Lecture Notes in Computer Science 793

Edited by G. Goos and J. Hartmanis

T. Aaron Gulliver   Norman P. Secord (Eds.)

Information Theory and Applications

Third Canadian Workshop
Rockland, Ontario, Canada
May 30 - June 2, 1993
Proceedings

Springer-Verlag
Berlin Heidelberg New York
London Paris Tokyo
Hong Kong Barcelona
Budapest

Series Editors

Gerhard Goos
Universität Karlsruhe
Postfach 69 80
Vincenz-Priessnitz-Straße 1
D-76131 Karlsruhe, Germany

Juris Hartmanis
Cornell University
Department of Computer Science
4130 Upson Hall
Ithaca, NY 14853, USA

Volume Editors

T. Aaron Gulliver
Department of Systems and Computer Engineering, Carleton University
1125 Colonel By Drive, Ottawa, Ontario, Canada K1S 5B6

Norman P. Secord
Communications Research Centre
3701 Carling Avenue, P.O. Box 11490, Station "H"
Ottawa, Ontario, Canada K2H 8S2

CR Subject Classification (1991): E.4, B.4, I.6, G.3, E.3, F.2, G.2

ISBN 3-540-57936-2 Springer-Verlag Berlin Heidelberg New York
ISBN 0-387-57936-2 Springer-Verlag New York Berlin Heidelberg

CIP data applied for

Typesetting: Camera-ready by author
SPIN: 10132045 45/3140-543210 - Printed on acid-free paper

Preface

The 1993 Canadian Workshop on Information Theory was held in Rockland, Ontario from May 30 to June 2. This was the third workshop to be held under the auspices of the Canadian Society of Information Theory, the previous two being in Ste. Jovite, Quebec in 1987 and Sidney, British Columbia in 1989. The aim of the workshops has been to provide an informal setting for Canadian researchers and graduate students to meet and exchange ideas. A wider call for papers was sent out for the 1993 workshop and we were pleased to welcome a number of participants from the United States and France among the 71 attendees. There were three plenary sessions given at the workshop, of which two are represented here. In addition, there were 32 regular presentations, and 24 of these appear as papers in this volume.

This is the first proceedings to be published from the workshop. Participants were asked to submit a manuscript following the workshop and each paper has been subject to peer review. An extensive list of reviewers appears at the end of this preface. The papers have been loosely grouped into four sections. A brief summary of the papers in each section is given below.

Coding and Cryptography

This section begins with the paper by the plenary lecturer, Prof. I. F. Blake. Prof. Blake and his co-authors, Gao and Lambert, present numerous new results in the construction of irreducible trinomials over finite fields. Of particular interest is the comprehensive table listing all irreducible trinomials of degree less than 2000 over F_2. Polemi and Sakkalis discuss algorithms for the identification and resolution of singularities in an algebraic curve over a finite field. They also discuss the computational complexity of these algorithms. Zarowski has developed a Schur form of the Berlekamp-Massey algorithm amenable to parallel implementation on a linear systolic array. He illustrates how to create a divisionless form of the Schur Berlekamp-Massey algorithm. Sablatash presents a comprehensive survey of the proposed all-digital HDTV systems with emphasis on the coding and modulation methods used. The paper by Drolet discusses the design of a VLSI chip able to perform finite field arithmetic over extension fields of F_2 of large order. Hendessi and Arof present the design of a pipelined VLSI chip that is capable of breaking the DES encryption standard by an exhaustive key search.

Coding and Modulation for Fading Channels

The second section of the volume also begins with the work of a plenary lecturer. Prof. S. Lin presents new results in the area of multilevel coded modulation in his paper co-authored with Rajpal and Rhee. The aim of this work is to develop methods of constructing very powerful codes with a simple structure and low decoding complexity. Lin and Pottie discuss coding strategies for a multiple-access frequency-hopped system employing antenna diversity. Boudreau presents a unified view of optimum noncoherent detection and various forms of differential detection employed for MPSK signals. Fattouche and Zaghloul show how a Hilbert transform based estimator can be used to remove the error floor present with differential detection of PSK signals in flat fading channels. Ferland and Chouinard study the use of DES encryption in a mobile communications environment with BCH block coding. Wautier, Dany and Mourot present a non-iterative block processing algorithm that is used with CAZAC training sequences to estimate the channel in a TDMA mobile environment. Trabelsi and Yongaçoḡlu present a new and improved approximation for the probability of packet success in a DS/CDMA multiple access system.

Decoding Techniques

Markman and Anderson discuss the use of the Pe-criterion in the analysis of sequential decoding for the binary symmetric channel. They extend this criterion to channels with intersymbol interference. Lodge, Young and Guinand show that by applying iterative soft decision decoding to concatenated convolutional codes, significant improvements in performance can be obtained over a single decoding. Roy and Fortier present a modification of the concept of a strongly-connected trellis for use in the Viterbi decoding of trellis coded modulation on a massively parallel computer. Chan and Haccoun use puncturing, a common technique with convolutional codes, to provide variable bandwidth efficiency with a single trellis coded modulation scheme. Sorokine, Kschichang and Durand have developed an efficient algorithm for the iterative construction of a trellis diagram for a binary block code. These results are used to derive a sequential stack algorithm for decoding. Le-Ngoc, Jia and Benyamin-Seeyar discuss the use of primitive elements of a prime field as multipliers for (T,U) permutation decoding. This increases the decoding capability of this technique.

Networks and Information Theory

Ghazi-Moghaddam, Lambadaris and Hayes introduce a constraint on the admission of overflow packets in hybrid switches that enhances the resource sharing. This bounds the observed delay in the overflow traffic by restricting their access. Zeytinoḡlu and Hsu discuss the transmission of wavelet transform coded digital audio signals over ATM packet networks. The coding algorithm is designed to be robust with respect to packet losses due to congestion. Kamoun and Ali propose the superposition of N interrupted Poisson processes as a model for a bursty, correlated arrival process such as at an ATM multiplexer. A statistical characterization is given of the traffic generated by such a process. Ross and Taylor examine the use of multiuser sequence detection to improve the performance of CDMA random-access systems. Huang and Leung propose a method of transfer function identification using a chaotic sequence generated with a nonlinear recurrence equation. The sequence follows a simple quadratic function in phase space so that the transfer function may be identified in this space rather than in the conventional time domain. Inkol and Saper present an algorithm for classifying pulsed signals having a phase or frequency modulated carrier. Khandani, Kabal and Dubois apply a dynamic programming approach to the problem of fixed-rate entropy coding of a memoryless source.

We would like to thank the following reviewers for their valuable time. Without their assistance, this volume would not have been possible.

Adnan Abu-Dayya
Norman C. Beaulieu
Anader Benyamin-Seeyar
Vijay K. Bhargava
Richard E. Blahut
Lorne Campbell
John C. Cartledge
James K. Cavers
Hervé Chabanne
Jean-Yves Chouinard
David C. Coll
Stewart Crozier
Vladimir Cuperman
Ingrid Daubechies
David E. Dodds
Ivan Fair
E. Barry Felstead
G. David Forney, Jr.
Nicolas D. Georganas
Georgios B. Giannakis
Anwar Hasan
Morris W. Hirsch
Paul Ho
Samir Kallel
Tadao Kasami
Torleiv Kløve
Frank R. Kschischang
James S. Lehnert
Tho Le-Ngoc
Rudolf Lidl
Shu Lin
John Lodge
Jon Mark
Lloyd J. Mason
P. Takis Mathiopoulos
Michael Moher
Brian Mortimer
Tetsa Papantoni-Kazakos
Miroslaw Pawlak
E. Roy Pike
Gregory J. Pottie
Michael B. Pursley
Amy Reibman
Jing-Fei Ren
Christopher Riordan
David Ruelle
Michael Sablatash
Yousef R. Shayan
Asrar Sheikh
Stanley S. Simmons
Elvino Sousa
Wayne E. Stark
Gordon L. Stüber
Desmond P. Taylor
Sergio Verdú
Qiang Wang
Victor K. Wei
Abbas Yongaçoğlu
Oyvind Ytrehus

We are grateful to the following organizations for their support of the workshop:

- Natural Sciences and Engineering Research Council of Canada
- Communications Research Centre
- Carleton University
- IEEE Information Theory Society
- IEEE Region 7
- Ottawa Carleton Research Institute
- Telecommunications Research Institute of Ontario

February 1994

T. Aaron Gulliver
Norman P. Secord

Table of Contents

Coding and Cryptography

Coding and Modulation for Fading Channels

Decoding Techniques

Networks and Information Theory

Constructive Problems for Irreducible Polynomials over Finite Fields *

Ian F. Blake, Shuhong Gao and Robert Lambert

University of Waterloo,
Waterloo, Ontario,
Canada N2L 3G1

Abstract. This paper discusses the techniques used in searching for irreducible trinomials in finite fields. We first collect some specific constructions of irreducible trinomials, then we show how to get new irreducible trinomials from given ones. We also make some comments on the irreducibility testing algorithms and on a primitivity testing algorithm although no experimantal results on primitive polynomials are reported on. Finally, updated tables of irreducible trinomials over F_2 are included.

1 Introduction

The subject of primitive polynomials over finite fields has been of interest over the past few decades, primarily due to their use in the generation of linear feedback shift register sequences which find application in communication systems, cryptographic systems and random number generation for Monte Carlo simulation. The use of trinomials, polynomials with only three nonzero coefficients, is attractive from the point of view of either software or hardware implementation, and the generation of such polynomials over the field of two elements, F_2, has received particular attention. The establishment of the primitivity of a binary irreducible polynomial of degree n generally requires the factorization of the integer $2^n - 1$. The values of n for which this factorization is known become increasingly sparse as n increases beyond a few hundred.

This article addresses the lesser question of the construction of irreducible trinomials over F_2, surveying some of the results available as well as considering some related questions. The next section contains some standard preliminary results on polynomials and bases over finite fields. Section 3 discusses specific irreducible trinomials and Section 4 considers some composition and recursive constructions for irreducible polynomials with particular attention to trinomials. Section 6 collects the known results on irreducible trinomials over F_2. Sections 5 and 7 contain some comments on irreducibility testing and primitivity testing algorithms. Some experimental results are reported on in the tables in the hope they may suggest new lines of investigation.

* This work was supported by a grant from the Information Technology Research Centre, a Centre of Excellence of the Province of Ontario

2 Preliminary Results

Let F_q denote the finite field with q elements and $F_q[x]$ the set of polynomials over F_q. The polynomial $f(x) \in F_q[x]$ is irreducible if it cannot be represented as a nontrivial product of two polynomials in $F_q[x]$ of lower degree. When $f(0) \neq 0$, the exponent of $f(x)$ is the least positive integer e such that $f(x)|x^e - 1$ and $f(x)$ is called primitive if it has degree n and exponent $q^n - 1$. It is sometimes convenient to refer to the index of $f(x)$, defined as $(q^n - 1)/e$.

Denote by $I_{n,q}[x]$ the set of irreducible polynomials of degree n over F_q. If $f(x) \in I_{n,q}[x]$ has a root α then

$$\underline{\alpha}_{poly} = \{1, \alpha, \cdots, \alpha^{n-1}\}$$

is a polynomial basis of F_{q^n} over F_q. The number of distinct ordered bases of F_{q^n} over F_q is

$$\prod_{i=0}^{n-1} (q^n - q^i) = q^{n(n-1)/2} \prod_{i=1}^{n} (q^i - 1)$$

which is also the order of $GL(n,q)$, the group of nonsingular matrices over F_q. The number of polynomial bases of F_{q^n} over F_q, up to conjugacy, is simply the number of irreducible polynomials of degree n over F_q,

$$|I_{n,q}[x]| = \frac{1}{n} \sum_{d|n} \mu(n/d) q^d$$

where μ is the Möbius function on Z_+.

If α is a root of $f(x) \in I_{n,q}[x]$ then

$$f(x) = \prod_{i=0}^{n-1} (x - \alpha^{q^i})$$

and if the roots of $f(x)$ are linearly independent they form a normal basis

$$\underline{\alpha}_{nor} = \{\alpha, \alpha^q, \cdots, \alpha^{q^{n-1}}\}.$$

The number of (unordered) normal bases of F_{q^n} over F_q is $|C(n,q)|/n$ where $C(n,q)$ is the set of $n \times n$ nonsingular circulant matrices over F_q. This number [16] is

$$\frac{1}{n} \Phi_q(x^n - 1)$$

where $\Phi_q(f(x))$ is the number of polynomials over F_q of degree less than $\deg f(x)$ which are relatively prime to $f(x)$. If the distinct irreducible factors of $f(x)$ have degrees $n_1, n_2, \cdots, n_r$ then

$$\Phi_q(f(x)) = q^n (1 - q^{-n_1})(1 - q^{-n_2}) \cdots (1 - q^{-n_r}).$$

If $f(x) \in I_{n,q}[x]$ and is primitive with linearly independent roots over F_q, then the roots form a primitive normal basis of F_{q^n} over F_n. The existence of such bases for all prime powers q and positive integers n has recently been established [11]. One might also consider dual and self dual bases and their enumeration but this goes beyond the interests of this article.

Interest here is restricted to the construction of irreducible polynomials over F_2. For computational purposes one might ask for a list of low weight (number of non-zero coefficients) primitive polynomials with independent roots of all degrees up to, say, $1,000$. As noted however, the only method of establishing primitivity of polynomials in $I_{n,2}[x]$ requires the factorization of $2^n - 1$ which is largely unknown for n beyond a few hundred. Except for a new and efficient primitivity testing algorithm given in Section 7, the harder problem of constructing primitive polynomials is ignored here.

The determination of all irreducible trinomials over F_2 for all degrees less than $1,000$ has been known for some time {[24],[25]}. In this set of tables for most of the degrees n for which the factorization of $2^n - 1$ is known, the irreducible polynomials have been tested for primitivity. With the use of symbolic computation packages such as Macsyma, Maple, Galois or AXIOM, it is not difficult to determine irreducible polynomials of quite high degree. However many analytical and useful results on trinomials have been established and the purpose of this article is to review the techniques by which such tables can be constructed and examine some of the interesting related results.

3 Irreducible Binomials and Trinomials

There are a few results on specific irreducible binomials and trinomials. We briefly mention them here. The reference [16] is used as a convenient reference for well established results. The existence of irreducible binomials is completely established as the following theorem shows.

Theorem 1 *[16] Let $a \in F_q^*$ with order e. Then the binomial $x^t - a$ is irreducible in $F_q[x]$ if and only if the integer $t \geq 2$ satisfies the following conditions:*

(i) each prime factor of t divides e but not $(q-1)/e$, and
(ii) if $4|t$ then $4|(q-1)$.

Corollary 2 *Let r be a prime factor of $q-1$ and $a \in F_q$. Assume that $q \equiv 1 (mod\ 4)$ if $r = 2$. Then $x^{r^k} - a$ is irreducible over F_q for every integer $k \geq 0$ if and only if a is not an r-th residue, i.e., a is not of the form η^r for some $\eta \in F_q$.*

When $r = 2$, the case that $q = 3(\bmod\ 4)$ is excluded by Corollary 2. In fact, $x^{2^k} - a$ is reducible for all $a \in F_q$ and for all $k > 1$ in this case. However, the next theorem shows how to construct an irreducible trinomial.

Theorem 3 *[1] Let $p \equiv 3(\bmod\ 4)$ be a prime and v the largest integer such that $2^v|(p+1)$. Define a_v recursively by the formula*

$$a_i = \begin{cases} \pm(\frac{a_{i-1}+1}{2})^{(p+1)/4}, & for\ 2 \leq i \leq v-1 \\ \pm(\frac{a_{i-1}-1}{2})^{(p+1)/4}, & for\ i = v, \end{cases}$$

with the initial $a_1 = 0$, and at each step one can choose either $+$ sign or $-$ sign. Then

$$x^{2^k} - 2a_v x^{2^{k-1}} - 1$$

is irreducible over F_p for every integer $k \geq 1$.

Now if $q = p^m \equiv 3 \pmod 4$, then m must be odd. Note that if $f(x) \in I_{n,q}[x]$ then $f(x) \in I_{n,q^k}[x]$ if and only if $\gcd(k, n) = 1$. We see that $x^{2^k} - 2a_v x^{2^{k-1}} - 1$ is also irreducible over F_q for every integer $k \geq 1$. This proves that when q is odd there is always an irreducible binomial or trinomial of degree 2^k over F_q. However, when $q = 2$, by the above argument there is no irreducible binomial of degree 2^k if $k > 2$ and we will see by Theorem 12 that there also is no irreducible trinomial.

Another classical result on trinomials is the following [16]:

Theorem 4 *The trinomial $x^p - x - b$, $b \in F_q$ where q is a prime power p^m is in $I_{p,q}[x]$ if and only if $Tr_{q|p}(b) \neq 0$.*

Corollary 5 *For $a, b \in F_q^*$, the trinomial $x^p - ax - b \in I_{p,q}[x]$ if and only if $a = A^{p-1}$ for some $A \in F_q$ and $Tr_{q|p}(b/A^p) \neq 0$.*

The next theorem is conjectured by Chowla [4], proved independently by Cohen [5] and Ree [19]:

Theorem 6 *For every $n \geq 2$, the number of $a \in F_p$ such that $x^n + x + a$ is irreducible over F_p is $p/n + O(p^{1/2})$ as p approaches ∞.*

These results are of interest but do not bear directly on the problem of finding irreducible trinomials over F_2.

4 Composition and Recursive Constructions of Irreducible Polynomials

There are several recursive constructions of irreducible polynomials (e.g. see [16]) and some of these lead to new irreducible trinomials from given ones. Two results are included here. We first note the following standard result.

Theorem 7 *[12] Let $f(x) = \sum_i f_i x^i \in I_{n,q}[x]$ have exponent e, and let $f^{[q]}(x) = \sum_i f_i x^{q^i - 1}$. Then all the irreducible factors of $f^{[q]}(x)$ are of degree e.*

Corollary 8 *A polynomial $f(x) \in I_{n,q}[x]$ is primitive if and only if $f^{[q]}(x)$ is irreducible.*

This Corollary will be applied to the case of trinomials in Section 6.

Notice that if $P(x)$ in the theorem is a trinomial then so is $P(x^t)$. The next theorem tells us when $P(x^t)$ is also irreducible.

Theorem 9 *[12] Let t be a positive integer and $P(x) \in I_{n,q}[x]$ with exponent e. Then $P(x^t)$ is irreducible over F_q if and only if*

(i) each prime factor of t divides e but not $(q^n - 1)/e$, and
(ii) if $4|t$ then $4|(q^n - 1)$.

Corollary 10 *Let $P(x) \in I_{n,q}[x]$ with exponent e and let r be a prime factor of e such that r does not divide $(q^n - 1)/e$. Assume that $q^n \equiv 1 (mod\ 4)$ if $r = 2$. Then $P(x^{r^k})$ is irreducible over F_q for every integer $k \geq 1$.*

As an application, we determine the irreducible self-reciprocal trinomials over F_2, i.e., polynomials of the form $x^{2t} + x^t + 1$. Since $x^2 + x + 1$ has exponent $e = 3$, one must have $t = 3^k$. For $t = 3^k$, it follows from Corollary 10 that $x^{2t} + x^t + 1$ is in fact irreducible over F_2 as $(2^2 - 1)/e = 1$. Hence all the irreducible self-reciprocal trinomials over F_2 are of the form $x^{2\cdot 3^k} + x^{3^k} + 1$.

5 Irreducibility Testing

A basic problem in constructing irreducible polynomials is to test when a given polynomial is irreducible. Generally, there are two algorithms to do this. One is derived from the Berlekamp algorithm for factorization of polynomials over finite fields. The other is from the algorithm of distinct degree factorization of polynomials. We outline the two algorithms as follows.

Suppose that $f(x)$ is a polynomial of degree n over F_q whose irreducibility is to be tested.

Algorithm 1 (Based on Berlekamp algorithm)

(1). Check if $\gcd(f(x), f'(x)) = 1$. If not, then $f(x)$ has repeated factors and is thus reducible.
(2). Compute $(x^i)^q = \sum_{j=0}^{n-1} t_{ij}x^j$ modulo $f(x)$ for $i = 0, 1, 2, \cdots, n-1$, where $t_{ij} \in F_q$, and form the $n \times n$ matrix $B = (t_{ij})$.
(3). Compute the rank of the matrix $B - I_n$ where I_n is the identity matrix of order n. $f(x)$ is irreducible if and only if the rank is $n - 1$.

Algorithm 2 (Based on distinct degree factorization)

(1). Check if $x^{q^n} - x \equiv 0 \mod f(x)$. If not then $f(x)$ is not an irreducible polynomial of degree n.
(2). For each prime divisor k of n, check if $\gcd(f(x), x^{q^{n/k}} - x) = 1$. If not for some k, then $f(x)$ has an irreducible factor of degree as a divisor of n/k and is thus reducible.

In a practical implementation of Algorithm 2, one might do the first step last, since on the way to computing x^{q^n} one may first get $x^{q^{n/k}}$. For an efficient algorithm to compute these quantities, one is referred to [7].

Note that $\gcd(f(x), x^{q^m} - x) = 1$ if and only if $f(x)$ has no irreducible factor of degree a divisor of m. When m is small, it is easy to compute $\gcd(f(x), x^{q^m} - x)$. This fact suggests that in implementing irreducibility testing algorithms one should check first if the tested polynomial has small factors, since a full irreducibility test for large degrees is much more expensive. Actually, Theorem 11 below, due to A. Odlyzko (private communication), shows that most of the polynomials have small factors, thus they are discarded at the early stage of the testing.

Theorem 11 *Denote by $a(n,k,q)$ the number of polynomials of degree n in $F_q[x]$ that have no factors of degree less than k. Then*

$$\lim_{n\to\infty}\frac{a(n,k,q)}{q^n}=\prod_{i=1}^{k-1}\left(1-\frac{1}{q^i}\right)^{d_i}=g(k,q),$$

where d_i denotes the number of irreducible polynomials of degree i in $F_q[x]$.

Proof. We first derive the generating function

$$g_k(z)=\sum_{n=0}^{\infty}a(n,k,q)z^n.$$

Note that any polynomial $f\in F_q[x]$ with no factors of degree less than k is of the form

$$f=f_kf_{k+1}\cdots,$$

where f_i has only irreducible factors of degree exactly i, for $i\geq k$. Note that there are $\binom{d_i+m-1}{m}$ polynomials $f_i\in F_q[x]$ of degree im with only irreducible factors of degree i, the generating function for these polynomials is

$$\sum_{m=0}^{\infty}\binom{d_i+m-1}{m}z^{im}=(1-z^i)^{-d_i}.$$

Thus

$$g_k(z)=\prod_{i=k}^{\infty}(1-z^i)^{-d_i}.$$

As $g_1(z)$ is the generating function for all the polynomials in $F_q[x]$, we have

$$g_1(z)=\sum_{n=0}^{\infty}q^nz^n=\frac{1}{1-zq}.$$

Therefore

$$g_k(z)=\frac{\prod_{i=1}^{k-1}(1-z^i)^{d_i}}{1-zq}.$$

Now $g_k(z)$ can be viewed as a function of the complex variable z, and $g_k(z)$ is analytic everywhere in the complex plane except at $z=1/q$ which is a pole of order 1. We apply Theorem 10.4 in [18]. Let R be a fixed real number such that $R>1/q$. Since $g_k(z)$ is continuous on the circle $|z|=R$, there is a real number w such that

$$\max_{|z|=R}|g_k(z)|\leq w.$$

Note that the residue of $g_k(z)$ at $z = 1/q$ is

$$-\frac{1}{q}\prod_{i=1}^{k-1}(1-\frac{1}{q^i})^{d_i} = -\frac{1}{q}g(k,q).$$

By Theorem 10.4 in [18], we have

$$\left|a(n,k,q) + (-\frac{1}{q}g(k,q))(\frac{1}{q})^{-n-1}\right| \le wR^{-n} + (R-\frac{1}{q})^{-1}R^{-n}\frac{1}{q}g(k,q).$$

Therefore

$$\left|\frac{a(n,k,q)}{q^n} - g(k,q)\right| \le (w + (R-\frac{1}{q})^{-1}\frac{1}{q}g(k,q))(qR)^{-n}.$$

But $(qR)^{-n} \to 0$ as $n \to \infty$, our result follows immediately.

When n is large, the number $1 - g(k,q)$ is the approximate ratio of polynomials of degree n in $F_q[x]$ with at least one factor of degree less than k with respect to all the polynomials of degree n in $F_q[x]$. Table 1 shows how this number changes for small q when k increases. In the table, $\bar{g}(k,q)$ denotes $(1-g(k,q))\cdot 100$, which is the percentage

k	1	2	3	4	5	6	7	8	9	10	11	12	13	14	15
$\bar{g}(k,2)$	0	75	81	86	88	90	92	93	93	94	95	95	96	96	96
$\bar{g}(k,3)$	0	70	79	85	88	90	91	93	93	94	95	95	96	96	96
$\bar{g}(k,4)$	0	68	79	84	88	90	91	93	93	94	95	95	96	96	96
$\bar{g}(k,5)$	0	67	78	84	88	90	91	93	93	94	95	95	96	96	96
$\bar{g}(k,7)$	0	66	78	84	88	90	91	93	93	94	95	95	96	96	96
$\bar{g}(k,8)$	0	66	78	84	88	90	91	93	93	94	95	95	96	96	96
$\bar{g}(k,9)$	0	65	78	84	88	90	91	93	93	94	95	95	96	96	96

Table 1.: Percentages of polynomials in $F_q[x]$ with small factors.

of the polynomials in $F_q[x]$ that have factors of degree less than k. Entries are rounded to the nearest integer. The data shows that about 90 percent of the polynomials in $F_q[x]$ have factors of degrees ≤ 6, and that more than 95 percent have factors of degrees ≤ 15! The result indicates the importance of testing for small factors before a full irreducibility test is carried out. It is also noted that Odlyzko [17] develops expressions for completely smooth polynomials (all factors less than a given degree), as appropriate for the discrete logarithm problem.

6 Known Results for Trinomials

As we have mentioned, much attention has been paid to irreducible (and primitive) trinomials over F_2 due to their usefulness in practice. In this section, we summarize the known tables about them and discuss very briefly how they are established. We denote $x^n + x^k + 1$ by $T_{n,k}(x)$.

The table of all irreducible trinomials over F_2 for all degrees less than $1,000$ is given by {[24],[25]}, where information about exponents is also given when possible. In Table 6, we list all the irreducible trinomials over F_2 of degrees up to 2000. It is possible to obtain such polynomials for much higher degrees but our immediate purpose is experimental, to use them to consider distribution patterns and attempt to formulate conjectures.

For trinomials, we still use the algorithms in section 5 to establish the irreducibility. However, trinomials in many cases can be seen to be reducible from the following theorem due to Swan [23].

Theorem 12 *$T_{n,k}(x)$ has an even number of irreducible factors if and only if*

(a) *n is even, k is odd, $n \neq 2k$ and $nk/2 \equiv 0$ or $1 \bmod 4$,*
(b) *n is odd, k is even not dividing $2n$ and $n \equiv \pm 3 \mod 8$,*
(c) *n is odd, k is even, $k|2n$, and $n \equiv \pm 1 \mod 8$.*

If n and k are both odd then the tests can be applied to $T_{n,n-k}(x)$.

We look at some consequences of this theorem. Firstly, by (a), we see that when $n \equiv 0 \mod 8$, $T_{n,k}(x)$ is always reducible, hence there is no irreducible trinomial of degree n over F_2. Secondly, both $T_{n,1}(x)$ and $T_{n,2}(x)$ cannot be irreducible simultaneously when $n > 3$, since if n is even then $T_{n,2}(x)$ is always reducible over F_2 and if $n > 3$ is odd then $T_{n,2}(x)$ is reducible by (c) when $n \equiv \pm 1 \mod 8$ and $T_{n,1}(x)$ is reducible by (b) when $n \equiv \pm 3 \mod 8$. Further, when $n \equiv \pm 3 \mod 8$, $T_{n,k}(x)$ is possibly irreducible only if k or $n - k$ is of the form $2d$ where d is a divisor of n. Suppose that $n = md$ and $k = 2d$. Then by Theorem 9, for $T_{n,k}(x)$ to be irreducible, $T_{m,2}(x)$ has to be irreducible and each prime divisor of d must divide e but not $(2^m - 1)/e$ where e is the exponent of $T_{m,2}(x)$. Also we must have $m \equiv \pm 3 \mod 8$, as if $m \equiv \pm 1 \mod 8$ then $T_{m,2}(x)$ is reducible by (c). Therefore, to find all the trinomials of degree $n \equiv \pm 3 \mod 8$ over F_2, one just need to check if $T_{n,2d}(x)$ is irreducible for each divisor d of n such that $n/d \equiv \pm 3 \mod 8$.

Table 2 [26] lists all integers $n \leq 30,000$ such that $T_{n,1}(x)$ is irreducible over F_2. It follows from Corollary 8 that for all the values of n in this table for which $T_{n,1}(x)$ is primitive (index 1), $T_{2^n-1,1}(x)$ is irreducible. It is also noted that from part (a) of Swan's theorem, $x^n + x + 1$ is always reducible for $n \equiv 2 \mod 8$. Table 3 reproduced from [6] gives all integers $n \leq 60,000$ such that $T_{n,2}(x)$ is irreducible over F_2. It is quite remarkable to observe that for $n \leq 60,000$ when $T_{n,2}(x)$ is irreducible one can easily check by using Table 3, and the condition above on the irreducibility of $T_{md,2d}(x)$, that there is no k other than 2 or $n - 2$ such that $T_{n,k}(x)$ is irreducible, except for $n = 21$, and we would conjecture that this is always so.

Conjecture. For $n > 21$, if $T_{n,2}(x)$ is irreducible over F_2 then $T_{n,k}(x)$ is reducible over F_2 for all $k \neq 2, n - 2$.

The case when $2^n - 1$ is prime has attracted special attention, since in this case all irreducible polynomials of degree n over F_2 are primitive. Primes of the form $2^p - 1$

n	Index	n	Index	n	Index
2	1	60	1	2380	?
3	1	63	1	3310	?
4	1	127	1	4495	?
6	1	153	1	6321	?
7	1	172	3.5	7447	?
9	7	303	?	10198	?
15	1	471	?	11425	?
22	1	532	1	21846	?
28	3.5	865	?	24369	?
30	$3^2.11$	900	?	27286	?
46	3	1366	?	28713	?

Table 2.: Values of $n \leq 30,000$ for which $T_{n,1}(x)$ is irreducible.

n	Index	n	Index	n	Index
3	1	123	1	20307	?
5	1	333	1	34115	?
11	1	845	?	47283	?
21	1	4125	?	50621	?
29	1	10437	?	57341	?
35	1	10469	?		
93	1	14211	?		

Table 3.: Values of $n \leq 60,000$ for which $T_{n,2}(x)$ is irreducible.

are referred to as Mersenne primes, where p is the Mersenne exponent. Suppose that p is a Mersenne exponent. Note that a polynomial $f(x)$ in $F_2[x]$ of degree p is irreducible if and only $x^{2^p} \equiv x \bmod f(x)$. So the irreducibility test is reduced to computing $x^{2^p} \bmod f(x)$. Table 4 is reproduced from [10], the latest contribution on this problem, considered by many researchers. The question mark in the last entry indicates that the search for Mersenne exponents for p in the interval $139268 < p < 216090$ has not been exhaustive.

We should remark that when $\gcd(x^n + x^k + 1, x^{2^m-1} - 1)$ is computed one can reduce n and k modulo $2^m - 1$ first then do the gcd. This is efficient in the situation where m is small, for example, when testing if $x^n + x^k + 1$ has factors of small degrees. This strategy can also be applied to other high degree polynomials.

Finally, we mention an interesting phenomenon for trinomials with Mersenne exponents. By Corollary 8, if $x^n + x + 1$ is irreducible over F_2 and $2^n - 1$ is a prime then $x^{2^n-1} + x + 1$ is irreducible in $F_2[x]$. From this it follows that $T_{2^n-1,1}(x)$ is irreducible for $n = 2, 3, 7, 127, 2^{127} - 1$. Let $A = 2^{127} - 1$. Then the next integer in the sequence is $B = 2^A - 1$ and we know that it is not prime. Since we do not know the complete factorization of B, we cannot determine if $T_{A,1}(x)$ is primitive and the irreducibility of

p	k
2	1
3	1
5	2
7	1,3
13	none
17	3,5,6
19	none
31	3,6,7,13
61	none
89	38
107	none
127	1,7,15,30,63
521	32,48,158,168
607	105,147,273
1279	216,418
2203	none
2281	715,915,1029
3217	67,576
4253	none
4423	271,369,370,649,1393,1419,2098
9689	84,471,1836,2444,4187
9941	none
11213	none
19937	881,7083,9842
21701	none
23209	1530,6619,9739
44497	8575,21034
86243	none
110503	25230,53719
132049	7000,33912,41469,52549,54454
216091?	none

Table 4.: Primitive trinomials for the known 31 Mersenne exponents.

$T_{B,1}(x)$ is thus unknown. However, Lenstra and Schoof [11] conjecture that $T_{B,1}(x)$ is indeed irreducible over F_2.

7 On Primitivity Testing

Suppose that one needs to find a primitive polynomial of degree n over F_q. In practice, one usually chooses at random a polynomial and then tests if it is primitive. If one needs to find a primitive polynomial of certain pattern, say with lowest weight, there appears to be no better search strategy other than exhaustive search. However, it is interesting to note that Shparlinskii [22] proves that there is a primitive polynomial of degree n

over F_2 of weight $w(f) \leq n/4 + o(n)$. If q is small relative to n, say $q = 2$, and if the weight of the polynomial is not of concern, Shparlinskii [22] and Shoup [21] show independently that one can restrict the search to a subset of polynomials over F_q in which the number of polynomials is polynomial in n. So for small q, the problem of constructing primitive polynomials is reduced to finding an efficient algorithm to test primitivity.

Let $f(x)$ be an irreducible polynomial of degree n over F_q. To prove that $f(x)$ is primitive, we have to show that $f(x)$ has exponent $q^n - 1$. As far as the authors know, the best algorithm for this is given by the following lemma.

Lemma 13 *Suppose that*

$$q^n - 1 = p_1^{r_1} p_2^{r_2} \cdots p_k^{r_k}, \tag{1}$$

where $p_1, p_2, \ldots, p_k$ are distinct primes. Then an irreducible polynomial $f(x)$ over F_q is primitive if and only if

$$x^{(q^n-1)/p_i} \not\equiv 1 \bmod f(x) \tag{2}$$

for $i = 1, 2, \ldots, k$.

An important prerequisite for this algorithm is the factorization of the integer $q^n - 1$. There are extensive tables [2] for the factorization of numbers of this form for small n and for small q, especially for the case $q = 2$. An updated version of these tables can be obtained from the authors of [2].

When the factorization of $q^n - 1$ is not known, we cannot test the primitivity of $f(x)$ in a reasonable time. It would be interesting to see if there is any efficient way to test primitivity without knowing the factorization of $q^n - 1$. The following problem is obviously difficult but may not be impossible.

Problem. Given the field F_q and an irreducible polynomial $f(x)$ of degree n over F_q, find an algorithm to compute the exponent of $f(x)$ in time polynomial in n and $\log q$. (The complete factorization of $q^n - 1$ is not assumed to be known.)

There are some short cuts in computing the quantities in (2). Since $f(x)$ is irreducible, we have

$$x^{(q^n-1)/(q-1)} = x x^q \cdots x^{q^{n-1}} \equiv (-1)^n f(0) \bmod f(x).$$

Thus if $p_i | (q-1)$ then

$$x^{(q^n-1)/p_i} = (x^{(q^n-1)/(q-1)})^{(q-1)/p_i} \equiv ((-1)^n f(0))^{(q-1)/p_i} \bmod f(x).$$

This suggests that we should distinguish the p_i in (1) that are divisors of $q - 1$ from those that are not. Suppose that $p_1, p_2, \cdots, p_\ell$ are divisors of $q - 1$ and $p_{\ell+1}, \cdots, p_k$ are not. Then for i from 1 to ℓ we just need to check if $((-1)^n f(0))^{(q-1)/p_i} = 1$ in F_q. We could save more computation by precomputing

$$a = ((-1)^n f(0))^{(q-1)/Q},$$
$$A \equiv x^{(q^n-1)/R} \bmod f(x)$$

where $Q = p_1 \cdots p_\ell$ and $R = p_{\ell+1} \cdots p_k$. Then we just need to compute

$$a^{Q/p_i}, \quad i = 1, \dots, \ell, \tag{3}$$

and

$$x^{R/p_j} \bmod f(x), \quad j = \ell + 1, \dots, k. \tag{4}$$

The quantities in (3) and (4) have similar structure and we show in the next paragraph an efficient way to compute them.

Let $v_1, \dots, v_k$ be k distinct integers and $V = v_1 \cdots v_k$ be their product. We would like to efficiently compute the k powers a^{V/v_i}, $i = 1, \dots, k$, where a is an element in some ring. Let $a_1 = a$ and $a_i = a^{v_1 \cdots v_{i-1}}$ for $k = 2, \dots, k$. Then

$$a^{V/v_i} = a_i^{v_{i+1} \cdots v_k}.$$

Note that $a_{i+1} = (a_i)^{v_i}$, so the a_i's can be computed recursively. This suggests that we should compute $a_1, a_2, \dots, a_k$ recursively one by one starting at a_1, and when a_i is computed we compute $a_i^{v_{i+1} \cdots v_k}$ immediately (and perhaps a_{i+1} at the same time). At the end, all the k powers a^{V/v_i} are computed. Only one a need be stored at any time.

To see how much computation one could save by this algorithm, we assume that computing an mth power in a ring needs $c \log m$ multiplications where c is a constant. Then computing the k quantities a^{V/v_i} independently needs

$$\sum_{i=1}^{k} c \log(V/v_i) = c(k-1) \sum_{i=1}^{k} \log v_i \tag{5}$$

multiplications. However, the number of multiplications needed in the above algorithm is

$$\sum_{i=1}^{k-1} c \log(v_{i+1} \cdots v_k) + \sum_{i=1}^{k} c \log v_i = c(\log v_1 + 2 \log v_2 + \cdots + k \log v_k), \tag{6}$$

which is much smaller than (5). In addition, (6) is minimized by re-ordering v_i such that $v_1 > v_2 > \cdots > v_k$.

8 Experimental Results for Trinomials

As noted earlier, some experimental results are included to consider certain questions on the distribution of trinomials over F_2. Table 6 lists all the irreducible trinomials $T_{n,k}(x)$ of degree less than $2,000$ for $k \leq n/2$, extending the previous tables [24, 25]. Define $N_2(n)$ to be the number of irreducible trinomials $T_{n,k}(x)$ over F_2 of degree n, $k \leq n/2$. Table 5 gives these numbers for $1 \leq n \leq 1,999$. The purpose for considering these numbers was to observe their behaviour with n. A casual look at the table might lead one to conclude, heuristically, that they do not appear to change with n although further analysis would be needed to substantiate this observation. One would in fact not expect them to by a simple heuristic assumption on the distribution of irreducible trinomials among the set of all irreducible polynomials.

Figures 1,2 and 3 consider the number of degrees r, for various ranges, for which there exists s irreducible trinomials $T_{r,k}(x),\ k \leq r/2$. Again the change in the distribution with the range is of interest, but the evidence given is too limited to draw conclusions. It is hoped to extend the range available to 5,000 in the near future which might confirm the suspicion that these quantities do not change appreciably with range and to allow a statement with greater statistical precision.

References

1. I.F. Blake, S. Gao and R.C. Mullin, "Explicit factorization of $x^{2^k}+1$ over F_p with prime $p \equiv 3 \bmod 4$", *AAECC*, **4** (1993), pp. 89–94.
2. J. Brillhart, D.H. Lehmer, J.L. Selfridge, B. Tuckerman and S.S. Wagstaff, "Factorizations of $b^n \pm 1, b = 2,3,5,6,7,10,11,12$ Up to High Powers", Vol. **22** of *Contemporary Mathematics*, AMS, 1988, 2nd edition.
3. L. Carlitz, "Factorization of a special polynomial over a finite field", *Pac. J. Math.*, **32** (1970) pp. 603-614.
4. S. Chowla, "A note on the construction of finite Galois fields $GF(p^n)$", *J. Math. Anal. Appl.*, **15** (1966), pp. 53–54. in 335-344.
5. S.D. Cohen, "The distribution of polynomials over finite fields", *Acta Arith.*, **17** 1970, pp. 255–271.
6. H. Fredricksen and R. Wisniewski, "On trinomials x^n+x^2+1 and $x^{8l\pm3}+x^k+1$ irreducible over $GF(2)$", *Information and Control*, **50** (1981) pp. 58-63.
7. J. von zur Gathen and V. Shoup, "Computing Frobenius maps and factoring polynomials", *Computational Complexity*, **2** (1992), pp. 187–224.
8. S.W. Golomb, *Shift Register Sequences*, Holden-Day Inc., 1967.
9. T. Hansen and G.L. Mullen, "Primitive polynomials over finite fields", *Math. Comp.*, **59** (1992) 639-643.
10. J.R. Heringa, H.W.J. Blőte and A, Compagner, "New primitive trinomials of Mersenne-exponent degrees for random-number generation", *Int'l J. Modern Physics C*, **3** (1992) 561-564.
11. H.W. Lenstra and R.J. Schoof, "Primitive normal bases for finite fields", *Math. Comp.*, **48** (1987) pp. 217-231.
12. R. Lidl and H. Niederreiter, *Finite Fields*, Cambridge University Press, 1987.
13. R.W. Marsh, W.H. Mills, R.L. Ward, H. Rumsey Jr. and L.R. Welch, "Round trinomials", *Pac. J. Math.*, **96** (1981) pp. 175-192. self-reciprocal 43-53.
14. W.H. Mills, "The degrees of the factors of certain polynomials over finite fields", *Proc. Amer. Math. Soc.*, **25** (1970) pp. 860-863.
15. W.H. Mills and N. Zierler, "On a conjecture of Golomb", *Pac. J. Math.*, **28** (1969) pp. 635-640.
16. A.J. Menezes, editor, *Applications of Finite fields*, Kluwer Academic Publishers, 1993.
17. A. Odlyzko, Discrete logarithms in finite fields and their cryptographic significance, *Proc. Eurocrypt '84*, pp. 224-314.
18. A. Odlyzko, Asymptotic enumeration methods, manuscript, 1993.
19. R. Ree, "Proof of a conjecture of S. Chowla", *J. of Number Theory*, **3** (1971), pp. 210–212.
20. E.R. Rodemich and H. Rumsey Jr., "Primitive polynomials of high degree", *Math. Comp.*, **22** (1968) pp.863-865.
21. V. Shoup, "Searching for primitive roots in finite fields", *Math. Comp.*, **58** (1992), pp. 369–380.

22. I.E. Shparlinskii, "On some problems in the theory of finite fields", *Russian Math. Surveys*, **46** (1991), pp. 199-240; or *Uspekhi Mat. Nauk*, **46** (1991), pp. 165–200.
23. R.G. Swan, "Factorization of polynomials over finite fields". *Pac. J. Math.*, **12** (1962) 1099-1106.
24. N. Zierler and J. Brillhart, "On primitive trinomials (Mod 2)", *Information and Control*, **13** (1968) pp. 541-554.
25. N. Zierler and J. Brillhart, "On primitive trinomials(Mod 2), II", *Information and Control*, **14** (1969) pp. 566-569.
26. N. Zierler, "On $x^n + x + 1$ over $GF(2)$", *Information and Control*, **16** (1970) pp. 502-505.
27. N. Zierler, "Primitive trinomials whose degree is a Mersenne exponent", *Information and Control*, **15** (1969) pp. 67-69.
28. N. Zierler, "On the theorem of Gleason and Marsh", *Proc. Amer. Math. Soc.*, **9** (1958), pp. 236-237.

Table 5. $N_2(n)$ for $1 \leq n \leq 1999$

	0	1	2	3	4	5	6	7	8	9
0	0	0	1	1	1	1	2	2	0	2
10	1	1	2	0	1	3	0	3	3	0
20	2	2	1	2	0	2	0	0	4	1
30	2	4	0	2	1	1	3	0	0	3
40	0	2	1	0	1	0	1	4	0	4
50	0	0	4	0	3	2	0	4	1	0
60	6	0	1	5	0	2	1	0	2	0
70	0	5	0	3	1	0	1	0	0	2
80	0	3	0	0	7	0	1	1	0	1
90	1	0	1	1	1	2	0	4	2	0
100	5	0	2	4	0	10	1	0	5	0
110	1	2	0	3	0	0	0	0	2	2
120	0	2	0	1	4	0	2	5	0	3
130	1	0	2	0	1	4	0	3	0	0
140	4	0	1	0	0	2	1	2	1	0
150	2	12	0	2	1	1	6	0	0	3
160	0	3	3	0	0	0	1	4	0	4
170	2	0	3	0	2	4	0	3	2	0
180	6	0	1	1	0	3	2	0	0	0
190	0	4	0	3	1	0	3	0	2	2
200	0	4	1	0	2	0	0	1	0	7
210	1	0	1	0	1	5	0	5	5	0
220	3	0	0	5	0	5	0	0	1	0
230	0	3	0	1	2	0	1	0	2	2
240	0	1	1	0	1	0	0	2	0	2
250	1	0	12	1	0	3	0	5	2	0
260	4	0	0	1	0	2	1	0	2	0
270	3	2	0	11	3	0	2	0	1	10
280	0	2	3	0	4	0	2	3	0	3
290	0	0	2	0	4	5	0	5	0	0
300	10	0	1	1	0	1	0	0	1	0
310	1	0	0	2	1	0	2	0	1	3
320	0	6	1	0	6	0	0	2	0	2
330	1	0	2	1	0	0	0	5	0	0
340	2	0	2	4	0	3	1	0	1	0
350	1	5	0	6	2	0	0	0	1	2
360	0	0	2	0	2	0	1	2	0	2
370	2	0	3	0	0	4	0	2	4	0
380	3	0	1	3	0	7	1	0	1	0
390	3	2	0	3	1	0	6	0	0	6
400	0	2	1	0	2	0	2	2	0	1
410	0	0	1	0	2	2	0	3	1	0
420	9	0	2	1	0	6	1	0	1	0
430	0	2	0	4	0	0	1	0	1	5
440	0	7	0	0	1	0	2	2	0	2
450	3	0	0	0	0	3	0	6	1	0
460	2	0	1	3	0	6	0	0	7	0
470	3	3	0	1	2	0	4	0	1	5
480	0	3	0	0	1	0	3	2	0	1
490	1	0	1	0	2	4	0	3	1	0

	0	1	2	3	4	5	6	7	8	9
500	7	0	0	3	0	2	3	0	2	0
510	2	6	0	4	2	0	2	0	3	1
520	0	4	3	0	2	0	1	6	0	3
530	0	0	3	0	2	0	0	1	1	0
540	12	0	0	5	0	1	0	0	0	0
550	1	2	0	7	0	0	2	0	1	4
560	0	3	0	0	1	0	2	5	0	2
570	2	0	0	0	1	2	0	3	0	0
580	1	0	2	1	0	6	0	0	8	0
590	1	0	0	4	4	0	1	0	0	2
600	0	2	1	0	1	0	1	3	0	5
610	1	0	3	0	2	3	0	1	1	0
620	4	0	1	8	0	2	1	0	2	0
630	0	1	0	2	2	0	4	0	0	8
640	0	5	1	0	0	0	1	4	0	5
650	1	1	2	0	4	2	0	5	2	0
660	6	0	4	4	0	6	0	0	1	0
670	2	3	0	5	0	0	3	0	0	2
680	0	0	2	0	1	0	1	3	0	5
690	1	0	2	0	1	2	0	2	2	0
700	4	0	4	0	0	9	0	0	4	0
710	0	2	0	2	4	0	3	0	1	5
720	0	6	1	0	1	0	2	3	0	2
730	1	0	1	0	0	12	0	2	1	0
740	4	0	2	8	0	4	1	0	3	0
750	1	3	0	1	2	0	13	0	2	5
760	0	4	2	0	0	0	0	2	0	3
770	0	0	2	0	2	3	0	4	1	0
780	9	0	1	4	0	4	0	0	0	0
790	0	3	0	1	2	0	0	0	2	1
800	0	2	0	0	3	0	2	3	0	5
810	5	0	8	0	2	3	0	3	1	0
820	1	0	4	4	0	1	1	0	2	0
830	0	2	0	5	1	0	0	0	1	3
840	0	2	1	0	1	1	3	4	0	1
850	2	0	3	0	0	7	0	6	2	0
860	2	1	1	0	0	5	2	0	3	0
870	1	1	0	1	0	0	2	0	0	6
880	0	4	5	0	1	0	0	3	0	5
890	1	0	1	0	2	1	0	2	1	0
900	14	0	1	8	0	2	2	0	1	0
910	0	3	0	3	0	0	2	0	1	5
920	0	4	0	0	5	0	1	3	0	0
930	2	0	6	0	0	1	0	2	1	0
940	0	0	1	3	0	9	0	0	3	0
950	0	3	0	3	2	0	2	0	0	5
960	0	3	0	0	5	0	1	4	0	5
970	0	0	9	0	0	3	0	5	0	1
980	0	0	3	2	0	1	1	0	1	0
990	2	5	0	2	1	0	2	0	3	1

	0	1	2	3	4	5	6	7	8	9
1000	0	4	0	0	0	0	0	5	0	2
1010	1	0	2	0	1	5	0	0	0	0
1020	3	0	1	3	0	3	3	0	2	2
1030	1	4	0	4	1	0	1	0	0	5
1040	0	1	1	0	3	0	0	3	0	7
1050	2	0	2	0	1	2	0	3	1	0
1060	2	0	3	4	0	2	0	0	0	0
1070	0	5	0	0	0	0	0	0	2	3
1080	0	6	1	0	1	1	3	5	0	6
1090	2	0	10	0	2	4	0	3	3	0
1100	8	0	1	6	0	6	1	0	1	0
1110	2	5	0	5	0	0	6	0	0	2
1120	0	5	1	0	0	0	3	5	0	3
1130	1	0	0	0	4	4	0	3	2	0
1140	5	0	2	0	0	2	2	0	2	0
1150	0	2	0	5	1	0	2	0	2	3
1160	0	3	0	0	3	0	2	2	0	3
1170	2	0	0	0	1	1	0	7	2	0
1180	2	0	3	3	0	3	1	0	10	0
1190	1	2	0	1	0	0	2	0	1	1
1200	0	3	1	0	5	0	1	3	0	4
1210	3	0	2	0	2	2	0	1	3	0
1220	2	0	0	3	0	2	1	0	2	0
1230	1	7	0	6	1	0	6	0	1	10
1240	0	3	4	0	0	0	3	4	0	2
1250	0	0	4	0	0	1	0	3	0	0
1260	15	0	0	2	0	6	2	0	1	0
1270	1	4	0	3	0	0	3	0	3	2
1280	0	4	1	0	2	0	2	2	0	5
1290	0	0	0	0	1	1	0	3	1	0
1300	6	0	2	0	0	2	1	0	2	0
1310	1	2	0	3	3	0	0	0	0	1
1320	0	2	0	0	2	0	1	5	0	2
1330	0	0	4	0	1	2	0	4	3	0
1340	5	0	0	2	0	1	0	0	1	0
1350	3	7	0	2	1	0	2	0	1	5
1360	0	4	1	0	3	0	1	2	0	3
1370	0	0	1	0	1	7	0	2	0	0
1380	2	0	0	1	0	6	3	0	2	0
1390	2	5	0	3	0	0	2	0	1	6
1400	0	4	2	0	11	0	0	8	0	2
1410	3	0	2	0	3	3	0	3	0	0
1420	4	0	4	6	0	5	1	0	5	0
1430	2	5	0	1	2	0	1	0	1	0
1440	0	3	1	0	1	0	1	5	0	7
1450	0	0	1	0	1	3	0	3	3	0
1460	3	0	0	2	0	2	1	0	3	0
1470	6	3	0	5	0	0	4	0	1	5
1480	0	4	1	0	0	0	1	4	0	1
1490	1	0	1	0	0	4	0	6	0	0

	0	1	2	3	4	5	6	7	8	9
1500	11	0	0	3	0	3	0	0	2	0
1510	1	4	0	3	1	0	1	0	3	5
1520	0	4	0	0	3	0	1	5	0	5
1530	3	0	0	0	1	1	0	3	0	0
1540	2	0	1	3	0	4	0	0	4	0
1550	2	2	0	6	2	0	1	0	1	2
1560	0	4	3	0	2	0	6	2	0	5
1570	1	0	2	0	0	11	0	7	2	0
1580	3	0	2	7	0	1	2	0	3	0
1590	3	3	0	1	0	0	5	0	0	2
1600	0	3	1	0	1	0	2	5	0	2
1610	0	0	1	0	0	3	0	12	1	0
1620	14	0	0	4	0	6	0	0	1	0
1630	1	1	0	5	2	0	3	0	4	2
1640	0	4	1	0	0	0	0	4	0	2
1650	4	0	1	0	0	1	0	3	0	0
1660	4	0	0	3	0	4	1	0	1	0
1670	0	3	0	3	1	0	2	0	2	7
1680	0	4	0	0	0	0	0	4	0	3
1690	0	0	4	0	1	1	0	4	2	0
1700	5	0	0	3	0	5	0	0	3	0
1710	1	2	0	4	0	0	2	0	0	3
1720	0	3	2	0	2	0	1	0	0	5
1730	0	0	0	0	1	5	0	1	1	0
1740	3	0	1	4	0	3	2	0	1	0
1750	2	1	0	1	0	0	1	0	0	1
1760	0	0	0	0	14	0	0	2	0	3
1770	2	0	1	0	2	2	0	4	2	0
1780	1	0	5	3	0	7	0	0	1	0
1790	1	4	0	2	0	0	0	0	2	3
1800	0	1	1	0	1	0	3	4	0	2
1810	3	0	0	0	1	4	0	1	2	0
1820	5	0	0	3	0	2	0	0	2	0
1830	1	4	0	3	0	0	2	0	2	1
1840	0	3	0	0	6	0	1	6	0	2
1850	0	0	0	0	1	6	0	5	0	0
1860	10	0	2	5	0	4	3	0	0	0
1870	2	3	0	2	0	0	0	0	2	3
1880	0	2	0	0	2	0	1	4	0	2
1890	3	0	0	0	0	1	0	0	0	0
1900	3	0	2	2	0	4	1	0	1	0
1910	0	13	0	3	0	0	0	0	3	5
1920	0	1	0	0	1	0	1	3	0	2
1930	0	0	2	0	1	4	0	3	1	0
1940	2	0	0	6	0	6	1	0	3	0
1950	1	2	0	16	0	0	2	0	1	2
1960	0	3	4	0	3	0	1	7	0	3
1970	0	0	0	0	5	1	0	2	0	0
1980	11	0	0	3	0	6	1	0	1	0
1990	2	2	0	4	1	0	1	0	0	3

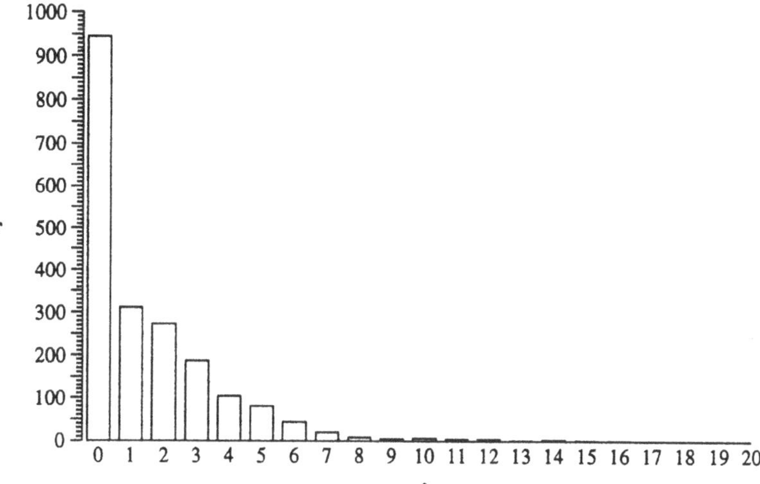

Figure 1. Number of degrees r, $1 \leq r \leq 2000$, for which there exists s irreducible trinomials $T_{r,k}(x)$, $k \leq r/2$.

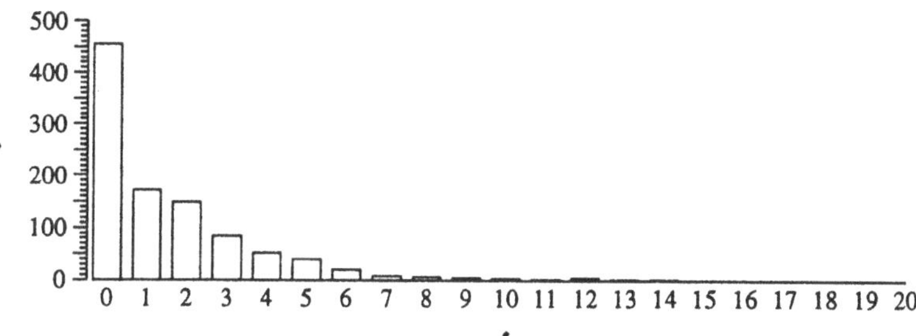

Figure 2. Number of degrees r, $1 \leq r \leq 1000$, for which there exists s irreducible trinomials $T_{r,k}(x)$, $k \leq r/2$.

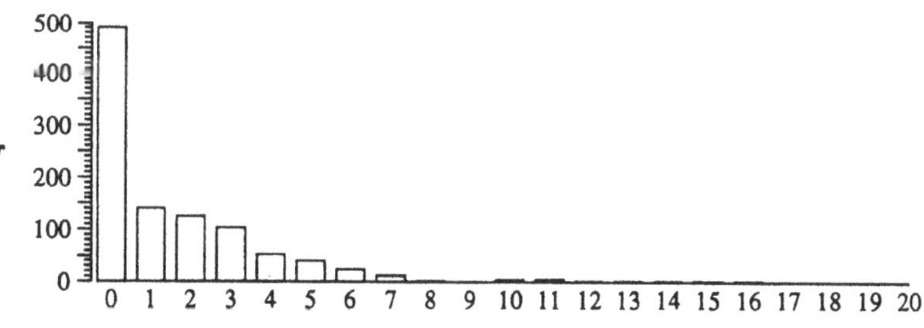

Figure 3. Number of degrees r, $1001 \leq r \leq 2000$, for which there exists s irreducible trinomials $T_{r,k}(x)$, $k \leq r/2$.

Table 6. Irreducible trinomials $T_{n,k}(x)$, $k \leq n/2$, for $1 \leq n \leq 2000$

2: 1
3: 1
4: 1
5: 2
6: 1, 3
7: 1, 3
9: 1, 4
10: 3
11: 2
12: 3, 5
14: 5
15: 1, 4, 7
17: 3, 5, 6
18: 3, 7, 9
20: 3, 5
21: 2, 7
22: 1
23: 5, 9
25: 3, 7
28: 1, 3, 9, 13
29: 2
30: 1, 9
31: 3, 6, 7, 13
33: 10, 13
34: 7
35: 2
36: 9, 11, 15
39: 4, 8, 14
41: 3, 20
42: 7
44: 5
46: 1
47: 5, 14, 20, 21
49: 9, 12, 15, 22
52: 3, 7, 19, 21
54: 9, 21, 27
55: 7, 24
57: 4, 7, 22, 25
58: 19
60: 1, 9, 11, 15, 17, 23
62: 29
63: 1, 5, 11, 28, 31
65: 18, 32
66: 3
68: 9, 33
71: 6, 9, 18, 20, 35
73: 25, 28, 31
74: 35
76: 21
79: 9, 19
81: 4, 16, 35
84: 5, 9, 11, 13, 27, 35, 39
86: 21
87: 13
89: 38
90: 27
92: 21
93: 2
94: 21
95: 11, 17
97: 6, 12, 33, 34
98: 11, 27
100: 15, 19, 25, 37, 49
102: 29, 37
103: 9, 13, 30, 31
105: 4, 7, 8, 16, 17, 28, 37, 43, 49, 52
106: 15
108: 17, 27, 31, 33, 45
110: 33
111: 10, 49
113: 9, 15, 30
118: 33, 45
119: 8, 38
121: 18, 30
123: 2
124: 19, 37, 45, 55
126: 21, 49
127: 1, 7, 15, 30, 63
129: 5, 31, 46
130: 3
132: 17, 29
134: 57
135: 11, 16, 22, 29
137: 21, 35, 57
140: 15, 29, 45, 65
142: 21
145: 52, 69
146: 71
147: 14, 49
148: 27
150: 53, 73
151: 3, 9, 15, 31, 39, 43, 46, 51, 63, 66, 67, 70
153: 1, 8
154: 15
155: 62
156: 9, 11, 57, 61, 63, 65
159: 31, 34, 40
161: 18, 39, 60
162: 27, 63, 81
166: 37
167: 6, 35, 59, 77
169: 34, 42, 57, 84
170: 11, 23
172: 1, 7, 81
174: 13, 57
175: 6, 16, 18, 57
177: 8, 22, 88
178: 31, 87
180: 3, 27, 33, 45, 55, 69
182: 81
183: 56
185: 24, 41, 69
186: 11, 79
191: 9, 18, 51, 71
193: 15, 73, 85
194: 87
196: 3, 33, 67
198: 9, 65
199: 34, 67
201: 14, 17, 59, 79
202: 55
204: 27, 99
207: 43
209: 6, 8, 14, 45, 47, 50, 62
210: 7
212: 105
214: 73
215: 23, 51, 63, 77, 101
217: 45, 64, 66, 82, 85
218: 11, 15, 71, 83, 99
220: 7, 33, 49
223: 33, 34, 64, 70, 91
225: 32, 74, 88, 97, 109
228: 113
231: 26, 34, 91
233: 74
234: 31, 103
236: 5
238: 73, 117
239: 36, 81
241: 70
242: 95
244: 111
247: 82, 102
249: 35, 86
250: 103
252: 15, 27, 33, 39, 53, 59, 67, 77, 81, 105, 109, 117
253: 46
255: 52, 56, 82
257: 12, 41, 48, 51, 65
258: 71, 83
260: 15, 35, 95, 105
263: 93
265: 42, 127
266: 47
268: 25, 61
270: 53, 81, 133
271: 58, 70
273: 23, 28, 53, 55, 56, 67, 88, 92, 98, 110, 113
274: 67, 99, 135
276: 63, 91
278: 5
279: 5, 10, 38, 40, 41, 59, 73, 76, 80, 125
281: 93, 99
282: 35, 43, 63
284: 53, 99, 119, 141
286: 69, 73
287: 71, 116, 125
289: 21, 36, 84
292: 37, 97
294: 33, 49, 61, 81
295: 48, 112, 123, 142, 147

297: 5, 83, 103, 122, 137
300: 5, 7, 45, 55, 57, 73, 75, 91, 111, 147
302: 41
303: 1
305: 102
308: 15
310: 93
313: 79, 121
314: 15
316: 63, 135
318: 45
319: 36, 52, 129
321: 31, 41, 56, 76, 82, 155
322: 67
324: 51, 81, 93, 99, 135, 149
327: 34, 152
329: 50, 54
330: 99
332: 89, 123
333: 2
337: 55, 57, 135, 139, 147
340: 45, 165
342: 125, 133
343: 75, 135, 138, 159
345: 22, 37, 106
346: 63
348: 103
350: 53
351: 34, 55, 79, 116, 134
353: 69, 95, 138, 143, 153, 173
354: 99, 135
358: 57
359: 68, 117
362: 63, 107
364: 9, 67
366: 29
367: 21, 171
369: 91, 110
370: 139, 183
372: 111, 135, 165
375: 16, 64, 149, 182
377: 41, 75
378: 43, 63, 107, 147
380: 47, 63, 105
382: 81
383: 90, 108, 135
385: 6, 24, 51, 54, 111, 142, 159
386: 83
388: 159
390: 9, 49, 89
391: 28, 31
393: 7, 62, 91
394: 135
396: 25, 51, 87, 109, 169, 175
399: 26, 49, 86, 109, 154, 181
401: 152, 170
402: 171
404: 65, 189
406: 141, 157
407: 71, 105
409: 87
412: 147
414: 13, 53
415: 102, 163
417: 107, 113, 155
418: 199
420: 7, 45, 65, 77, 87, 127, 135, 161, 195
422: 149, 177
423: 25
425: 12, 21, 42, 66, 111, 191
426: 63
428: 105
431: 120, 200
433: 33, 61, 118, 153
436: 165
438: 65
439: 49, 133, 145, 156, 171
441: 7, 31, 35, 127, 196, 212, 217
444: 81
446: 105, 153
447: 73, 83
449: 134, 167
450: 47, 79, 159
455: 38, 62, 74
457: 16, 61, 123, 210, 217, 226
458: 203
460: 19, 61
462: 73
463: 93, 168, 214
465: 31, 59, 103, 124, 158, 217
468: 27, 33, 143, 171, 183, 189, 195
470: 9, 149, 177
471: 1, 119, 127
473: 200
474: 191, 215
476: 9, 15, 129, 141
478: 121
479: 104, 105, 122, 158, 224
481: 138, 201, 231
484: 105
486: 81, 189, 243
487: 94, 127
489: 83
490: 219
492: 7
494: 17, 137
495: 76, 89, 118, 226
497: 78, 216, 228
498: 155
500: 27, 75, 95, 125, 185, 243, 245
503: 3, 26, 248
505: 156, 174
506: 23, 95, 135
508: 9, 109
510: 69, 197
511: 10, 15, 31, 160, 202, 216
513: 26, 85, 175, 242
514: 67, 103
516: 21, 91
518: 33, 45, 113
519: 79
521: 32, 48, 158, 168
522: 39, 171, 259
524: 167, 195
526: 97
527: 47, 123, 147, 152, 198, 239
529: 42, 114, 157
532: 1, 37, 81
534: 161, 261
537: 94
538: 195
540: 9, 11, 81, 99, 113, 135, 155, 165, 179, 191, 207, 211
543: 16, 28, 58, 203, 235
545: 122
550: 193
551: 135, 240
553: 39, 57, 94, 99, 109, 255, 258
556: 153, 273
558: 73
559: 34, 70, 148, 210
561: 71, 109, 155
564: 163
566: 153, 273
567: 28, 112, 143, 245, 275
569: 77, 210
570: 67, 143
574: 13
575: 146, 258
577: 25, 27, 231
580: 237
582: 85, 261
583: 130
585: 88, 121, 151, 157, 232, 256
588: 35, 77, 91, 99, 151, 201, 245, 253
590: 93
593: 86, 108, 119, 177
594: 19, 27, 35, 195
596: 273
599: 30, 210
601: 201, 202
602: 215
604: 105
606: 165
607: 105, 147, 273
609: 31, 91, 128, 181, 233
610: 127
612: 81, 157, 297
614: 45, 177
615: 211, 232, 230
617: 200
618: 295
620: 9, 93, 95, 185
622: 297
623: 68, 87, 128, 185, 230, 251, 296, 311
625: 133, 156
626: 251
628: 223, 289
631: 307
633: 101, 292
634: 39, 315
636: 217, 269, 311, 315

967: 36, 130, 210, 400
969: 31, 74, 308, 404, 446
972: 7, 115, 153, 155, 243, 279, 297, 377, 405
975: 19, 121, 133
977: 15, 63, 207, 374, 480
979: 178
982: 177, 277, 481
983: 230, 342
985: 222
986: 3
988: 121
990: 161, 297
991: 39, 171, 193, 466, 472
993: 62, 367
994: 223
996: 65, 369
998: 101, 417, 477
999: 59
1001: 17, 54, 354, 422
1007: 75, 96, 351, 386, 405
1009: 55, 148
1010: 99
1012: 115, 301
1014: 385
1015: 186, 258, 447, 466, 484
1020: 135, 461, 495
1022: 317
1023: 7, 43, 127
1025: 294, 306, 383
1026: 35, 375, 399
1028: 119, 203
1029: 98, 343
1030: 93
1031: 68, 116, 287, 371
1033: 108, 330, 340, 498
1034: 75
1036: 411
1039: 21, 88, 279, 364, 418
1041: 412
1042: 439
1044: 41, 71, 361
1047: 10, 430, 470
1049: 141, 227, 290, 293, 327, 357, 390
1050: 159, 371
1052: 291, 357
1054: 105
1055: 24, 243
1057: 198, 331, 438
1058: 27
1060: 439, 525
1062: 49, 297, 405
1063: 168, 208, 285, 370
1065: 463, 476
1071: 7, 50, 56, 281, 436
1078: 361, 445
1079: 230, 282, 342
1081: 24, 235, 318, 348, 423, 525
1082: 407
1084: 189
1085: 62
1086: 189, 217, 321
1087: 112, 240, 445, 457, 481
1089: 91, 148, 206, 283, 400, 466
1090: 79, 243
1092: 23, 65, 77, 113, 143, 169, 201, 353, 455, 479
1094: 57, 261
1095: 139, 383, 476, 538
1097: 14, 86, 303
1098: 83, 87, 503
1100: 35, 93, 165, 209, 245, 407, 497, 519
1102: 117
1103: 65, 71, 164, 189, 434, 465
1105: 21, 57, 96, 321, 363, 439
1106: 195
1108: 327
1110: 417, 549
1111: 13, 40, 238, 321, 531
1113: 107, 217, 238, 280, 310
1116: 59, 227, 333, 405, 479, 495
1119: 283, 329
1121: 62, 107, 113, 176, 203
1122: 427
1126: 105, 309, 561
1127: 27, 171, 317, 465, 512
1129: 103, 135, 208
1130: 551
1134: 129, 189, 321, 441
1135: 9, 36, 106, 298
1137: 277, 313, 458
1138: 31, 183
1140: 141, 189, 413, 419, 539
1142: 357, 461
1145: 227, 506
1146: 131, 243
1148: 23, 221
1151: 90, 125
1153: 241, 268, 415, 499, 541
1154: 75
1156: 307, 513
1158: 245, 249
1159: 66, 129, 424
1161: 365, 409, 551
1164: 19, 85, 161
1166: 189, 513
1167: 133, 265
1169: 114, 173, 207
1170: 27, 267
1174: 133
1175: 476
1177: 16, 64, 70, 126, 186, 288, 412
1178: 375, 491
1180: 25, 99
1182: 77, 405, 425
1183: 87, 108, 136
1185: 134, 317, 386
1186: 171
1188: 75, 153, 251, 261, 287, 327, 395, 413, 507, 525
1190: 233
1191: 196, 212
1193: 173
1196: 281, 519
1198: 405
1199: 114
1201: 171, 360, 388
1202: 287
1204: 43, 129, 387, 459, 559
1206: 513
1207: 273, 423, 511
1209: 118, 145, 290, 356
1210: 243, 363, 463
1212: 203, 567
1214: 257, 605
1215: 302, 322
1217: 393
1218: 91, 215, 471
1220: 413, 555
1223: 255, 549, 588
1225: 234, 547
1226: 167
1228: 27, 193
1230: 433
1231: 105, 265, 355, 390, 453, 496, 517
1233: 151, 172, 247, 256, 368, 520
1234: 427
1236: 49, 119, 151, 301, 317, 441
1238: 153
1239: 4, 10, 56, 58, 146, 154, 193, 466, 610, 616
1241: 54, 165, 501
1242: 203, 395, 403, 479
1246: 25, 69, 421
1247: 14, 27, 90, 585
1249: 187, 237
1252: 97, 265, 409, 577
1255: 589
1257: 289, 347, 595
1260: 21, 135, 195, 231, 261, 295, 335, 381, 385, 405, 421, 483, 545, 585, 589
1263: 77, 346
1265: 119, 338, 371, 467, 552, 576
1266: 7, 447
1268: 345
1270: 333
1271: 17, 53, 380, 450
1273: 168, 357, 495
1276: 217, 427, 541
1278: 189, 385, 637
1279: 216, 418
1281: 229, 392, 547, 596
1282: 231
1284: 223, 315
1286: 153, 441
1287: 470, 622
1289: 99, 204, 212, 242, 609
1294: 201
1295: 38
1297: 198, 337, 565
1298: 399
1300: 75, 175, 217, 475, 525, 607

1634: 227, 683
1636: 37, 493, 663
1638: 173, 217, 397, 721
1639: 427, 709
1641: 287, 319, 382, 637
1642: 231
1647: 310, 469, 494, 757
1649: 434, 660
1650: 579, 583, 727, 803
1652: 45
1655: 53
1657: 16, 105, 520
1660: 37, 445, 615, 763
1663: 99, 319, 808
1665: 176, 349, 376, 488
1666: 271
1668: 459
1671: 202, 292, 406
1673: 90, 134, 605
1674: 755
1676: 363, 413
1678: 129, 709
1679: 20, 42, 204, 279, 479, 537, 668
1681: 135, 373, 610, 802
1687: 31, 139, 391, 648
1689: 758, 805, 826
1692: 359, 489, 625, 835
1694: 501
1695: 29
1697: 201, 420, 768, 822
1698: 459, 767
1700: 225, 311, 441, 567, 825
1703: 161, 285, 590
1705: 52, 162, 217, 633, 744
1708: 93, 211, 303
1710: 201
1711: 178, 465
1713: 250, 332, 425, 626
1716: 221, 377
1719: 113, 268, 611
1721: 300, 531, 684
1722: 39, 839
1724: 261, 749
1726: 753
1729: 94, 118, 216, 820, 862
1734: 461
1735: 418, 651, 727, 853, 861
1737: 403
1738: 367
1740: 259, 515, 711
1742: 869
1743: 173, 302, 506, 602
1745: 369, 702, 764
1746: 255, 783
1748: 567
1750: 457, 481
1751: 482
1753: 775
1756: 99
1759: 165
1764: 105, 199, 231, 273, 297, 371, 413, 453, 469, 539, 603, 611, 735, 759
1767: 250, 623
1769: 327, 560, 882
1770: 279, 607
1772: 371
1774: 117, 829
1775: 486, 771
1777: 217, 544, 801, 859
1778: 635, 815
1780: 457
1782: 57, 81, 105, 245, 585
1783: 439, 520, 559
1785: 214, 364, 392, 452, 574, 676, 724
1788: 819
1790: 593
1791: 190, 349, 512, 682
1793: 114, 254
1798: 69, 685
1799: 312, 321, 374
1801: 502
1802: 843
1804: 747
1806: 101, 497, 581
1807: 123, 438, 525, 700
1809: 521, 734
1810: 171, 283, 427
1814: 545
1815: 163, 698, 736, 856
1817: 479
1818: 495, 871
1820: 11, 45, 335, 359, 779
1823: 684, 735, 893
1825: 9, 759
1828: 273, 343
1830: 381
1831: 51, 99, 511, 771
1833: 518, 670, 881
1836: 243, 891
1838: 53, 465
1839: 836
1841: 66, 710, 873
1844: 339, 375, 447, 557, 623, 705
1846: 901
1847: 180, 257, 596, 614, 768, 920
1849: 49, 412
1854: 885
1855: 39, 136, 607, 633, 807, 837
1857: 688, 709, 785, 788, 809
1860: 13, 27, 31, 279, 341, 527, 555, 713, 761, 871
1862: 149, 797
1863: 260, 293, 527, 592, 685
1865: 53, 521, 738, 762
1866: 11, 367, 891
1870: 121, 253
1871: 261, 345, 806
1873: 199, 763
1878: 253, 341
1879: 174, 613, 852
1881: 370, 892
1884: 669, 867
1886: 833
1887: 353, 392, 652, 800
1889: 29, 497
1890: 371, 667, 931
1895: 873
1900: 235, 297, 525
1902: 733, 945
1903: 778, 930
1905: 344, 358, 424, 946
1906: 931
1908: 945
1911: 67, 161, 172, 196, 304, 371, 392, 469, 616, 644, 686, 770, 791
1913: 462, 530, 639
1918: 477, 705, 793
1919: 105, 344, 680, 720, 848
1921: 468
1924: 327
1926: 357
1927: 25, 151, 168
1929: 31, 553
1932: 277, 637
1934: 413
1935: 103, 236, 382, 811
1937: 231, 297, 930
1938: 747
1940: 113, 177
1943: 11, 60, 158, 519, 528, 960
1945: 91, 288, 294, 328, 358, 652
1946: 51
1948: 603, 709, 919
1950: 9
1951: 121, 759
1953: 17, 35, 70, 178, 266, 280, 287, 413, 532, 560, 605, 650, 746, 842, 875, 925
1956: 279, 291
1958: 89
1959: 371, 409
1961: 771, 783, 965
1962: 99, 135, 639, 747
1964: 21, 113, 735
1966: 801
1967: 26, 197, 236, 449, 596, 809, 866
1969: 175, 546, 585
1974: 165, 245, 301, 365, 745
1975: 841
1977: 230, 740
1980: 33, 63, 125, 297, 307, 363, 435, 545, 759, 845, 875
1983: 113, 161, 967
1985: 311, 584, 597, 668, 672, 774
1986: 891
1988: 555
1990: 133, 217
1991: 546, 896
1993: 103, 124, 717, 762
1994: 15
1996: 307
1999: 367, 585, 732
2000:

Singular Algebraic Curves over Finite Fields

Despina Polemi[1] and Takis Sakkalis[2]

[1] Department of Mathematics, State University of New York, Farmingdale, NY 11735, polemid@snyfarva.bitnet

[2] Department of Mathematical Sciences, Oakland University, Rochester, MI 48309 sakkalis@vela.acs.oakland.edu

ABSTRACT. This paper presents algorithms for the identification and resolution of rational and non-rational singularities (by means of *blowings-up*) of a projective plane curve $C : F(x_1, x_2, x_3) = 0$ with coefficients in a finite field $\mathbf{k}$. As a result, the *genus* of the curve is computed. In addition the running time of the algorithms are also analyzed.

KEY WORDS AND PHRASES. finite field, rational points, non-rational points, singular points, blowing-up process, genus.

1 Introduction.

Algebraic curves are an essential tool in such areas of mathematics as algebraic geometry, computer aided geometric design, coding theory and cryptography. An algebraic curve can be given by its *implicit* equation or a *parametric* (*explicit*) equation. The representation of choice is determined by the operations one wants to perform with the curve. Singular points play a fundamental role in the theory of algebraic curves, in particular, in problems like the rational parametrization of a curve, Sendra & Winkler [14] and the topology of real curves, Sakkalis [12].

Computational techniques dealing with algebraic curves with singular points, defined over an algebraically closed field, have been considered by Abhyankar & Bajaj [1] and more recently by Sakkalis & Farouki [13]. On the other hand, the resolution (*desingularization*) of singularities of an algebraic curve is a well known task when the coefficient field is algebraically closed (Fulton [3], Walker [18], Hartshorne [4]).

The problem of desingularization of algebraic curves over finite fields is an important one in the theory of error-correcting codes, and it has been considered by several authors. Le Brigand & Risler in [7] resolve the singularities of a curve using the blowing-up process. However, in that paper it is assumed that the points on the curve are all rational over the ground field (*rationality assumption*). The same assumption is adopted in Vladut & Manin [17], where they concentrate on the algorithmic aspect of the blowing-up process applied to modular curves.

The concept of the Cremona transformations along with the blowing-up process is used by Polemi et. al. in [9], [11] to algorithmically resolve the singularities of a curve, where the rationality assumption is deleted. Another desingularization technique, which does not require the rationality assumption, is based on the fact that desingularization is equivalent to the functorial nature of the integral closure. This last method is included in the IBM computer algebra system AXIOM (SCRATCHPAD) [2], [10].

The deficiency of singularities on the curve from its maximum allowable limit is measured by a birational invariant of the curve, called *genus*. Effectively computing the genus of a singular curve defined over a finite field is another important task in the theory of error correcting codes, since curves of small genus with many rational points give "good" algebraic geometric Goppa codes.

In this paper we are interested in algebraic singular curves defined over finite fields. Our objective is to develop robust and efficient algorithms that deal with curves of this type. More precisely, if $C : F(x_1, x_2, x_3) = 0$ is such a curve, in Section 2 we present an algorithm (Algorithm S-M) that identifies the rational and nonrational singularities of C and computes their multiplicities. In Section 3 we reconsider the method of the *blowing-up* process algorithmically (Algorithm B-U). Furthermore, we resolve the singularities of the curve using successive applications of blowings-up (Algorithm RES). In addition, we analyze the number of steps of the mentioned algorithms as a function in n, the degree of the curve C, and the cardinality q of the finite field. It is interesting to note that our procedures do not depend on the rationality assumption. Therefore, they can be applied to any algebraic curve over a finite field whose points are not necessarily rational over that field. We conclude with Section 4, where we compute the genus of C using the quantities return from our above mentioned algorithms.

2 Singularities over a finite field.

For an extensive exposition of the concepts that arise in this section, we refer the reader to Silverman ([15], Chapters 1-2), Tsfasman ([16], Ch. 2.3) and Moreno ([8], §5.4).

We denote by $\mathbf{k} = \mathbf{F}_q$ a finite field with $q = p^e$ elements, where p is a prime number, and its algebraic closure by $\overline{\mathbf{k}}$. Let $F(X,Y,Z)$ be a homogeneous polynomial of degree n with coefficients in the finite field $\mathbf{k}$ so that $F(X,Y,Z)$ is absolutely irreducible over $\mathbf{k}$, that is F is irreducible over $\overline{\mathbf{k}}$.

Definition 1. In the projective plane $\mathbf{P}^2(\overline{\mathbf{k}})$ we call the set of points

$$C = \{ (x_1, x_2, x_3) \in \mathbf{P}^2(\overline{\mathbf{k}}) : \quad F(x_1, x_2, x_3) = 0 \},$$

the projective plane algebraic curve defined over $\mathbf{k}$.

The projective plane $\mathbf{P}^2(\overline{\mathbf{k}})$ can be thought as the union of three non-disjoint subsets

$$U_i = \{ (x_1, x_2, x_3) \in \mathbf{P}^2(\overline{\mathbf{k}}): \quad x_i \neq 0 \} \quad i = 1, 2, 3.$$

Each U_i is isomorphic to the affine space $\mathbf{A}^2(\overline{\mathbf{k}})$, and is called an *affine neighborhood* of $\mathbf{P}^2(\overline{\mathbf{k}})$.

Definition 2. The set of points in the intersection $C \cap U_i$ is called the affine representation of C in the affine neighborhood U_i.

Each affine representation of the curve C is designated by a polynomial function in the field of rational functions $\mathbf{k}(C)$ of the curve C.

Let $f(x, y) \in \mathbf{k}(C)$ be an affine representation of C in some affine neighborhood U_i and let $P = (a, b)$ in $C \cap U_i$. We say that P is a *singular point* (or a *singularity*) of C if both formal partial derivatives f_x and f_y equal zero at P; otherwise it is a *nonsingular* point. If P is a singular point, then the Taylor expansion of the polynomial $f(x, y)$ about the singular point $(x, y) = (a, b)$ is

$$f(\xi, \eta) = f_m(\xi, \eta) + \dots + f_n(\xi, \eta), \tag{1}$$

where $\xi = x-a$, $\eta = y-b$, and $f_i(\xi, \eta)$ are homogeneous forms of ascending degree i with $m \geq 2$. In that case m is called the *multiplicity* of the singularity P and it is denoted by $m_P(C)$. Furthermore, if P is a singularity and the discriminant of $f_m(\xi, \eta)$ is non-zero, then we call P an *ordinary* singularity otherwise the singularity is called *nonordinary*. Moreover if $f_m(\xi, \eta)$ factors in $\mathbf{k}$ (or in $\overline{\mathbf{k}}$) as

$$f_m(\xi, \eta) = \prod_{i=1}^{m} (\alpha_i \xi + \beta_i \eta)^{r_i}, \tag{2}$$

then $\alpha_i \xi + \beta_i \eta = 0$ are the *tangents* of the curve C at the singularity P. If a_i, $b_i \in \mathbf{k}$ then $\alpha_i \xi + \beta_i \eta = 0$ is called a *rational tangent*, otherwise it is called nonrational. We note that a point $P \in C$ is nonsingular if and only if its multiplicity $m_P(C) = 1$.

Definition 3. Let $Q \in C$. Then, the *degree* $d(Q)$ of Q is a positive integer with the property that Q is conjugate to precisely $d(Q)$ points $(x_1, x_2, x_3) \in C$ under the action of the Galois group $Gal(\overline{\mathbf{k}}/\mathbf{k})$.

If $d(Q) = 1$ then Q is a $\mathbf{k}$-*rational* point. Otherwise Q is *nonrational* of degree $d(Q) > 1$. In particular, $d(Q) = [\mathbf{k}(Q) : \mathbf{k}]$ where $\mathbf{k}(Q)$ is the normal extension of $\mathbf{k}$ generated by the coordinates of Q. The point Q becomes rational in $\mathbf{k}_{d(Q)}$, a finite extension of the ground field $\mathbf{k}$ of degree $d(Q)$.

The following theorem provides the means for constructing a finite extension field $\mathbf{k}_s$ of $\mathbf{k}$, so that all singular points of C are rational in $\mathbf{k}_s$.

Theorem 4. *Let $\{P_i\}$ be the nonrational singular points of degree $d(P_i) > 1$ on the curve $C : F(X, Y, Z) = 0$ defined over $\mathbf{k} = \mathbf{F}_q$. Then, there exists an integer s with the property that in the finite extension $\mathbf{k}_s = \mathbf{F}_{q^s}$ of $\mathbf{k}$ the points P_i become $\mathbf{k}_s$-rational.*

Proof. The singular points of C in the affine neighborhood U_1 are the solutions of the system of equations

$$f(x, y) = f_x = f_y = 0, \tag{3}$$

where f_x, f_y denote the partial derivatives of $f(x,y)$ with respect to x, y respectively. Since the polynomial $F(X,Y,Z)$ is absolutely irreducible, we see that $f(x,y)$ is absolutely irreducible as well. Let t be an indeterminate, and consider the polynomial

$$a(x,t) = \mathrm{Res}_y(f, tf_x + f_y) .$$

We observe that $a(x,t) \neq 0$, for if not, f and tf_x+f_y would have a common factor, $r(x,y)$ say, of positive degree in y. But then $r(x,y)$ divides $(t_1-t_2)f_x$ for t_1, t_2 in $\mathbf{k}$. The latter contradicts the fact that f and f_x have no common factor, since f is irreducible. We write $a(x,t) = \sum_{i=0}^m a_i(x)t^i$ and let $A(x) = \mathrm{GCD}(a_0, \cdots, a_m)$. Let (x_0, y_0) be a singular point of C. Then $f(x_0,y_0) = tf_x(x_0,y_0)+f_y(x_0,y_0) = 0$ for all t. Thus, $a(x_0,t) \equiv 0$ which implies that x_0 is a root of $A(x)$. Similarly, by working over the affine neighborhoods U_2 and U_3, we can find polynomials $B(y)$ and $D(z)$. Note that if the polynomials $A(x), B(y), D(z)$ are all constants the curve C is nonsingular. Let s be the least common multiple of the degrees of the irreducible factors (over $\mathbf{k}$) of the polynomial $E(u) = A(u)B(u)D(u)$, and let K be the splitting field of $E(u)$ over $\mathbf{k}$. Then $K = \mathbf{k}_s = \mathbf{F}_{q^s}$. Finally, we note that every singular point of C has become $\mathbf{k}_s$-rational. □

We will now describe a method that identifies the $\mathbf{k}_s$-rational singular points of C and their multiplicities. First, we will find the singular points of C in the affine neighborhood U_1. Define

$$p(x) = \mathrm{Res}_y(f, f_x),\ q(y) = \mathrm{Res}_x(f, f_y) ,$$

and note that the irreducibility of $f(x,y)$ implies that $p(x)q(y) \neq 0$. As we noted previously, the singular points of C, in the affine neighborhood U_1, are precisely the solutions of system (3). Let then (x_0, y_0) be such a singular point. Then obviously $p(x_0) = q(y_0) = 0$. In order to solve the above system we employ the following algorithm. We first find the roots of the polynomials $p(x)$ and $q(y)$ in $\mathbf{k}_s$. Let $(a,b) \in \mathbf{k}_s \times \mathbf{k}_s$ with $p(a) = q(b) = 0$. Then, we test whether (a,b) is a solution of the system (3), and thus a singular point of C.

Now let (a,b) be a singular point and let $g(x,y) = f(x+a, y+b)$. Expand the polynomial $g(x,y)$ around the point $(0,0)$ to obtain

$$g(x,y) = g_m(x,y) + g_{m+1}(x,y) + \ldots + g_n(x,y),$$

where g_i are homogeneous polynomials of ascending degree i, and the lowest degree m is the multiplicity of the singularity. The discriminant of $g_m(x,y)$ yields the type of the singularity (a,b). Finally, by repeating the above procedure over the affine neighborhoods U_2 and U_3 we identify all singularities of the curve C. We summarize in the following:

Algorithm S-M * Singularities-Multiplicities *
Input: An absolutely irreducible projective plane curve C, described by the homogeneous polynomial $F(X,Y,Z)$ of degree n, with coefficients in $\mathbf{k}$. The

fixed integer s.
Output: The $\mathbf{k}_s$-rational singular points of C and their multiplicities.

Step 1: Let $f(x,y) := F(X,Y,1)$ be the affine representation of C in U_1. Compute the partial derivatives f_x and f_y of the polynomial $f(x,y)$ with respect to x and y.
Step 2: Identify the distinct $\mathbf{k}_s$-rational singular points of the curve C by solving the system of equations $f(x,y) = f_x = f_y = 0$ over the extended ground field $\mathbf{k}_s$.
Step 3: Let P be a singular point. Translate P to the origin by applying an affine linear transformation, and thus find the type of the singularity P.
Step 4: Repeat steps 1-3 for the affine neighborhoods U_2 and U_3, where C is designated by the polynomials $f(x,z) := F(X,1,Z)$ and $f(y,z) := F(1,Y,Z)$, respectively.

We will now compute the *time complexity* of the Algorithm S-M. Time complexity is thought as the upper bound for the number of bit operations needed to perform an arithmetic task. For our estimates we use the *bitwise computational model* which is a simplified RAM model of logarithmic cost. We also use the computational results of Koblitz [6] and Knuth [5]. We do not consider the shift operations, memory access, etc.

Proposition 5. *The running time of the Algorithm S-M is $O(q^s n^6 \log^3 q^s)$, where n is the degree of the curve C and q^s is the cardinality of the extended ground field $\mathbf{k}_s = \mathbf{F}_{q^s}$ with $q = p^e$.*

Proof. The elements of the finite field $\mathbf{k}$ are polynomials with coefficients in $\mathbf{F}_p$ regarded modulo a primitive polynomial of degree e. The passage from the field $\mathbf{k}$ to the field $\mathbf{k}_s$ requires $O(e^3 t^3 \log^2 p)$, where $t = s/e$ (see [17], p.2640). Let $C(i)$ denote the time complexity of the ith step.

For the first step we compute the time needed to calculate the partial derivatives f_x and f_y. Since $f(x,y) \in [\mathbf{k}(y)](x)$ is considered as a polynomial in x, it has at most $O(n)$ terms, and thus this operation involves at most $O(n)$ multiplications of integers (i.e. the exponents of x) by polynomials in y of degrees at most $O(n)$ (i.e. the coefficients). Hence, the total number of multiplications is $O(n^2)$. Each multiplication of an integer by a field element requires $O(e\log^2 p)$ operations. Hence $C(1)$ is $O(n^2 e\log^2 p)$, which can be replaced by $O(n^2 \log^2 q)$.

For the second step we note that the time needed to compute the polynomials $p(x)$ and $q(y)$ is $O(n^3)$. We also note that the degrees of $p(x)$ and $q(y)$ are $O(n^2)$. The evaluation of a polynomial of degree $\leq n^2$ at an $a \in \mathbf{k}_s$ requires $O(n^2 \log^3 q^s)$ operations. Thus, the total number of operations required to find whether the point $x = a$ is a zero of $p(x)$ is $O(q^s n^2 \log^3 q^s)$. For each root a of $p(x)$, we solve $f(a,y) = 0$; this takes $O(q^s n^3 \log^3 q^s)$. Since there are $O(n^2)$ roots of $p(x)$, $C(2)$ is of the order $O(q^s n^5 \log^3 q^s)$.

For the third step, we apply the affine transformation $x \longmapsto x+a$ and $y \longmapsto y+b$ when $P=(a,b)$ and $a,\, b \neq 0$. This requires the evaluation of $f(x+a,y+b)$, that is we need to estimate the time for computing the representation:

$$(x+a)^i(y+b)^j = \sum_{k,l} x^k y^l P_{kl}(a,b)\,,$$

where $i+j$ are $\leq n$. Since the complexity of multiplying two polynomials in $\mathbf{F}_q[T]$ of degree N and M is $O(NM\log^3 q)$, we conclude that the time needed to obtain the expansions $(x+a)^i$ and $(y+b)^j$ is at most $O(n^2\log^3 q^s)$. Hence, $a_{ij}(x+a)^i(y+b)^j$ can be computed in time $O(n^2\log^3 q^s)$. Since $f(x,y)$ itself has $O(n^2)$ terms, the complexity for transforming P to the origin is $O(n^4\log^3 q^s)$.

To determine whether the singularity is ordinary or not, we compute the discriminant of f_m. The latter requires $O(n^3)$ time. Thus the total complexity of Step 3 is $O(n^4 log^3 q^s)$. Since there are $O(n^2)$ $\mathbf{k}_s$-rational singular points of C, we get that the running time of the algorithm is $O(q^s n^6\log^3 q^s)$. □

Corollary 6. *If* $s=1$ *then the running time of the Algorithm S-M is* $O(qn^6\log^3 q)$.

Example 1A. Consider the curve C described by the homogeneous form

$$F(X,Y,Z) = X^{21} + X^5Y^{16} + Y^4X^9Z^8 + XY^{12}Z^8 + YX^{10}Z^{10} + \\ + X^2Y^9Z^{10} + Y^3X^4Z^{14} + Y^7Z^{14} + XZ^{20} + YZ^{20} + Z^{21}$$

of degree 21, defined over the finite field $\mathbf{k} = \mathbf{F}_q$ with $q=2$ elements.

The affine representation of C in the affine neighborhood U_2 with local coordinates $x = \frac{X}{Y}$, $z = \frac{Z}{Y}$ is

$$f(x,z) = x^{21} + x^5 + x^9z^8 + xz^8 + x^{10}z^{10} + x^2z^{10} + x^4z^{14} + z^{14} + xz^{20} + z^{21} \quad .$$

The point $P_1 = (0,0) \in C \cap U_2$ is a $\mathbf{k}$-rational singular point of multiplicity $m_{P_1}(C) = 5$.

The affine representation of C in the affine neighborhood U_3 with local coordinates $y = \frac{Y}{X}$, $z = \frac{Z}{X}$ is

$$f(y,z) = 1 + y^{16} + y^4z^8 + y^{12}z^8 + yz^{10} + y^9z^{10} + y^3z^{14} + y^7z^{14} + z^{20} + yz^{20} + z^{21} \quad .$$

The point $P_2 = (1,0) \in C \cap U_3$ is a singular point. We transform P_2 into the origin by applying the linear transformation $y \mapsto y'+1$, $z \mapsto z$ to obtain the polynomial

$f(y',z) =$

$$= z^{21} + y'^{16} + z^8y'^{12} + z^8y'^8 + z^{20}y' + z^{10}y'^9 + z^{10}y'^8 + z^{14}y'^7 + z^{14}y'^6 + z^{14}y'^5 + z^{14}y'^4 \quad .$$

Thus, the multiplicity of the curve C at the point P_2 is $m_{P_2}(C) = 16$.

3 Blowing-up Process.

In this section we discuss the *blowing-up* process, a method which resolves the singularities of an algebraic curve. A useful reference for this topic is Tsfasman ([16], ch. 2.5).

Definition 7. The *blowing-up* of a point P_1 in the projective plane $\mathbf{P}^2 = \mathbf{P}^2(\overline{\mathbf{k}})$ is a birational morphism $\pi_1 : S_1 \to \mathbf{P}^2$, where

i. S_1 is a projective smooth surface.
ii. $\pi_1^*(\mathbf{P}^2 - P_1) \simeq S_1 - E_1$.
iii. $\pi_1^*(P_1) = E_1 \simeq \mathbf{P}^1(T_1)$,

where T_1 is the tangent space of $\mathbf{P}^2$ at P_1. The morphism π_1^* is the inverse of the morphism π_1, which is defined everywhere except at the point P_1. The projective line E_1, called the exceptional line, is the image of the point P_1 under the morphism π_1^*.

Let $C : F(X,Y,Z) = 0$ be an absolutely irreducible projective plane curve of degree n, defined over $\mathbf{k}$. Let $f(x,y) = 0$ be a polynomial equation, which represents the curve C around its singular point $P_1 = (0,0)$ with multiplicity m_1.

The blowing-up of P_1 in $\mathbf{P}^2$ induces the blowing-up of P_1 in the curve $C : f(x,y) = 0$. In particular, we obtain the birational morphism $\tilde{\pi}_1 : C_1 \to C$, where C_1 is an affine projective curve, as follows: Using equation (1) we can rewrite the polynomial $f(x,y)$ as

$$f(x,y) = f_{m_1}(x,y) + f_{m_1+1}(x,y) + ... + f_n(x,y).$$

The affine quadratic transformation $Q_1 : x = u\ , y = us$ applied on $C : f(x,y) = 0$ yields

$$f(u,us) = u^{m_1}[\,f_{m_1}(1,s) + uf_{m_1+1}(u,s) + ... + u^{n-m_1}f_n(u,s)\,].$$

The exceptional line is $E_1 : u = 0$ and the curve $C_1 : f_1(u,s) = 0$, where $f_1(u,s) = f_{m_1}(1,s) + uf_{m_1+1}(u,s) + ... + u^{n-m_1}f_n(u,s)$. The points of the intersection $E_1 \cap C_1$ of the exceptional line E_1 and the curve C_1 are the points P_{1i} *infinitely near* P_1 in the first neighborhood.

Proposition 8 ([3], p.165). *With the above notation, the cardinality of the set* $\tilde{\pi}_1^*(P_1) = \{P_{1i} \in C_1\}$ *is less than or equal to* m_1, *and the multiplicities* $m_{P_{1i}}(C_1) \le r_i$.

Proof. The points P_{1i} belong to the set

$$\tilde{\pi}_1^*(P_1) = \{\,P_{1i} = (0,s) : f_{m_1}(1,s) = 0\,\}.$$

Since $\mathbf{k}$ is a finite field, the cardinality of the set $\tilde{\pi}_1^*(P_1)$ is less than or equal to m_1.

Using basic facts on *intersection indices*, ([3], p.74), we have $m_{P_{1i}}(\tilde{C}_1) \leq I(P_{1i}, f_{m_1}(1,s) \cap u)$. The polynomial $f_m(1,s)$ can be factored in $\mathbf{k}$ (or in $\bar{\mathbf{k}}$) as $f_m(1,s) = \prod_{i=1}^{m_1}(a_i + s)^{r_i}$. Thus

$$m_{P_{1i}}(\tilde{C}_1) \leq r_i \sum_{i=1}^{m_1} I(P_{1i}, (a_i + s) \cap u) = r_i \quad \square$$

We will now describe the blowing-up process algorithmically.

Algorithm B-U * Blowing-Up Process *
Input: An absolutely irreducible projective plane curve $C : F(X,Y,Z) = 0$ of degree n defined over $\mathbf{k}$, the singular point $P_1 = (0,0,1)$ together with its multiplicity m_1.
Output: An affine representation of the curve C_1 over C, the points P_{1i}, and their multiplicities.
Method:

Step 1: Let $C : f(x,y) = 0$ be the affine representation of C around a singular point P_1. The quadratic transformation $Q_1 : x = u,\ y = us$ (or $Q_2 : x = vt,\ y = v$) transforms the curve C into the curve $C_1 : f_1(u,s) = 0$, where

$$f(u, us) = u^{m_1}(f_1(u,s)).$$

Step 2: Find the points P_{1i} by intersecting the exceptional line $E_1 : u = 0$ with the curve C_1. Compute the multiplicities of the points P_{1i}.

Proposition 9. *The running time of the Algorithm B-U is $O(m_1 q^s n^4 \log^3 q^s)$, where n is the degree of the curve C, and q^s the cardinality of the extended ground field $\mathbf{k}_s$.*

Proof. Let $C(i)$ denote the time complexity of the ith step. Since a quadratic transformation requires $O(n^2)$ time (Abhyankar [1], p.277), then $C(1)$ is $O(n^2)$.

For the second step we find the roots of the polynomial $f_1(0,s) = 0$ over $\mathbf{k}_s$. This requires $O(q^s n \log^3 q^s)$. On the other hand, it takes $O(n^4 \log^3 q^s)$ to transform each point P_{1i} to the origin. Thus the transformation of all $O(m_1)$ points P_{1i} to the origin requires $O(m_1 n^4 \log^3 q^s)$. Thus $C(2) = O(m_1 q^s n^4 \log^3 q^s)$, which is also the running time of the Algorithm B-U. $\square$

Corollary 10. *If $P_1 = (0,0,1)$ is a $\mathbf{k}$-rational singular point of multiplicity m_1, then the running time of the Algorithm B-U is $O(m_1 q n^4 \log^3 q)$.*

We *resolve* the singularity P_1 by successive applications of blowings-up until we obtain a curve $\tilde{C}$ with only simple points; we call $\tilde{C}$ the *nonsingular model of C over P_1*.

Notation. The composition $\pi_N \circ \cdots \circ \pi_1$ yields the birational morphism $\pi : \tilde{S} \longrightarrow \mathbf{P}^2$ from the smooth surface $\tilde{S}$ to the projective plane $\mathbf{P}^2$. The composition $\tilde{\pi}_N \circ \cdots \circ \tilde{\pi}_1$ yields the morphism $\tilde{\pi} : \tilde{C} \longrightarrow C$ from the nonsingular model $\tilde{C}$ to the singular curve C. The intersection $E_i \cap C_i$ of the exceptional divisor E_i and the curve C_i yields the points *infinitely near* P_1 in the ith neighborhood.

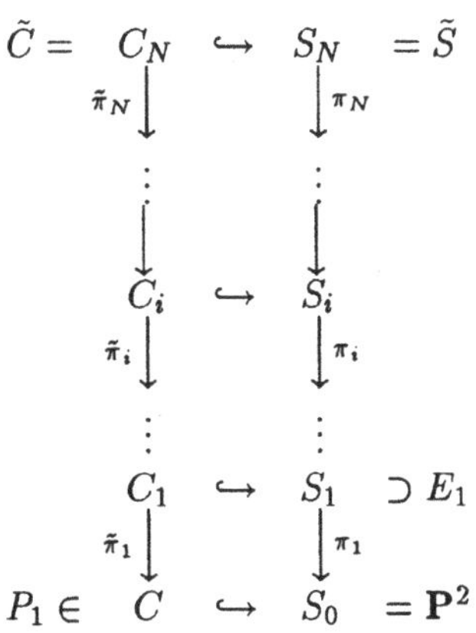

Figure 1.

The collection of all the points infinitely near P_1 in all neighborhoods forms the *singularity tree* of P_1. The point P_1 is the *root* of the tree and the points infinitely near P_1 are the *nodes*. We keep a count at each node equal to the multiplicity of the transformed curve at that point.

Theorem 11 Resolution. *Let C be an irreducible curve in the surface $S_0 = \mathbf{P}^2$. Then there exists a finite sequence of blowings-up $S_N \rightarrow S_{N-1} \rightarrow \ldots \rightarrow S_1 \rightarrow S_0$ such that the strict transform $C_N = \tilde{C}$ of C on S_N is nonsingular.*

Proof. The existence and construction of the blowings-up are described in the Algorithm RES. Since the genus g_i of the successive curves C_i is decreasing for each $i = 0, ..., N$, and since the genus is a nonnegative integer, we conclude that the resolution process terminates. (We refer the reader to §4 of this paper for computing the genus explicitly). □

Remark. The surface $S_1 = \mathbf{P}^2 \times \mathbf{P}^1$ where $S_N = \mathbf{P}^2 \times \mathbf{P}^1 ... \times \mathbf{P}^1$. Using Segre's embedding, the high dimensional surface S_N can be embedded in the projective surface $\mathbf{P}^N$ (see Hartshorne [4], p.13). Thus the nonsingular model $\tilde{C}$ can become a projective curve.

Let us now describe the resolution process algorithmically.

Algorithm RES: * Resolution process *

Input: An absolutely irreducible projective plane curve $C : F(X,Y,Z)$ of degree n defined over $\mathbf{k}$. The $\mathbf{k}_s$-rational singular points $P_i \in C$ (transformed at the

origin), along with their multiplicities m_i. The affine representations of C around all P_i.
Output: Singularity trees of all singular points $P_i \in C$.
Method:

Step 1: Apply Algorithm B-U to one of the singular points of C, e.g. the point P_1 of multiplicity m_1.
Step 2: If C_1 is a smooth curve in the surface S_1, i.e. $m_{P_{1i}}(C_1) = 1$, then $C_1 = \tilde{C}$ is a non-singular model of the curve C over P_1. If not, then apply Algorithm B-U to those points $P_{1i} \in C_1$ which are singular.
Step 3: Repeat Step 2 until all the points infinitely near P_1 at any neighborhood are simple. Construct the singularity trees of all singular points $P_i \in C$.

Proposition 12. *The running time of the Algorithm RES is $O(mq^s n^{10} \log^3 q^s)$, where $m = \sum_i m_i$, n is the degree of the curve C, and q^s is the cardinality of the extended ground field $\mathbf{k}_s$.*

Proof. Since the singularities are $\mathbf{k}_s$-rational, no further field extensions are required to carry out the Algorithm RES. The resolution of each of singular point requires at most $O(n^2)$ blowings-up (Abhyankar [1]). Each blowing-up requires $O(m_1 q^s n^4 \log^3 q^s)$ (see Proposition 9). The degree of the original curve C will grow at worst by a factor of $O(n^2)$ (Abhyankar [1]). Thus the construction of one singularity tree takes $O(m_1 q^s n^8 \log^3 q^s)$. Hence the overall time bound for the construction of all singularity trees around all $O(n^2)$ singularities P_i is $O(mq^s n^{10} \log^3 q^s)$, where $m = \sum_i m_i$, which is the time complexity of the Algorithm RES. □

Corollary 13. *If all singular points $P_i \in C$ of multiplicity m_i are $\mathbf{k}$-rational, then the running time of the Algorithm RES is $O(mqn^{10} \log^3 q)$, where $m = \sum_i m_i$.*

Example 1B. Refer to Example 1A. We first apply Algorithm RES to the singular point P_1 of C.

BLOWING-UP I of the point P_1 with local coordinates, x and z of multiplicity $m_{P_1}(C) = 5$. The equation of the local exceptional line E_1 is $z = 0$. The affine equation of the new curve C_1 is $f_1(x_1, z) = x_1^5 + x_1^{21} z^{16} + x_1^9 z^{12} + x_1 z^4 + \cdots$. The first node lying over P_1 is the point P_{11} with local coordinates, x_1 and z of multiplicity $m_{P_{11}}(C) = 5$.
BLOWING-UP II of the point P_{11}. The equation of the local exceptional line E_2 is $z = 0$. The affine equation of the new curve C_2 is $f_2(x_2, z) = z^5(x_2^5 + x_2 + z^4 + \cdots) = 0$. The nodes lying above P_{11} are P_{111} with local coordinates x_2 and z of multiplicity $m_{P_{111}}(C_2) = 1$, and the point P_{112} with local coordinates, $x_2 + 1 = x'$ and z and multiplicity $m_{P_{112}}(C_2) = 4$.

BLOWING-UP III of the point P_{112}. The equation of the local exceptional line E_2 is $z=0$. The affine equation of the new curve C_3 is $f_3(x_3,z)=x_3^4+z^2x_3^2+\cdots=0$. The node that is lying above P_{112} is P_{1121} with local coordinates x_3 and z of multiplicity $m_{P_{1121}}(C_3)=4$.

BLOWING-UP IV of the point P_{1121}. The equation of the local exceptional line E_4 is $z=0$. The affine equation of the new curve C_4 is $f_4(x_4,z)=x_4+x_4^2+\cdots=0$. The nodes lying over the blown up point, P_{1121} are the points P_{11211} with local coordinates x_4 and z of multiplicity $m_{P_{11211}}(C_4)=2$, and the point P_{11212} with local coordinates, $x_4+1=x'$ and z of multiplicity $m_{P_{11212}}(C_4)=2$.

BLOWING-UP V of the point P_{11212}. The equation of the local exceptional line E_5 is $z=0$. The affine equation of the new curve C_5 is $f_5(x_1^{'},z)=x_1^{'2}+z+z^2+zx_1^{'}+\cdots=0$. The node lying above P_{11212} is the point P_{112121} with local coordinates $x_1^{'}$ and z of multiplicity, $m_{P_{112121}}(C_5)=1$.

BLOWING-UP VI of the point P_{11211}. The equation of the local exceptional line E_5 is $z=0$. The affine equation of the new curve C_5 is $f_5(x_5,z)=x_5^2+1+z^2+z+\cdots=0$. The point, above P_{11211} is the point P_{112111} with local coordinates $x_5+1=x''$ and z and multiplicity $m_{P_{112111}}(C_5)=1$.

We will now apply Algorithm RES to the singular point P_2:

BLOWING-UP 1 around the point P_2: The equation of the local exceptional line E_1 is $z=0$. The affine equation of the new curve C_1 is $f_1(y_1,z)=y_1^{16}+y_1^8+z^5+\cdots=0$. Thus the nodes lying above the point $P_2=(0,0)$ are: $P_{21}=(1,0)$ with local coordinates y_1 and z of multiplicity $m_{P_{21}}(f_1(y_1,z))=5$, and the point P_{22} of multiplicity $m_{P_{22}}(f_1(y_1,z))=m_{(0,0)}(f_1(y_1^{'},z))=5$.

BLOWING-UP 2 around the point P_{22}: The equation of the local exceptional line E_2 the affine equation of the new curve C_2 and the infinitely near points are described . The equation of the local exceptional line E_2 is $y_1'=0$. The affine equation of the new curve C_2 is $f_2(y_1',\tilde{z})=\tilde{z}^5+y_1'^3+\cdots=0$. Thus the node lying above P_{22} is the point $P_{221}=(0,0)$ of multiplicity $m_{P_{221}}(C_2)=3$.

BLOWING-UP 3 around the point P_{221}. Preceding as above we have the nodes P_{2211} of multiplicity $m_{P_{2211}}(C_3)=1$, and the point P_{2212} where $m_{P_{2212}}(C_3)=2$ lying above the blown up point P_{221}.

BLOWING-UP 4 around the point P_{2212}: The point, lying above P_{2212} is the point P_{22121} of multiplicity $m_{P_{22121}}(C_4)=1$.

BLOWING-UP 5 around the point P_{21} with local coordinates y_1 and z and multiplicity $m_{P_{21}}(C_1)=5$. The equation of the local exceptional line E_2 is $y_1=0$. The affine equation of the new curve C_2 is $f_2(y_1,z_1)=z_1^5+y_1^3+y_1^{11}+\cdots+z_1^2y$. Thus the node lying above P_{21} is the point P_{211} with local coordinates y_1 and z_1 and multiplicity $m_{P_{211}}=3$.

BLOWING-UP 6 around the point P_{211}. The equation of the local exceptional line E_3 is $z_1=0$. The affine equation of the new curve C_3 is $f_3(y_2,z_1)=$

$y_2^3 + y_2 + z_1^2 + \cdots = 0$. The nodes lying above P_{211} are the points P_{2111} with local coordinates y_2 and z_1 of multiplicity $m_{P_{2111}}(C_3) = 1$, and the point P_{2112} with local coordinates $y_2+1 = y'$ and z_1 of multiplicity $m_{P_{2112}}(C_3) = 2$.

BLOWING-UP 7 around the point P_{2112}. The equation of the local exceptional line E_4 is $y' = 0$. The affine equation of the new curve C_4 is $f_4(y', z_2) = y' + 1 + z_2^2 + \cdots = 0$. Thus the node lying above P_{2112} is the point P_{21121} with local coordinates y' and $z_2 + 1 = z'$ and multiplicity $m_{P_{21121}}(C_4) = 1$.

The singularity trees of P_1 and P_2 and the multiplicity of the successive curves at each node follow:

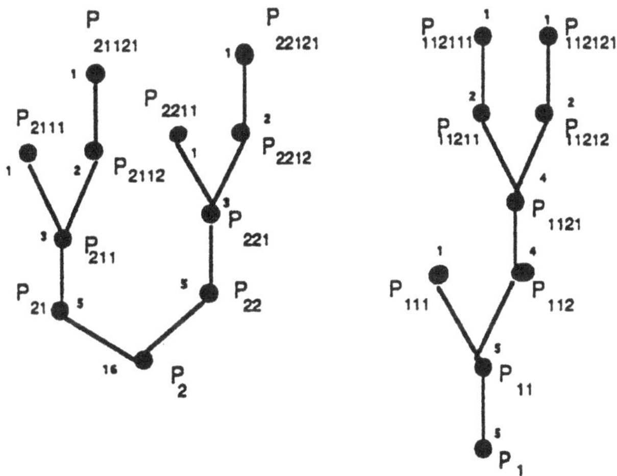

4 Genus Calculation.

In this section we calculate the genus of an algebraic curve using Algorithms S-M, and RES described in the former sections. Let C be an absolutely irreducible singular projective plane curve defined over a finite field $\mathbf{k} = \mathbf{F}_q$ where $q = p^e$ for some prime number p. We assume that the singular points $P \in C$ are $\mathbf{k}_s$-rational, for some finite field extension $\mathbf{k}_s$ of $\mathbf{k}$.

An important invariant associated with every point $P \in C$ is the integer δ_P, which measures how singular C is at P. For an exact definition of δ_P we refer the reader to Hartshorne's book ([4], p.298).

Lemma 14. *The integer δ_P can be computed as*

$$\delta_P = \sum_i \frac{1}{2} m_i(m_i - 1),$$

taken over all infinitely near points P_i of P including P.

Proof. For the proof see Hartshorne's book ([4], p. 393).

Definition 15. The genus g of the singular plane curve C with degree n is

$$g = \frac{(n-1)(n-2)}{2} - \sum_{P} \delta_P,$$

where the sum is taken over all singular points $P \in C$.

Proposition 16. *The genus g of the singular plane curve C with degree n can be computed in $O((n^6 + n^{10}m)T)$ time, where $T = q^s log^3 q^s$, m_P denotes the multiplicity of C at each singular point P, $m = \sum_P m_P$, and q^s is the cardinality of the extended ground field $\mathbf{k}_s$.*

Proof. The time taken to compute the genus g of the curve C is bound by the time to compute the integer δ_P. The latter is bound by the time to compute the singular points $P \in C$ along with their multiplicities, plus the time to compute the nodes and multiplicities m_i in each singularity tree of $P \in C$. But these are the quantities return from Algorithms S-M and RES which require $O(q^s n^6 \log^3 q^s)$ and $O(mq^s n^{10} \log^3 q^s)$ time respectively (see Propositions 5 and 12).

Therefore the integer δ can be computed in $O(q^s n^6 \log^3 q^s + mq^s n^{10} \log^3 q^s) = O((n^6 + n^{10}m)T)$ time, where $T = q^s log^3 q^s$. □

Corollary 17. *If all singular points $P \in C$ are $\mathbf{k}$-rational, then the genus g of the singular plane curve C with degree n can be computed in $O((n^6 + n^{10}m)T)$ time, where $T = qlog^3 q$.*

Example 1C. Refer to Examples 1A, 1B, and Figure 2. The integers $\delta_{P_1} = 5.4 + 5.4 + 1.0 + 4.3 + 4.3 + 2.1 + 2.1 + 1.0 + 1.0 = 68$ and $\delta_{P_2} = 16.15 + 5.4 + 3.2 + 1.0 + 2.1 + 1.0 + 5.4 + 3.2 + 1.0 + 2.1 + 1.0 = 296$ yield the integer $\delta = \delta_{P_1} + \delta_{P_2} = 364$. Thus the genus of C is $g = \frac{(21-1)(21-2)}{2} - \frac{1}{2}364 = 8$.

Remarks.
1. In [11], another resolution method is described for any algebraic curve (not necessarily plane). We can also generalize our algorithm RES to any algebraic curve using methods similar to those in [11].
2. The genus of a curve has also been computed in [2], [11], [10] based on other desingularization methods.

References

1. S. Abhyankar and C. Bajaj. Computations with algebraic curves. *Lecture Notes in Computer Science, Springer-Verlag,* 358, July 1988.
2. M. Bronstein, M. Hassner, A. Vasquez, and C.J. Williamson. Computer algebra algorithms for the construction of error correcting codes on algebraic curves. *IEEE Proceedings on Information Theory,* June 1991.

3. D. Le Brigand and J.J. Risler. Algorithm de Brill-Noether et codes de Goppa. *Bull. Soc. math. France*, 116:231–253, 1988.
4. D. Polemi. *The Brill-Noether Theorem for Finite Fields and an Algorithm for Finding Algebraic Geometric Goppa Codes.* Proceedings of the IEEE International Symposium on Information Theory, p.38, Budapest, Hungary, 1991.
5. D. Polemi, C. Moreno, and O. Moreno. A construction of a.g. Goppa codes from singular curves. *submitted for publication*, 1992.
6. T. Sakkalis. The topological configuration of a real algebraic curve. *Bull. Austr. Math. Soc.*, 43:37–50, 1991.
7. T. Sakkalis and R. Farouki. Singular points of algebraic curves. *Journal of Symbolic Computation*, 9:405–421, 1990.
8. S. Vladut and Y. Manin. Linear codes and modular curves. *Journal of Soviet Mathematics*, 30, no. 6:2611–2643, 1985.

A Divisionless Form of the Schur Berlekamp-Massey Algorithm

Christopher J. Zarowski

Department of Electrical Engineering, Queen's University
Kingston, Ontario, Canada K7L 3N6

Abstract. A procedure exists for mapping the Berlekamp-Massey algorithm (BMA) into so-called Schur form. This procedure will be shown here to be applicable to the development of a divisionless Schur BMA. When the BMA is combined with the Schur BMA the result is an efficient parallel algorithm for the computation of the error-locator polynomial which is used in the decoding of Reed-Solomon, or, more generally, Bose-Chaudhuri-Hocquenghem codes.

1 Introduction

The Berlekamp-Massey algorithm (BMA) is important in the computationally efficient decoding of Reed-Solomon (RS), and, more generally, Bose-Chaudhuri-Hocquenghem (BCH) block codes. The BMA computes the error-locator polynomial, the zeros of which yield the locations of errors in the received word. Some variations of the BMA compute the error-evaluator polynomial as well. A sequential implementation of the BMA has an asymptotic time complexity of $O(t^2)$, where t is the number of errors that the code is designed to correct. In [1] a so-called Schur form of the BMA was derived. This algorithm is called the Schur BMA. (The Schur BMA is also known as the two-pass BMA in [2].) It is analogous to the Schur algorithm for Toeplitz matrices [3], and so is amenable to parallel implementation on a linear array of processors, as shown in [4]. Specifically, an array of $2t$ processors is needed containing $3t - 1$ multipliers. The algorithm of [1] employs finite field division, and so the processor array requires one divider as well. The machine in [4] will compute the error-locator polynomial (but not the error evaluator polynomial) in $O(t)$ time. This includes array initialization, and the time taken to read out the final solution. Note that the direct implementation of the BMA on a linear processor array without the aid of the Schur BMA would have a time complexity of $O(t \log t)$.

According to [1] Schur Berlekamp-Massey algorithms are derived by mapping from classical Berlekamp-Massey algorithms. Thus, if the BMA requires division then the Schur BMA will require it as well. Therefore, it will now be shown that the divisionless BMA in [5] may be used to construct a divisionless Schur BMA using the mapping procedure in [1].

2 Derivation of the Divisionless Schur BMA

Algorithm 1 (below) summarizes the divisionless BMA found in [5]. The syndrome sequence is denoted by

$$S_0 , \ldots , S_{r-1} , \quad r = 2t . \tag{2.1}$$

We recall that this sequence is obtained by evaluating the received word, expressed as a polynomial, at the zeros of the code generator polynomial (see, for example, [6] for details). The error-locator polynomial is given by $\sigma^{(k)}(z) = (1/\alpha_0^{(k)})\alpha^{(k)}(z)$, where $\alpha^{(k)}(z) = \alpha_0^{(k)} + \alpha_1^{(k)}z + \cdots + \alpha_{l_k}^{(k)}z^{l_k}$, and where l_k is the length of the shortest linear feedback shift register (LFSR) that synthesizes $S_0 , \ldots , S_k$. This LFSR is parametrized by $\sigma^{(k)}(z)$. As well, $\beta^{(k)}(z) = \beta_0^{(k)} + \beta_1^{(k)}z + \cdots + \beta_{l_k+1}^{(k)}z^{l_k+1}$ is an auxiliary polynomial introduced into the computation for the purpose of eliminating the division operations. We see that this has the effect of doubling the number of multiplications needed to compute the error-locator polynomial (compare with the BMA in Fig. 11 of [1]).

It is important to note that the only statement in the divisionless BMA of Algorithm 1 that makes it unsuitable for efficient implementation on a linear processor array is

$$d_k = \sum_{i=0}^{l_k} \alpha_i^{(k)} S_{k-i} . \tag{2.2}$$

It is this statement that causes a parallel implementation of Algorithm 1 to have a time complexity of $O(t \log t)$ instead of the more desirable $O(t)$. The computation in (2.2) can be viewed as an explicit inner product between a sequence of syndromes, and error-locator polynomial coefficients. The mapping procedure to follow replaces the above explicit inner product computation with an implicit computation that does not spoil the parallelism that is otherwise inherent in the divisionless BMA.

$\alpha^{(0)}(z) := 1 ;\ \beta^{(0)}(z) := z ;$
$l_0 := 0 ;\ \delta_0 := 1 ;$
For $k := 0$ to $r - 1$ do begin
  $d_k := \sum_{i=0}^{l_k} \alpha_i^{(k)} S_{k-i} ;$
  $\alpha^{(k+1)}(z) := \delta_k\alpha^{(k)}(z) - d_k\beta^{(k)}(z) ;$

```
If (d_k = 0) or (k < 2l_k) then begin
  β^(k+1)(z) := zβ^(k)(z) ;
  l_{k+1} := l_k ;
  δ_{k+1} := δ_k ;
  end;
If (d_k ≠ 0) and (k ≥ 2l_k) then begin
  β^(k+1)(z) := zα^(k)(z) ;
  l_{k+1} := k − l_k + 1 ;
  δ_{k+1} := d_k ;
  end;
end;
```

Algorithm 1. The divisionless Berlekamp-Massey algorithm.

To obtain the divisionless Schur BMA we begin by considering the Hankel system of equations

$$[\alpha_{l_k}^{(k)}\ \alpha_{l_k-1}^{(k)}\ \cdots\ \alpha_1^{(k)}\ \alpha_0^{(k)}\ 0 \cdots 0]$$

$$\left[\begin{array}{cccc|ccc} S_0 & S_1 & \cdots & S_{r-l_{r-1}-1} & S_{r-l_{r-1}} & \cdots & S_{r-1} \\ S_1 & S_2 & \cdots & S_{r-l_{r-1}} & S_{r-l_{r-1}+1} & \cdots & 0 \\ \vdots & \vdots & & \vdots & \vdots & & \vdots \\ S_{l_{r-1}} & S_{l_{r-1}+1} & \cdots & S_{r-1} & 0 & \cdots & 0 \end{array}\right]$$

$$= \quad [u_0^{(k)}\ u_1^{(k)}\ \cdots\ u_{r-1}^{(k)}\ u_{r-1}^{(k)}] \tag{2.3}$$

which defines the Schur variables, $u_j^{(k)}$. Equation (2.3) is directly analogous to Equation (77) in [1]. The rationale for (2.3) is also considered in detail in [1]. Thus, from (2.3) we have

$$u_j^{(k)} = \sum_{i=0}^{l_k} \alpha_i^{(k)} S_{j+l_k-i} \tag{2.4}$$

for $j = k - l_k, \ldots, r - l_k - 1$. Equation (2.4) is used to map recursions in terms of polynomials $\alpha^{(k)}(z)$ into recursions in terms of Schur variables. It is also necessary to define the auxiliary Schur variables

$$v_j^{(k)} = \sum_{i=0}^{l_k+1} \beta_i^{(k)} S_{j+l_k-i} \tag{2.5}$$

for $j = k - l_k + 1, \ldots, r - l_k$. This is because we need to map $\beta^{(k)}(z)$ as well as $\alpha^{(k)}(z)$. Equation (2.5) arises as a consequence of applying (2.4) to Algorithm 1, as will be seen below.

We may now consider the details of the mapping process. First of all, note that (2.4) allows (2.2) to be immediately rewritten as

$$d_k = u^{(k)}_{k - l_k} \quad . \tag{2.6}$$

Thus, d_k may be computed without the need to evaluate the sum in (2.2) directly. That is, an explicit inner product computation has been replaced by an implicit inner product computation.

The main recursion of Algorithm 1 is

$$\alpha^{(k+1)}(z) = \delta_k \alpha^{(k)}(z) - d_k \beta^{(k)}(z) \quad ,$$

and this may be rewritten in terms of the individual polynomial coefficients as

$$\alpha_i^{(k+1)} = \delta_k \alpha_i^{(k)} - d_k \beta_i^{(k)} \quad (i = 0 , \dots , l_{k+1}) \quad . \tag{2.7}$$

From (2.4) (with k replaced by $k + 1$), and (2.7)

$$\begin{aligned} u_j^{(k+1)} &= \sum_{i=0}^{l_{k+1}} \alpha_i^{(k+1)} S_{j + l_{k+1} - i} \\ &= \sum_{i=0}^{l_{k+1}} [\delta_k \alpha_i^{(k)} - d_k \beta_i^{(k)}] S_{j + l_{k+1} - i} \\ &= \delta_k \sum_{i=0}^{l_{k+1}} \alpha_i^{(k)} S_{j + l_{k+1} - i} - d_k \sum_{i=0}^{l_{k+1}} \beta_i^{(k)} S_{j + l_{k+1} - i} \quad . \end{aligned} \tag{2.8}$$

We must rewrite the summations in the last equality of (2.8) in terms of Schur, and auxiliary Schur variables as defined in (2.4), and (2.5), respectively. Since $l_{k+1} \geq l_k$, and $\alpha_i^{(k)} = 0$ for $i > l_k$, we have via (2.4)

$$\sum_{i=0}^{l_{k+1}} \alpha_i^{(k)} S_{j + l_{k+1} - i} = \sum_{i=0}^{l_k} \alpha_i^{(k)} S_{j + l_{k+1} - i} = u^{(k)}_{j + l_{k+1} - l_k} \quad . \tag{2.9a}$$

Similarly, $\beta_i^{(k)} = 0$ for $i > l_k + 1$, and so via (2.5)

$$\sum_{i=0}^{l_{k+1}} \beta_i^{(k)} S_{j + l_{k+1} - i} = \sum_{i=0}^{l_k} \beta_i^{(k)} S_{j + l_{k+1} - i} = v^{(k)}_{j + l_{k+1} - l_k} \quad . \tag{2.9b}$$

Substituting (2.9a,b) into (2.8) yields the main recursion for the divisionless Schur BMA, which is

$$u_j^{(k+1)} = \delta_k u^{(k)}_{j + l_{k+1} - l_k} - d_k v^{(k)}_{j + l_{k+1} - l_k} \quad . \tag{2.10}$$

From Algorithm 1, for $d_k = 0$ with $k < 2l_k$ we have the recursion $\beta^{(k+1)}(z) = z\beta^{(k)}(z)$. Following the procedure which gave us (2.10) we see that this recursion maps to

$$v_j^{(k+1)} = v_{j + l_{k+1} - l_k - 1}^{(k)} \quad . \tag{2.11}$$

Similarly, for $d_k \neq 0$ with $k \geq 2l_k$ the recursion $\beta^{(k+1)}(z) = z\alpha^{(k)}(z)$ maps to

$$v_j^{(k+1)} = u_{j + l_{k+1} - l_k - 1}^{(k)} \quad . \tag{2.12}$$

Finally, the initialization statements of Algorithm 1, i.e.,

$$\alpha^{(0)}(z) = 1 \,, \quad \beta^{(0)}(z) = z$$

map to, respectively,

$$u_j^{(0)} = S_j \; (j = 0, \ldots, r-1) \,, \quad v_j^{(0)} = S_{j-1} \; (j = 1, \ldots, r) \;, \tag{2.13}$$

again with the aid of (2.4) and (2.5).

The complete divisionless Schur BMA appears as Algorithm 2 (below). The algorithm was verified by implementing it as a computer program in the Pascal language.

$u_j^{(0)} := S_j \; (j = 0, \ldots, r-1)$;
$v_j^{(0)} := S_{j-1} \; (j = 1, \ldots, r)$;
$l_0 := 0$;
$\delta_0 := 1$;
For $k := 0$ to $r - 1$ do begin
    $d_k := u_{k - l_k}^{(k)}$;
    If ($d_k = 0$) or ($k < 2l_k$) then begin
        $l_{k+1} := l_k$;
        $\delta_{k+1} := \delta_k$;
        $v_j^{(k+1)} := v_{j + l_{k+1} - l_k - 1}^{(k)}$
        $(j = k - l_{k+1} + 2, \ldots, r - l_{k+1})$;
        end;
    If ($d_k \neq 0$) and ($k \geq 2l_k$) then begin
        $l_{k+1} := k - l_k + 1$;
        $\delta_{k+1} := d_k$;
        $v_j^{(k+1)} := u_{j + l_{k+1} - l_k - 1}^{(k)}$
        $(j = k - l_{k+1} + 2, \ldots, r - l_{k+1})$;
        end;
    $u_j^{(k+1)} := \delta_k u_{j + l_{k+1} - l_k}^{(k)} - d_k v_{j + l_{k+1} - l_k}^{(k)}$
    $(j = k + 1 - l_{k+1}, \ldots, r - l_{k+1} - 1)$;
    end;

Algorithm 2. The divisionless Schur Berlekamp-Massey algorithm.

As the reader can see, the derivation of the divisionless Schur BMA is very short and straightforward, especially when compared with the effort required to obtain the divisionless microlevel Euclid algorithm (MLE) in [7], for example. A working parallel computer for the calculation of the error-locator polynomial without division would execute both Algorithms 1 and 2 concurrently. This would be done along the lines discussed in [4]. A discussion of the merits of a parallel implementation of the BMA/Schur BMA relative to alternative decoders (such as those based on the MLE) appears in [4] as well.

3 Conclusions and Suggestions for Further Work

The mapping procedure of [1] has been shown to be effective in the development of a divisionless Schur BMA for the computation of the error-locator polynomial. It is anticipated that the algorithm shall have an efficient implementation on a linearly connected parallel processor array. This expectation is reasonable since the Schur BMA has been shown to have such an efficient implementation in [4]. The time complexity is $O(t)$, on an array of $O(t)$ processors.

It would be of interest to consider the application of the mapping procedure to the problem of finding Schur forms of BMA variants that compute the error-evaluator polynomial in addition to the error-locator polynomial. For example, one might consider mapping the BMA variant on p. 153 of [8]. As well, note that the divisionless Schur BMA considered here, and the Schur BMA of [1] are frequency domain algorithms as defined in [8]. It may be desirable to seek a time-domain Schur BMA, and a time-domain divisionless Schur BMA, perhaps by beginning with the time-domain algorithms in [8]. Such algorithms might be used to construct a parallel version of the sequential time-domain decoder considered in [9]. A parallel-pipelined decoder appears in [10]. This decoder can handle erasures. It would be desirable to incorporate such a feature into decoder structures that employ the Schur BMA.

References

[1] C. J. Zarowski, "Schur Algorithms for Hermitian Toeplitz, and Hankel Matrices with Singular Leading Principal Submatrices," IEEE Trans. on Signal Proc., vol. 39, Nov. 1991, pp. 2464-2480.

[2] H.-M. Zhang, P. Duhamel, "On the Methods for Solving Yule-Walker Equations," IEEE Trans. on Signal Proc., vol. 40, Dec. 1992, pp. 2987-3000.

[3] S.-Y. Kung, Y. H. Hu, "A Highly Concurrent Algorithm and Pipelined Architecture for Solving Toeplitz Systems," IEEE Trans. on Acoust., Speech, and Signal Proc., vol. ASSP-31, Feb. 1983, pp. 66-75.

[4] C. J. Zarowski, "Parallel Implementation of the Schur Berlekamp-Massey Algorithm on a Linearly Connected Processor Array," to appear in the IEEE Transactions on Computers (accepted September 1993).

[5] X. Youzhi, "Implementation of Berlekamp-Massey Algorithm Without Inversion," IEE Proc.-I, vol. 138, June 1991, pp. 138-140.

[6] R. Blahut, *Theory and Practice of Error Control Codes.* Reading, Massachusetts: Addison-Wesley, 1983.

[7] T. Citron, "Algorithms and Architectures for Error Correcting Codes," Ph. D. dissertation, Stanford University, 1986.

[8] R. Blahut, "A Universal Reed-Solomon Decoder," IBM J. Res. Dev., vol. 28, Mar. 1984, pp. 150-158.

[9] Y. Shayan, T. Le-Ngoc, V. Bhargava, "A Versatile Time-Domain Reed-Solomon Decoder," IEEE J. on Sel. Areas in Comm., vol. 8, Oct. 1990, pp. 1535-1542.

[10] H. M. Shao, I. S. Reed, "On the VLSI Design of a Pipeline Reed-Solomon Decoder Using Systolic Arrays," IEEE Trans. on Comp., vol. 37, Oct. 1988, pp. 1273-1280.

Error-Control Coding, Modulation and Equalization for All-Digital Advanced Television: State of the Art and Future Possibilities

Mike Sablatash

Communications Research Centre
3701 Carling Avenue
P.O. Box 11490, Station H
Ottawa, Ontario, K2G 4G6, Canada

Abstract. Data rates, channels and channel transmission encoder-decoder and modulator-demodulator pairs for the four all-digital HDTV systems proposed in each of the United States and in Europe for terrestrial broadcasting are described, as well as for advanced all-digital television systems for satellite broadcasting. Recent research directions and trends to (1) provide sufficient rejection of co-channel interference from NTSC carriers, (2) achieve low BER, high data rate and reasonable implementation complexity and cost, (3) create a signal which minimizes degradation due to interferences and noise, and minimizes interference into other systems, (4) provide bandwidth efficiency, and (5) provide graceful degradation, coverage range and robustness are succinctly described for the terrestrial HDTV systems. For the satellite ATV systems items for improvement and recent research directions and trends to realize them are also described. A concluding discussion and topics for future research end the paper.

1 Introduction

All-digital high-definition television (HDTV) systems for terrestrial broadcasting have been proposed in the United States [1-4], as well as an all-digital advanced television (ATV) system for satellite broadcasting [16]. All-digital HDTV systems have also been proposed in Europe for terrestrial [5-13] and satellite [14-15] broadcasting.

All of these HDTV or advanced television (ATV) systems employ fairly modern methods of error control - where the term "error control" is meant to cover both error control in the system to minimize the effects of interferences and noise into the system, as well as error control to minimize errors in other systems, caused by that system.

The main goals of this paper are to describe and discuss the theory and methods for error control used by these HDTV (and one ATV) systems. It will identify similarities and differences in their error control, modulation and equalization methods and discuss possible improvements to them. It will also succinctly describe recent research directions and trends in error control which give much more attention to advanced and relatively new theory and techniques in information, coding and communication theory. Finally, the author's views are provided about the best choices for error control, modulation and equalization schemes.

The specific HDTV (and ATV) systems which have been proposed - and, in most cases, built and tested in various ways - are introduced next.

First, the 4 all-digital HDTV systems proposed and tested for terrestrial broadcasting [1-4] in the United States are: (1) Digicipher [1], (2) Digital Spectrum Compatible (DSC-HDTV) [2], (3) Advanced Digital Television (ADTV) [3], and (4) Channel Compatible Digicipher (CCDC) HDTV System [4].

These systems compress a raw digitized source image data rate of about 1 Gbps to between 12.59 and 18.88 Mbps. After digitized audio and control data are added, as well as a considerable number of bits for error control because of the sensitivity to errors of the highly compressed data, the transmission data rates lie between 19.2 and 26.43 Mbps. The Digicipher, ADTV and CCDC systems all employ a choice (which can be switched automatically) of 16 or 32 quadrature amplitude modulation (QAM), and a concatenation of an outer Reed-Solomon (RS) code with an inner trellis-coded modulation (TCM) employing a convolutional encoder. The DSC-HDTV system selects 2-vestigial sideband (VSB) or 4-VSB and uses an RS code with t=10 and erasure correcting. The Digicipher and CCDC RS codes have t=5, and the ADTV system uses an RS code with t=10. All of these systems incorporate adaptive equalization. The DSC-HDTV uses a modified duo-binary scheme and a precoder/post-comb, and post-coder, to reduce National Television Systems Committee (NTSC) interference.

In May 1993 an alliance of HDTV proponents was formed to collaborate in creating a single, unified HDTV system, with the intention of incorporating the best aspects of all the proposed systems, the best ideas, knowledge and experiences of all the proponents, and some relatively recent research results. This Grand Alliance has announced in July 1993, that it will use one of the following coding and modulation methods: 4-VSB, 6-VSB (trellis-coded version of 4-VSB), 32 QAM, and 32 spectrally-shaped QAM (SS-QAM). It will also investigate coded orthogonal frequency division multiplexing (COFDM).

The European proposals (some of which have been tested) for all-digital terrestrial broadcasting [5-13] are: (1) the SPECTRE (Special Purpose Extra Channels for Terrestrial Radiocommunication Enhancements) HDTV system [6,7,12], (2) the DIAMOND (Digital Scalable Modulation for New Broadcasting) HDTV system [10,11], (3) STERNE (Système de télévision en radiodiffusion numérique) HDTV system [8,17], and (4) HD-DIVINE digital terrestrial HDTV [5,9,13].

All four of these European all-digital terrestrial broadcasting systems use coded orthogonal frequency division multiplexing (COFDM) [18,25].

The SPECTRE system employs a (255,239) RS code. The modem uses software running in real time to execute the FFT for modulating and assembling about 400 carriers into the OFDM. The modem can be programmed to perform a variety of digital modulation schemes. Quadrature phase-shift keying (QPSK) and 8-PSK had been implemented by June 1992, but 16-PSK and 16-QAM are also possible [6,7,12].

For the DIAMOND system design it was concluded that to reach high efficiency (30/60 Mbit/s in UHF/VHF) one needs a guard interval, a precise digital clock recovery and potential use of dual polarization. No indication of the use of coding with OFDM is revealed in [10,11]. However, in [6] it is noted that a BCH code is used.

The STERNE system uses a convolutional code concatenated with trellis coded QAM, and frequency and time interleaving [8,17]. Soft decision decoding is used. For broadcasting networks, use of an RS code has been suggested in concatenation with TCM.

The HD-DIVINE HDTV system uses a (224,208) shortened RS code, and an adaptive equalizer.

It is noted in [13] that the proposed European satellite broadcasting system, the Italian Telettra Eureka 256 project [14,15], was a pioneering achievement of the first rank. QPSK and 8-PSK modulation with various concatenations of convolutional, BCH and trellis codes were investigated for satellite transmission of HDTV with great success.

In the United States the GI-Digicipher I NTSCTV for multiple video channels per carrier has been developed for satellite transmission [16]. It has a transmission BW of 24 MHz, an information data rate of 26.9 Mbps, a QPSK symbol rate of 19.5 MHz, and a maximum of 10 NTSC video channels. It uses a concatenation of a rate ¾, 64-state punctured convolutional code and a (125,115) RS code. This system is commercially available.

Researchers have recently turned their attention to applying advanced and relatively recent new theory and technology in information, coding and communication theory to HDTV error-control [26-31]. In addition the applications of wavelets to channel coding has opened up new possibilities for investigation [32-35].

This paper emphasizes the theory and techniques of channel coding, modulation, and equalization and other error control schemes for terrestrial broadcasting of HDTV in the United States and Europe. The greatest emphasis is placed on the systems proposed for use in the United States.

2 Descriptions of Data Rates, Channels and Channel Transmission Encoder-Decoder and Modulator-Demodulator Pairs

From the available information supplied in the submissions of the four all-digital HDTV systems for terrestrial broadcasting in the U.S., one can examine the similarities and differences among the video sources. What is spectacular is the enormous reduction in the digitized data rate of the raw source from about 1 Gbps to the compressed and compacted video data rates of between 17.47 and 18.88 Mbps for 32-QAM and 12.59 and 13.60 Mbps for 16-QAM, and 16.92 Mbps (max.) for DSC-HDTV. The video data rate for 32-QAM is 5/4 that for 16-QAM, and the additional data rates for digitized audio, control, auxiliary data and error control is the same for 16-QAM and 32-QAM. Clearly, such highly compressed data will require considerable protection from errors.

The sources of such errors for over-the-air and cable TV broadcasting are summarized as follows. For over-the-air broadcast, sources of undesired in-band radiations include: multipath co-channel interference, cosmic noise, and man-made impulse and intustrial noise. Sources of undesired out of band radiations include: adjacent channel TV, taboo TV channels, FM radio (88-108 MHz), other services (mobile, CB), man-made industrial and military interferences, antenna characteristics, transmission line losses and reflections, random noise, R.F. response, timing and non-linear processing, I.F. and tuner generated beats, Nyquist slope sound traps, and imperfections and quadrature distortion in the quasi-synchronous or envelope modulator.

For cable transmission, sources of undesired radiations include terrestrial broadcast signals, satellite and MAC signals, heterodyners, demodulators and VSB-AM modulators in the headend, random noise, non-linear distortion and multipath in cables and repeaters, phase jitter, filtering, descrambling and defects in the set-top converter and de-scrambler, and in the demodulation, recording and modulation processes in the VCRs.

Table 1shows the origins and constituents of the transmission data and the symbol rates for the four systems and modulation methods. The total data rate is obtained by adding the data rates for digitized audio and for control and auxiliary data to the video data rate. The transmission data rate is obtained from the total data rate by adding the data rate for error control to it. The transmission symbol rate is obtained from the transmission data rate by dividing the transmission data rate by the number of bits per transmission symbol (i.e., by 5 for 32 QAM and by 4 for 16 QAM). It is obvious that a considerable additional data transmission rate is required for error control, as noted earlier.

Table 1. Simulcast All-Digital System Differences

Transmission Encoder

System and Modulation Method	Video Data Rate Mbps	Total Data Rate Mbps	Transmission Data Rate Mbps	Transmission Symbol Rate Msps
Digicipher (32-QAM)	17.47	18.22	24.39	4.88
Digicipher (16-QAM)	12.59	13.34	19.51	4.88
DSC-HDTV				
(2-VSB)	8.46 (min)		11.14 (min)	10.76
(4-VSB)	16.92 (max)		21.04(max)	10.76
ADTV				
(32-QAM)	17.73	14.8(SP) 3.7(HP)	19.20(SP) 4.8 (HP) 24 (total)	4.80 3.84 (HP) 0.96 (SP)
(16-QAM)	12.75	11.84(SP) 2.96(HP)	15.36(SP) 3.84(HP) 19.2 (total)	4.80
CCDC				
(32-QAM)	18.88	19.89	26.43	5.286
(16-QAM)	13.60	14.60	21.15	5.286

A summary of techniques for signal transmission encoding and decoding and modulating-demodulating can be readily observed from Table 2.

Table 2. Similarities and Differences in Signal Transmission Encoding and Decoding and Modulating-Demodulating for North American Proposals

	Digicipher	DSC-HDTV	ADTV	CCDC
Modulation	16-&32-QAM	2/4-VSB	16-&32-QAM	16-&32-QAM
RS FEC	(116,106,5) 16-QAM	(167,147,10)	(148,127,10) 16-&32-QAM	(121,116,5) 16-QAM
	(155,106,5) 32-QAM	Erasure Correcting		168,158,5) 32-QAM
TCM (inner) Rates	3/4 (16-QAM) 4/5 (32-QAM)		0.9 (16-&32-QAM)	3/4(16-QAM) 4/5 (32-QAM)
Convolutional Encoder Constraint Length	7			7
Interleaver-Deinterleaver		Inter- and intra-Segment	Cell Interleaving	
Depth #1	445μsec			7 Data Lines
Depth#2	102μsec			31 Symbols
Equalizer Type	LMS	LMS	Fractinally Spaced	Fractionally Spaced
Equalizer Taps	256	80 FF, 200 FB	64	256
Echo Range (μsecs)	-2 to 24	-2 to 24	±4	-2 to 24
Precoder/Post Comb Filtering		Reduces CCI; 2:1 C/N Threshold Loss		

Figure 1 shows a generic block diagram for HDTV systems for terrestrial broadcasting.

Fig.1. Generic HDTV Transmission System for Terrestrial Broadcasting

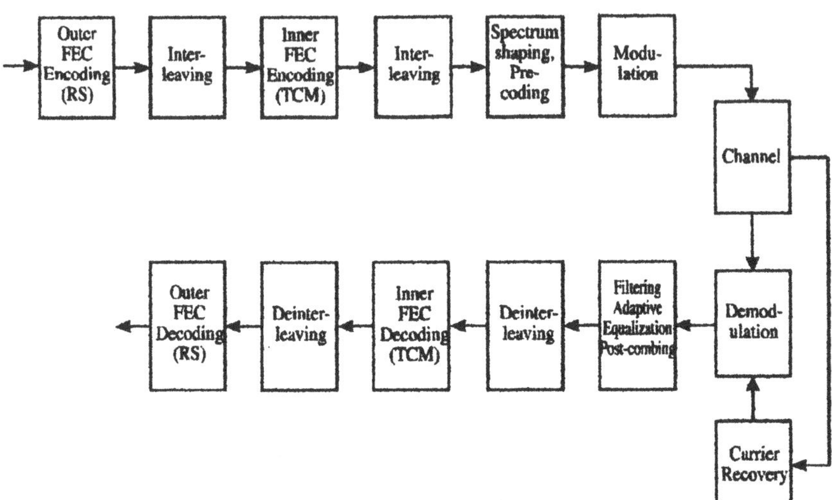

Detailed information about the four European all-digital HDTV systems is not as uniformly available because they are not all at an equal stage of development and, possibly, for reasons of confidentiality. Some further details for the (DIAMOND) Thomson-CSF/LER TV/HDTV follow. Transmission BW, 8MHz, NTSC channel; data rate, 34 Mbps; modulation, 64-QAM/OFDM; integration interval (T_s), 70.4µs ($1/T_s$=14.205 KHz); guard interval (T_g), 8.8 µs; reference symbol sent every 15 symbols; number of carriers, 512 (491 useful); channel coding, BCH; and noise threshold, C/N (8 MHz)=28 dB.

Thomson-CSF has agreed that using TCM would improve the C/N ratio.

It should be noted that this system worked well with interfering co-channel TV. The system FEC overhead was 18%. A 6 MHz, 20 Mbps version has been tested in the U.S.A.

The pioneering work for the RAI/Telettra Eureka 256 project using different concatenated error-correcting codes for satellite transmission of all-digital HDTV, and experimental transmissions are described in [14] and [15]. Two solutions which looked particularly promising are (1) QPSK modulation combined with a convolutional code of rate ¾ concatenated with a BCH (255,239) FEC block code, and (2) 8-PSK modulation with rate ⅔ trellis coding, concatenated with an interleaved BCH (255,239) block code. The latter has a higher spectral efficiency, but an increased modem complexity.

Some specifications for multiple (video) channels per carrier (MCPC) for the GI-Digicipher I NTSCTV for satellite broadcasting follow. Transmission BW, 24MHz, data rate, 26.9 Mbps (coded data rate=39.0 Mbps); modulation, QPSK; channel coding, concatenated convolutional coding with RS - rate 3/4 (punctured), 64-state convolutional code, and RS (125,115) code; noise threshold,for one error event per 15 minutes, C/N at 19.5 MHz is 7 dB; QPSK symbol rate, 19.5 MHz; maximum number of NTSC video channels, 10; receiver filter (analog), 4th order Butterworth; and transmit filter (FIR), optimized for D/A, elliptic (analog) filter and receive filter. This system is particularly interesting because it uses a punctured convolutional code which enables different levels of protection to be used for different parts of the data stream.

3 Items for Improvement of Terrestrial and Satellite Broadcast All-Digital HDTV Systems

The following items for improvement have been the subjects of research and development studies and papers on improving the terrestrial broadcast of all-digital HDTV systems.

1. The provision of sufficient rejection of co-channel interference (CCI) from NTSC carriers. This topic has been addressed in the DSC-HDTV system [2] by using a modified duobinary coding scheme. This scheme creates spectral nulls after the post-comb filter in the receiver that are close to the NTSC video, colour and audio carrier frequencies [2,41].

There are several drawbacks to the DSC-HDTV approach [26]. One of these is its use of a hard-decision decoder, since the slicer is placed before the postcoder, resulting in incompatibility with maximum likelihood decoding. Another drawback is that such an algorithm cannot easily be made adaptive. Finally, a smooth trade-off between noise and interference performance cannot be made.

In the ADTV system [3] the low frequency and high frequency portions of the spectrum are separated before transmission so that the NTSC picture carrier falls in the region between these portions. One obvious drawback to this scheme is the inefficient spectrum use that will result when NTSC stations are eventually taken off the air. Such a scheme is also transmitter dependent, sacrifices channel capacity for interference immunity, and cannot be disabled if NTSC interference is not a problem.

Using coded orthogonal frequency multiplexing (COFDM), or simply OFDM, carriers affected by CCI can be eliminated at the expense of a small reduction in overall channel bandwidth efficiency [25]. Again, the drawback to this is inefficient spectrum use.

2. Achieving low BER, high data rate and reasonable implementation complexity and cost have always been items for improvement, and progress in their realization continues to be made.

3. A partial extension of the last item is creating a signal which minimizes degradations due to interferences and noise, and minimizes interference into other systems. A major problem with the NTSC system is the fact that little channel coding was implemented to achieve this goal. By applying modern communication theory principles a great deal can now be done to achieve this and other goals listed here.

4. Bandwidth efficiency is clearly an item for improvement because of the demands on a crowded spectrum by the innovation of new services.

5. Graceful degradation scalability, coverage range and robustness are items for which improvements would be welcome.

For satellite broadcast all-digital ATV systems the items for improvement are as follows.

1. Achieving maximum energy efficiency for the given bandwidth and satellite channel characteristic. This involves optimizing modulation and coding for a satellite broadcasting system with given bandwidth limitations, interference levels, and channel distortion.

2. Minimizing BER and maximizing data rate with acceptable implementation complexity and cost.
3. Using medium-class DBS satellites and small (≈50 cm) receiving antennas.
4. Allowing flexibility to provide a large number of multimedia services.

The next section identifies categories and discusses recent research directions and trends to realize these improvements.

4 Recent Research Directions and Trends

Since about the middle of 1991, quite a few outstanding and well-known researchers with expertise in information theory, error-control coding, modulation and equalization have turned their attention and efforts to finding new results to improve the performance of all-digital HDTV transmission [26-31,38-40], in addition to those researchers who, starting in 1990, have pursued research into the use of COFDM for all-digital HDTV transmission. The first research work and results have almost all been published in very well-known and popular IEEE conference proceedings so only brief descriptions are provided in this and the next Section.

4.1 Research with the Main Goal of Minimizing Degradation due to CCI

One of the most pernicious and difficult problems in designing signal transmission encoder-decoder and modulator-demodulator pairs is sufficient rejection of the CCI due to the NTSC picture carrier, colour sub-carrier and sound carrier. Two approaches to minimize the effects of CCI are described next.

Joint decision-feedback and trellis-coded modulation. A modified duo-binary scheme for partial response signalling which creates spectral nulls very close to the NTSC video, colour and audio carrier frequencies is described in [2] and [41]. It consists of a digital precoder inserted between the RS encoder and modulator in the channel encoder, and a post-comb filter, modulo-4 interpreter and slicer, between the receiver demodulator and filter, and a de-interleaver, and an error correcting decoder and deformatter at the receiver end. The input to the precoder is an M-level signal which is precoded with a modulo-M adder and a k-symbol delay element. By choosing the delay element appropriately, spectral nulls can be positioned close to the NTSC video, colour and audio carrier frequencies. In the receiver a modified slicer maps the (2M-1)-level signal at the output of the comb filter into an M-level signal at the cost of an SNR reduction of 3 dB. As discussed in Section 3, there are several drawbacks to the DSC-HDTV approach.

In [26] a form of joint decision feedback equalization (DFE) and trellis-coded modulation is used to combat co-channel interference. DFE is normally not used in coded systems because delay-free decisions are required from the decoder. It is explained how a DFE can be used in this application by exploiting the cyclostationary properties of the interference. A receiver structure is derived which is compatible with coded modulation. The algorithm is shown to be more efficient than previous methods. Simulation results demonstrated significant gains over TCM alone and over linear equalizers [26]. Advantages of the technique are the adaptive processing, smooth trade-off of noise performance against interference performance, and its capability to make use of powerful coding schemes.

Design of signal constellations and associated precoders for partial-response channels. The major transmission issue of co-channel interference from NTSC into HDTV signals is addressed in [30] through the design of signal constellations and associated precoders for partial-response channels. An equivalent partial-response (PR) channel $1\text{-}Z^{k}$ arises in over-the-air broadcasting of HDTV signals when a comb filter is used in an HDTV receiver to reduce NTSC co-channel interference. It is shown that besides PAM and square (SQ) QAM, generalized square (GSQ) and generalized hexagonal (GHX) constellations can be used with this precoding technique. These GSQ and GHX constellations typically have smaller peak-to-average power ratios than those for PAM and SQ constellations of comparable BW efficiencies. Different TCM schemes may be applied to different classes of bits from an HDTV video/audio source encoder for graceful degradation. The resulting signals are then multiplexed together in the time domain before they are processed by a combined precoder and constellation mapper. As an example, for DSC-HDTV based on VSB/PAM, a 1D 4-PAM constellation for TCM with 8 states and rate ½, and with BW efficiency of 1.0 and nominal coding gain of 10.00 dB may be used for a more important class of bits, and a 4D, 4-PAM constellation for TCM with 8 states and rate ⅔, and with BW efficiency of 1.75 and nominal coding gain of 6.02 dB for a less important class.

It can easily be shown that if a coded modulation scheme originally designed for an AWGN channel can be perfectly used for an inter-symbol interference channel with a Tomlinson/Harashima precoder, then it can also be perfectly used for a general PR channel with a precoder. The word "perfectly" means that in addition to the transmitter constellation C not being altered by the precoder, the MSED between different sequences of symbols $\{\tilde{R}_n\}$ received from the PR channel in the absence of additive noise $\{N_n\}$ is equal to the MSED between different sequences of symbols $\{S_n\}$ at the output of the translated constellation mapper. The latter is also the MSED of the coded modulation scheme when applied to an AWGN channel [30].

A table of characteristics of some simple TCM schemes that can be perfectly used in the HDTV transceiver with comb filtering is given in [30].

The use of a variation of the Tomlinson/Harashima precoder greatly expands the list of constellations for such applications. The significance of this expansion is twofold. First, there can be a better match between the bit rate of an HDTV signal and the size of the constellation and, therefore, a better error rate performance. Second, these constellations can be used jointly to provide graceful degradation in the received signal quality. One may use conventional codes such as RS codes, TCM and their concatenations for a baseband HDTV transceiver for which the PR channel has transfer function $1\text{-}Z^{k}$. The results for this channel are extended to a more general PR channel, $1-\Sigma_{i=1}^{J} c_i Z^{-i}$, with integer coefficients c_i.

An interleaver/deinterleaver enhances the decoder performance in the presence of bursty noise which can come from the TV channel, comb filter, or adaptive equalizer.

The theory and techniques in [30] appear to be very promising to minimize NTSC interference into HDTV.

4.2 Research to Achieve Low BER, High Data Rate and Reasonable Implementation Complexity and Cost, and Minimize Interference into Other Systems

Coded orthogonal frequency division multiplexing. The theory of orthogonal frequency division multiplexing (OFDM) and coded orthogonal frequency division multiplexing (COFDM) is given in [10-12,17-25] with applications to digital audio and television broadcasting. Reference [19] has a good list of references through which the theory can be traced to its beginnings. Since about 1990 [6] COFDM has been the basis for each of the all-digital HDTV systems developed or under development in Europe. It was successfully developed for digital sound broadcasting in France, starting in about 1985 [19, 20, 22]. In 1991 interest in using COFDM for all-digital television transmission in the U.S.A. began, and in October 1992 a workshop centred around this technique was held at MIT. The Grand Alliance intends to investigate COFDM. J.R. Forrest, an eminent researcher in the UK has recently stated that "there seems little doubt that OFDM is the most powerful and cost-effective" [36].

The general principles of OFDM and COFDM are independent of applications such as digital audio broadcasting (DAB) and HDTV. Most of this section describes such general principles. There is considerable current study and research to determine whether COFDM can be successfully used in the HDTV environment. For example, it may be possible to use it for transmission over the television channel bandwidths in Europe, but this appears to be more difficult over the 6 MHz BW North American television bandwidths. Interference from NTSC carriers also poses a special problem for HDTV transmissions by COFDM. A great hope, of course, is that COFDM with its many desirable features, will prove to be applicable to North American HDTV transmission.

In the presence of inter-symbol interference caused by the transmission channel, the orthogonality between signals is no longer maintained. This can be solved by preceding each elementary signal in the time domain by a guard interval which can be used to "absorb" the ISI [19]. The loss due to mismatching between the emitted signals and the pulse response of the receiving filter corresponds to a loss of 10 log $(T_s+\Delta)/T_s$ which, in practice, can be kept below 1 dB. (Here Δ is the length of guard interval and T_s the duration of the elementary signal. If the channel impulse response has a finite duration less than Δ, and if the channel impulse response varies slowly compared with the duration $T_s+\Delta$, multipath propagation echoes will add constructively to the signal, eliminating (except for second-order effects) the need for a ghost canceller.

The OFDM signal can be written as

$$x(t)=\sum_{j}\sum_{k=0}^{N-1} I_{jk}\cdot g_k(t-jT)$$

where the I_{jk} is a set of complex numbers taken from a finite alphabet representing the emitted discrete input data signal, $g_k(t)=\exp(j2\pi f_k t)$ for $-\Delta\le t\le T_s$ is the base pulse, $f_k=f+k/T_s$ with $0\le k\le N-1$ is the set of N carrier frequencies, Δ is the guard interval length, T_s is the effective symbol length and, finally, $T=T_s+\Delta$ is the physical symbol length. If $\Delta=T_s/4$ there is a 1 dB loss.

The necessity for adding coding has been discussed in [19,21] for sound broadcasting to mobile receivers, and in [20] for digital video transmission. A concatenation of an RS block code and a convolutional code with TCM is quite effective for mobile communication channels.

Incorporating some of the descriptions and principles above, and adding those emanating from the use of coding provides the following principles of COFDM [37].

1. The first principle consists of splitting the information to be transmitted into a given number of modulated carriers with individual low bit rates, so that the corresponding symbol time becomes larger than the delay spread of the channel. Then, provided that a guard interval is inserted between successive symbols, the channel frequency selectivity no longer generates inter-symbol interference. Nevertheless, some of the carriers are enhanced by constructive interference, while others suffer from destructive interference. Therefore, if only this principle was applied some information would be well received and other information would be destroyed by the "local fadings".

2. The second principle systematically exploits the multipaths between the transmitter and the receiver by using the fact that signals sufficiently separated in frequency and time cannot be identically affected by the channel. Therefore, the COFDM system includes linking of elementary signals (information modulating a given carrier during a given symbol time) transmitted at distant locations of the time-frequency domain. This is achieved by convolutional coding associated with soft decision Viterbi decoding, in conjunction with frequency and time interleaving.

The diversity provided by interleaving plays an important role in the system. The Viterbi decoding efficiency is then at a maximum because successive samples presented at its input are affected by independent distortions. Even when the receiver is not moving the diversity in the frequency domain is sufficient to ensure correct behaviour of the system. It means that for this system, multipath is a form of diversity and should be considered as an advantage. Of course, as for any system able to benefit from multipath, the larger the transmission bandwidth, the better the ruggedness of the system.

In other words, the time-frequency diversity caused by interleaving allows the Viterbi decoder to integrate local fading phenomena over the whole signal bandwidth and over the time interleaving depth: the system performance is then related to an "average signal-to-noise ratio" criterion. (This is from [37], intended for DAB. The same principle should hold for ATV, but parameters and designs will likely be very different.)

This signal-to-noise ratio will increase as soon as the received signal power is augmented by echoes that cannot combine destructively: this is the case when the echoes are separated by a minimum delay equal to the inverse of the signal bandwidth.

COFDM systems have been described [10,17,19] which constructively combine (by a power sum) the multipath echoes. Although this seems to be effective for DAB, further investigation is required to show whether this will likely be useful for ATV.

The symbols overlap in the frequency domain (but are orthogonal). They are separated by the guard interval in the time domain. The functions in the frequency domain are sin $(f)/(f)$ functions in the case of a rectangular window.

This system concept involves inherent multipath handling by using the guard interval in the time domain, and high spectrum efficiency by separating the symbols only in the time domain with a guard interval of about 20%.

A single-frequency network is a set of transmitters spread throughout a given territory, temporarily synchronized, and transmitting the same signal at the same frequency. This allows for a very large coverage (a whole country or parts of a country).

This time-frequency arrangement matches the channel characteristics, and results in:

1. Separation of the symbols in the time domain where the dispersion of the channel is very large (basic to single-frequency network applications); and
2. Overlapping symbols in the frequency domain where dispersion of the channel is less critical (Doppler spread).

As suggested above, the COFDM system allows for single-frequency networking because it can benefit from echoes which do not differ from the transmitted signal (but is only a delayed version of it), and because these echoes can simply be "man-made echoes".

This technique can also be used locally to recover a shadowed area by direct reamplification using a co-channel retransmitter often called a "Gap-filler".

Time diversity in COFDM is achieved through convolutional coding plus time-interleaving.

Frequency diversity and space diversity are achieved by using a guard interval, convolutional coding and frequency interleaving.

When gap-filling or single-frequency networking are used, the space diversity becomes available on the network side. This means that spatially distributed transmitters contribute by addition of powers to the received signal. As the position of these transmitters is not concentrated in only one direction, this feature is very useful to avoid a complete shadowing when a natural obstacle such as a building or a hill masks a given direction of the horizontal plane.

On the other hand, when the receiver is close to a given transmitter of the network, this space diversity is no longer available; however, it is not needed in this case because the vicinity of the transmitter implies high signal-to-noise ratio.

The simultaneous contribution of these various diversity techniques makes the COFDM system very robust and efficient for digital audio broadcasting of high quality sound with a continuity of service inside the coverage area. This has been verified by simulation studies as well as by numerous experiments in Europe and in America. It also shows great promise for transmission of all-digital television.

OFDM is robust to white noise. It may be of interest for certain applications - such as military - that when there are jammers QAM has a 20 dB advantage over OFDM, but OFDM has a 4 dB advantage over QAM if the carrier is switched off. In the presence of impulse noise OFDM has a 10 dB advantage over QAM.

Trellis-coded QAM appears to be a good solution for frequency-selective channels. A study cited in [17] considers 4 source bits, encoded into 6 bits for 64-QAM. The source bit rate in a 6 MHz channel is

$$4 \ X \ \textit{effective bandwidth} \ X \ \frac{\textit{symbol length}}{\textit{GI} + \textit{symbol length}} \ X \ \textit{reference carrier ratio},$$

where the effective bandwidth is estimated at 5.5 MHz, the reference carrier ratio is 0.95, and the symbol length/ (GI + symbol length) is $\frac{7}{8}$. The result is a source bit rate of 18.3 Mbps.

The COFDM performances in the presence of narrow-band interferers are are follows:

1. If a given sub-carrier is spoiled short bursts of error are corrected by the inner code.
2. Because of the known position of the interferer a weighting of the reliability at the Viterbi decoder input can be used.
3. If more than one carrier is affected the frequency interleaving splits the error bursts.

The COFDM performances in the presence of impulsive noise are as follows:

1. In the presence of low power impulsive noise a dilution of the disturbing power over about 1000 carriers saves 30 dB.
2. In the presence of high power impulsive noise time interleaving splits the whole erroneous symbol over time.

The COFDM performances in a co-channel environment are as follows.

1. COFDM is slightly annoying. It is white noise-like and power efficient.
2. COFDM is resistant to interferers because: (a) disturbing energy is time and frequency averaged, and (b) the channel analysis allows the soft decision decoder to better correct the most likely errors.

COFDM permits tailoring of its spectrum by switching off sub-carriers where strong co-channel interference occurs, such as at the NTSC picture and colour sub-carriers.

In the presence of adjacent channels, the COFDM FFT process actually implements an additional digited rectangular filtering, which results in a wider useful bandwidth and better adjacent signal rejection.

According to [17] coding plus OFDM not only solves the problem of echoes but also gets rid of all other transmission impairments without extra complexity through synergistic interaction of: (1) the FFT; (2) time/frequency interleaving; and (3) the soft decision optimal decoder.

COFDM in a dense network offers the following advantages: (1) homogeneous received power, (2) reduced co-channel interfering power, (3) no "cliff" effect (i.e., very rapid increase in BER as a function of decreasing SNR, (4) network scalability, (5) coverage accuracy, (6) possibility of omnidirectional aerials, (7) reduced frequency re-use distance, and (8) artificially time-variant channel.

Corresponding to time diversity the COFDM feature is coding plus time interleaving. Corresponding to frequency diversity the COFDM feature is coding plus frequency interleaving. Corresponding to space diversity the COFDM feature is coding plus a guard interval.

Recent work at NHK in Japan has investigated trellis-coded OFDM and its performance for terrestrial digital TV broadcasting [38]. This work showed that by using TCM-16QAM-OFDM data of about 12.5 Mbit/secs can be transmitted in a 6 MHz BW with little degradation in the bit error rate performance compared with a conventional 4 OFDM system. A compression to 10 Mbits/sec was predicted in the near future, and then the transmission capacity of the TCM-16QAM-OFDM can be used with parity bits of a FEC code with a rate of about 70% to 80%. Their studies led to TCM-16QAM-OFDM as a better choice than TCM-32QAM-OFDM and TCM-64QAM-OFDM.

Very recent work [46] at Philips Research Labs in The Netherlands has shown that the application of RS codes in conjuction with error and erasure decoding offers an excellent alternative to CCI notches for an OFDM-based broadcast scheme.

One drawback of COFDM is that its envelope is not constant, which leads to the necessity to work the transmitter amplifier 3-5 dB under the clipping point, but work is going on to decrease this penalty.

Other drawbacks include: (1) implementation complexity (IFFT/FFT), (2) overhead due to the guard interval and reference waveform, (3) performance depends on channel estimation (tilt and gain adjust), and (4) use of lower symbol rate makes the system vulnerable to phase noise.

A summary of some main advantages are: (1) interference rejection by putting "holes" in the spectrum (although this reduces the spectrum utilization), (2) performance flexibility: carriers can be modulated at different data rates, and with different modulations, and (3) a possible SNR gain due to constructive adding of echoes.

Practical, direct incorporation of modulation schemes based on QPSK modulation into QAM-based modulation schemes. In [29] it is demonstrated how modulation schemes based on QPSK modulation can be directly incorporated into QAM-based modulation schemes. It is shown that this leads directly to an easily implementable structure which is both efficient in BW and data reliability.

Concatenated codes are known for having the capability of providing an effective and practical approach to achieving low BER, high data rate and reasonable implementation complexity. It is argued in [29] that the correct solution to the concatenated coding problem for HDTV transmission is to simply extend the codes developed for QPSK to QAM modulation.

For non-concatenated codes, a trellis code based on a binary code at rate ⅔ is usually best, a fact that follows from study of the asymptotic coding gain of a trellis code. However, for concatenated coding systems employing QAM-based trellis coded modulation, optimization of coding gain is achieved by analysis of the number of nearest neighbours in contrast to non-concatenated systems, for which asymptotic coding gain governs. Analysis is provided in [29] to support this by demonstrating the effect of the number of nearest neighbours on the error rate.

Analysis shows that a four-way partition with the number of states much greater than two is a very efficient method of trellis coding in a concatenated coding system. Such a four-way partition of QAM is a natural extension of QPSK modulation. It is, therefore, a simple matter to incorporate any good QPSK code into a trellis coding scheme for QAM modulation. A concatenated coding scheme based on QPSK trellis codes and symbol error correcting coding is proposed in [29]. A specific example employing a standard rate ½ 64-state code was shown to perform better than a rate ⅔ 16-state code Theoretical and simulated performance curves are given for symbol error rate versus E_s/N_0 for 16-QAM, R=½, 64-state; 16-QAM, R=⅔, 16-state; 32-QAM, R=½, 64-state; 32-QAM, R=⅔, 16-state; 64-QAM, R=½, 64-state; and 64-QAM, R=⅔, 16-state TCM using the above two (R=½ and R=⅔) convolutional codes. The above TCM scheme is concatenated with RS (120,110) for 16-QAM, RS (160,150) for 32-QAM and RS (200,190) for 64-QAM, and theoretical and simulated performance curves are given for RS block error rate versus E_s/N_0. These performance curves demonstrate high error control capability.

The distinct implementation advantage over other trellis coding schemes is emphasized, due to the fact that a standard off-the-shelf Viterbi decoder can be used in the trellis decoder rather than a custom part.

New soft-decision algorithms: trellis decoding of block codes and approximate A Posteriori probability. There has been considerable recent research work on using trellis decoding of block codes. Kuroda [39] has analyzed and simulated trellis decoding of majority-logic decodable codes. The particular code which he has studied is the (273,191) difference set cyclic code, which is a majority-logic decodable projective geometry code used in Japanese teletext and FM multiplex broadcasting systems.

The soft decision *a posteriori* probability algorithm (APP) was proposed by J.L. Massey for majority-logic decodable codes. It is superior to conventional hard decision decoding but it is difficult to adapt this algorithm to multi-phase shift keying, which is important for digital modulation schemes for coded modulation.

Kuroda uses a new soft decision decoding algorithm by using a trellis diagram in the majority-logic decoding stage. When Hamming distance is adopted as a branch metric this algorithm becomes a variable threshold algorithm. This algorithm is easy to apply to MPSK modulation systems. The simulation results of the (273,191) code show that the error correction performance of this new algorithm is 1.0 dB superior to the APP algorithm and 1.5 dB superior to the variable threshold hard decision decoding.

This algorithm can be adapted as a decoding algorithm of coded modulation using the (273,191) code. Kuroda's simulation results show that the performance of the (273,191) coded modulation system is better than that of the Ungerboeck code for large signal-to-noise ratios.

Approximate *a posteriori* probability (APP) decoding is applied to digital video transmission (this includes HDTV) in [31]. Threshold decodable block codes with large block length and APP soft decision decoding are proposed. The method works very well due to the weight function associated with APP soft decision decoding of threshold decodable codes. When the number of components in the parity equations is large it is demonstrated that the new method gives excellent error performance, whereas there is a substantial degradation in the performance of the least reliable symbol approximation presented by Tanaka, *et al.* and others. An analysis of the effects of feedback is given. The results of this analysis show that if the performance criterion is word error rate rather than BER, feedback of previously decoded bits is essential to obtain all possible coding gain from the soft decision decoder.

The performance of the (1057,813,34) threshold-decodable code (rate=0.77) is compared with the following concatenated system with the same rate: a punctured convolutional code with rate 5/6 punctured from a rate ½ code having generator polynomials (in octal representation) 171 and 133 as the inner code, and a (255,239,16) RS outer code, and infinite depth interleaving. The concatenated code performance (block error rate) is approximately 0.4 dB better than the performance of a (1057,813,34) threshold-decodable code at a block error rate of 10^{-6} [31].

Thus, the concatenated code provides superior performance and the threshold-decodable block code provides a much simpler implementation while suffering a slight degradation in performance.

Error performance was examined for large as well as short block length codes, and it was shown to be better than LRS (least reliable symbol) approximation to APP decoding. When the channel is quantized, the new algorithm has nearly equal performance to the exact APP decoding, even for codes with a large number of components in the parity equations where the LRS approximation suffers a substantial degradation in performance.

This paper is the first to apply APP decoding to digital video, and the results show considerable promise for this application.

Recent work in France has resulted in a turbo-code approaching the Shannon limit. In very recent unpublished work hierarchical channel coding and modulation of digital television using turbo-codes has been described [48].

Very recent recent work in Germany has described a 3-levelhierarchical digital television system which uses different iterative decoding algorithms for combined concatinated coding and multiresolution modulation [49].

Spread spectrum television broadcasting. In a very original and surprising approach W.F. Schreiber has proposed joint source-channel coding of HDTV and the use of a new form of spread-spectrum code-division multiplex (SS-CDM) channel coding for its transmission, together with progressive or "multiresolution" source coding [44]. It is claimed in [44] that by using this method the maximum image quality, reliability and spectrum for over-the-air terrestrial broadcasting will be obtained. In Schreiber's proposed scheme the coded information is divided into a large number of parallel data streams, each of which is spread to full channel BW, and becomes a component in the CDM scheme. Each has a different priority and therefore a different CNR. The resulting signal occupies one standard 6-MHz channel and is expected to have excellent interference performance.

It is asseverated in [44] that this proposed system meets the requirement of reliable performance in the presence of analog channel impairments, high spectrum efficiency, low cost, easy inter-operability, the ability to be upgraded over time in a non-disruptive manner, and the possibility of producing less expensive receivers of lower performance. This system involves subband coding, adaptive selection and modulation of subband samples, scrambling, hybrid analog/digital transmission, and spread-spectrum processing of both analog and digital data. Efficient spectrum utilization is achieved, in part, by a soft threshold in which each receiver produces an image of maximum possible quality given its particular reception conditions. Hence, the service area is substantially larger than that of systems that deliver the same quality to all receivers, as is the maximum quality that is obtained in regions of very good reception conditions. Encoders and decoders of various resolutions can work together without modification. Interference performance is as good as that of any other proposed system, so it is equally possible to give a second channel to each broadcaster who wishes to transmit HDTV. As NTSC stations are shut down, HDTV power levels can be raised, increasing image quality without modification of receivers.

Apparently it can achieve adequate quality without temporal processing, so that full motion rendition is retained and receivers can be somewhat less complex. Motion compensation can be readily used, enabling higher quality or narrower required bandwidth.

Channel Coding Using Wavelets. Wavelet matrices are obtained as the coefficients in equations for finding wavelet packets [32-34]. Input bits or blocks of information can be represented or coded by the rows of a wavelet matrix, shifted in an appropriate manner, and the output symbols obtained by adding the entries in each column of the matrix. A certain special case gives rise to fractal modulation [35]. This type of channel coding has not yet been investigated for HDTV transmission.

Fractal Modulation. As mentioned in the immediately preceding sub-section fractal modulation [35] arises from a special case of a wavelet matrix. Fractal modulation will, theoretically, enable transmissions over channels of unknown time duration and unknown BW. There are quite a few problems in implementation to be overcome, however, particularly how to do synchronization.

Fractal modulation is quite intriguing, but requires both theoretical and technical advances before it can be studied for digital HDTV transmission.

4.3 Research to Provide Graceful Degradation, Coverage Range and Robustness

Multiresolution joint source/channel coding. The theoretical optimality, established by Shannon [43], of the separation of source coding or redundancy removal from a source, from channel coding, which requires insertion of redundancy to minimize errors due to a noisy channel, holds only in the limit of infinitely complex and long codes and, more importantly, for a single-channel or point-to-point communication system [27]. On the other hand, Cover [42] showed that one could trade off channel capacity from the poor channels to the better ones, and that this trade-off can, in theory, be worthwhile. As noted in [27] these ideas point out the efficiency of using a multiresolution or embedded approach to digital broadcasting.

As noted in [27] this justifies the choice of a multiresolution (MR) source coding scheme to represent a source compacted by a hierarchy of resolutions, to which a "matched" MR transmission scheme can be designed to produce an efficient end-to-end design.

In [27] the design of an end-to-end MR system is proposed which includes an MR channel coding scheme. In this way a graceful degradation can be provided and spectral efficiency for digital broadcasting improved [27].

The use of MR joint source-channel coding in broadcasting HDTV is shown in [27] to be an efficient alternative to single-resolution schemes, which suffer from a sharp threshold effect (the digital "cliff") in the fringes of the broadcast area. The MR approach improves the coverage and robustness of the transmission scheme. The alternatives available for MR transmission through embedded modulation and error correction codes are examined. It is shown how multiresolution trellis-coded modulation (TCM) can be used to increase the coverage range. The performance of coding schemes and simulations are presented. The trade-off involved in the choice of low- and high-coverage areas as well as the comparative costs and complexities of different multiresolution transmission alternatives are discussed.

In [27] the superiority of an embedded MR transmission scheme over independent transmissions of the MR source resolutions is shown, and the trade-offs in robustness and broadcast area coverage of low- and high-resolutions between embedded MR and SR digital systems for QAM constellations are pointed out, and the benefits of using joint MR source and channel coding are highlighted.

Punctured convolutional codes. Rate-compatible punctured convolutional codes (RCPC) have proven to be very appropriate for use in concatenation with inner TCM, and with inner and outer interleavers, to provide unequal error protection (UEP) for transmission over frequency-selective land-mobile fading channels [23,24]. Slow frequency hopping was also used with OFDM in the studies described in [23,24], in which digital audio (MUSICAM) broadcasting is given as an example.

For matching HDTV source signals of unequal significance to channels of varying rate versus CNR RCPC codes appear to have considerable potential. Thus, RCPC codes are very interesting for joint source-channel coding. References [23] and [24] should be studied and the theory and techniques described therein adapted to digital HDTV transmission.

4.4 Brief Discussion of Recent Research on Satellite and Packet Transmission

Satellite transmission. In [28] various digital transmission techniques for digital satellite broadcasting have been analyzed, and transmission parameters optimized for a 27 MHz standard TV transponder. Numerical results show that the scheme with the best performance of those studied is rate ⅔ trellis-coded 8 PSK modulation concatenated with an RS (255,239) block code. For a net data rate of 40 Mbps, a BER of 10^{-10} can be achieved with E_{b0}/N_0=9.5 dB, including channel, interference, and demodulator impairments. QPSK, in combination with the RS (255,239) block code is a close second, with E_{b0}/N_0=10 dB.

By link budget evaluation it was shown that medium-class DBS satellites allow users equipped with 60 cm antennas to receive digital DBS carriers.

This new concept of digital direct satellite broadcasting (D-DBS) allows unprecedented flexibility by providing a large number of audio-visual services. An optimization procedure to obtain the system parameters is described and the results presented. Channel distortion and uplink/downlink interference effects are taken into account by using a computer simulation. A link budget analysis shows how a medium power direct-to-home TV satellite can provide multimedia services to users with small (60 cm) antennas.

ATM-based B-ISDN coding and modulation for HDTV transmission. Section V of [45] describes transmission of HDTV images compressed by wavelets on a B-ISDN ATM network. A model for the output bit rate of compressed HDTV images derived through spectral analysis and autoregressive models was used to simulate and ATM multiplexer. Curves of cell loss probability versus buffer length are shown for both bursty and non-bursty traffic with the aid of the simulated ATM multiplexer.

5 Concluding Discussion and Future Research

Beginning with the pioneering work on using concatenated error control codes for satellite transmission for the RAI/Teletra 256 project by Cominetti [14,15] during the late 1980's, the investigation of more and more sophisticated theory and techniques for error control modulation and equalization has grown at an increasing rate. The June 1990 Digicipher all-digital HDTV proposal [1] was the first system for terrestrial broadcasting to use concatenated TCM and RS codes. It also employed adaptive equalization. The Zenith and

AT&T system used a cleverly modified duo-binary scheme with a pre-coder at the transmitter end and a post-coder at the receiver end to deal with one of the most difficult problems - interference from NTSC carriers - by creating spectral nulls at the carriers in the receiver.

The two other U.S. proposals [3,4] used concatenated TCM and RS codes, and spectral shaping to minimize interference. The modulation methods for three of the systems were 16- and 32-QAM, while for the Zenith and AT&T system it was 2- and 4-VSB.

The Grand Alliance has announced that its system will use one of 4-VSB, 6-VSB (a trellis coded version of 4-VSB), 32-QAM or 32 SS-QAM. It will also investigate COFDM.

Since about 1989 or 1990 OFDM and COFDM have been the focus of European research and development into all-digital HDTV for terrestrial broadcasting. Since 1991 or 1992 a great deal of interest and investigation has been generated in the U.S.A. and Canada aimed at fully exploring the performances of variations and modifications of COFDM which would be appropriate for North American terrestrial broadcast all-digital HDTV.

The co-channel interference problem has generated increasing research activity, especially in the design of combined coding, modulation and some form of precoding and postcoding. Creating holes in the spectrum, as in the ADTV system, or by eliminating COFDM carriers, to solve the co-channel interference problem will probably not be acceptable.

What are some reasonable possibilities? The author offers his <u>personal</u> views on this matter.

A combination of COFDM with partial response (precoding) techniques could eliminate making spectral holes in the COFDM. The combination of precoding and error control methods investigated by Wei [30] deserves much more in-depth work.

The work on APP decoding [31], combined with precoding should be investigated. The new work on turbo-codes [48] and iterative decoding for combined concatenated coding and multiresolution modulation [49] is particularly promising. Both of these are applied to OFDM systems.

The marriage of COFDM and combined DFE and TCM could lead to the solution of the co-channel interference problem, as well as excellent error performance.

If some systematic way were formed to adapt to source and source coding changes, joint source-channel coding could take advantage of Cover's theory [42] to provide graceful degradation. In this connection rate-compatible punctured convolutional codes are obvious contenders [23,24].

Completely new approaches are possible using wavelet matrices and fractal modulation, but research on using these for HDTV has not even started, to the best knowledge of the author.

Error-control techniques have opened up new vistas in satellite broadcasting of HDTV [28].

We are just awakening to the rich possibilities offered by the theory for a truly reliable, spectrum efficient and technologically and economically feasible transmission system for all-digital HDTV for terrestrial broadcasting.

References

1. Digicipher HDTV system description, General Instruments Corporation, Videocipher Division, 6262 Lusk Boulevard, San Diego, California 92121. Submitted on behalf of the American Television Alliance to the FCC Advisory Committee on Advanced Television Systems, Working Party 1, Systems Analysis (Aug. 22, 1991).
2. Digital Spectrum Compatible technical details, Zenith Electronics Corporation and AT&T. Submitted to the FCC Advisory Committee on Advanced Television Systems, Working Party 1, Systems Analysis (Sept. 23, 1991).
3. Advanced Digital Television system description, Advanced Television Research Consortium, consisting of Davis Sarnoff Research Centre, NBC, North American Philips and Thomson Consumer Electronics. Submitted to the FCC Advisory Committee on Advanced Television Systems, Working Party 1, Systems Analysis (Jan. 20, 1992).
4. Channel Compatible Digicipher HDTV System (CCDC), MIT. Submitted on behalf of the American Television Alliance to the FCC Advisory Committee on Advanced Television Systems, Working Party 1, Systems Analysis (May 14, 1992).
5. E. Stare: Development of a prototype system for digital terrestrial HDTV, TELE Magazine, No. 2, 1-6 (1992).
6. A.G. Mason, N.K. Lodge: Digital terrestrial television development in the SPECTRE project, Proc. of 1992 Int. Broadcasting Conf., .359-366 (June 1992).
7. N.K. Lodge, A.G. Mason" A rugged and flexible digital modulation scheme for terrestrial high definition television, Proc. of 1992 NAB HDTV World Conf., Las Vegas, Nevada, 174-180 (April 13-16, 1992).
8. P. Bernard: 'STERNE': the CCETT proposal for digital television broadcasting, Proc. of 1992 International Broadcasting Conf., 372-374 (., June 1992).
9. J. Weber: HD-DIVINE. Viewgraphs presented at MIT Workshop on High Data Rate Digital Broadcasting, on behalf of Telecom Denmark (Oct. 26-27, 1992).
10. T. de Couasnon: OFDM efficiency: guard intervals, clock tracking and cross polarization. Viewgraphs presented at MIT Workshop on High Data Rate Digital Broadcasting, for Thomson-CSF Labs. Électroniques de Rennes (Oct. 26-27, 1992).
11. T. de Couasnon" Thomson field trials. Viewgraphs presented at MIT Workshop on High Data Rate Digital Broadcasting, on behalf of Thomson-CSF Laboratoires Électroniques de Rennes (Oct. 26-27, 1992).
12. N.K. Lodge, A.G. Mason: A rugged and flexible digital modulation scheme for terrestrial high definition television, EBU Review - Technical, No. 253, 20-26 (Autumn 1992).
13. P. Appelquist, O. Franchesi: Development of a terrestrial digital HDTV system - the HD-Divine project, EBU Review - Tech., No. 253, 40-47 (Autumn 1992).
14. M. Cominetti: Perspectives and evolution of HDTV by satellites, Proc. of Third Int. Workshop on HDTV, vol. II, Torino, Italy (Aug. 30-31-Sept. 1, 1989).
15. G. Barbieri, M. Cominetti: Experimental point-to-multipoint digital HDTV transmission via satellite during the Football World Cup 1990, EBU Review - Technical, No. 244, .230-237 (Dec. 1990).

16. S.A. Lery: GI-Digicipher I NTSCTV. Viewgraphs presented in a tutorial, Channel coding and modulation applied to HDTV and multichannel NTSC, 1993 Consumer Electronics Conference (June 1993).

17. J.B. Le Floch, B. Sauer: Equalization, Coding and Modulation for ATV. Viewgraphs presented at MIT Workshop on High Data Rate Digital Broadcasting, on behalf of France Telecom, CCETT (Oct. 26-27, 1992).

18. R.V. Paiement: Evaluation of single-carrier and multi-carrier modulation techniques for ATV. Paper presented at MIT Workshop on High Data Rate Digital Broadcasting, on behalf of the Communications Research Centre, (Oct. 26-27, 1992).

19. M. Alard, R. Lassalle: Principles of modulation and channel coding for digital broadcasting for mobile receivers, EBU Review - Technical, No. 224, 168-190 (Aug., 1987).

20. J.F. Hélard, B. Le Floch: Trellis-coded orthogonal frequency division multiplexing for digital video transmission, Proc. of Globecom, '91, .23.5.1-23.5.7 (Dec. 1991).

21. B. Sauer, D. Castelain, G. Degoulet, M. Rivière, B. Le Floch: Digital terrestrial broadcasting of audiovisual signals, Proc. of SPECTRUM 20/20 Symp., Toronto, Canada, 19 pages (Oct. 1992).

22. J.C. Rault, D. Castelain, B. Le Floch: The coded orthogonal frequency division multiplexing (COFDM) technique and its application to digital radio broadcasting towards mobile receivers, Proc. of Globecom '89 (Dec. 1989).

23. P. Hoeher: TCM on frequency - selective land-mobile fading channels, Proc. of 5th Tirrenia Int. Workshop on Digital Communs., Italy (Sept. 1991).

24. P. Hoeher, J. Hagenauer, E. Offer, Ch. Rapp: Performance of an RCPC-coded OFDM-based digital audio broadcasting (DAB) system, Proc. of IEEE Globecom '91, .2.1.1-2.1.7 (Dec. 1991).

25. J.J. Gledhill, S.V. Anikhindi, P.A. Avon: The transmission of digital television in the UHF band using orthogonal frequency division multiplexing, Proc. of the 6th Int. Conf. on Digital Processing of Signals in Communs., IEEE Conf., Publ. No. 340, 175-180 (Sept. 1991).

26. J.J. Nicolas, J.S. Lim, "Equalization and interference rejection for the terrestrial broadcast of digital HDTV", Proc. ICASSP '93, vol. IV, Minneapolis, Minnesota, .IV-176 - IV-179 (April 26-30, 1993).

27. K. Ramchandran, A. Ortega, K.M. Uz, M. Vetterli: Multiresolution broadcast for digital HDTV using joint source/channel coding, IEEE Journal on Selected Areas in Communications, vol. 11, 6-23 (Jan. 1993).

28. R. de Gaudenzi, C.Elia, R. Viola: Analysis of satellite broadcasting systems for digital HDTV, IEEE J Selected Areas in Communs, vol. 11, .99-110 (Jan. 1993).

29. C. Heegard, S.A. Lery, W.H. Paik: Practical coding for QAM transmission of HDTV, IEEE J. on Selected Areas in Communs, vol. 11, 111-118 (Jan. 1993).

30. L.-F. Wei: Precoding technique for partial-response channels with applications to HDTV transmission, IEEE Journal on Selected Areas in Communications, vol. 11, 127-135 (Jan. 1993).

31. S.A. Raghavan, Y. Hebron, I. Gurantz: On the application of approximate APP decoding to digital video transmission, IEEE Journal on Selected Areas in Communications, vol. 11, 136-152 (Jan. 1993).

32. M.A. Tzannes, M.C. Tzannes: Constant envelope bit-by-bit channel coding using wavelets, Proc. of the Canadian Conference on Electrical and Computer Engineering, vol. 1, MM2.5.1-MM2.5.4 (Sept. 13-16, 1992).
33. M.A. Tzannes, M.C. Tzannes: Block biorthogonal channel coding using wavelets, Proc. MILCOM '92.
34. M.A. Tzannes: Channel coding: Wavelets and communication systems> Viewgraphs presented at a workshop on wavelets and their applications, Aware Inc., One Memorial Drive, Cambridge, MA 02142 (July 23-24, 1992).
35. G.W. Wornell, A.V. Oppenheim: Wavelet-based representations for a class of self-similar signals with application to fractal modulation, IEEE Trans. on Information Theory, vol. 38, .785-800 (March 1992).
36. F. Forrest: Key issues in HDTV/ATV systems, EBU Review - Technical, vol.253, 20-26 (Autumn 1992).
37. The COFDM system intended for digital audio broadcasting takes benefit from time, frequency and space diversity. Documents, CCIR Study Groups, Period 1990-1194, from Collection of CCETT Contributions.
38. M. Saito, S. Moriyami, O. Yamada: A digital modulation method for terrestrial digital TV broadcasting using trellis coded OFDM and its performance, Proc. Globecom '92 (Dec. 1992).
39. T. Kuroda: A study on the coded modulation scheme using majority logic decodable codes. Unpublished paper from NHK Science and Tech. Rese. Labs, distributed at a seminar presented at the Communications Res. Centre, Ottawa, Can. (Feb. 1993).
40. R.J. Siracusa, K. Josejah, J. Zdepski, D. Raychaudhuri: Flexible and robust packet transport for digital HDTV, IEEE Journal on Selected Areas in Communs, vol.11, 88-98 (Jan. 1993).
41. C. Eilers, P. Fockens: The DSC-HDTV interference rejection system, IEEE Trans. Consumer Electronics, vol.38, 101-107 (June 1992).
42. T. Cover: Broadcast channels, IEEE Trans. Inf. Theory, vol.IT-18, .2-14 (Jan. 1972).
43. C.E. Shannon: A mathematical theory of communication, Bell System Tech. J., vol.27, 379-423 (1948).
44. W.F. Schreiber: Spread-spectrum television broadcasting, SMPTE J., vol.101. 538-549 (Aug. 1992).
45. Argenti, F., G. Benelli, A. Mecocci: Source coding and transmission of HDTV images compressed with the wavelet transform, IEEE J. on Selected Areas in Communications. vol.11, 46-58 (Jan. 1993).
46. P. de Bot, C.P.M.J. Baggen: Reed-Solomon cades for ODFM broadcasting over frequency-selective channels using error and erasure decoding. Proc. 1993 Int. Tirrenia Workshop on Digital Communications, Tirrenia, Italy (Sept. 1993).
47. C. Berron, A. Gl;avieux, P. Thitimajshimo: Near Shannon limit error-correcting coding and decoding: turbo-codes. Proc. ICC '93, 1064-1070 (June 1993).
48. C. Berrow, P. Combelles: Digital television: Hierarchial channel coding using turbo-codes. To appear, Proc. ICC '94.
49. L. Papke, K. Fazel: Different iterative decoding algorithms for combined concatinated coding and multiresolution modulation. To appear, Proc. ICC '94.

A Finite Field Arithmetic Unit VLSI Chip

Germain Drolet

Dept. of Electrical & Computer Engineering
Royal Military College
Kingston, Ontario
K7K 5L0

Abstract. This paper presents a circuit operating on fields $\mathbb{F}_2[X]/\langle f(X)\rangle$ where $f(X)$ is a binary irreducible polynomial of degree $m \geq 2$, and $\mathbb{F}_2 = \mathrm{GF}(2)$. This circuit is able to perform back to back multiplications and inversions for any such $f(X)$ and any value of m within a specified range, m being possibly large. It is assumed that the elements of the field are expressed as polynomials in X of degree less than m (polynomial basis). The circuit consists mainly of a Serial Input–Serial Output multiplier which is similar to the one published by Yeh, Reed, Truong in 1984. An element of the field is inverted by raising it to the power $2^m - 2$, and so the outputs of the multiplier are fed back into its inputs. Even though the circuit can operate on any size of field within the specified range, it is better suited for large fields; circuitry achieving better performance can be designed for small fields (Parallel Input–Parallel Output).

1. Introduction

In 1990, Shayan, Le–Ngoc, Bhargava designed a versatile Reed–Solomon decoder circuit, *i.e.* one that can be programmed to decode Reed–Solomon codes of any dimension, but with fixed block length. The VLSI chip that they constructed operates on Reed–Solomon codes of block length 31, that is over the alphabet $\mathrm{GF}(2^5)$ [9]. Some attempt has been made to construct a more versatile chip in which the block length, and hence the alphabet size, are no longer fixed [4]. One of the difficulties in doing so, is to design a circuit performing arithmetic operations on fields $\mathrm{GF}(2^m)$ where m is allowed to vary from one application to the other; most arithmetic circuits operate on fields of a fixed size.

In 1984, Yeh, Reed, Truong proposed two circuits performing multiplications on extension fields of varying degree over GF(2). One circuit is Serial Input–Serial Output, and the other is Parallel Input–Parallel Output [11]. The first circuit has since been improved by Zhou [12]. All of these use polynomial basis representation of the elements of the finite fields. If representation of the elements in more than one basis is allowed, *e.g.* using a polynomial basis together with its dual basis, then better circuits and better performance can be obtained [3]. This might not be convenient however, requiring transformations from one basis to the other. Green implemented Yeh–Reed–Truong's second circuit, since his application dealt with fields $\mathrm{GF}(2^m)$ of relatively small size, $3 \leq m \leq 8$, and required parallel output. However, the inverter circuit he investigated turned out to be more complex, requiring an inverting time that grows exponentially with m. It is impractical when $m > 8$ [4].

In this paper, we look at a circuit operating on fields $\mathbb{F}_2[X]/\langle f(X)\rangle$ where $f(X)$ is a binary irreducible polynomial of degree $m \geq 2$, and $\mathbb{F}_2 = \mathrm{GF}(2)$. This circuit must be able to perform back to back multiplications and inversions for any such $f(X)$ and any value of m within a specified range, m being possibly large. We assume that the elements of the field are expressed as polynomials in X of degree less than m (polynomial basis). The circuit consists mainly of a Serial Input–Serial Output multiplier which is similar to Yeh–Reed–Truong's. An element of the field is inverted by raising it to the power $2^m - 2$, and so the outputs of the multiplier are fed back into its inputs. Even though the circuit can operate on any size of field within the specified range, it is better suited for large fields. This makes the chip more interesting for applications to the field of *cryptography* than *coding for the correction of errors*. Circuitry achieving better performances has been designed for small fields (Parallel Input–Parallel Output) [7], but impractical circuit complexity makes it unsuitable to applications with large extension fields of GF(2).

Simultaneous versatility (circuits are independent of the irreducible polynomial) and large field size requirements are obtained by Hasan–Bhargava [5] and Araki *et al.* [1]. Their performances differ slightly:

circuit complexity:	$\mathcal{O}(m)$ for [1]
	$\mathcal{O}(m^2)$ for [5]
time complexity:	$\mathcal{O}(m^2)$ for [1]
	$\mathcal{O}(m)$ for [5]
computation time:	not equal for [1]
	equal for [5]

The performances of the circuit presented in this paper are closer to Araki's, the only difference being that multiplications/inversions of any elements all require the same time when the value of m is fixed. It also has the advantage over any other circuit known to the author, that it can be cascaded to allow operations on larger fields, notably m as large as 100.

Section II reviews the basics of Yeh–Reed–Truong's serial multiplier, and section III presents a modification to Yeh–Reed–Truong's multiplier and its integration into the arithmetic unit. Section IV describes a VLSI chip based on this circuitry. The chip can stand alone or in a cascade to operate on very large fields.

2. Review of Yeh–Reed–Truong Serial Multiplier

Consider the following binary polynomials:

$$a(X) = \sum_{i=0}^{m-1} a_i X^i,$$

$$b(X) = \sum_{i=0}^{m-1} b_i X^i,$$

$$c(X) = \sum_{i=0}^{m-1} c_i X^i,$$

$$f(X) = \sum_{i=0}^{m} f_i X^i,$$

where we assume $f_m = 1$ and $f(X) \neq X^m$, that is $f(X)$ has at least one non-zero coefficient other than f_m. Yeh, Reed and Truong [11] derived an electronic circuit that calculates the unique binary polynomial $d(X)$ of degree at most m–1 such that

$$d(X) \equiv a(X)\ b(X) + c(X) \pmod{f(X)}.$$

This simply represents multiplication and addition in a ring of the form $\mathbb{F}_2[X]/\langle f(X) \rangle$, where $\mathbb{F}_2 = \mathrm{GF}(2)$ is the binary finite field. This circuit will, in particular, perform operations in finite fields $\mathrm{GF}(2^m)$ when $f(X)$ is an irreducible polynomial in $\mathbb{F}_2[X]$. Indeed when $f(X)$ is irreducible, the ideal $\langle f(X) \rangle$is maximal in $\mathbb{F}_2[X]$ and then $\mathbb{F}_2[X]/\langle f(X) \rangle$ is a field isomorphic to $\mathrm{GF}(2^m)$ [6]. In the remainder of this paper, $f(X)$ need not be irreducible unless stated.

$d(X)$ is simply calculated in the way one would normally do by hand. That is,

$$d(X) = \sum_{i=0}^{m-1} b_i\ X^i a(X) \quad + \ c(X)$$

$$= \sum_{i=0}^{m-1} b_i a^{(i)}(X) \ + \ c(X) \qquad (1)$$

in which $X^i a(X)$ is reduced modulo $f(X)$, and where

$$a^{(i)}(X) \equiv X^i a(X) \pmod{f(X)}$$

$$\deg a^{(i)}(X) < m \ .$$

Thus "at each step $j=1, 2, \ldots, m$" $X^{j-1} a(X)$ is multiplied by X modulo $f(X)$ and added to the partial sum $c^{(j-1)}(X) = \sum_{i=0}^{j-2} b_i a^{(i)}(X) + c(X)$, yielding $c^{(j)}(X)$. The initial conditions of this recurrence ($j-1=0$) are $c^{(0)}(X) = c(X)$ and $a^{(0)}(X) = a(X)$. After m repetitions the partial sum will contain the reduction modulo $f(X)$ of $a(X)\,b(X) + c(X)$.

In this paper we assume that m is a relatively large integer, on the order of say 100. In such cases the coefficients of all polynomials must be fed serially into an electronic circuit. Feeding them in parallel would require one pin for each coefficient of each polynomial, hence there would be far too many pins for a chip.

The circuit consists of a cascade of m identical cells, each calculating and adding one term to the sum (1) [as per previous paragraph]. One such cell is presented in figure 1, in which the coefficients of all polynomials are fed in serially and ordered by increasing powers of X. The starred polynomials are reciprocal: *e.g.* $a(X)^* = X^{m-1} a(X^{-1})$. Inputting $a(X)^*$ in the circuit is equivalent to ordering the coefficients of $a(X)$ (fed into the circuit) by decreasing powers of X. In other words the coefficients of starred polynomials fed into the multiplier are ordered by decreasing powers of X, while the coefficients of non–starred polynomials are ordered by increasing powers of X. This convention differs from that chosen by Yeh, Reed, Truong in [11] and will become more convenient later, simplifying some expressions. It does not affect the operation of the circuit since with both conventions the same sequences of coefficients enter the circuit. With the present ordering, the coefficients will form a polynomial which is the reciprocal of the one obtained with the ordering of Yeh, Reed, Truong; this is indicated by the stars in figure 1.

In this circuit:

– The coefficient f_m of $f(X)$ need not be fed since it is assumed to be 1 always. Therefore the coefficients fed in are those of the polynomial

$$f_{m-1}X^{m-1} + f_{m-2}X^{m-2} + \ldots + f_1X + f_0 \ ,$$

that is the reduction of X^m modulo $f(X)$.

– The term $b_{j-1} a^{(j-1)}(X)$ is added to the partial sum $c^{(j-1)}(X)$ to yield $c^{(j)}(X)$.

– $a^{(j-1)}(X)$ is multiplied by X by feeding $a^{(j-1)}(X)^*$ through one fewer delay el-

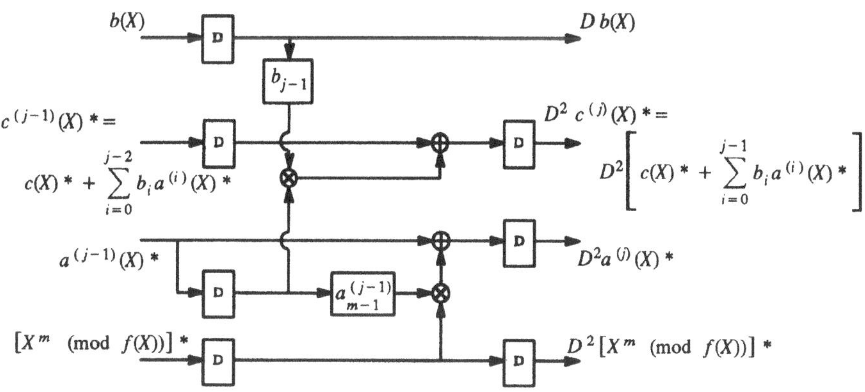

Fig. 1. Yeh–Reed–Truong j–th Multiplier Cell

ement. This causes coefficients of $a^{(j-1)}(X)$ to be aligned with coefficients of the next power of the polynomials $c^{(j-1)}(X)$ and $X^m \pmod{f(X)}$. It is then reduced modulo $f(X)$ by adding the coefficients of X^m modulo $f(X)$ when $a^{(j-1)}_{m-1}$ equals 1 (leading coefficient of $a^{(j-1)}(X)$).

– The consequence of passing $b(X)$ through a single delay element (instead of two as for all other polynomials) is to "push" $b(X)$ faster through the cascade. This is needed because at the instant that all signals enter the first stage of the cascade, the coefficient b_0 must be loaded in the memory cell. However when the signals enter the second stage of the cascade, the coefficient b_1 must be loaded in the memory cell. This indicates that the coefficients of $b(X)$ must travel faster than those of the other polynomials. Similarly when all signals enter the j–th stage of the cascade, the coefficient b_{j-1} must be loaded in the memory cell.

It is clear that by cascading m such circuits with initial values $c^{(0)}(X) = c(X)$, $a^{(0)}(X) = a(X)$, the output of the m–th stage is

$$D^{2m} c^{(m)}(X) * = D^{2m} d(X) * = D^{2m} \left[\sum_{i=0}^{m-1} b_i a^{(i)}(X) + c(X) \right]^*,$$

that is $a(X)\, b(X) + c(X) \pmod{f(X)}$ ordered by decreasing powers of X and delayed by $2m$ clock cycles.

3. Extended Yeh–Reed–Truong Multiplier cell

When $f(X)$ is irreducible in $\mathbb{F}_2[X]$, the inverse of $0 \neq a(X) \in \mathbb{F}_2[X]/\langle f(X) \rangle$ is $a(X)^{-1} = a(X)^{2^m-2}$ and is best computed by:

$$a(X)^{2^m-2} = \underbrace{\left[\left[\left[a(X)^2 a(X)\right]^2 a(X)\right]^2 \ldots\; a(X)\right]^2}_{m\,-\,1 \text{ times}} . \tag{2}$$

This computation is easily performed by feeding the output of the multiplier back into one input of the multiplier while the other input selects alternatively the output of the multiplier or the original element to be inverted. $2(m-1) - 1$ multiplications have to be performed, yielding the inverse $a(X)^{-1}$ of $a(X)$.

The use of the Yeh–Reed–Truong multiplier to perform the computation (2) poses a problem during the squaring cycle, for then a given output $y(X)$ must be fed back into both inputs of the multiplier. But we recall that in order to calculate $[y(X)\,y(X) \pmod{f(X)}]^*$, $y(X)$ and $y(X)^*$ are the two necessary inputs to the multiplier. Before $y(X)$ can be fed back into the multiplier, one has to wait for the last coefficient of $y(X)^*$ to come out of the multiplier, hence introducing large delays into the computation time of (2). In order to resolve this deadlock we will modify the basic cell of the Yeh–Reed–Truong multiplier such that both outputs $y(X)$ and $y(X)^*$ are simultaneously calculated.

Extra circuitry is introduced to compute

$$D^{2m}[\,a(X)\,b(X) + c(X) \pmod{f(X)}\,] .$$

It is achieved in a similar fashion to the Yeh–Reed–Truong multiplier cell, applying equation (1). When $a^{(i)}(X)$ is fed into the cell, its multiplication by X is accomplished by passing it through one more delay element than the other signals; $a^{(i)}(X)$ becomes delayed by one more clock cycle. The circuit performing this is shown in figure 2, where we see that $X^m \pmod{f(X)}$ is added to the delayed $a^{(i)}(X)$ when $a_{m-1}^{(j-1)}$ equals 1. $a_{m-1}^{(j-1)}$ is the last coefficient entering the cell on input line $a(X)$, but is the first coefficient entering through input line $a(X)^*$. In fact, this is the coefficient that is stored in the memory cell of the j–th Yeh–Reed–Truong multiplier cell. The same applies to coefficient b_{j-1} of $b(X)$. Consequently the additional circuitry does not stand alone but is designed to work in conjunction with the Yeh–Reed–Truong multiplier cell. Also in the resulting extended multiplier cell, all signals except $b(X)$ are duplicated, with coefficients ordered by both increasing and decreasing powers of X. The following equivalent expressions are used in the notation of Figure 2 (these are easily shown and the proof is omitted):

$$[X^m \pmod{f(X)}]^* \equiv X^{-1} \pmod{f^R(X)} ,$$

$$\left[Xa^{(j-1)}(X) \pmod{f(X)}\right]^* = X^{-1}a^{(j-1)}(X)^* \pmod{f^R(X)} ,$$

where $f^R(X) = X^{\deg f(X)} f(X^{-1})$ is the reciprocal of $f(X)$, and X^{-1} is the inverse of X in the ring $\mathbb{F}_2[X]/\langle f^R(X)\rangle$(notice that given the assumptions $f_m = 1$ and $f(X) \neq X^m$, X is invertible in $\mathbb{F}_2[X]/\langle f^R(X)\rangle$). This new notation is used to emphasize that the previously defined $a(X)^*$ is in general different than $a^R(X)$; $a(X)^* = a^R(X)$ if and only if $\deg a(X) = m - 1$.

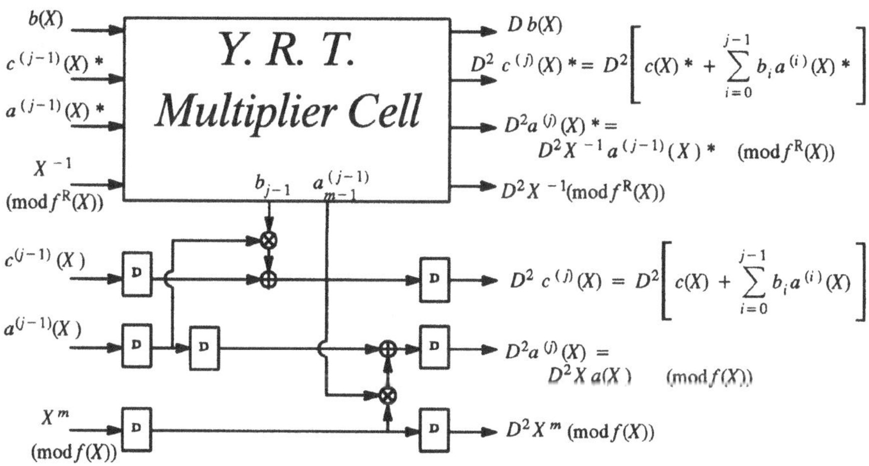

Fig. 2. j–th Yeh–Reed–Truong extended Multiplier Cell

For example, a cascade of $m = 4$ such cells would yield a multiplier for $GF(2^4)$ when $f(X)$ is irreducible of degree 4. The output polynomials would be delayed by 8 clock cycles. Moreover, feedback can be introduced to this cascade in order to calculate the inverse of an element in GF(2^4). The resulting arithmetic unit and the necessary connections are shown in figure 3.

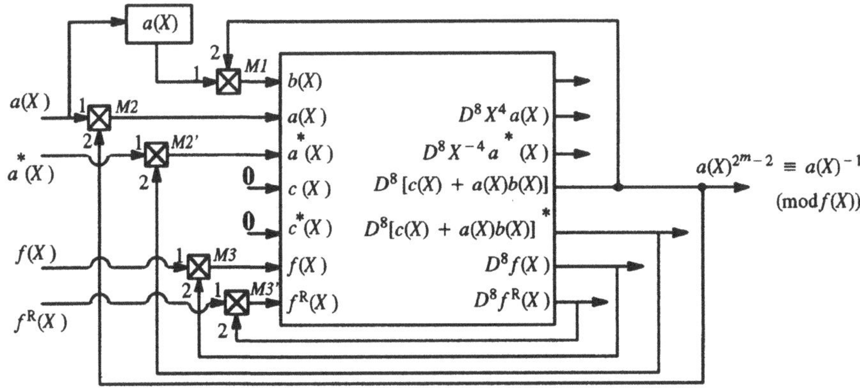

Fig. 3. Use of Yeh–Reed–Truong Multiplier Cells into an Arithmetic Unit

In the main block representing the cascade of 4 Yeh–Reed–Truong extended multiplier cells, the factor D^8 represents the afore mentioned delay of 8 clock cycles between input polynomials and output polynomials. The multiplexers *M2, M2', M3, M3'* will select input 1 during the first multiplication and select input 2 for all subsequent multiplications. Multiplexer *M1* will select input 1 for the first and second multiplications. After that, it will select input 2 and 1 alternatively to compute $a(X)^{2^m-2}$ as per equation (2). It successively calculates $a(X)^2$, $a(X)^3$, $a(X)^6$, $a(X)^7$, $a(X)^{14}$, ... until power $2^m - 2$ is reached: $2(m-1)-1$ multiplications are required. Note that this is only valid when $f(X)$ is irreducible. If $f(X)$ is not irreducible, $a(X)$ may not have an inverse in the ring $\mathbb{F}_2[X]/\langle f(X)\rangle$, and if it does its inverse will not be given by $a(X)^{2^m-2}$ in general.

Finally we notice that a cascade of M Yeh–Reed–Truong extended multiplier cells performs operations in the extension field of degree M over GF(2) when $f(X)$ is irreducible of degree M. But it can also perform operations in extension fields of degree $2 \leq m \leq M$ if the last $(M-m)$ Yeh–Reed–Truong extended multiplier cells are disabled. One control signal, RECEIVE_ENABLE, is used to mark the beginning and end of input coefficients of the polynomials and to disable these last $(M-m)$ cells. This increases the versatility of the circuit. The processing time is however not optimal if $m < M$. Indeed, even though the computation is finished at the output of the m–th cell, the result must be passed through all M Yeh–Reed–Truong extended multiplier cells. So

$2(M - m)$ clock cycles could be taken off the overall delay by using a shorten cascade. The throughput remains the same but the pipe takes longer to fill.

Multiplications and inversions on finite fields $\mathbb{F}_2[X]/\langle f(X)\rangle$ of different size and different irreducible polynomial $f(X)$ can be performed back to back by the circuit.

4. VLSI Implementation

The Yeh–Reed–Truong extended multiplier cell contains 137 gates and 400 gates are required for overhead (control, status and acknowledgement). An arithmetic unit circuit with 20 cascaded Yeh–Reed–Truong extended multiplier cells contains roughly 3000 gates (12000 transistors) and is manufactured on a single chip through the *Canadian Microelectronic Corporation* using *Northern–Telecom* 3µm CMOS technology. Extra connections are added to the chip so that the Yeh–Reed–Truong extended multiplier cells of many chips can be cascaded. Operations in large fields can be achieved in this fashion. For example, a cascade of five chips can be wired to operate in extension fields over GF(2) of degree $2 \leq m \leq 100$, with no required programing of the chips. The chip is fully synchronous and requires a clock signal CLK as well as a square wave, 2xCLK, of twice the clock frequency. The latter is used by two flip–flops inside each Yeh–Reed–Truong extended multiplier cell, and need not be in phase with CLK. It is integrated in a 48–pin package. 25 pins are required by the basic multiplier, 11 more pins are needed to make the chip "cascadable", and extra pins are added for testability of the chip.

The chip has two modes of operation: *open–loop* (for multiplications) and *closed–loop* (for inversions) indicated by a status pin STATUS. Both modes are Serial Input–Serial Output. The *open–loop* mode is systolic and each multiplication requires $2m$ clock cycles. Once the pipe is filled one coefficient of the product is obtained at each clock cycle. Leading coefficients of output polynomials are indicated by a pulse on pin $\overline{\text{STROBE}}$ (active low) for synchronization.

The *closed–loop* mode is not systolic and each inversion requires $2m(2m-3) = \mathcal{O}(m^2)$ clock cycles $[2(m-1)-1$ multiplications]. Various powers of $a(X)$ will successively appear at the output and $a(X)^{2^M-2}$ is indicated by a *low*–pulse on pin STROBE, where M is the total number of Yeh–Reed–Truong extended multiplier cells in the cascade. If the inversion is performed in a field extension of degree $m < M$, the chip will still raise $a(X)$ to the power $2^M - 2$. During this computation, the inverse $a(X)^{2^m-2}$ will come out and will not be acknowledged by the chip. The user must count the powers and "intercept" the inverse. Input lines $c(X)$, $c(X)^*$ must remain *low* during an inversion and input line $b(X)$ is ignored. When $\overline{\text{STROBE}}$ gives the pulse indicating the end of an inversion, the chip returns into the *open–loop* mode. In order to perform a succession of inversions, the chip must constantly be brought back into the *closed–loop* mode. This is done by raising a control MODE pin for at least one clock cycle.

5. Conclusion

The circuitry developed in this paper extends the Serial Input–Serial Output multiplier cell of Yeh–Reed–Truong. A cascade of the new cell calculates the reduction modulo $f(X)$ of $a(X)\,b(X) + c(X)$ and outputs its coefficients ordered by both increasing and decreasing powers of X. This multiplier is recursively used by feeding back its outputs on its inputs to invert an element. It is better suited for operations on large extension fields of GF(2).

The circuit is very versatile in the sense that it can perform operations on any field extension of degree $2 \leq m \leq M$ over GF(2). These fields are isomorphic to $\mathbb{F}_2[X] / \langle f(X) \rangle$, where $f(X)$ is a binary irreducible polynomial of degree m. When $m > 2$, many such polynomials exist [2]. The circuit can perform operations no matter which one is chosen. Multiplications and inversions on finite fields of different size and/or different irreducible polynomial $f(X)$ can be performed back to back by the circuit. $f(X)$ only needs to be fed in together with the elements to operate on; no programming of the circuit is required.

The circuit can also be used to multiply in rings isomorphic to $\mathbb{F}_2[X] / \langle f(X) \rangle$, where $f(X)$ is not necessarily irreducible. In such cases the *closed–loop* mode (for inversion) will simply raise an element to the power $2^M - 2$, M being the number of cascaded multiplier cells. Even if the original element is invertible in the ring, the result will not be its inverse in general.

The VLSI arithmetic unit chip constructed from this circuitry operates on extension fields of degree no more than 20 over GF(2). It can, however, be cascaded to operate on larger extension fields. Its main disadvantage is that almost all input polynomials must be duplicated, coefficients of which being ordered by both increasing and decreasing powers of X. An improved version of the present circuit is being investigated following Zhou's approach [12]. It is expected that similar operations can be performed with a smaller gate count. A smaller technology will be used, yielding higher throughputs.

References

1. K. Araki, I. Fujita, M. Morisue, "Fast Inverter over Finite Filed Based on Euclid's Algorithm", *Trans. IEICE*, vol. E 72, pp 1230–1234, Nov 1989.

2. L. Childs, *A Concrete Introduction to Higher Algebra*, Undergraduate Text in Mathematics, Springer–Verlag, New–York, 1979.

3. M. Diab, "Systolic Architectures for Multiplication over Finite Field GF(2^m) ", *Applied Algebra, Algebraic Algorithms, and Error Correcting Codes*, Proceedings of the 8th International Conference, AAECC–8, Tokyo, Japan, August 1990, Springer–Verlag, pp 329–340.

4. B.K. Green, *Design of a Single–Chip Universal Reed–Solomon Decoder*, Master's Thesis presented to the Dept. of Elec. and Comp. Eng., Royal Military College, Kingston, Ontario, May 1992.

5. M.A. Hasan, V.K. Bhargava, "Bit–Serial Systolic Divider and Multiplier for GF(2^m) ", *IEEE Trans. Comput.*, vol C–41, no 8, pp 972–980, Aug 1992.

6. T.S. Hungerford, *Algebra*, Graduate Text in Mathematics 73, Springer–Verlag, New–York, 1984.

7. M. Kovac, N. Ranganathan, M. Varanasi, "SIGMA: A VLSI Systolic Array Implementation of a Galois Field GF(2^m) Based Multiplication and Division Algorithm", *IEEE Trans. on VLSI Systems*, vol 1, no 1, pp 22–30, Mar 1993.

8. P.A. Scott, S.E. Tavares, L.E. Peppard, "AS Fast VLSI Multiplier for GF(2^m) ", *IEEE J. Selected Areas Commun*, vol 4, no 1, pp 62–66, Jan 1986.

9. Y.R. Shayan, T. Le–Ngoc, V.K. Bhargava, "A Versatile Time–Domain Reed–Solomon Decoder", *IEEE J Selected Areas Commun*, vol 8, no 8, pp 1535–1542, Oct 1990.

10. C.C. Wang, T.K. Truong, H.M. Shao, L.J. Deutsch, J.K. Omura, I.S. Reed, "VLSI Architecture for Computing Multiplications and Inverses in GF(2^m) ", *IEEE Trans. Comput.*, vol C–34, no 8, pp 709–717, Aug 1985.

11. C.S. Yeh, I.S. Reed and T.K. Truong, "Systolic Multipliers for Finite Fields GF(2^m) ", *IEEE Trans. Comput.*, vol C–33, no 4, pp 357–360, Apr 1984.

12. B.B. Zhou, "A new Bit–serial Systolic Multiplier over GF(2^m) ", *IEEE Trans. Comput.*, vol C–37, no 6, pp 749–751, June 1988.

A Successful Attack Against the DES

Faramarz Hendessi and Mohammad R. Aref

Department of Electrical Engineering, Isfahan University of Technology, Isfahan, IRAN

Abstract. It is shown that the Data Encryption Standard (DES) function is divided into 68 subfunctions, and therefore the exhaustive key search attack could be done by a pipelining method. A chip is designed whose purpose is to attack the DES and a searching machine is detailed based on it. It is shown that the DES could be broken, easily and cheaply, by this machine.

1 Introduction

The publication of the Data Encryption Standard algorithm in 1975 throw a challenge to the world to break it [1] and many comments, both for it and against it, resulted. Amongst them, the Hellman's machine [2] is a method of importance. This machine is designed to exhaust all the 2^{56} keys at a rate 10^{12} keys per second and would, therefore, require about one day for the entire search. The machine itself would consist of about 10^6 DES chips, each searching a different portion of the key space at a rate of about one key per microsecond. The estimates on the cost of such machine vary and range from few to tens of millions of dollars.

The idea of searching machine is further developed in this paper. We show that the DES algorithm could be carried out to a pipelining problem. Contrary to Diffie and Hellman, we don't use the DES chip in the searching machine. Rather we design a new chip, called DES Breaker (DESB), with purpose of using the pipelining idea. It is shown that the searching machine using DESB would be much cheaper than Hellman's machine.

2 Encryption algorithm

The Data Encryption Standard was designed to encipher blocks of data consisting of 64 bits under the control of a 56-bit key. A block to be enciphered is subject to the initial permutation IP, then to a complex key-dependent computation and finally to the inverse of the initial permutation IP^{-1}. The key-dependent computation consists of 16 rounds (iterations) of a calculation that is described in terms of the cipher function f (figure 1). Using the notation defined in the appendix[1], the calculation of the ith round is shown by the following formula:

$$\begin{cases} \underline{R_i} = \underline{L_{i-1}} \oplus f\left(\underline{k_i}, \underline{R_{i-1}}\right) & \\ \underline{L_i} = \underline{R_{i-1}} & i = 1, 2, .., 16 \end{cases} \tag{1}$$

Where underlines variables represent 32–bit and 48–bit vectors, $\left(\underline{R_0}, \underline{L_0}\right)$ is the permuted input block, and $\underline{k_i}$, called the ith subkey, is a function of the key and

[1] It is suggested that the appendix be read before the article.

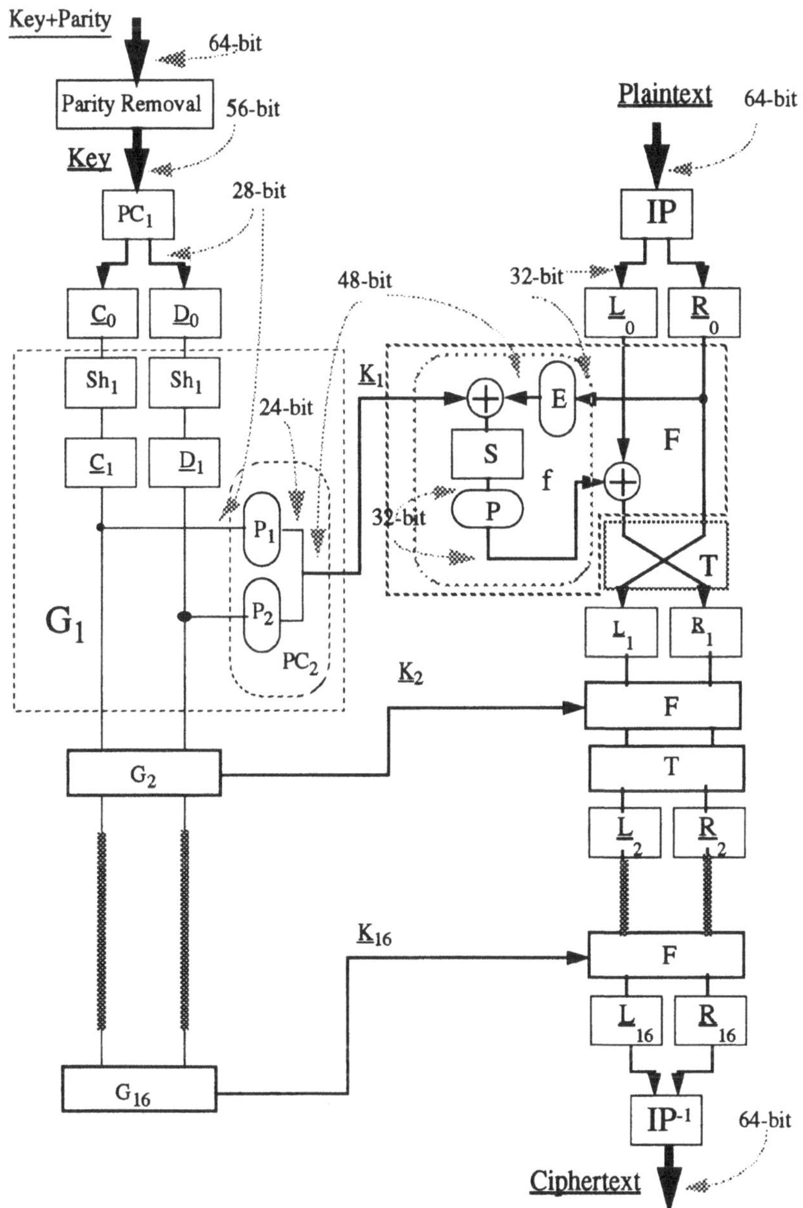

Fig. 1. The encryption algorithm in DES.

is generated in the ith round by the following procedure:

$$\begin{cases} \underline{C_i} = Sh_j\left(\underline{C_{i-1}}\right) \\ \underline{D_i} = Sh_j\left(\underline{D_{i-1}}\right) \\ \underline{k_i} = \left(P1(\underline{C_i}), P2(\underline{D_i})\right) & \begin{cases} j=1 & i=1,2,9,16 \\ j=2 & else \end{cases} \end{cases} \tag{2}$$

Where $(\underline{C_0}, \underline{D_0})$ is the permuted key by $PC1$, and $Sh_j(.)$ denotes j bits circular left shift of the argument block [Appendix].

3 Divided encryption algorithm

To bring the algorithm into the pipelining problem, the algorithm should be divisible into sub-algorithms, and this section shows that the DES is so.

Let $\underline{f_i}$ be the output of the f function in the ith round. Also let $\underline{L_{-1}} = \underline{R_0}$, $\underline{L_{17}} = \underline{R_{16}}$,, and $\underline{f_0} = \underline{0}$. Using these definitions (1) could be rewritten as follows:

$$\begin{cases} \underline{L_{i+1}} = \underline{L_{i-1}} \oplus \underline{f_i} \\ \underline{f_{i+1}} = f\left(\underline{k_{i+1}}, \underline{L_{i+1}}\right) = f\left(\underline{k_{i+1}}, \underline{L_{i-1}} \oplus \underline{f_i}\right) \end{cases} \tag{3}$$

Considering the structure of the f function (A-3), the above formula could be rearranged as :

$$\begin{cases} \underline{L_{i+1}} = \underline{L_{i-1}} \oplus \underline{f_i} \\ \underline{f_{i+1}} = P\left(S\left(E\left(\underline{L_{i-1}} \oplus \underline{f_i}\right) \oplus \underline{k_{i+1}}\right)\right) \end{cases} \tag{4}$$

Combining (2) and (4) yields following recursive structure of the encryption algorithm:

$$\begin{cases} \left(\underline{L_0}, \underline{L_{-1}}\right) = IP(\underline{Plaintext}) \\ \left(\underline{C_0}, \underline{D_0}\right) = PC1(\underline{Key}) \\ \underline{f_0} = \underline{0} \end{cases} \tag{5}$$

$$\begin{cases} \underline{C_{i+1}} = Sh_j\left(\underline{C_i}\right) \\ \underline{D_{i+1}} = Sh_j\left(\underline{D_i}\right) \\ \underline{f_{i+1}} = P\left(S\left(E\left(\underline{L_{i-1}} \oplus \underline{f_i}\right) \oplus \left(P1(Sh_j(\underline{C_i})), P2(Sh_j(\underline{D_i}))\right)\right)\right) \\ \underline{L_{i+1}} = \underline{L_{i-1}} \oplus \underline{f_i} \qquad \begin{cases} j=1 & i=0,1,8,15 \\ j=2 & else \end{cases} \end{cases} \tag{6}$$

$$i = 0, 1, .., 16$$

$$\underline{Ciphertext} = IP^{-1}\left(\underline{L_{17}}, \underline{L_{16}}\right) \tag{7}$$

Equation (6) could be calculated for $(i+1)$th round if $\underline{L_{i-1}}, \underline{f_i}, \underline{C_i}$, and $\underline{D_i}$ have been specified at the same time.

Considering the structure of XOR function, $\underline{X} \oplus \underline{Y}$ and $\overline{\underline{X} \oplus \underline{Y}}$, for any $\underline{X}$ and $\underline{Y}$, can be calculated simultaneously. Thus to increase parallelism, every vector and its complement are considered at the same time. Therefore letting

$$\underline{w_i} = \left(\underline{L_i}, \underline{L_{i-1}}, \overline{\underline{L_{i-1}}}, \underline{f_i}, \overline{\underline{f_i}}, \underline{C_i}, \overline{\underline{C_i}}, \underline{D_i}, \overline{\underline{D_i}}\right) \tag{8}$$

be an input, equation (6) can be represented in the following recursive form:

$$\underline{w_{i+1}} = M_j\left(\underline{w_i}\right) \qquad \begin{cases} j=1 & i=0,1,8,15 \\ j=2 & else \end{cases} \tag{9}$$

Where $M_{j(.)}$ is the Boolean function defined by (6).

To calculate $\underline{w_0}$ from $\underline{Plaintext}$ and $\underline{key}$, and to reach $\underline{Ciphertext}$ from $\underline{w_{16}}$, we also define two functions $B0(.)$ and $B1(.)$ as follows:

$$\underline{w_0} = B0\left(\underline{Plaintext}, \underline{key}\right) \tag{10}$$

$$\underline{Ciphertext} = B1\left(\underline{w_{16}}\right) \tag{11}$$

The encryption algorithm is, therefore, divided into 18 tasks.

$$DES_{\underline{key}}(\underline{Plaintext}) = B1 M_1 M_2^6 M_1 M_2^6 M_1^2 B0\left(\underline{Plaintext}, \underline{key}\right) \tag{12}$$

where M_j^n means that n repeated M_j transformations are performed on the input vector.

On the other hand, $\underline{w_{i+1}}$ is calculated in four steps. In step one, $\underline{L_{i-1}} \oplus \underline{f_i}$, which is $\underline{L_{i+1}}$, and $\overline{\underline{L_{i-1}} \oplus \underline{f_i}}$ are calculated. In the second step $\underline{P1(Sh_j(\underline{C_i}))}, \underline{P2(Sh_j(\underline{D_i}))}, \overline{\underline{P1(Sh_j(\underline{C_i}))}}, \overline{\underline{P2(Sh_j(\underline{D_i}))}}, E\left(\underline{L_{i+1}}\right) \oplus \underline{k_{i+1}}$ and $\overline{E\left(\underline{L_{i+1}}\right) \oplus \underline{k_{i+1}}}$ are calculated. Consider that $P1, P2, Sh_j$, and E need no time to be executed [Appendix]. In the third and fourth steps $P\left(S\left(E\left(\underline{L_{i+1}}\right) \oplus \underline{k_{i+1}}\right)\right)$ and $\overline{P\left(S\left(E\left(\underline{L_{i+1}}\right) \oplus \underline{k_{i+1}}\right)\right)}$ are calculated respectively. Thus M_j is divided into four subfunctions defined as follows:

$$\underline{q_i} = A0\left(\underline{w_i}\right), \qquad \underline{u_i} = A1_j\left(\underline{q_i}\right), \qquad \underline{v_i} = A2\left(\underline{u_i}\right), \qquad \underline{w_{i+1}} = A3\left(\underline{v_i}\right) \tag{13}$$

where $\underline{q_i}, \underline{u_i}$, and $\underline{v_i}$ are defined as

$$\begin{cases} \underline{q_i} = \left(\underline{L_i}, \underline{L_{i+1}}, \overline{\underline{L_{i+1}}}, \underline{C_i}, \overline{\underline{C_i}}, \underline{D_i}, \overline{\underline{D_i}}\right) \\ \underline{u_i} = \left(\underline{L_{i+1}}, \underline{L_i}, E\left(\underline{L_{i+1}}\right) \oplus \underline{k_{i+1}}, \overline{E\left(\underline{L_{i+1}}\right) \oplus \underline{k_{i+1}}}, \underline{C_{i+1}}, \underline{D_{i+1}}\right) \\ \underline{v_i} = \left(\underline{L_{i+1}}, \underline{L_i}, \overline{S\left(E\left(\underline{L_{i+1}}\right) \oplus \underline{k_{i+1}}\right)}, \underline{C_{i+1}}, \underline{D_{i+1}}\right) \end{cases} \tag{14}$$

Thereby the M_j could be represented as

$$M_j = A0 A1_j A2 A3 \tag{15}$$

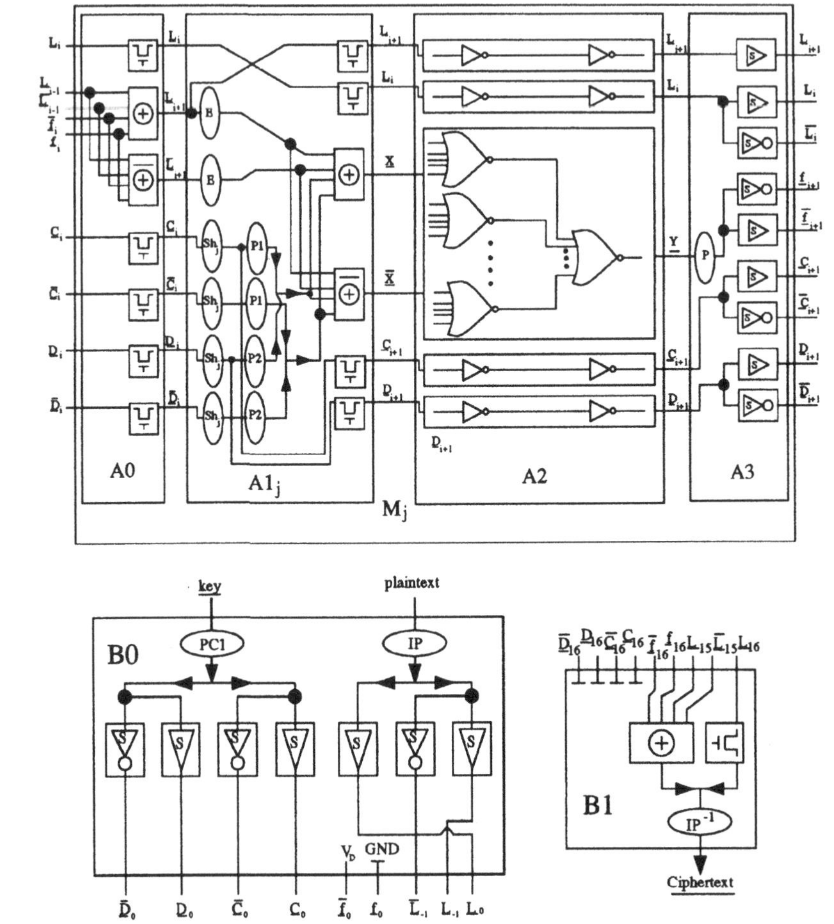

Fig. 2. Structures of M_j, Bo, and $B1$.

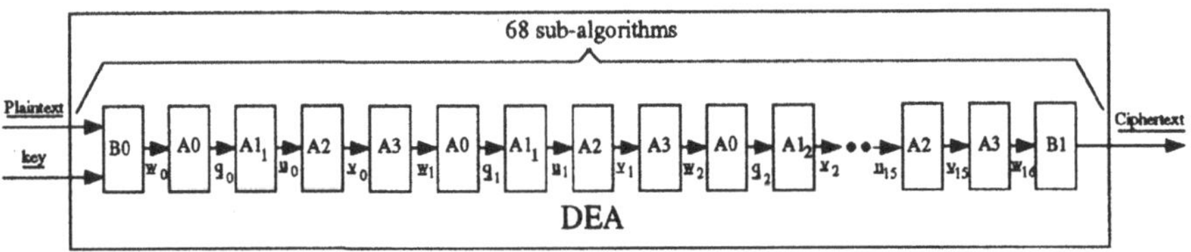

Fig. 3. Structure of DEA.

Figure 2 shows the block diagram of the structures of M_j, Bo, and $B1$. As it is apparent in the figure, there exist some delays to synchronize the output on every parallel path. Since the functions $IP, IP^{-1}, PC1, P, E, Sh_j, P1$, and $P2$ could be implemented just by permuting the wires, no delay has been predicted for them [Appendix].

Eventually, considering (12) and (15), the Divided Encryption Algorithm (DEA) of the DES consists of 68 tasks (figure 3). It is apparent from figure 2 that the maximum task delay is equal to the delay of two serial NORs (NOTs) and due to A2. This delay is equal about 1.5 nano seconds in the 5-micron-NMOS-VLSI implementation [3].

4 Exhaustive key search by pipelining method

Figure 4 shows the flow chart of the pipelining attack to the DES. $\underline{PL_0}, \underline{CI_0}$, and $\underline{CI_1}$ are chosen such that

$$\underline{CI_0} = DES_{\underline{key}}(\underline{PL_0}) \quad and \quad \underline{CI_1} = DES_{\underline{key}}(\overline{\underline{PL_0}}) \tag{16}$$

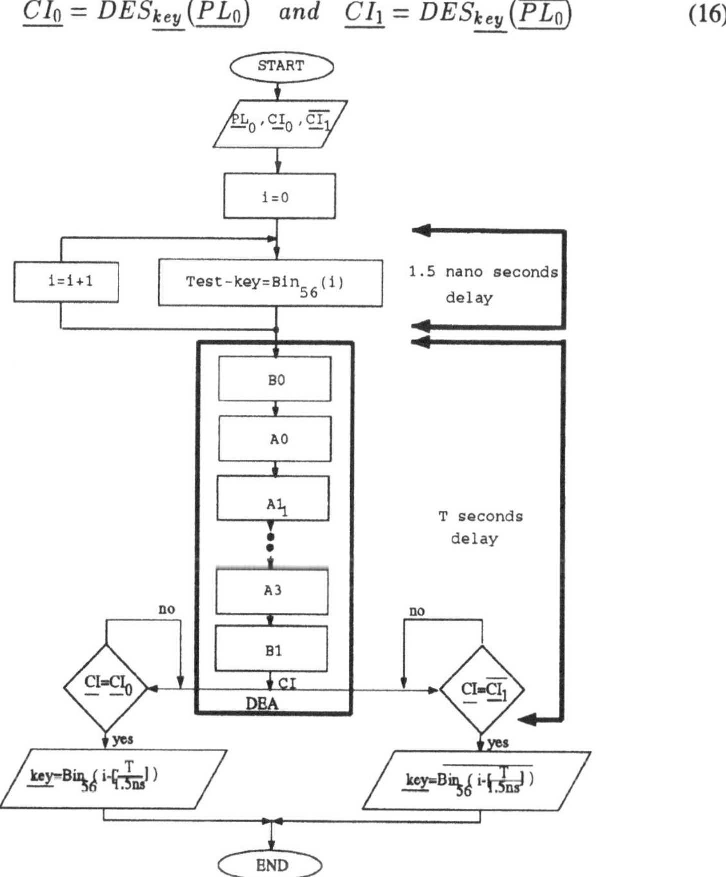

Fig. 4. Flow chart of the pipelining attack to the DES.

Thus if the output of the DEA, shown $\underline{CI}$, is equal to $\underline{CI_0}$, the key is equal to the tested key, and due to the following property of the DES, called complementary property, if $\underline{CI}$ is equal to $\overline{CI_1}$, the key is equal to the complementary of the tested key.

$$DES_{\overline{\underline{key}}}(\underline{Plaintext}) = \overline{DES_{\underline{key}}(\underline{Plaintext})} \tag{17}$$

The tested key is changed independent of the encryption time with a delay about 1.5ns. Thus $\frac{T}{1.5ns}$ is subtracted from counter number at the end. Where T is the encryption time plus comparison time. On the other hand, since the average number of the tested keys is 2^{54} [2], the average searching time will be 312.75 days.

5 The DES Breaker (DESB) chip

Figure 5 shows a block diagram of a chip which uses 8 DEAs for the pipelining attack. The attack is started by installing $\underline{PL_0}, \underline{CI_0}$ and $\overline{CI_1}$ in the three serial input parallel output shift registers. Then the 53 bits of the test key are changed by an external counter. Any time when the output of one of the functions is equal to $\underline{CI_0}$ or $\overline{CI_1}$, the output m_0 will be one. In that time, the output $\underline{D}$ shows the last 3 bits of the tested key, and the output $m1$ shows that whether the key must be complemented or not.

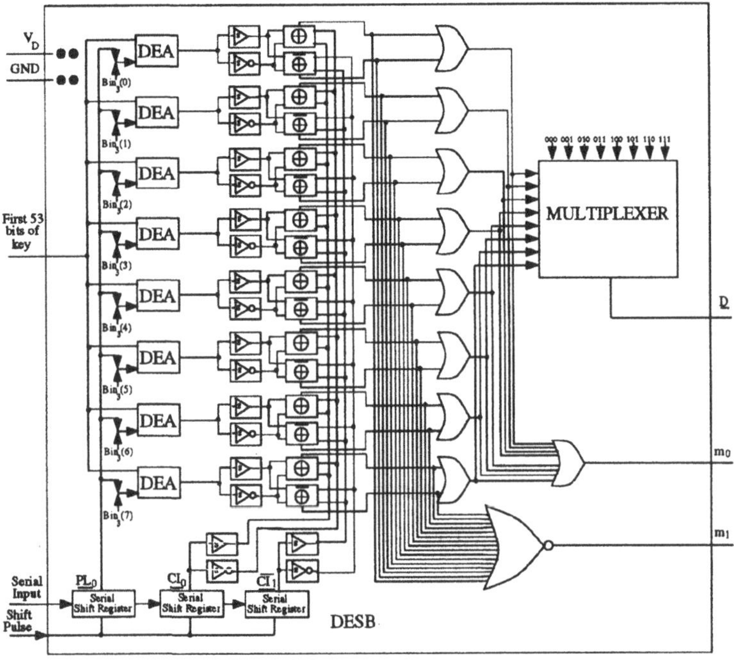

Fig. 5. Structure of DESB.

Including one pin for shift pulse of the shift registers and two pins for supplying the chip, DESB chip will have 62 pins. On the other hand it is shown in [4] that NMOS-VLSI implementation of this chip consists of 770000 transistors. Therefore DESB chip could be implemented by today's technology.

6 Searching machine

If n parallel breaker chips are employed to attack to the DES, the attack time decreases to $\frac{312.76}{8n}$ days. For example with n=32 this time will be just 1.2 days.

On the other hand, since the DES chips have been used since 1977, many individuals have been experimenting with the chip. Assume at least 1000 of these individuals are interested to have a searching machine to break the DES. Thus at least 32000 DESB chips will be produced. Due to this large demand, the chip could be produced at a cost less than $40 a piece [4]. Including an additional cost increase of 100 percent to account for the other required circuits, the pipelining attack to the DES by a searching machine which consists of 32 breaker chips costs almost $2560. This value is 0.000116 of the estimated price by Hellman, and shows that the DES is not secure any more.

7 Conclusion

It has been shown that the exhaustive attack can succeed if DESB is used instead of the DES. DESB has been designed to use pipelining method against the DES. It has been shown that DESB chip could be implemented by today's technology at a cost less than $40 a piece. It has also been shown that a searching machine, consists of 32 DESB chips, breaks the DES in 1.2 days by a cost of $2500. Therefore the DES with 56–bit key is not secure anymore if it is used in the electronic codebook mode.

Appendix

A. Notation

To better understand the DES structure, it is necessary to define the following functions:

1–The permutation functions IP, IP^{-1}, and P:

— IP yields a 64-bit output block by permuting the bits of the 64-bit input block.

— IP^{-1} denotes the inverse permutation of IP.

— P permutes a 32-bit input block to yield a 32-bit output block.

The permutation tables could be found in [5], [6], and [7].

2– The permuted choice function $PC1, PC2, P1$, and $P2$:

— $PC1$ removes 8 parity bits of the input block and permutes the remaining 56 bits.

— $PC2$ chooses 48 bits of the 56-bit input block and permutes them.

— $P1$, and $P2$, defined as follows, divide $PC2$ into two parallel parts.

$$P1(x_0, x_1, .., x_{27}) = (x_{13}, x_{16}, x_{10}, x_{23}, x_0, x_4, x_2, x_{27}, x_{14}, x_5, x_{20}, x_9, x_{22}, x_{18}, x_{11}, x_3, x_{25}, x_7, x_{15}, x_6, x_{26}, x_{19}, x_{12}, x_1) \quad \text{(A-1)}$$

$$P2(x_0, x_1, .., x_{27}) = (x_{12}, x_{23}, x_2, x_8, x_{18}, x_{26}, x_1, x_{11}, x_{22}, x_{16}, x_4, x_{19}, x_{15}, x_{20}, x_{10}, x_{27}, x_5, x_{24}, x_{17}, x_{13}, x_{21}, x_7, x_0, x_3) \quad \text{(A-2)}$$

where x_i is specified the ith bit of the input block $\underline{X}$.

More details could be found in [5], [6], and [7].

3– The expander function E:

— E takes a 32–bit block as input and yields a 48–bit block as output by repeating some bits.

The method of the repetition could be found in [5], [6], and [7].

4– The function S:

— S yields a 32–bit output from a 48–bit input by using 8 distinct functions called S-boxes. Each S-box takes a block of 6 bits as input and yields a block of 4 bits as output.

The details about S-boxes could be found in [4], [5], [6], [7], [8], [9].

5– The function f:

— f operates on two blocks, one of 32 bits and one of 48 bits, and produces a block of 32 bits. The f function is defined by using the S, E, and P functions as follows

$$f(\underline{X}, \underline{Y}) = P(S(E(\underline{X}) \oplus \underline{Y})) \quad \text{(A-3)}$$

where $\oplus$ denotes the XOR operation which applies modulo-2 addition bit-by-bit over the whole vectors $E(\underline{X})$ and $\underline{Y}$.

6– The shift function Sh_j:

— Sh_j shifts the bits of the 28-bit input block j bits to the left circularly. For example

$$\begin{aligned} Sh_1(x_0, x_1, .., x_{27}) &= (x_1, x_2, x_3, .., x_{27}, x_0) \\ Sh_2(x_0, x_1, .., x_{27}) &= (x_2, x_3, .., x_{27}, x_0, x_1) \end{aligned} \quad \text{(A-4)}$$

7– The function Bin_m:

— Bin_m yields m-bit binary representation of input integer. For example $Bin_3(2) = 010$.

B. VLSI Implementation

The mentioned functions in part A could be implemented by NMOS-VLSI technology as follows:

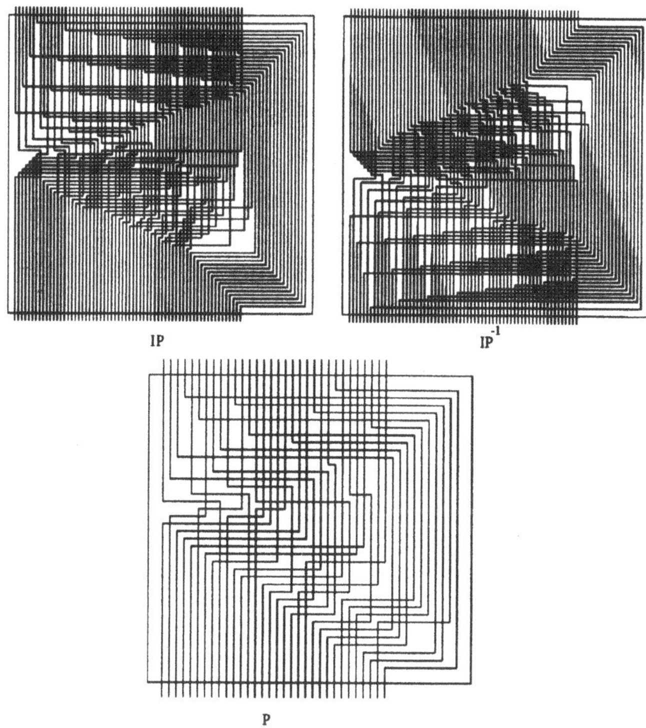

Fig. B-1. Implementations of P, IP, and IP^{-1}.

1– *The permutation functions* IP, IP^{-1}, *and* P: These functions could be implemented just by permuting some wires (figure B-1). Since no gate is needed to implement them, they need no time to be executed.

2– *The permuted choice function* $PC1, P1,$ *and* $P2$: These functions could be also implemented by exchanging the wires (figure B-2), and so, they don't need any time to be executed either.

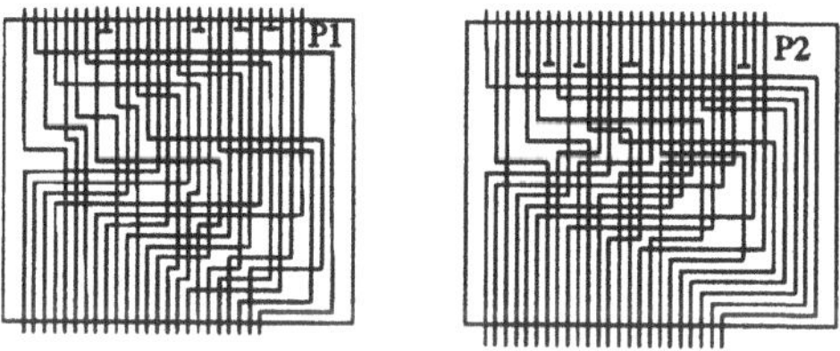

Fig. B-2. Implementations of $P1$, $P2$ and $PC1$. (Continued . . .)

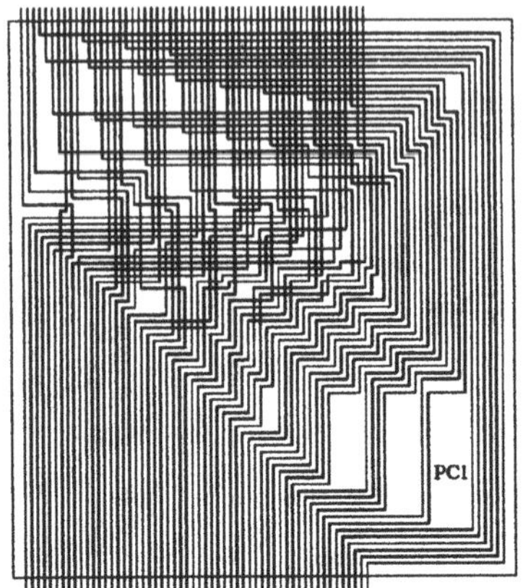

Fig. B-2. Implementations of $P1$, $P2$ and $PC1$.

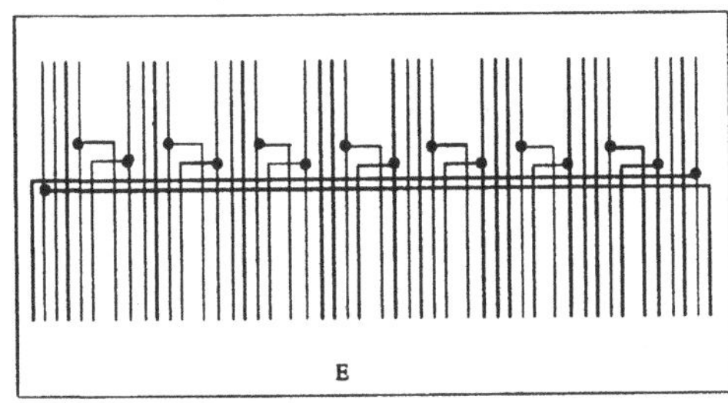

Fig. B-3. Implementation of E.

3– *The expander function E:* Figure B-3 shows the implementation of the E function. As it is apparent, no delay and no gate are necessary to implement this function either.

4– *The function S:* A Boolean function could be easily found for any output bit of this function [9]. Thus the S function is implemented by using 32 Boolean functions. If the NOR representation of the Boolean functions are used, and if it is assumed that the input block and its complement are accessible at the same time, the S function and its complement could be implemented by 3808 transistors [4].

5– *The shift function* Sh_j*:* It is easy to show that this function is a kind of permutation function [4], and so, it could be implemented by permuting some wires. Figure B-4 shows the implementations of Sh_1 and Sh_2.

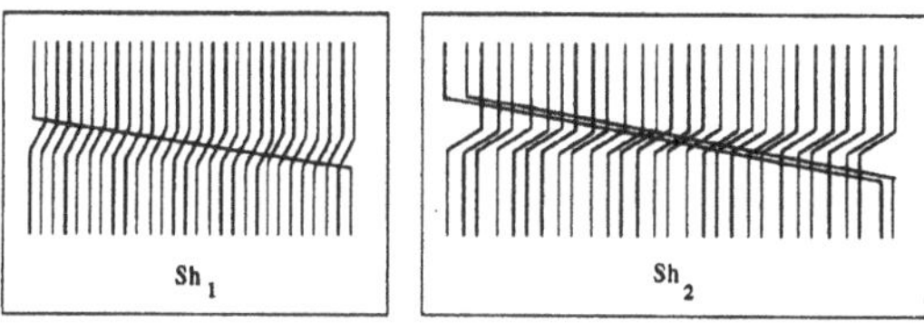

Fig. B-4. Implementations of Sh_1 and Sh_2.

6– XOR operation: The Boolean representation of $\underline{X} \oplus \underline{Y}$ is implemented as figure B-5. As it is apparent in the figure, if the inputs and their complements are accessible at the same time, $\overline{\underline{X} \oplus \underline{Y}}$ could be also implemented at the same time.

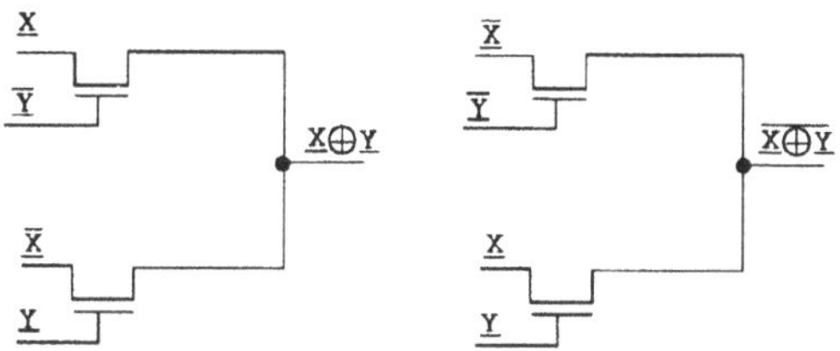

Fig. B-5. Implementations of $\underline{X} \oplus \underline{Y}$ and $\overline{\underline{X} \oplus \underline{Y}}$.

Acknowledgment

The authors strongly acknowledge professor A. U. H. Sheikh[2] for his useful hints and his efforts for the revision of the paper, and professor J. Y. Chouinard[3] for his interest and valuable advice to the authors. In addition, we are grateful to Jahad-Daneshgahi center of Isfahan Univ. of Tech. for its generous financial support.

Reference

[1]: National Bureau of standards," Encryption algorithm for computer data protection: Requests for comments," Federal Register, Vol. 40. No. 12134,1975.

[2]: Diffie, W. and Hellman, M. E. ," Exhaustive cryptanalysis of the NBS Data Encryption Standard," computer magazine, 10, 6, 74–84, 1977.

[3]: Punckuell, D. A. and Eshraghian, K. , *Basic VLSI Design (principles and applications)* , Prentice-Hall of Australia, 1985.

[4]: Hendessi, F. , *Cryptanalysis of the Data Encryption Standard*, Master Thesis, Department of Electrical Engineering, Isfahan Univ. of Tech., Isfahan, Iran, 1989.

---

2 Professor Sheikh is with the Department of Systems and Computer Engineering at Carleton University

3 Professor Chouinard is with Electrical Engineering Department at University of Ottawa.

[5]: National Bureau of standards," Specification for the Data Encryption Standard," Federal Register, Vol. 46, January 1977.

[6]: Meyer, C. H. and Matyas, S. M. , *Cryptography: A new Dimension Computer Data Security,* John Wiley, New York, 1982.

[7]: Konheim, A. G. , *Cryptography: A Primer,* John Wiley, New York, 1981.

[8]: Brickell, E. F. , Moore, J. H. , and Purtil, M. R. ," Structure in the S-boxes of the DES (Extended abstract) ," Advanced in cryptology, Proc. of Crypto.'86, LNCS, Vol. 263, springer-verlag, 1987, pp. 3–8.

[9]: Schaumller-Binchl, " Cryptanalysis of the Data Encryption Standard, Method of Formal Coding," Proc. of Crypto.'82, LNCS, Vol. 149, springer-verlag, 1983, pp. 235–259.

Low-complexity and High-Performance Multilevel Coded Modulation for the AWGN and Rayleigh Fading Channels*

Shu Lin, Sandeep Rajpal and Do Jun Rhee

Department of Electrical Engineering
University of Hawaii at Manoa
Honolulu, Hawaii 96822, U.S.A.

Abstract. The multilevel coding method is a powerful technique for constructing bandwidth efficient modulation codes systematically with arbitrarily large distance parameters from Hamming distance component (**block** or **convolutional**) codes in conjunction with proper bits-to-signal mapping through signal-set partitioning. The multilevel modulation codes constructed by this method allow the use of multi-stage decoding procedures that provide good trade-off between performance and decoding complexity.

In this paper, constructions of multilevel modulation codes for both the AWGN and the Rayleigh fading channels are discussed. Distance parameters, such as the minimum squared Euclidean distance, minimum symbol distance and minimum product distance, which determine the error performance of a multilevel modulation code are expressed in terms of the minimum Hamming distances of the component codes. Error performances of some good multilevel modulation codes for either the AWGN or the Rayleigh fading channels are given. These codes achieve high performance with low decoding complexity.

1 Introduction

One of the dramatic developments in bandwidth-efficient communications over the past twelve years is the introduction and rapid applications of combined coding and bandwidth efficient modulation, known as **coded modulation**, for reliable data transmission. Using coded modulation, reliable data transmission can be obtained without compromising bandwidth efficiency. The first coded modulation scheme was introduced by Ungerboeck in 1976[1], and later published in 1982[2]. In this scheme, trellis(or convolutional) codes were combined with various types of modulation signal sets to form modulation codes by proper bits-to-signal mapping through signal set partitioning. This scheme is now known as the **trellis-coded modulation**(TCM). Since the publication of Ungerboeck's

* This research was supported by NSF Grants NCR-911540, BCS-9020435 and NASA Grant NAG 5-931

prize winning paper, there has been a great deal of research on construction of TCM codes and many good TCM codes have been constructed[3-19].

Combining block coding and modulation, now known as **block-coded modulation** (BCM), originated from Imai and Hirakawa's paper published in 1977[20], in which they proposed a multilevel method of combining binary block component codes with a channel signal set to form a block modulation code. This multilevel method was later extended and used by others for constructing block modulation codes[21-35].

The multilevel method is a very powerful technique for constructing bandwidth efficient modulation codes systematically with arbitrarily large **distance parameters** from Hamming distance component **(block** or **convolutional)** codes in conjunction with proper bits-to-signal mapping through signal set partitioning. Particularly, it provides the flexibility to coordinate the distance parameters of a code such that the best performance for a given channel can be attained. Furthermore, the multilevel modulation codes constructed by this method allow the use of multi-stage decoding procedures that provide good **trade-off** between error performance and decoding complexity.

In this paper, constructions of multilevel bandwidth efficient modulation codes for both the AWGN and Rayleigh fading channels are presented. Distance parameters, such as the minimum squared Euclidean distance, minimum symbol distance, and minimum product distance, which determine the error performance of a multilevel modulation code are expressed in terms of the minimum Hamming distance of the component codes. Guidelines for constructing good multilevel modulation codes for either the AWGN channel or the Rayleigh fading channel are presented. Error performances of some good multilevel modulation codes for either the AWGN or the Rayleigh fading channel are given. These codes achieve high performance with low decoding complexity.

The organization of this paper is as follows. In Section 2, a brief review of the multilevel method for constructing multilevel modulation codes is given. Distance parameters of a multilevel modulation code are defined and expressed in terms of the minimum Hamming distances of the component codes. Guidelines for constructing good modulation codes for either the AWGN or the Rayleigh fading channel are presented. In Section 3, some good multilevel modulation codes and their error performances using multi-stage decoding are given. These codes achieve significant coding gains over the uncoded reference systems with low decoding complexity. In Section 4, a multilevel product coded modulation scheme is presented. Modulation codes constructed by this scheme achieve high-performance with reduced decoding complexity.

2 Multilevel Coded Modulation

In a coded modulation system, information sequences are encoded into signal sequences over a certain modulation signal set, e.g., an MPSK signal set or a QAM signal set. These signal sequences form a modulation code. In the following, we first define some important distance parameters of a modulation code and then give a brief review of the basic multilevel method for constructing modulation codes.

Let s be a point $(X(s), Y(s))$ in a two dimensional Euclidean space R^2. let s and s' be two points in R^2. The **squared Euclidean distance** between s and s', denoted $d^2(s, s')$, is defined as follows:

$$d^2(s, s') \triangleq (X(s) - X(s'))^2 + (Y(s) - Y(s'))^2. \tag{2.1}$$

Let $\mathbf{v} = (s_1, s_2, \cdots, s_n)$ and $\mathbf{v}' = (s_1', s_2', \cdots s_n')$ be two n-tuples over R^2. The squared Euclidean distance between $\mathbf{v}$ and $\mathbf{v}'$, denoted $d^2(\mathbf{v}, \mathbf{v}')$, is defined as follows:

$$\begin{aligned} d^2(\mathbf{v}, \mathbf{v}') &\triangleq \sum_{j=1}^{n} (X(s_j) - X(s_j'))^2 + (Y(s_j) - Y(s_j'))^2 \\ &= \sum_{j=1}^{n} d^2(s_j, s_j') \end{aligned} \tag{2.2}$$

Let C be a modulation code of length n with signals from a certain modulation signal space S. The error performance of C depends on several distance parameters. Let $d^2(\mathbf{x}, \hat{\mathbf{x}})$ denote the **squared Euclidean distance** between two code sequences, $\mathbf{x}$ and $\hat{\mathbf{x}}$, in C. The **minimum squared Euclidean distance** of C, denoted $d_E^2[C]$, is defined as follows:

$$d_E^2[C] \triangleq \min\{d^2(\mathbf{x}, \hat{\mathbf{x}}) : \mathbf{x}, \hat{\mathbf{x}} \in C \text{ and } \mathbf{x} \neq \hat{\mathbf{x}}\} \tag{2.3}$$

The **symbol (Hamming) distance** between two code sequences $\mathbf{x}$ and $\hat{\mathbf{x}}$, denoted $\delta_H(\mathbf{x}, \hat{\mathbf{x}})$, is the number of different symbols between the two sequences. The **minimum symbol distance** of C, denoted $\delta_H[C]$, is defined as the minimum symbol distance between any two code sequences in the code. The **product distance** between $\mathbf{x}$ and $\hat{\mathbf{x}}$ denoted by $\Delta_p^2(\mathbf{x}, \hat{\mathbf{x}})$ is defined as follows:

$$\Delta_p^2(\mathbf{x}, \hat{\mathbf{x}}) = \prod_{k=1, x_k \neq \hat{x}_k}^{n} d^2(x_k, \hat{x}_k) \tag{2.4}$$

where $d^2(x_k, \hat{x}_k)$ is the squared Euclidean distance between k-th signals, x_k and $\hat{x}_k$, of $\mathbf{x}$ and $\hat{\mathbf{x}}$. The **minimum product distance** of C, denoted $\Delta_p^2[C]$, is the minimum product distance between any two code sequences with symbol distance $\delta_H[C]$ in the code.

For the AWGN channel, the error performance of a code depends primarily on its minimum squared Euclidean distance and path multiplicity[2]. For the fading channel, the error performance of a code depends primarily on its minimum symbol distance, minimum product distance, and path multiplicity [36-38]. It depends on the minimum squared Euclidean distance in a lesser degree.

The basic multilevel method for constructing modulation codes consists of five steps: (1) selection of a modulation signal set S; (2) labeling of signal points by strings of labeling symbols through signal set partitioning; (3) selection of component codes; (4) combining component codes into a code over a signal label set; and (5) label-to-signal mapping to form a multilevel modulation code.

The construction method and the derivation of distance properties of a multilevel modulation code are best explained by an example. Consider the two-dimensional 8-PSK signal constellation with unit energy as shown in Figure 1. Label each of the 8 signal points by a string of 3 bits, abc, where a is the first labeling bit and c is the last labeling bit. The signal labeling is achieved by set partitioning process as shown in Figure 2. The signal set S is partitioned into a chain of partitions. The first partition consists of two disjoint subsets which are labeled by "0" and "1". The second partition consists of 4 disjoint subsets which are labeled by "00", "01", "10" and "11" respectively. The third partition consists of 8 disjoints subsets, each consists of only one 8-PSK signal point which is uniquely labeled by a string of 3 bits as shown in Figure 2.

The partition is carried out in such a way that, as the partition level increases, the **intra-set distance**[2] (the minimum squared Euclidean distance among signal points) of a set in the partition increases. The intra set distances at the 3 partition levels in Figure 2 are: $d_1 = 0.586$, $d_2 = 2.0$, and $d_3 = 4.0$ respectively. From Figure 2, we see that each subset in the first partition is a QPSK signal set and each subset in the second partition is a BPSK signal set. The above partitioning process is called the **binary partition**.

The labeling strings formed from the above partitioning process have the following important properties: (1) two signal points with labels different at the first bit position are at a squared Euclidean distance at least $d_1 = 0.586$ apart; (2) two signal points with labels identical at the first bit position but different at the second bit position are separated by a squared Euclidean distance at least $d_2 = 2.0$ apart; and (3) two signal points with labels identical at the first two bit positions but different at the last bit position are at a squared Euclidean distance $d_3 = 4.0$ apart. The monotonically increasing property of the intra-set distances d_1, d_2 and d_3 is a key to the construction of bandwidth efficient modulation codes.

Suppose block component codes are used for the code construction. For $1 \le i \le 3$, let A_i be a binary (n, k_i, δ_i) linear block code of length n, dimension k_i, and minimum Hamming distance δ_i. Let

$$\begin{aligned}\mathbf{v}^{(1)} &= (v_1^{(1)}, v_2^{(1)}, \cdots, v_j^{(1)}, \cdots, v_n^{(1)}) \\ \mathbf{v}^{(2)} &= (v_1^{(2)}, v_2^{(2)}, \cdots, v_j^{(2)}, \cdots, v_n^{(2)}) \\ \mathbf{v}^{(3)} &= (v_1^{(3)}, v_2^{(3)}, \cdots, v_j^{(3)}, \cdots, v_n^{(3)})\end{aligned} \tag{2.5}$$

be three codewords in A_1, A_2 and A_3 respectively. We form the following sequence:

$$\mathbf{v}^{(1)} * \mathbf{v}^{(2)} * \mathbf{v}^{(3)} \triangleq (v_1^{(1)}v_1^{(2)}v_1^{(3)}\ ,\ v_2^{(1)}v_2^{(2)}v_2^{(3)}\ ,\ \cdots\ ,v_j^{(1)}v_j^{(2)}v_j^{(3)}\ ,\\ \cdots\ ,v_n^{(1)}v_n^{(2)}v_n^{(3)}) \tag{2.6}$$

For $1 \leq j \leq n$, we take $v_j^{(1)}v_j^{(2)}v_j^{(3)}$ as the label for a signal point in the 8-PSK signal constellation as shown in Figure 1. Let $\lambda(\cdot)$ be the mapping which maps the label $v_j^{(1)}v_j^{(2)}v_j^{(3)}$ into its corresponding signal point s_j, i.e., $\lambda(v_j^{(1)}v_j^{(2)}v_j^{(3)}) = s_j$. Then

$$\lambda(\mathbf{v}^{(1)} * \mathbf{v}^{(2)} * \mathbf{v}^{(3)}) \triangleq (\lambda(v_1^{(1)}v_1^{(2)}v_1^{(3)})\ ,\ \lambda(v_2^{(1)}v_2^{(2)}v_2^{(3)})\ ,\ \cdots\ ,\lambda(v_j^{(1)}v_j^{(2)}v_j^{(3)})\ ,\\ \cdots\ ,\lambda(v_n^{(1)}v_n^{(2)}v_n^{(3)})) \tag{2.7}$$

is a sequence of n 8-PSK signals. Let

$$\begin{aligned} C &\triangleq \lambda[A_1 * A_2 * A_3] \\ &= \{\lambda(\mathbf{v}^{(1)} * \mathbf{v}^{(2)} * \mathbf{v}^{(3)}) : \mathbf{v}^{(1)} \in A_1,\ \mathbf{v}^{(2)} \in A_2, \text{and } \mathbf{v}^{(3)} \in A_3\} \end{aligned} \tag{2.8}$$

Then C is a 3-level block 8-PSK modulation code of length n and dimension $k = k_1 + k_2 + k_3$. Since $k_1 + k_2 + k_3$ information bits are encoded into a code sequence of n 8-PSK signals, the spectral efficiency is

$$\eta[C] = (k_1 + k_2 + k_3)/n \ \text{ bits/symbol}$$

In the above construction, each component code contributes one level of labeling.

The distance parameters of the above 8-PSK modulation code can be expressed in terms of the minimum Hamming distances of its component codes and are given in the following distance theorem.

Distance Theorem: Let $d_E^2[C]$, $\delta_H[C]$ and $\Delta_p^2[C]$ denote the minimum squared Euclidean distance, minimum symbol distance and minimum product distance of the 3-level 8-PSK code, $C = \lambda[A_1 * A_2 * A_3]$, respectively. Then $d_E^2[C], \delta_H[C]$ and $\Delta_p^2[C]$ are given as follows:

(1) $d_E^2[C] = \min\{\delta_i d_i : 1 \leq i \leq 3\}$ (2.9)
(2) $\delta_H[C] = \min\{\delta_i : 1 \leq i \leq 3\}$, (2.10)
(3) Let k be the smallest integer in the index set $I = \{1, 2, 3\}$ for which $\delta_k = \delta_H[C]$. Then

$$\Delta_p^2[C] = (d_k)^{\delta_k} \tag{2.11}$$

where d_1, d_2, and d_3 are the intra-set distances of the partition chain 8-PSK/QPSK/ BPSK.

The proof of (2.9) can be found in [20,22,23] and the proofs of (2.10) and (2.11) can be found in [39].

From the expressions for the minimum squared Euclidean distance and product distances of a multilevel modulation code given by (2.9) and (2.11), it is clear why the intra-set distances should be kept as large as possible during the signal set partitioning and labeling process. The distance theorem provides general guidelines for constructing good multilevel modulation codes for both the AWGN and fading channels. For the AWGN channel, the error performance of a modulation code depends mainly on its minimum squared Euclidean distance. In this case, expression (2.9) should be used as a guideline for code construction. For a given minimum squared Euclidean distance, the component codes should be chosen to maximize the spectral efficiency and minimize the decoding complexity and path multiplicity. For the fading channel, the error performance of a modulation code depends strongly on its minimum symbol and product distances. Both these distances should be as large as possible, and they play different roles in determining the error performance of a code. At low SNR, the minimum product distance is more important; whereas at high SNR, the minimum symbol distance becomes more important[36-38]. In designing modulation codes for the fading channel, expressions of (2.10) and (2.11) should be used as the design guidelines.

The above multilevel code construction method and distance theorem can be generalized to any signal constellation, MPSK or QAM, in a straightforward manner. The component codes may be either block or convolutional codes, binary or non-binary.

3 Some 3-level 8-PSK Modulation Codes and Their Error Performances

In this section, some good 3-level 8-PSK modulation codes for the AWGN and Rayleigh fading channels are constructed using the distance theorem as the general guideline. The error performance of these codes based on multi-stage soft-decision decoding [29,40] are given. In the code construction, the component codes are chosen to have simple trellis structure so that the Viterbi decoding algorithm can be used at each decoding stage.

In the following, the notation (n, k, d) denotes a linear block code of length n, dimension k and minimum Hamming distance d.

Code-I: A 3-level 8-PSK Block Modulation Code for the AWGN Channel

Suppose we want to design a 3-level 8-PSK modulation code for the AWGN channel, with minimum squared Euclidean distance 8 and spectral efficiency around 2 bits/symbol. From expression (2.9), we find that the three component codes must have minimum Hamming distances at least, 14, 4, and 2 respectively. In choosing the component codes to achieve the required minimum squared Euclidean distance and spectral efficiency, the overall decoding complexity, phase symmetry, and other factors must also be taken into consideration. One possible choice of the component codes is: (1) A_1 = (32, 6, 16), the first order Reed-Muller(RM) code of length 32; (2) A_2 = (32, 26, 4), the 3rd-order RM code of length 32; and (3) A_3 = (32, 31, 2), the even parity check code of length 32. With this choice, the resultant 3-level 8-PSK code of length 32, $C(1) = \lambda[A_1 * A_2 * A_3]$, has the following parameters: $\eta[C] = 63/32 = 1.969$ bits/symbol, $d_E^2[C] = 8, \delta_H[C] = 2$, and $\Delta_p^2[C] = 16$. This code has a spectral efficiency almost the same as the uncoded QPSK, 2 bits/symbol.

The reason to choose RM codes as component codes in the code construction is that they have **relatively simple trellis diagrams**[18,41] and hence can be decoded with the soft-decision Viterbi decoding algorithm.

The (32, 6, 16) RM code has a simple 16-state and 4-section trellis diagram[18]. This trellis diagram is **loosely connected** and consists of 8 **parallel** and **structurally identical** 2-state sub-trellis diagrams without cross connections between them[41]. As a result, 8 identical and very simple 2-state Viterbi decoders can be built to decode this code in parallel at the first stage of decoding of the 8-PSK code $C(1)$. This parallel structure of the trellis diagram not only simplifies the decoding complexity but also speeds up the decoding process.

The (32,26,4) RM code has a 16-state and 4-section trellis diagram which consists of 2 parallel and structurally identical 8-state sub-trellis diagrams without cross connections between them. The parallel structure in the trellis allows us to devise 2 identical and simple 8-state Viterbi decoders to decode the (32, 26, 4) RM code in parallel at the second stage of decoding of the 8-PSK code C(1).

The even parity check $(32, 31, 2)$ code has a 2-state trellis diagram. Therefore, the decoding of this component code is also very simple. We see that the overall decoding complexity for the 8-PSK code $C(1)$ is quite simple.

The bit error performance of $C(1)$ over the AWGN channel is shown in Figure 3. We see that the code achieves a 3.5 dB real coding gain over the uncoded QPSK at the bit-error-rate(BER) 10^{-6}. Since the code has a minimum symbol distance 2 and a large minimum product distance 16, it also performs very well over the Rayleigh fading channel as shown in Figure 4. The code achieves an 10.61 dB real coding gain over the uncoded QPSK at the BER 10^{-3}. The simulation is carried out assuming independent fading, no channel state information and squared Euclidean distance as the decoding metric.

Overall, $C(1)$ provides good performance for both the AWGN and Rayleigh fading channels with relatively simple decoding complexity. Furthermore, it is invariant under 45° degree phase rotation [34] which is important for synchronization purpose.

Code-II: A 3-level 8-PSK Block Modulation Code for the Rayleigh Fading Channel

Now, consider the design of a 3-level 8-PSK block modulation code for the Rayleigh fading channel. Suppose we want to construct a code with minimum symbol distance 4 and minimum product distance greater than 1. From expression (2.10) in the distance theorem, we find that the smallest minimum Hamming distance of the component codes must be 4. From expression (2.11) in the distance theorem, we find that for the minimum product distance to be greater than 1, the first component code must not be the code with the smallest minimum Hamming distance. In this case, either the second or the third component code should be chosen to have the smallest minimum Hamming distance. A possible choice of the component codes is: (1) $A_1 = (32, 16, 8)$, the 2nd-order RM code of length 32; (2) $A_2 = (32, 26, 4)$, the 3rd-order RM code of length 32; and (3) $A_3 = A_2 = (32, 26, 4)$. The resultant 3-level 8-PSK code, denoted $C(2) = \lambda[A_1 * A_2 * A_3]$, has the following parameters: $\eta[C] = 2.125$ bits/symbol, $d_E^2[C] = 4.688, \delta_H[C] = 4$, and $\Delta_p^2[C] = 16$. The code has a minimum product distance 16 and spectral efficiency more than 2 bits per symbol.

The first component code has a 64-state and 4-section trellis diagram which consists of 8 **parallel** and **structurally identical** 8-state sub-trellis diagrams without cross connections. As a result, 8 identical 8-state Viterbi decoders can be devised to decode this code at the first stage of decoding of $C(2)$. The second and third component codes are both the 3rd-order RM code of length 32 whose decoding complexity has already been discussed in the case of Code-I. Again, we see that the overall decoding complexity for $C(2)$ is quite simple.

The error performance of this code over the Rayleigh fading channel is shown in Figure 4. The simulation is carried out assuming independent fading, no channel state information and squared Euclidean distance as the decoding metric. The code achieves a 12.33 dB real coding gain over the uncoded QPSK at the BER 10^{-3} with bandwidth reduction (or higher bandwidth efficiency). This code outperforms $C(1)$ over the Rayleigh fading channel because it has larger minimum symbol distance. This code is invariant under 90° phase rotation.

Code-III: A 3-level 8-PSK TCM Code for the Rayleigh Fading Channel

In the following, we consider the construction of a two-level 8-PSK modulation code using two component codes, a convolutional code and a block code. The convolutional component code, denoted A_1, is a rate-1/2 code of constraint length 5 with generator matrix $G(D) = [1 + D + D^3, 1 + D^2 + D^4]$. This code has **minimum free branch distance** 5 [42] and a 16-state trellis diagram. The block component code, denoted A_2, is the (32, 26, 4) RM code. The trellis structure of this code was discussed in the construction of C(1).

At each time unit, the two code bits at the output of the convolutional code encoder form the first two label bits for an 8-PSK signal point, the block component code contributes the third label bit as shown in Figure 5. The resultant two-level 8-PSK code, denoted $C(3) = \lambda[A_1 * A_2]$, is a TCM code with the following parameters: $\eta[C] = 1.8125$ bits/symbol, minimum symbol distance $\delta_H[C] = 4$, and minimum product distance $\Delta_p^2[C] = 256$.

The error performance of $C(3)$ over the Rayleigh fading channel with two-stage decoding is shown in Figure 4. It achieves an impressive 14.33 dB real coding gain over the uncoded QPSK at the BER 10^{-3}. The coding gain is achieved at the expense of a 9.375 % bandwidth expansion. The decoding complexity of $C(3)$ is reasonably simple.

If the best free Hamming distance convolutional code of constraint length 6 (with a 32-state trellis diagram) [43] is used as the convolutional component code A_1, the resultant two-level 8-PSK TCM code, denoted $C(4)$, does not perform as well as $C(3)$ as shown in Figure 4. $C(3)$ slightly outperforms $C(4)$ with less decoding complexity.

Other multilevel TCM codes can be constructed in the same manner.

4 Product Modulation Codes

In this section, multilevel coded modulation and product coding technique are combined to form product coded modulation systems to achieve high-performance with reduced decoding complexity.

4.1 Two-dimensional Product Codes

Let C_1 be an (N_1, K_1) binary block code and C_2 be an (N_2, K_2) binary block code. The product of C_1 and C_2, denoted $C_1 \times C_2$, is formed in three steps. A message of K_1K_2 bits is first arranged in a $K_2 \times K_1$ array of K_2 rows and K_1 columns. Each row of this array is encoded into an N_1-bit codeword in C_1. This row encoding results in a $K_2 \times N_1$ array of K_2 rows and N_1 columns. Then each column of this second array is encoded into an N_2-bit codeword in C_2. As a result of the two-step encoding, we obtain an $N_2 \times N_1$ code array. This code array is then transmitted column by column (or row by row). The collection of all the distinct code arrays form a two-dimensional product code[43]. If the minimum Hamming distances of C_1 and C_2 are δ_1 and δ_2 respectively, then the minimum Hamming distance of their product $C_1 \times C_2$ is $\delta_1 \times \delta_2$. C_1 and C_2 are called the horizontal and vertical component codes of the product code respectively.

4.2 Construction of Product Modulation Codes

For $1 \leq i \leq 3$, let $C_{i,1} = (N, k_{i,1}, \delta_{i,1})$ be a binary block code with minimum Hamming distance $\delta_{i,1}$ and $C_{i,2} = (n, k_{i,2}, \delta_{i,2})$ be a binary block code with minimum Hamming distance $\delta_{i,2}$. Now we form 3 product codes, $P_1 = C_{1,1} \times$

$C_{1,2}$, $P_2 = C_{2,1} \times C_{2,2}$ and $P_3 = C_{3,1} \times C_{3,2}$. Let A, B, and C be three code arrays from P_1, P_2 and P_3 respectively. Let

$$\begin{aligned} \mathbf{a}_j &= (a_{j,1}\ ,\ a_{j,2}\ ,\ \cdots\ , a_{j,n}\ ) \\ \mathbf{b}_j &= (b_{j,1}\ ,\ b_{j,2}\ ,\ \cdots\ , b_{j,n}\ ) \\ \mathbf{c}_j &= (c_{j,1}\ ,\ c_{j,2}\ ,\ \cdots\ , c_{j,n}\ ) \end{aligned} \tag{4.1}$$

be the j-th columns of the code arrays, A, B, and C respectively. We form the following sequence :

$$\mathbf{a}_j * \mathbf{b}_j * \mathbf{c}_j = (a_{j,1}b_{j,1}c_{j,1}\ ,\ a_{j,2}b_{j,2}c_{j,2}\ ,\ \cdots\ , a_{j,n}b_{j,n}c_{j,n}) \tag{4.2}$$

For $1 \leq \ell \leq n$, we take $a_{j,\ell}b_{j,\ell}c_{j,\ell}$ as the label for a signal point in the 8-PSK signal constellation as shown in Figure 1. Let $\lambda(\cdot)$ be the mapping which maps the label $a_{j,\ell}b_{j,\ell}c_{j,\ell}$ into its corresponding signal point $s_{j,\ell}$, i.e., $\lambda(a_{j,\ell}b_{j,\ell}c_{j,\ell}) = s_{j,\ell}$. Then

$$\begin{aligned} \lambda(\mathbf{a}_j * \mathbf{b}_j * \mathbf{c}_j) &= (\lambda(a_{j,1}b_{j,1}c_{j,1})\ ,\ \lambda(a_{j,2}b_{j,2}c_{j,2})\ , \cdots,\ \lambda(a_{j,n}b_{j,n}c_{j,n})) \\ &= (s_{j,1}, s_{j,2},\ \cdots\ , s_{j,n}) \end{aligned} \tag{4.3}$$

is a sequence of n 8-PSK signals. For $1 \leq j \leq N$, combining the corresponding columns $\mathbf{a_j}, \mathbf{b_j}$, and $\mathbf{c_j}$ of code arrays A, B and C by the above bits-to-signal mapping, we obtain an $n \times N$ array of 8-PSK signals, denoted $\lambda(A * B * C)$. This array is then transmitted column by column.

Note that, for $1 \leq j \leq N$, the columns $\mathbf{a_j}, \mathbf{b_j}$, and $\mathbf{c_j}$ are codewords from the vertical codes $C_{1,2}$, $C_{2,2}$ and $C_{3,2}$ respectively. Then, for $1 \leq j \leq N$,

$$\begin{aligned} \Lambda &\triangleq \lambda[C_{1,2} * C_{2,2} * C_{3,2}] \\ &= \{\lambda(\mathbf{a}_j * \mathbf{b}_j * \mathbf{c}_j)\ :\ \mathbf{a}_j \in C_{1,2}\ ,\ \mathbf{b}_j \in C_{2,2} \text{ and } \mathbf{c}_j \in C_{3,2}\} \end{aligned} \tag{4.4}$$

forms a 3-level 8-PSK modulation code of length n, dimension $k = k_{1,2} + k_{2,2} + k_{3,2}$, and minimum squared Euclidean distance $d_E^2[\Lambda] = \min\{0.586 \times \delta_{1,2}\ ,\ 2 \times \delta_{2,2}\ ,\ 4 \times \delta_{3,2}\}$. Consequently, the following collection of distinct arrays of 8-PSK signals,

$$\Omega = \{\lambda(A * B * C)\ :\ A \in P_1\ ,\ B \in P_2 \text{ and } C \in P_3\} \tag{4.5}$$

form a **product** 8-PSK modulation code of length nN, dimension $k_{1,1} \times k_{1,2} + k_{2,1} \times k_{2,2} + k_{3,1} \times k_{3,2}$, and minimum squared Euclidean distance

$$d_E^2[\Omega] = \min\{0.586 \times \delta_{1,1} \times \delta_{1,2}\ ,\ 2 \times \delta_{2,1} \times \delta_{2,2}\ ,\ 4 \times \delta_{3,1} \times \delta_{3,2}\} \tag{4.6}$$

From (4.6), we see that we can construct product 8-PSK codes with **arbitrarily** large minimum squared Euclidean distance by choosing the horizontal and vertical component codes properly. If the vertical component codes are chosen using expressions (2.10) and (2.11) as the guidelines, good codes with large minimum symbol and product distances can be constructed.

In the above construction of product modulation code, binary linear block codes are used as the horizontal codes for forming the product codes. In fact, we can use nonbinary codes of length N with symbols from GF(2^m) as the horizontal codes for the product codes. In this case, the horizontal encoding must be done first followed by the column encoding. This is to ensure that the columns of the $n \times N$ signal array are codewords in Λ.

4.3 Decoding

One obvious way (though impractical) of decoding the proposed product modulation code is to compare each received code sequence with all the possible code sequences and find the closest one in terms of minimum squared Euclidean distance. The decoding complexity associated with this technique would be simply enormous. We will focus on a suboptimal decoding procedure which allows decoding of the codes with reduced decoding complexity while maintaining good performance.

Consider the decoding of an 8-PSK product modulation code Ω. Recall that each codeword in Ω can be written in the form $\mathbf{V} = (\mathbf{v}_1, \mathbf{v}_2, \cdots, \mathbf{v}_N)$ with $\mathbf{v}_i \in \Lambda$ for $1 \leq i \leq N$. Let $\mathbf{R} = (\mathbf{r}_1, \mathbf{r}_2, \cdots, \mathbf{r}_N)$ be the received sequence. The decoding is performed in 3 stages with P_1, P_2 and P_3 decoded in sequence. At the first stage of decoding, each column $\mathbf{a}_j$ of P_1 is decoded based on the received sequence $\mathbf{r}_j$, using a multi-stage soft-decision decoding algorithm for Λ[40]. After the column decoding of P_1, the horizontal code $C_{1,1}$ of P_1 then uses the decoded estimates of columns to decode the rows of P_1 using hard-decision decoding. The horizontal row decoding is advantageous since it corrects additional errors in the columns and thus helps reduce the error propagation into the next stage of decoding. In fact, if the horizontal codes are chosen powerful enough, the error propagation effect is negligible, thereby eliminating the major problem of multi-stage decoding and improving the error performance. Thus, after the horizontal row decoding of P_1 we have new estimates of the columns (the corrected estimates). These new estimates along with the received sequence are passed to the second decoding stage and are used to decode the columns of P_2. The decoding of P_2 and P_3 follows the same procedure as that of P_1.

4.4 Examples

Example 4.1: In this example, we will construct a **product** modulation code over 8-PSK for the AWGN channel.

Choose $C_{1,2}$ to be the $(8,1,8)$ repetition code with minimum distance 8, $C_{2,2}$ to be the $(8,7,2)$ even parity code with minimum distance 2, and $C_{3,2}$ to be the trivial $(8,8,1)$ code with minimum distance 1. These three vertical codes are used to form Λ. Also, let $C_{1,1}$ be the $(255,187)$ BCH code with error correcting capability $t = 9$. Two Reed-Solomon (RS) codes are used at the second level to form $C_{2,1}$. The first one is the 12-error-correcting $(127,103)$ RS code over

$GF(2^7)$. The second one is the 12-error-correcting $(128, 103)$ extended RS code over $GF(2^7)$. $C_{3,1}$ is chosen to be the six-error-correcting $(255, 243)$ RS code over $GF(2^8)$. The overall spectral efficiency of Ω is $\eta = 1.75147$ bits/symbol and the phase invariance is 45°. The $(8, 1, 8)$ repetition code associated with the first labeling bit of Λ has a 2-state trellis which is used to decode the columns of P_1 at the first stage of decoding with the Viterbi algorithm. Also, the $(8, 7, 2)$ even parity code associated with second level of Λ has a very simple 2-state trellis which is used to form the decoded estimates of the columns of P_2 (i.e. the second stage of decoding). The decoding complexity associated with decoding the third level of Λ is very small and hence will be ignored. The total decoding complexity of Λ due to the soft-decision multi-stage decoding is therefore $2 + 2 = 4$ states. The decoding complexity of the horizontal codes of the product codes depends upon what kind of decoding algorithm is chosen for the BCH and RS codes.

Figure 6 shows the simulation for the probability of bit error. As can be seen from Figure 6, the suboptimum decoding allows one to achieve 5.38 dB real coding gain over uncoded QPSK at the bit error rate of 10^{-6}. Also shown in Figure 6, is the bit-error-performance obtained for the single level concatenation of Λ with the 16 error correcting NASA standard $(255, 223)$ RS code over $GF(2^8)$ [44]. Both systems (i.e., the single-level concatenation and the proposed product modulation code) have the same spectral efficiency and comparable decoding complexity, however the performance of the proposed **product** modulation code is better than the single level concatenation. The proposed code outperforms the single-level concatenation code by about 0.28 dB at the bit error rate of 10^{-6}, due to the preferential error correcting capability of the proposed code.

Example 4.2: In this example, we will construct a **product** modulation code over 8-PSK for the Rayleigh fading channel.

Choose $C_{1,2}$ to be the $(8, 4, 4)$ RM code with minimum distance 4, $C_{2,2}$ to be the $(8, 7, 2)$ even parity code with minimum distance 2, and $C_{3,2}$ to be the even parity $(8, 7, 2)$ code with minimum distance 2. These three vertical codes are used to form Λ. Note, that Λ is designed specifically for the Rayleigh fading channel. The code has minimum symbol distance $\delta_H[\Lambda] = 2$ and the minimum product distance $\Delta_p^2[\Lambda] = 4$.

Also, let $C_{1,1}$ be the $(127, 106)$ BCH code with error correcting capability $t = 3$. $C_{2,1}$ is chosen to be the 9-error-correcting $(127, 109)$ RS code over $GF(2^7)$. $C_{3,1}$ is chosen to be the four-error-correcting $(127, 119)$ RS code over $GF(2^7)$. The overall spectral efficiency of Ω is $\eta = 1.988$ bits/symbol and the phase invariance is 90°. The $(8, 4, 4)$ RM code associated with the first labeling bit of Λ has a 4-state trellis which is used to decode the columns of P_1 at the first stage of decoding with the Viterbi algorithm. Also, the $(8, 7, 2)$ even parity code associated with second level of Λ has a very simple 2-state trellis which is used to form the decoded estimates of the columns of P_2 (i.e. the second stage of decoding). The decoding complexity associated with the third stage of decoding is comparable to that of the second stage. The total decoding complexity of Λ due to the soft-decision multi-stage decoding is therefore $4 + 2 + 2 = 8$ states. The decoding complexity of the horizontal codes of the product codes depends upon what kind of decoding algorithm is chosen for the BCH and RS codes.

Figure 7 shows the simulation for the probability of bit error. The simulation has been carried out assuming independent fading, no channel state information and minimum squared Euclidean distance as the decoding metric. As can be seen from Figure 7, the suboptimum decoding allows one to achieve 13.29 dB real coding gain at the bit error rate 10^{-3} and 40.94 dB real coding gain at the bit error rate of 10^{-6} over uncoded QPSK.

The proposed code outperforms all the codes available in literature, both in terms of performance and complexity.

The above construction of product modulation codes can be generalized to other signal constellations in a straight forward manner.

5 Conclusion

In this paper, we have shown that multilevel coding method is a powerful technique for constructing modulation codes for both the AWGN and fading channels. Some multilevel codes have been constructed and they achieve significant coding gains with relatively simple decoding complexity. The product coded modulation scheme based on the multilevel method provides a technique for constructing long powerful codes to achieve high performance with low decoding complexity.

References

[1] G. Ungerboeck and I. Csajka, "On Improving Data-link Performance by Increasing the Channel Alphabet and Introducing Sequence Coding," presented at 1976 *International Symposium on Information Theory*, Ronneby, Sweden, June 1976.

[2] G. Ungerboeck, "Channel Coding with Multilevel/Phase Signals," *IEEE Trans. on Information Theory*, Vol. IT-28, No. 1, pp. 55-67, January 1982.

[3] G. Ungerboeck, "Trellis-Coded Modulation with redundant Signal Set, Part II: State of the Art," *IEEE Communications Magazine*, Vol. 25, No. 2, pp. 12-21, February 1987.

[4] G.D. Forney, R.G. Gallager, G.R. Lang, F.M. Longstaff, and S.U. Quereshi, "Efficient Modulation for Band-limited Channels," *IEEE J. Selected Areas in Communications*, Vol. SAC-2, pp. 632-646, September 1984.

[5] A.R. Calderbank and J.E. Mazo, " A new description of trellis codes," *IEEE Trans. on Information Theory*, Vol. IT-30, pp. 784-791, Nov. 1984.

[6] A.R. Calderbank and N.J.A. Sloane, "Four-Dimensional Modulation with an Eight State Trellis Code," *AT&T Tech. Journal*, Vol. 64, No. 5, May-June 1985; also, "A new family of codes for dial-up voice lines," in *Proc. IEEE Global Telecomm. Conf.*, Nov. 1984, pp. 20.2.1-20.2.4.

[7] S.G. Wilson, H.A. Sleeper II, Paul J. Schottler, and Mark T. Lyons, "Rate 3/4 Convolutional Coding of 16-PSK: Code Design and Performance Study," *IEEE Trans. on Communications*, Vol. COM-32, pp. 1308-1315, December 1984.

[8] M. Oerder, " Rotationally invariant trellis codes for MPSK modulation," *1985 International Communication Conference Record* , pp. 552 - 556, Chicago, June 23-26, 1985.

[9] A.R. Calderbank and N.J.A. Sloane, " An eight-dimensional trellis code," *Proceedings of the IEEE*, Vol. 74, pp. 757-759, May 1986.

[10] S.G. Wilson, "Rate 5/6 Trellis-Coded 8PSK," *IEEE Trans. on Communications*, Vol. COM-34, October 1986.

[11] A.R. Calderbank and N.J.A. Sloane, "New Trellis Codes Based on Lattices and Cosets," *IEEE Trans. on Information Theory*, Vol. IT-33, pp. 177-195, March 1987.

[12] E. Zehavi and J.K. Wolf, "On the Performance Evaluation of Trellis Codes," *IEEE Trans. on Information Theory*, Vol. IT-33, pp. 196-202, March 1987.

[13] G.J. Pottie and D.P. Taylor, "An Approach to Ungerboeck Coding for Rectangular Signal Sets," *IEEE Trans. on Information Theory*, Vol. IT-33, pp. 285-289, March 1987.

[14] L.F. Wei, "Trellis-Coded Modulation with Multi-dimensional Constellations," *IEEE Trans. on Information Theory*, Vol. IT-33, pp. 483-501, July 1987.

[15] D. Divsalar and M.K. Simon, "Multiple Trellis Coded Modulation (MTCM)," *IEEE Trans. on Communications*, Vol. COM-36, pp. 410-419, April 1988.

[16] M. Rouanne and D.J. Costello, Jr., "A Lower Bound on the Minimum Euclidean Distance of Trellis Coded Modulation Schemes," *IEEE Trans. on Information Theory*, Vol. IT-34, pp. 1011-1020, September 1988, Part I.

[17] G.D. Forney, Jr., "Coset Codes I: Introduction and Geometrical Classification," *IEEE Trans. on Information Theory*, Vol. IT-34, pp. 1123-1151, September 1988, Part II.

[18] —, "Coset Codes II: Binary Lattices and Related Codes," *IEEE Trans. on Information Theory*, Vol. IT-34, pp. 1152-1187, September 1988, Part II.

[19] S.S. Pietrobon, et al., "Trellis-Coded Multi-Dimensional Phase Modulation," *IEEE Trans. on Information Theory*, Vol. IT-36, January 1990.

[20] H. Imai and S. Hirakawa, "A New Multilevel Coding Method Using Error Correcting Codes," *IEEE Trans. on Information Theory*, Vol. IT-23, No. 3, pp. 371-376, May 1977.

[21] E.L. Cusack, "Error Control Codes for QAM Signaling," *Electron. Letters*, Vol. 20, No. 2, pp. 62-63, 19 Jan 1984.

[22] V.V. Ginzburg, "Multidimensional Signals for a Continuous Channel," *Problemy Peredachi Informatsii*, Vol. 20, No. 1, pp. 28-46, 1984.

[23] S.I. Sayegh, "A Class of Optimum Block Codes in Signal Space," *IEEE Trans. on Communications*, Vol. COM-30, No. 10, pp. 1043-1045, October 1986.

[24] R.M. Tanner, "Algebraic Construction of Large Euclidean Distance Combined Coding Modulation Systems," *Abstract of Papers, 1986 IEEE International Symposium on Information Theory*, Ann Harbor, October 6-9, 1986.

[25] T. Kasami, T. Takata, T. Fujiwara and S. Lin, "Error Control Systems with Combined Block Coding and M-ary PSK Modulations," *The Proceedings of the 10th Symposium on Information Theory and Its Applications*, Enoshima, Japan, pp. 393-398, November 1987.

[26] —, "Construction of Block Codes for 2^ℓ-ary PSK and 2^ℓ-ary QASK Modulations," *Papers of IEICE of Japan*, Vol. IT87-126, pp. 49-54, March 1988.

[27] E. Biglieri and M. Elia, "Multidimensional Modulation and Coding,' *IEEE Trans. on Information Theory*, Vol. IT-34, pp. 803-809, July 1988.

[28] G.J. Pottie and D.P. Taylor, "Multi-Level Channel Codes Based on Partitioning," *IEEE Trans. on Information Theory*, Vol. IT-35, No. 1, pp. 87-98, January, 1989.

[29] A.R. Calderbank, "Multi-Level Codes and Multi-Stage Decoding," *IEEE Trans. on Communications*, Vol. COM-37, No. 3, pp. 222-229, March 1989.

[30] F.P. Kschischang, P.G. buda, and S. Pasupathy, "Block Coset Codes for M-ary Phase Shift Keying," *IEEE J. Selected Areas in Communications*, Vol. SAC-7, pp. 900-913, August 1989.

[31] T. Takata, S. Ujita, T. Fujiwara, T. Kasami and S. Lin, "Linear Structure and Error Performance Analysis of Block PSK Modulation Codes,"*IEICE Trans. on Communications, Electronics, Information and Systems of Japan*, Vol. J73-A, No. 2, pp. 314-321, February 1990.

[32] T. Kasami, T. Takata, T. Fujiwara, and S. Lin, "A Concatenated Coded Modulation Scheme for Error Control," *IEEE Trans. on Communications*, Vol. COM-38, No. 6, pp. 752-763, June 1990.

[33] —, "Cross-Over Construction for Multi-Level Block Modulation Codes ," *IEICE Trans. on Communications, Electronics, Information and Systems of Japan*, Vol. E73, No. 10, October 1990.

[34] —, "On Linear Structure and Phase Rotation Invariant Properties of Block 2^ℓ-PSK Modulation Codes," *IEEE Trans. on Information Theory*, Vol. IT-37, No. 1, January 1991.

[35] —, "On Multi-Level Block Modulation Codes," *IEEE Trans. on Information Theory*, Vol. IT-37, No. 4, July 1991.

[36] E. Biglieri, D. Divsalar, P.J. McLane and M.K. Simon, *Introduction to Trellis-Coded Modulation with Applications*, Macmillan, 1991.

[37] C. Schlegel and D.J. Costello, Jr., "Bandwidth Efficient Coding for Fading Channels: Code Construction and Performance Analysis", *IEEE Journal Select Areas Communications*, Vol. SAC-7, No. 9, pp. 1356-1368, Dec. 1989.

[38] B. Vucetic, "Concatenated Multilevel Block Coded Modulation", *IEEE Trans. on Communications*, Vol. 41, No. 1, January 1993.

[39] J. Wu and S. Lin, "Multilevel Trellis MPSK Modulation Codes for the Rayleigh Fading Channel," *IEEE Trans. on Communications*, Vol. 41, No. 9, pp. 1311-1318, July 1993.

[40] T. Takata, S. Ujita, T. Kasami and S. Lin, "Multistage Decoding of Multilevel Block M-PSK Modulation Codes and Its Performance Analysis," *IEEE Trans. on Information Theory*, Vol. 39, No. 4, July 1993.

[41] T. Kasami, T. Takata, T. Fujiwara and S. Lin, " On Structural Complexity of the L-Section Minimal Trellis Diagrams for Binary Linear Block Codes," *IEICE Trans. on Fundamentals of Electronics, Communications and Computer Sciences*, Vol. E76-A, No. 9, September 1993.

[42] S. Rajpal, D.J. Rhee and S. Lin, "Multi-Dimensional Trellis Coded Phase Modulation Using a Multilevel Concatenation Approach," Abstracts of the 1991 IEEE International Symposium on Information Theory, Budapest, Hungary, June 23-28, 1991.

[43] S. Lin and D.J. Costello, Jr., *Error Control Coding: Fundamentals and Applications*, Prentice-Hall, Englewood Cliffs, New Jersey, 1983.

[44] T. Kasami, T. Takata, T. Fujiwara, and S. Lin, "A Concatenated Coded Modulation Scheme for Error Control," *IEEE Trans. on Communications*, Vol. COM-38, No. 6, pp. 752-763, June 1990.

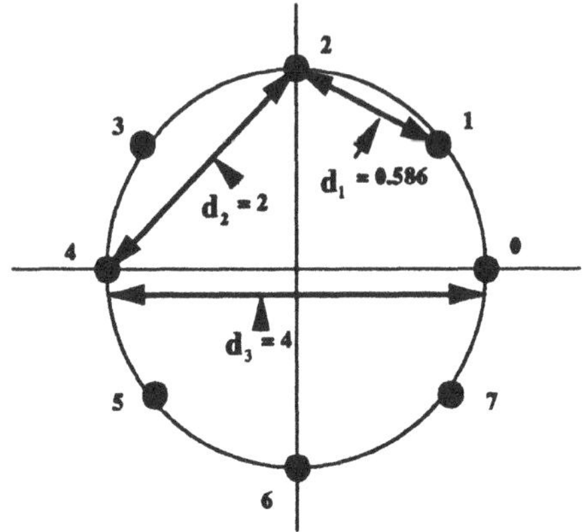

Figure 1 An 8-PSK signal constellation

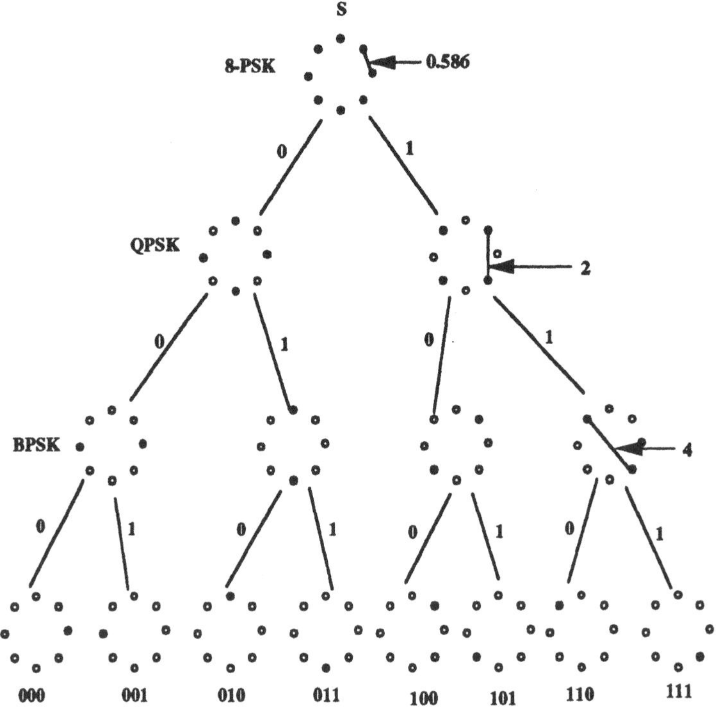

Figure 2 The 8-PSK signal set and its partition chain 8-PSK / QPSK / BPSK

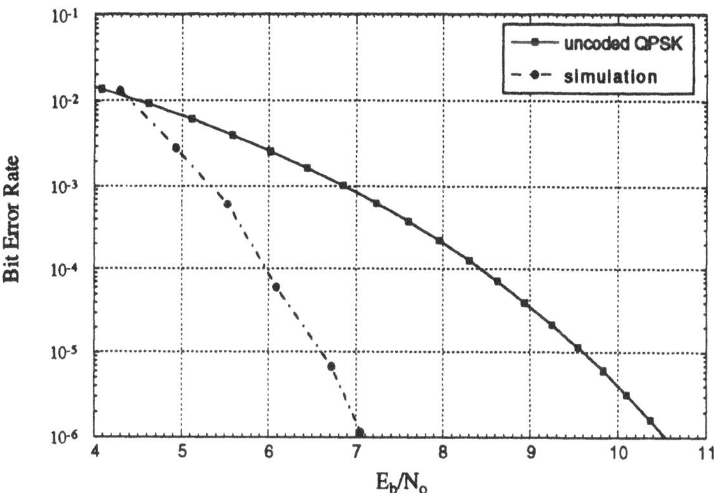

Figure 3 Bit-error performance of the basic 3-level modulation code C(1) over the AWGN channel

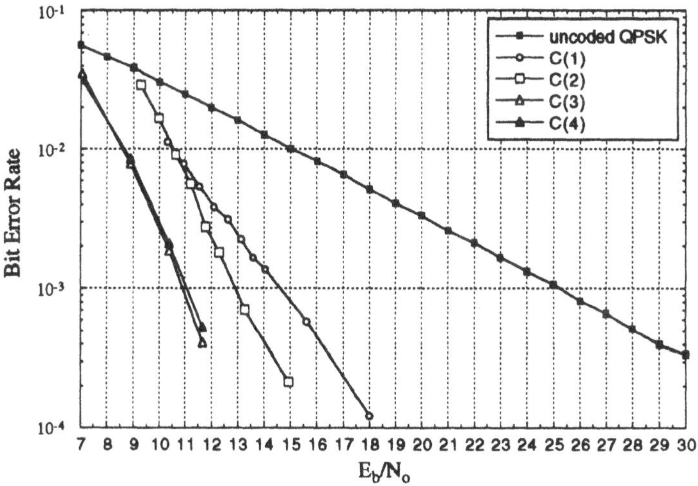

Figure 4 Bit-error performances of the multilevel modulation codes, C(1), C(2), C(3) and C(4) over the Rayleigh fading channel

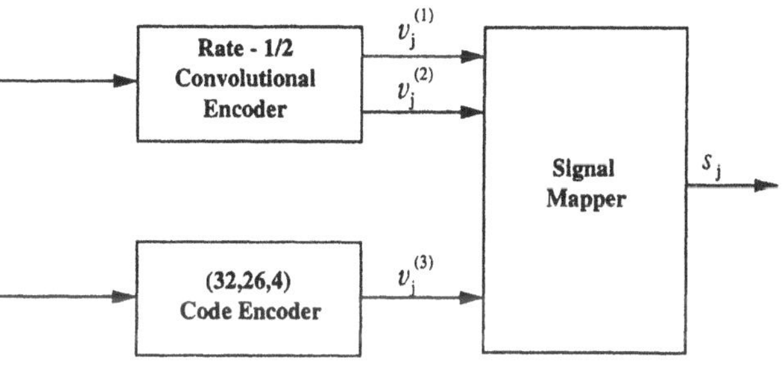

Figure 5 C(3) -Encoder

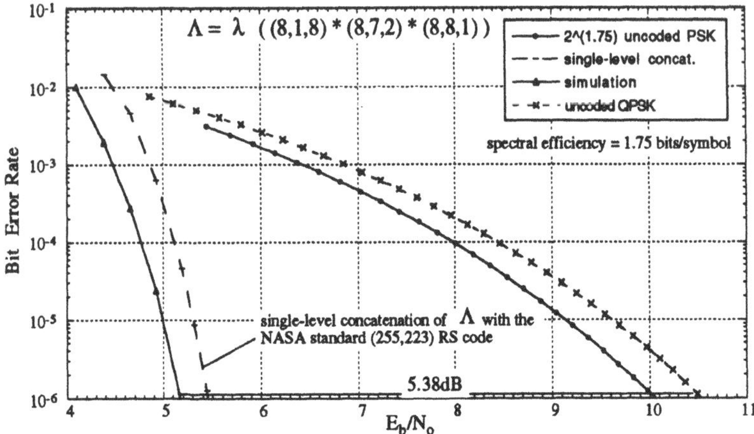

Figure 6 Bit-error performance of the product coded modulation scheme of Example 1 in Section 4 over the AWGN channel

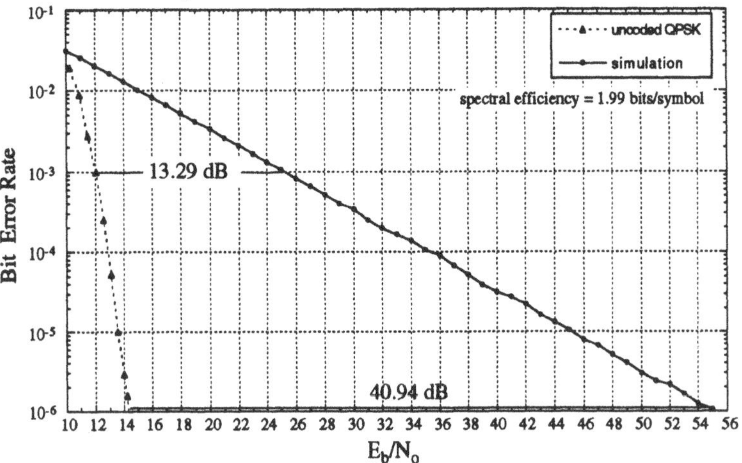

Figure 7 Bit-error performance of the product coded modulation scheme of Example 2 in Section 4 over the Rayleigh fading channel

Channel Coding for a Frequency-Hopped Wireless Transceiver[1]

Victor S. Lin and Gregory J. Pottie

Electrical Engineering Department
University of California, Los Angeles
405 Hilgard Ave.
Los Angeles, CA 90024

Abstract - We consider a multiple-access frequency-hopped system employing antenna diversity and error correction coding. A construction based on orthogonal Latin squares is used for the hopping patterns to improve code performance in the presence of interference. The measures of interest are the system performance, delay and decoder complexity, providing a basis for comparing BCH, Reed-Solomon (RS), and convolutional codes. It is shown that when binary non-coherent FSK modulation and hard decision detection are used, the BCH, RS, and convolutional codes that can meet the delay and bandwidth constraints have similar performance and comparable complexity. Soft decision decoding for the convolutional code and error-and-erasure correction decoding for the RS code are investigated in terms of the performance-complexity trade-offs for different decoding metrics. It is shown that improvement over hard decision decoding is significant when the right decoding metric is selected.

1. Introduction

The dominant impairments to transmission in cellular radio systems are multipath fading and interference from other users sharing the common spectrum. Multipath propagation results in severe fading, which can dramatically increase the signal to noise ratio (SNR) required for reliable operation. In a time-division multiple access (TDMA) system the users are assigned slots which are kept from frame to frame, with interfering users in nearby cells assigned slots in the same way. Because of factors such as geographic proximity and shadow fading, the result is that interference power also displays wide variations across slots, which further degrades the 90% to 99% worst case performance. The conventional solution of restricting the frequency re-use between cells significantly reduces capacity. Since the re-use distance must be made large enough to allow sufficient margin against multipath fading, a high performance price is paid.

In a direct sequence CDMA system, the problem of differing interference levels is mitigated by the despreading process. Effectively, for large enough spreading each

1. This work supported in part by AT&T Bell Laboratories Mathematical Sciences Research Center and ARPA contract DAAB07-92-R-C977.

user deals with the average interference levels, and thus the system may be engineered to deal with the average rather then worst case interference levels. Channel coding completes the task, and further reduces the SNR required for reliable operation. Unfortunately, wideband direct sequence systems suffer from two problems: (1) the power consumption of the receiver is very large and (2) synchronous access is not possible on the up-link, since the chip duration is very short. The latter problem results in decreased capacity, since the interference generated within the cell considerably exceeds the interference generated in neighboring cells. A frequency hopped system can alleviate both of these difficulties since the hopping rate is independent of the bandwidth occupied, and therefore can be set low enough to permit low-power receiver design and synchronous operation.

An ongoing research project at UCLA is the design of a low-power hand-held transceiver [1]. Data rates between 2 kb/s to 160 kb/s are to be supported, using a 26 MHz bandwidth. The maximum hop rate is 160 khops/s and the maximum range is 500 meters. Experimental characterization of the radio channel has found that the typical delay spread is less than 400 ns for both indoor and outdoor microcellular environments [2][3]. At the maximum hop rate, the hopping or baud interval (6.25 μs) is large relative to the delay spread; hence, the effect of ISI caused by multipath is negligible and equalization is not needed. In order to achieve a high capacity with low power consumption, it is essential that diversity be employed. Dual antenna diversity, frequency hopping, and channel coding will be implemented. The frequency hopped transceiver will consume less than 300 mW in active mode with two receiver branches, in comparison to well over 1 Watt of power for a DS system implemented in the same technology [4]. This is possible because of a careful design of the digital direct frequency synthesizer/digital-to-analog converter pairs [5]. In the following, we outline our progress in the design of the hopping patterns and selection of the channel codes, and the basic trade-offs in choice of metrics for decoding.

2. Latin Squares

Slow frequency hopping in combination with interleaving and channel coding is effective in dealing with variations in the signal to interference ratio among successive hops, provided the hopping patterns are selected properly. A number of recent papers [6][7][8] have examined slow-hopped systems. We will concentrate a simple construction introduced in [9] which is especially suitable for this application.

A latin square of order n is an n x n matrix with entries from a set R of n distinct elements, say R = {0,1,...,n-1} such that each row and column contains every element of R exactly once. Two Latin squares are said to be orthogonal if the n^2 ordered pairs (i,j), where i and j are the entries from the same position in the respective squares, exhaust the n^2 possibilities, i.e., every ordered pair occurs exactly once.

If n is prime, then there is a simple construction for a family of n-1 mutually orthogonal latin squares. For a = 1,...,n-1 we define an n x n matrix {a} by setting

$$\{a\}_{ij} \equiv ai + j \pmod{n}$$

It is easy to verify that each matrix is a latin square: for fixed i, the numbers ai+j,

$j=0,\ldots,n-1$ are distinct, and for fixed j, the numbers ai+j, $i=0,\ldots,n-1$ are distinct, in both cases assuming modulo arithmetic. It remains to verify that if $a \neq b$, then the matrices $\{a\}$ and $\{b\}$ are orthogonal. This amounts to showing that for all $c, d = 0,1,\ldots,n-1$, there is a unique solution (i,j) to the linear system

$$c \equiv ai + j$$
$$d \equiv bi + j$$

This follows from the fact that the determinant of this linear system is a-b, which is non-zero (mod n).

If n is prime, then the residues modulo n form the finite field of order n. The above construction extends to finite fields of prime power order in the natural way: now R is the set of field elements, and for every non-zero field element a, we define a matrix $\{a\}$ with the ij^{th} entry $\{a\}_{ij} = ai+j$, where addition and multiplication take place in the field.

The above construction may be used in a synchronous cellular radio system. That is, the timing of the slot boundaries must be aligned across cells, with guard bands potentially required for large cells to account for propagation delays from nearby cells. Each cell is assigned a matrix $\{a\}$ in the set of n-1 mutually orthogonal latin squares. Then up to n users can be accommodated in any cell, each user being assigned a field element. For example, suppose n=8, a=5, and a user is assigned the element 7. We might interpret the rows as frequency slots and the columns as time slots for a frequency hopped system. Then at time 0, frequency 3 is transmitted, at time 1 frequency 6 is transmitted, etc. Except for other cells which also are assigned matrix $\{5\}$ there will be at most one collision per user at any given time/frequency pair over the 8 slot duration of the frame. If a regular re-use pattern of 7 cells is used for the matrices, then the signals from users sharing the same hopping patterns will be heavily attenuated. In this way, each user suffers collisions from an ensemble of users occupying the nearby cells, rather than with one dominant interferer. Obviously, with a larger value of n, the re-use distance increases and the randomization of interference hits is more complete. Also, the rows might be reinterpreted as time slots in a frame, and the columns as frame positions within a superframe for a pure TDMA system.

The construction allows each mobile to quickly generate its own time/frequency hopping pattern given minimal information transfer from the base station. The only information required, given that n and the primitive polynomial p(x) are known, is the matrix number, a, and the field element, e. Then if i is the row number, the column corresponding to i is just $j \equiv e + ai \pmod{p(x)}$. Hence, the overhead involved in communicating the hopping pattern is negligible. Note that we cannot get every possible value of n using this construction. However, primes and prime powers are dense for moderate n, affording a wide range of choices.

A very similar construction has recently been proposed by Wyner [10] using a combinatorial configuration constructed by Roche [11] for use in wideband multi-tone systems, to assure that users in nearby cells interfere in at most one tone.

The effectiveness of the Latin squares construction was illustrated in [9] for a slow-hopped DQPSK system. Simulations were conducted in combination with rate-1/2 and rate-1/4 convolutional codes with 32 states, for a multi-cell system using the same propagation assumptions as in [12]. It was further assumed that Rayleigh fading was independent between frequency hops, and that channel parameters were estimated from the data. The results indicate that the system capacity without some randomization of interferers is quite poor, but with the Latin squares approach capacity increased beyond that reported for a DS-CDMA system in [12], with considerably lower decoding delay. The addition of dual antenna diversity further improved performance.

Simulations were also conducted to examine performance in flat fading conditions. The results in this case are particularly interesting in illustrating what we refer to as interferer diversity. Even though the desired signal level is now at the same level in each slot, the interference levels are independent, since a different set of interferers in encountered in each slot. Thus, even in flat-fading conditions, the interference is independently fading, leading to independent signal to interference ratios. Therefore channel coding continues to provide a significant improvement in performance, effectively averaging over the SNR variations while providing coding gain. This is of considerable interest in handheld transceiver applications, since fading can be flat over tens of MHz in some indoor situations. In these cases, frequency hopping in itself would not be sufficient to achieve diversity benefit; some randomization of interference is also required. Consequently, the Latin squares construction will be incorporated in the final design.

3. Channel Code Selection

Fig. 1 outlines the system model used in the investigation of alternative channel coding schemes. The input bit stream is encoded into the coded q-bit symbol sequence of a block code or a convolutional code and the resulting coded sequence with a rate of f_c symbols/s is fed to the interleaver. The interleaver rearranges the coded sequence in such a way that the fading channel with burst error characteristic is transformed into a channel having independent errors. The coded symbol at the output of the interleaver, to be transmitted with a rate of $f_c \times q$, is assigned to one of the two binary orthogonal FSK signals.

The frequency hopping binary FSK signal is output from a frequency hopper where the hopping local signal is multiplied to the binary FSK signal. A slow frequency hop scheme is assumed in which a hop is made in each slot interval consisting of a few symbols. It is assumed that the received signal experiences independent Rayleigh fading slot-by-slot, and that it is perturbed by additive white Gaussian noise. While this model does not yield a capacity estimate, it is sufficient to evaluate the relative effectiveness of coding schemes in a situation where we in fact expect the SNR to be independent from hop to hop as a result of the use of the Latin squares construction.

The receiver dehops the signal to obtain the received binary FSK signal. Noncoherent detection is used for the reception, because the handheld system is designed to allow fast hopping. The receiver employs dual antenna diversity with postdetection equal gain combining to achieve diversity. This was found to outperform selection diversity, and in any case when the channel SNR cannot be estimated with any accuracy for a slot (e.g., fast hopping), two complete receiver branches are required to make antenna diversity effective. Thus, equal gain combining comes essentially for free. We assume that the signals received at the two antennas have low correlation.

As a result of interleaving/deinterleaving and frequency hopping, errors within a code word appear to be independent. We assume that a convolutional interleaver is used because it results in less delay than block interleaving. In this section, we consider hard decision decoding for convolutional codes and error correction for block codes. Later, we examine soft decision decoding for convolutional codes and error-and-erasure correction for block codes.

In order to compare the various coding options in terms of BER performance, we assume ideal interleaving so that the channel is memoryless. For non-coherent binary FSK (NC-BFSK) modulation with ideal interleaving, the error rate performance over a frequency-nonselective, slowly fading channel is

$$P_2 = \frac{1}{2+\bar{\gamma}_b} \tag{1}$$

where is $\bar{\gamma}_b$ the average signal-to-noise ratio, defined as

$$\bar{\gamma}_b = \frac{\xi_b}{N_o} E(\alpha^2) \tag{2}$$

The term $E(\alpha^2)$ is the average value of the Rayleigh distributed envelope squared.

When L-th order antenna diversity technique is used, the performance of square-law-detected binary FSK is well approximated by [13]

$$P_2 \approx \left(\frac{1}{\bar{\gamma}_b}\right)^L \binom{2L-1}{L} \tag{3}$$

For dual antenna diversity, L is set equal to two.

3.1 Performance of Reed-Solomon and BCH Codes

With Reed-Solomon (RS) codes over $GF(2^q)$, q binary channel symbols are used to form one code symbol. Therefore, the probability of code symbol error is $P = 1 - (1 - P_2)^q$, where P_2 is given by (1) and (2), with and without antenna diversity, respectively. For M-ary ($M = 2^q$) block codes with error correction decoding, the decoded symbol error probability can be approximated as

$$P_{es} \approx \frac{1}{N} \sum_{i=t+1}^{N} i \binom{N}{i} P^i (1-P)^{N-i} \tag{4}$$

where $t = [(d-1)/2]$ is the number of errors that can be corrected by the code, d is the minimum distance of the code, N is the block length, and [x] is the largest integer less than or equal to x. The probability of a decoded bit error is

$$P_{eb} = \frac{2^{q-1}}{2^q - 1} P_{es} \tag{5}$$

For BCH codes, the probability of decoded bit error is just $P_{eb} = P_{es}$, where P_{es} is given by (4). Thus, by combining (4), (5) and (2) or (3), the probability of decoded bit error for block codes can be computed at any signal-to-noise ratio, with or without antenna diversity.

3.2 Performance of Convolutional Codes

Next, we consider the error rate performance of convolutional codes decoded using the Viterbi algorithm with a hard decision decoding metric. The decoded bit error probability of a rate b / n convolutional code can be upperbounded by

$$P_b < \frac{1}{b} \sum_{l=d}^{\infty} \beta_l D^l \tag{6}$$

where d is the minimum free distance of the code, and β_l is the total information weight of all paths of distance l from the all-zeros path. With hard decision decoding

$$D = \sqrt{4P_2(1-P_2)} \tag{7}$$

where P_2 is given by (1) and (2), with and without antenna diversity, respectively.

The weight and distance structure of many convolutional codes have been computed and tabulated in the literature. Thus, equations (6) and (7) can be used to evaluate the performance of a specific convolutional code.

To study the performance advantage of coded systems over an uncoded one, a system simulation was designed and completed. For decoding Reed-Solomon and BCH codes, the Berlekemp-Massey algorithm and Forney algorithm were implemented. The Viterbi algorithm was employed for decoding convolutional codes.

The simulation results were checked against the corresponding analytical results whenever possible in order to verify the accuracy of our simulation software during the initial phases of the simulation development. It was found that the assumption of ideal interleaving was good. Analytical expressions (5) and (6) produced fairly tight upper bounds. For each coding scheme, the difference between the performance curves obtained from the upper bound and the simulation is less than 1 dB for $10^{-2} < \text{BER} < 10^{-4}$. The simulation results were used to compute coding gains. Fig. 2 illustrates typical BER curves generated from the simulation data for a particular convolutional code with and without dual antenna diversity. Some of the coding gain data are summarized in Table 1. It shows that using the rate-1/2, 32 state (v=5) convolutional code resulted in near minimum SNR at the target BER, with and without antenna diversity. The coding gain achieved is 19.1 dB with antenna diversity, compared to 13.0 dB with a single antenna. This highly motivates the use of both channel coding and antenna diversity in the system architecture.

The code rate and code size for the codes in Table 1 were chosen based on their potential to meet code performance with decoding delay tolerable for voice transmission, practical code complexity, and channel bandwidth constraints.

3.3 Decoding Delay

To evaluate and compare the delay of the selected codes, we defined three different delay measurements: decoding delay, interleaving delay and overall delay. The delay quantities are specified in signaling interval units, Ts, to provide a normalized delay representation. Denoted τ_d, decoding delay is the waiting time for some number of encoded symbol to be received before starting the decoding process. The interleaving delay, τ_i, for a (I, J) convolutional interleaver is proportional to $(I-1) \times J$, where values of I and J are chosen to randomize the error bursts of the fading channel and depend on the coding scheme, the code parameter, and the number of channel symbols per frequency slot [14]. The overall delay is the sum of the decoding delay and interleaving delay, i.e. $\tau_c = \tau_d + \tau_i$.

To show how delay is determined by the system architecture, examine a system with BFSK modulation and a Reed-Solomon code over $GF(2^q)$. The code has a block length spanning $(2^q - 1) \times q$ channel symbols and thus, $\tau_d/Ts = (2^q - 1) \times q$. Assume there are $c \times q$ channel symbols per slot or equivalently, c code symbols per frequency slot. In order to ensure that all code symbols in a code word suffer from independent fading, successive code symbols must be interleaved across slots. This requires that $I = c$ and $J = (2^q - 1)$. Therefore, $\tau_i/Ts = (c - 1) \times (2^q - 1) \times q$ and $\tau_c/Ts = (2^q - 1) \times c \times q$.

Now, consider a BFSK system with c x q channel symbols per frequency slot. In this example the coding scheme is a rate-1/2 convolutional code with memory ν (no. of states = 2^ν). It has been shown that the full minimum distances of the 32 the 64 state codes are obtained with a truncation depth of 19 and 27, respectively [15]; these parameters were used in computing the delays given in Table 1. For the purpose of delay comparison, we assume that the full minimum distance is obtained with a truncation depth of $4.5 \times \nu$. This implies $\tau_d/Ts = 9 \times \nu$. Setting $I = c \times q$ and $J = 9 \times \nu$ as the interleaving parameters guarantees independent fading condition for the successive symbols generated by the encoder with a separation less than truncation depth. Thus, $\tau_i/Ts = (c \times q - 1) \times 9 \times \nu$, and $\tau_c/Ts = 9 \times \nu \times c \times q$.

It can also be shown that using BCH codes with block length N and bit-by-bit interleaving result in $\tau_d/Ts = N$, $\tau_i/Ts = N \times (c \times q - 1)$, and $\tau_c/Ts = N \times c \times q$. Note that for similar system parameters, the overall delays for the BCH and Reed-Solomon codes with equal block lengths are the same. The overall delay expressions derived above imply that a Reed-Solomon code over $GF(2^q)$ has comparable overall delay to a convolutional code with memory $\nu = 2^q/9$.

Codes satisfying the voice transmission delay requirement are listed in Table 1. To compute the delay in seconds, we assumed that the delay constraint is 20 ms, each slot consists of 6 BFSK signals for the length 63 block codes and 10 for remaining codes, and the channel transmission rate is 16 kbit/s.

Table 1: Code Performance and Delay Comparison

Code Type RS = Reed-Solomon CC = Convolutional	**Gain w/Single Antenna** (dB)	**Gain w/Dual Antenna** (dB)	**Coding Delay w/o Interleaving (Ts)**	**Overall Coding Delay (Ts)**
BCH (n=31, k=16, d=7)	11.5	17.9	31	310
BCH (n=63, k=36, d=11)	12.3	18.3	63	378
RS (n=31, k=15, d=17)	13.1	18.2	155	310
RS (n=63, k=33, d=31)	13.2	18.7	378	378
CC (r=1/2, v=5)	13.5	19.3	38	380
CC (r=1/2, v=6)	13.6	19.5	54	540

To ensure practical code complexity, code selections were limited to block codes with block length less than 255 and convolutional codes with 128 states or less. These are codes which are currently used in various digital communication systems. Channel bandwidth constraints were included in the code selections by limiting the code rate to approximately 1/2; this limits the channel bit rate to a value no larger than twice the user data rate. This limitation was imposed to avoid excessively large bandwidth requirements with high data rates and to decrease link vulnerability from frequency selective fading at high data rates.

3.4 Decoder Complexity

Code complexity was quantified in terms of multiplies per decoded symbol for the block codes and additions per decoded symbol for the convolutional code. We found that decoding a single t-error correcting BCH code word required approximately $10t^2 + 3tn$ multiplications, obtained by adding $10t^2$ multiplications for executing the Berlekemp-Massey algorithm to 2tn for evaluating the syndrome, and tn for Chien search. For a t-error correcting RS code, the number of multiplications required to decode a code word increases due to the multiplications in the Forney algorithm. However, the number of operations normalized per bit for RS codes is not necessarily higher than BCH codes when the code length and rate are fixed. In fact, our calculation showed that the (63, 33) RS codes required 26 multiplies per bit versus 33 multiplies per bit for the (63, 36) BCH code. For a 2^v convolutional code, a total of $3*2^v$ addition operations are necessary at each stage. This includes two additions to compute the cumulative path metrics of the paths merging at each state and a comparison (subtraction) to determine which incoming path survives. For convolutional codes, in addition, a trace back operation is required to complete the decoding. For the codes parameters listed in Table 1, the convolutional codes required more than twice the operations per bit than the comparable block codes. This fact was partly confirmed by the longer simulation times for convolutional codes.

The results of analysis and simulation indicate that for a system architecture employing slow frequency hop and dual antenna diversity to combat multipath fading, a BCH code performs as well as an RS code with a comparable code rate. Our investigation also shows that a rate-1/2, 32 state convolutional code can attain a relatively large coding gain at 10^{-3} BER while meeting the tight delay constraint for two-way speech transmission. In addition, convolutional codes offer the advantage of efficient soft decision decoding. The conclusion is that complexity and performance for the code candidates are similar for hard decision decoding; however, the dependence on soft decision decoding must be determined before deciding which code is the best candidate. We now examine some of the relevant trade-offs.

4. Decoding Metrics

For Rayleigh fading channels, soft decision decoding with perfect channel parameters can effectively double the diversity order available through coding [13]. However, for a real system where channel parameters are obtained by an imperfect estimator, the unreliable estimates could significantly degrade the performance benefits of soft decoding with side information. The reliability of the channel parameter estimator, the modulation scheme, and the channel condition are some of the factors that affect the formulation of a good soft decoding metric. Furthermore, the trade-offs in performance and complexity should be considered when selecting a decision metric for NC-BFSK systems because most soft decoding schemes require much higher receiver/decoder complexity than hard or erasure decoding.

4.1 Soft Decision Metrics for Fading AWGN Channels

The optimum soft decision metric is derived from the likelihood function of the decision variables. For a NC-BFSK system, in which one of two frequencies is transmitted with equal probability to a receiver and the transmitted signal is corrupted by additive white Gaussian noise with spectral density $N_0/2$, a model of the received signal is

$$r_i(t) = \alpha \sin(\omega_i t + \phi) + n(t); \qquad i = 0, 1 \tag{8}$$

We assume that the amplitude, α, is known and the phase, ϕ, is a random variable uniformly distributed in the interval $[0, 2\pi]$. It can be shown that when the constant terms are eliminated, the log-likelihood functions may be written

$$M_{ML} = \ln(p_i(r)) \sim \ln\left(I_o\left(\frac{2\alpha z_i}{N_o}\right)\right); \qquad i = 0, 1 \tag{9}$$

where z_i is the decision variable produced by a square law detector (the optimum demodulator for NC- FSK), and $I_0(\bullet)$ is the modified Bessel function [16].

The mapping of the decision variables into the maximum-likelihood (ML) soft metric involves a very complicated function, $\ln(I_0(\bullet))$. To implement the branch computation part of the Viterbi decoder, a look-up table will be required to transform the decision variables into branch metrics. Since the transforming function has a linear and a non-linear region, one way to reduce the size of the look-up table is by storing the values of the function over the non-linear region and use a linear approximation formula, which does not require costly memory storage, to compute the function over the linear region. Using this procedure, the size of the look-up table used in our study is about 2 Kilobytes.

Besides the ML decoding metric, we consider two suboptimum but less complicated soft decision metrics. The first is based on the Euclidean distance concept. The Euclidean distance metric has been shown to be the optimal soft metric for coherent PSK systems in AWGN channels, but it is not optimal for NC-BFSK systems. It may be written:

$$\mathrm{M_E} = (z_i - \alpha)^2 + (z_j)^2 \tag{10}$$

where $i = 0$ and $j = 1$ for the hypothesis that z_0 was received, and vise versa for the hypothesis that z_1 was received. The metric is simpler not only because it does not involve any complicated functions but also the fade magnitude of the signal is the only channel parameter appearing in the metric. The other metric we considered required even less processing and complexity by using a simple linear combining scheme:

$$\mathrm{M_{LC}} = z_i - z_j \tag{11}$$

where values of i and j are defined as in the previous metric. The linear combining metric does not require any channel state information.

4.2 Soft Decision Metrics for Multiple-Access Channels

FH/SS systems operating in a multiple-access channel will experience background thermal noise, as well as, interference from other users. We use the following model of the received signal for the multiple-access channel:

$$r_i(t) = \alpha \sin(\omega_i t + \phi) + \beta i(t) + n(t); \qquad i = 0, 1 \tag{12}$$

where the fade levels α and β are independent Rayleigh distributed random variables $i(t)$ and $n(t)$ are independent narrowband white Gaussian processes. Since interference is typically the dominant impairment in multiple-access environments, we can assume that the noise term is negligible and set the two-sided spectral density of $i(t)$ to $N_o/2$. Thus, when $E[\beta^2]$ is normalized to one, then

$$\mathrm{E}\left[\beta i(t) \beta i(\tau)\right] = \mathrm{E}\left[\beta^2\right] \mathrm{E}\left[i(t) i(\tau)\right] = \frac{N_o}{2} \delta(t - \tau) \tag{13}$$

Note that for the FH system under consideration, the only real difference between the fading AWGN channel and the multiple-access channel is that the power of the "noise" in the latter is changing from hop to hop.

To perform well for the multiple-access channels, metrics (9) and (10) were reformulated to account for the variation in noise power in each symbol interval. The reformulations are given by (14) and (15), respectively:

$$M_{L,1} = \ln\left[\mathrm{Io}\left(\frac{2\alpha z_i}{\beta^2 N_o}\right)\right] \tag{14}$$

$$M_{E,1} = \frac{(z_i - \alpha)^2 + (z_j)^2}{\beta^2 N_o} \tag{15}$$

For NC-BFSK, the relevant quantities for soft decision decoding are the decision variables for the two frequencies, the received signal power, and the noise power for each slot. We next present ways of obtaining the channel parameters for slow-hopped systems.

4.3 Estimation of Channel Parameters

The estimation of channel parameters from the data and with training sequences was investigated in [9], in the context of DQPSK. It was found that it is better from the point of view of capacity to form estimates directly from the data-bearing signals for short slots of 8-16 symbols, rather than appending a training sequence. In addition, it was found that for some combinations of cell loading and channel codes it was better to use an erasure-declaring mechanism than to use the soft metric proposed; but in any case performance was always better than using simple hard decisions for a properly chosen erasure threshold. Simulations were also conducted using the soft metric with perfect channel knowledge, revealing a very large gap in performance. Thus, channel state information can be very valuable in decoding.

For NC-BFSK signaling with square-law detection, the signal power for a slot can be estimated by accumulating the larger decision variable, z_i, for each received symbol in the slot. This type of estimation involves hard decision demodulation. While it is relatively easy to estimate the power of the desired signal for useful signal to interference ratios, it is more difficult to accurately estimate the interference power. For NC-BFSK, an orthogonal signaling scheme, one way to form an estimate of the noise (or interference) power for a slot is to accumulate the smaller decision variable, z_j, in the slot. The reliability of the signal power and noise power estimates depends on the sample size of the estimator, which is equal to the number of symbols per slot.

4.4 Erasure Metric

When the channel state information are not so reliable, performing error-and-erasure correction decoding is a way to increase code performance gains without incurring increased cost in system complexity. The mechanism we have chosen for erasing unreliable NC-BFSK symbols is based on a ratio threshold test, in which channel symbols having a signal envelope ratio (i.e. the ratio between the decision variables) below a certain threshold are erased. This erasure declaration mechanism does not use any channel state information; hence, requires very little additional complexity in comparison to hard decision decoding. In the branch metric computations, erasures are assigned a value half-way between the binary values for the expected symbols.

We also investigated error-and-erasure correction decoding for RS codes, which achieves some performance benefit with a trivial increase in decoding computation in comparison to error correction decoding [17]. The mechanism for declaring erasures is again based on a ratio threshold test, in which the code symbols having the lowest signal envelope ratio, z_0/z_1 (assuming that $z_0 > z_1$), is erased. When $z_0/z_1 > \theta > 1$, the decision corresponding to z_0 appears to have a good quality. This erasure declaration metric recognizes that the worst BFSK symbol in each q-bit code symbol is the weak link but it will inevitably fail to erase some symbols which are in error, and will erase some symbols which are not in error. There is an optimum range of values for the number of erasures, N_e, declared so that residual error correcting capability, N_c, is sufficient to correct the remaining errors in the received word. The optimum range of values was found during simulation.

4.5 Performance Evaluations

Performance evaluation by simulation for a NC-BFSK system over fading AWGN channels, as well as, multiple-access channels using soft decision decoding metrics given in (9), (10), (11), (14) and (15) was performed. Receivers with perfect knowledge of channel parameters α and N_0, and ones with estimated channel parameters were simulated. Table 2 summarizes the relevant E_b/N_0 data for a rate-1/2, 32-state convolutional coded system with dual antenna diversity.

Table 2: Eb/No at Pb = 10^{-3} for different decoding metrics for a dual antenna diversity system.

	Fading AWGN Channel		Multiple-Access Channel	
Metric Type	Perfect	Estimated	Perfect	Estimated
Maximum Likelihood	8.25	9.25	8.25	9.10
Linear Combining	8.50	8.50	10.0	10.0
Euclidean	8.25	9.0	8.75	9.50

The data in Table 2 shows that there is no single metric that out-performs all others in every scenario. For the fading AWGN channel, the linear combining metric is clearly a very good selection in terms of having a good complexity-performance trade-off. However, for the multiple-access channel, the maximum-likelihood (ML) metric might be the better choice since it holds a slight performance advantage over the other two metrics. In short, when fading interferers are the dominant impairment, the metrics using imperfect estimates of channel parameters performed better than the one using no channel parameters at all. For single antenna systems, the performance gap between the alternative metrics is even bigger.

For comparing soft decision decoding against hard decision decoding, the simulation showed that soft decoding with the ML metric performed 3.5 dB better than hard decision decoding for a rate-1/2, 32 state convolutional code for a single antenna system. When dual antenna diversity was employed, the improvement decreased to approximately 2 dB. Fig. 3 illustrates the improvements in SNR of soft decision decoding over hard decision decoding for SNR ranging from 8 to 20 dB.

Simulation of the error-and-erasure correction decoding for a RS code was also carried out and the results are included in Table 3, where performance of coded systems for different coding schemes and decoding methods are summarized.

Table 3: Code Performance for Different Decoding Methods

Code Type RS = Reed-Solomon CC = Convolutional	**Decoding method**	**Gain w/Single Antenna** (dB)	**Gain w/Dual Antenna** (dB)
RS (n=31, k=15, d=17)	Error Correction	13.1	18.2
	Error & Erasure Correction	14.9	19.5
CC (r=1/2, v=5)	Hard Decision Decoding	13.5	19.3
	Error & Erasure Decoding	15.0	20.0
	Soft Decision Decoding	17.0	21.5

The simulation data showed that a rate-1/2, 32 state convolutional code with soft decision decoding requires approximately 2 dB less signal-to-noise at 10^{-3} BER than a length 31, RS code with error-and-erasure correction decoding on a Rayleigh fading channel, both with and without dual antenna diversity. Thus, a convolutional code with ML soft decision decoding appears to be the most suitable code selection for our particular frequency hopped wireless architecture.

5. Conclusion

We have considered channel coding options for a frequency-hopped system employing NC-BFSK signaling. We have concluded that the hopping patterns should be selected so as to randomize the interference encountered in successive hops, and have observed the advantages offered by the Latin squares construction for synchronous systems. Alternative channel codes were compared on the basis of delay, complexity, and bandwidth efficiency, with the conclusion that when hard decision decoding is employed, the BCH, RS, and convolutional codes that can meet the delay and bandwidth constraints have similar performance and comparable complexity. At the desired bit error rate, the convolutional codes are slightly better since they required approximately one dB less signal-to-noise ratio than the best performing block codes. Furthermore, the performance gain obtained by soft decoding of the convolutional codes with the maximum-likelihood metric was shown to be more significant than the gain obtained by error-and-erasure correction decoding of block codes. This was true even with imperfect channel state estimates. The benefits of dual antenna diversity in combination with coding were demonstrated.

One way to form an estimate of the SNR for a slot is to accumulate the mean squared error between the received signal and hard decision demodulation. The larger the mse, the less reliable the slot. The selection of a soft decision metric depends on the reliability of the channel parameters estimator. That is, the number of levels of quantization to be used in subsequent decoding depends on the application. For example, with a very slowly changing channel, results could be accumulated over several slots, and many bits of soft decision information extracted. At the other extreme, for only a small number of data symbols per slot the best that can be expected is to be able to declare erasures. The same is true of fast hopped systems, where all that could be done is to monitor the received signal strength and declare erasures when it is below some empirically determined threshold. In the future, we hope to investigate algorithms for adapting the metric to changing channel conditions and applications.

References

[1] J. Min, A. Rofougaran, V. Lin, M. Jensen, A. Abidi, G. Pottie, and Y. Rahmat-Samii, "A Low-Power Handheld Frequency-Hopped Spread Spectrum Transceiver Hardware Architecture," Virginia Tech Symposium on Wireless Personal Communications, Blackburg, VA, June 1993.

[2] G. Zaharia, G. El Zein, J. Citerne, "Time delay measurements in the frequency domain for indoor radio propagation," IEEE Antennas and Propagation Society International Symposium. 1992 Digest., pp. 1388-91

[3] D.M.J. Devasirvatham, C. Banerjee, R.R. Murray, D.A. Rappaport, "Two-frequency radiowave propagation measurements in Brooklyn," 1st International Conference on Universal Personal Communications '92 Proceedings, pp. 01:05/1-5.

[4] C. Chien, P. Yang, E. Cohen, R. Jain, and H. Samueli, "A 12.7 Mchips/s all-digital BPSK direct sequence spread-spectrum IF transceiver in 1.2 μm CMOS," To appear in 1994 IEEE International Solid-State Circuits Conference.

[5] G. Chang, A. Rofougaran, M-K. Ku, A.A. Abidi, H. Samueli, "A low-power CMOS digitally synthesized 0-13 MHz sinewave generator," To appear in 1994 IEEE International Solid-State Circuits Conference.

[6] B. Gudmundson, J. Sköld, and J.K. Ugland, "A Comparison of CDMA and TDMA Systems," Proc. 1992 IEEE Vehic. Tech. Conference, Denver, May 11-13, 1992, pp. 732-735.

[7] H. Sasaoka, "Block coded 16-QAM/TDMA cellular radio system using cyclical slow frequency hopping," Proc. 1992 IEEE Vehic. Tech. Conference, pp. 405-408.

[8] N.R. Livneh, R. Meridan, M. Ritz, and G. Silbershatz, "Frequency hopping CDMA for digital radio," Proc. Int'l. Commsphere Symposium, Herzliya, Israel, Dec. 1991.

[9] G.J. Pottie and A.R. Calderbank, "Channel Coding Strategies for Cellular Radio," IEEE International Symposium on Information Theory, San Antonio, TX, Jan. 1993. (submitted to IEEE Transactions on Vehicular Technology).

[10] A.D. Wyner, "Multi-tone multiple-access for cellular systems,", AT&T Bell Laboratories Tech. Memo, BL011217-920811-11TM, Nov. 1991.

[11] J. Roche, "Families of binary matrices having nearly orthogonal rows," AT&T Bell Laboratories Tech. Memo, BL011217-920812-12TM
[12] K.S. Gilhousen, et. al., "On the Capacity of a Cellular CDMA System," IEEE Trans. Vehic. Tech., vol. 40, May 1991, pp. 303-312.
[13] J.G. Proakis, *Digital Communications.* 2nd Ed. New York: McGraw-Hill, 1989.
[14] A.J. Viterbi, J.K. Omura, *Principles of Digital Communication and Coding.* New York: McGraw-Hill, 1979.
[15] S. Lin and D.J. Costello, *Error Control Coding: Fundamentals and Applications.* Englewood Cliffs, NJ: Prentice-Hall, 1983.
[16] A.D. Whalen, *Detection of Signals in Noise.* San Diego: Academic Press, Inc., 1971.
[17] R.E. Blahut, *Theory and Practice of Error Control Codes.* New York: Addison-Wesley, 1983.

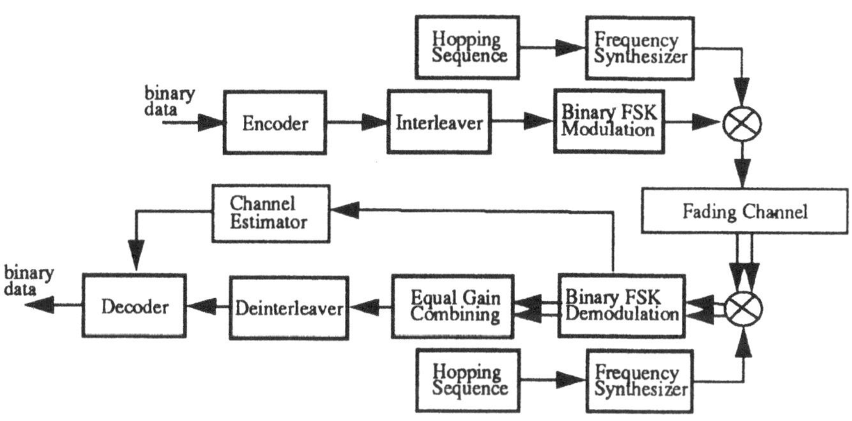

Fig. 1. System Model.

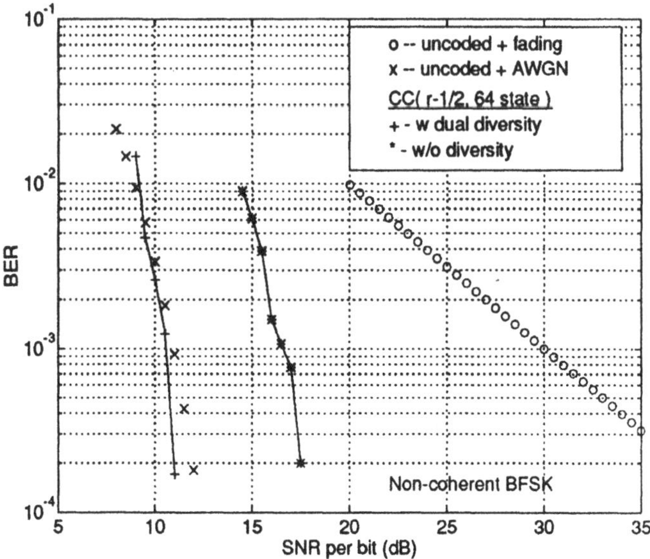

Fig. 2. BER vs. SNR for rate-1/2, 64 state code.

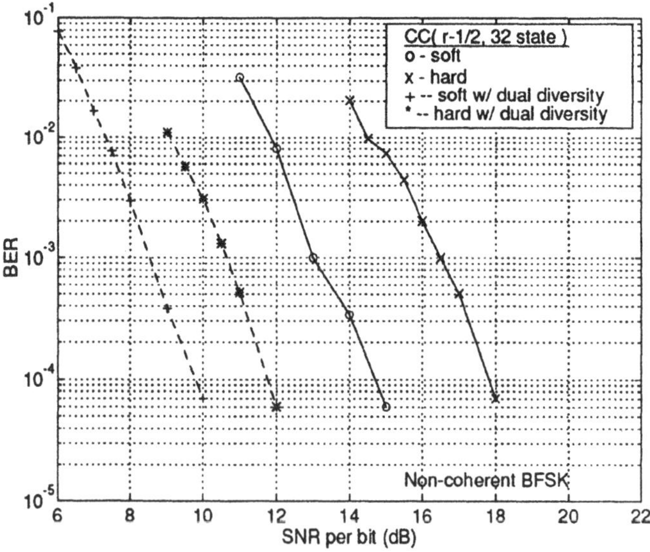

Fig. 3. BER vs. SNR for rate-1/2, 32 state code.

A Unified View of Noncoherent Detection and Differential Detection of Phase Modulated Signals

Daniel Boudreau

Communications Research Centre
Industry and Sciences Canada, Ottawa
Ontario, K2H 8S2
Tel.: (613) 990-6278
Fax: (613) 990-6339
email: dan.boudreau@crc.doc.ca

Abstract. This paper discusses, in a unified fashion, the optimum noncoherent detection and different forms of differential detection for MPSK signals. It links together Multiple-Symbol Differential Detection, Differential Detection with Nonredundant Error Correction as well as other forms of differential detection schemes using decision feedback. The unifying tool used in this article is a differential phase trellis, which shows explicitly the phase differences used in the detectors.

1 Introduction

Noncoherent detection of phase modulated signals, under the form of differential detection, has been the subject of many research efforts in recent years. The simplicity and the robustness of the resulting detector implementation makes differential detection very attractive for channels, such as fading channels, for which coherent detection is not practical. The traditional differential detection method is accomplished by comparing the received phase in a given symbol interval to that in the previous symbol interval and making a multilevel decision on the difference between these two phases [1]. In such a method, the use of the previous symbol as a phase reference introduces a loss in performance. It is of interest to obtain noncoherent methods that improve on the basic differential detector performance by observing and using the received signal through a window larger than two symbol intervals.

Starting with the maximum likelihood (ML) index over a finite window of received samples on an additive white Gaussian noise (AWGN) channel, a *multiple-symbol differential detector* (MSDD) may be obtained. It computes the received phase differences over more than two symbol intervals. This detection scheme fills the gap between ideal coherent and traditional two-symbol differential detection of MPSK signals. A much referenced paper on this subject is the one by Divsalar and Simon [1], although other authors have also derived such a detector [2], [4].

The idea of using a window of more than two samples (or more than one differential detector) for detecting phase modulated signals has been around for more than 20 years. Chow and Ko [5] proposed to use higher order differential detectors to improve the decisions for the one-symbol differential detection of binary DPSK, by using decision feedback. This idea was used again for the differential detection of Minimum Shift Keying (MSK) signals by Masamura et al. [6], and for M-ary DPSK by Samejima et al. [7]. It was referred to as differential detection with *nonredundant error correction* (NEC).

This paper discusses, in a unified fashion, the optimum noncoherent detection and the different forms of differential detection mentioned above for M-ary PSK signals. The optimum noncoherent detection of MPSK signals on an additive white Gaussian noise channel is first presented. It is used to obtain the MSDD scheme of Divsalar and Simon. This method is then interpreted in terms of correlations of the differential detector outputs with some of the possible phase differences obtained on a specific phase trellis.

The phase trellis interpretation is a convenient way to link MSDD and the other methods. It shows how to feed back previous decisions to improve the MSDD detector. Several feedback MSDD schemes can then be derived, among which the NEC methods of [6] and [7] find their places.

Computer simulation is used to compare the performance of these different related methods on an AWGN channel.

2 Optimum Noncoherent Detection on an Additive White Gaussian Noise Channel

Assume a general linear M-ary signaling system in which the baseband symbols are represented as $u_m(t)$, for $m = 1, 2, \cdots, M$ and $0 \leq t \leq T$. The symbols are filtered by the transmit filter and sent on the (complex baseband) AWGN channel. The received baseband signal $r(t)$ is filtered and sampled at the symbol rate, assuming that sufficient statistics are obtained at this point (e.g. the samples out of a filter matched to the transmit filter are used). The sampled received signal, at sampling instant kT, can then be expressed as

$$r(k) = \alpha e^{-j\phi} u_m(k) + z(k), \tag{1}$$

where $\alpha e^{-j\phi}$ is the channel complex gain and the $z(k)$'s are the samples of the complex white Gaussian noise process with variance N_o. Note that the sampling period T has been normalized to one.

For equiprobable transmitted symbols detected with random phase, the maximum likelihood index over a window of N samples, for the mth hypothesis, is given by [8]

$$\Lambda_m = \exp\left[-\frac{\alpha^2}{2N_o}\sum_{i=0}^{N-1}|u_m(k-i)|^2\right] I_0\left(\frac{\alpha}{N_0}\left|\sum_{i=0}^{N-1} r(k-i)u_m^*(k-i)\right|\right) \tag{2}$$

where $I_0(x)$ is the modified Bessel function of order zero.

Since $I_0(x)$ is a monotonically increasing function of its argument, for *constant modulus signals*, the index can be transformed into the correlation

$$\Lambda_m = \left| \sum_{i=0}^{N-1} r(k-i) u_m^*(k-i) \right|^2 \tag{3}$$

This expression is general for phase modulated signals. For equiprobable Continuous Phase Modulated (CPM) [9] signal sequences, the ML index is obtained by using the same form of correlation receiver [10], where $r(k)$ is a sampled version (at the Nyquist rate) of the received signal (without filtering) and $u_m(k)$ is a sampled version of the complex transmitted signal

$$u_m^{\mathrm{CPM}}(k) = e^{j\phi_m(k)} \tag{4}$$

with $\phi_m(k)$ the continuous phase.

3 Multiple-Symbol Differential Detection of MPSK Signals

For MPSK signals, the maximum likelihood index over a window of N symbols on an AWGN channel is of the form

$$\Lambda_m = \left| \sum_{i=0}^{N-1} r(k-i) e^{-j\phi_{k-i}^m} \right|^2 \qquad m = 1, 2, \ldots M^{N-1} \; , \tag{5}$$

where $e^{j\phi_k^m}$ is the mth possible transmitted symbol at sampling instant $t = kT$.

In order to avoid phase ambiguity, the phase information is differentially encoded at the transmitter and is expressed as

$$\phi_{k-i}^m = \phi_{k-N+1}^m + \sum_{n=0}^{N-i-2} \Delta\phi_{k-i-n}^m \tag{6}$$

with

$$\phi_{k-N+1}^m = \phi_0^m + \sum_{j=0}^{k-N} \Delta\phi_{k-j-N+1}^m \; , \tag{7}$$

where $\Delta\phi_k^m$ is the phase difference between instants k and $k-1$.

Equation (5) leads directly to the MSDD method for which the likelihood index is

$$\Lambda_m^{\mathrm{MSDD}} = \mathrm{Re}\left[\sum_{j=0}^{N-2} \sum_{\ell=0}^{N-2-j} r(k-j) r^*(k-\ell-j-1) e^{-j\theta_{(k-j)\ell}^m} \right] \tag{8}$$

$$m = 1, 2, \ldots M^{N-1}$$

where

$$\theta^m_{(k-j)\ell} = \sum_{i=0}^{\ell} \Delta\phi^m_{k-j-i} \ . \tag{9}$$

The variable $\theta^m_{(k-j)\ell}$ is seen to be the phase difference between instants $(k-\ell-j-1)$ and $(k-j)$. For $\ell = 0, 1, 2, \cdots, N-2$, the phase differences given in (9) can be represented by the outputs of a systematic rate $1/(N-1)$ convolutional encoder, as shown in Fig. 1.

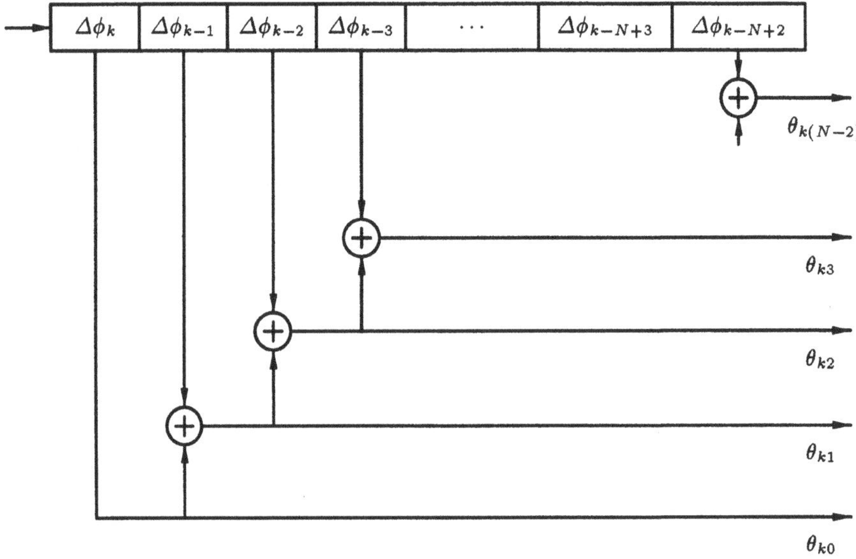

Fig. 1. Convolutional encoder representation of $\theta^m_{(k-j)\ell}$, for $j = 0$

The corresponding trellis diagram has M^{N-2} states. An example of this trellis is given in Fig. 2 for $M = 2$ and $N = 4$. The computation of the argument of the real operator in $\Lambda^{\mathrm{MSDD}}_m$ is then equal to the correlation of the *unquantized* differential detector outputs

$$\mathcal{D}_{(k-j)\ell} = r(k-j)r^*(k-\ell-j-1) \tag{10}$$

with the possible outputs of the trellis over $N-1$ branches, when the correlations with $\theta^m_{(k-j)\ell}$, $(N-1-j) \le \ell \le (N-2)$ are left out. The corresponding phase trellis is shown in Fig. 3 for $M = 2$ and $N = 4$. The detector considers the received samples for instants $k, k-1, k-2, k-3$, computes the phase differences $\mathcal{D}_{k0}, \mathcal{D}_{k1}, \mathcal{D}_{k2}, \mathcal{D}_{(k-1)0}, \mathcal{D}_{(k-1)1}, \mathcal{D}_{(k-2)0}$ and searches the trellis for the most likely path. This gives the decisions $\Delta\hat{\phi}_k$, $\Delta\hat{\phi}_{k-1}$ and $\Delta\hat{\phi}_{k-2}$. Once this is accomplished, a new block of N samples is input and the process repeats.

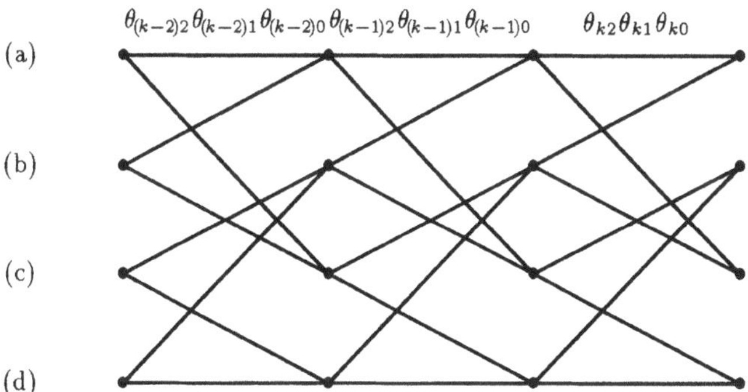

Fig. 2. A general phase trellis for $M = 2$ and a window of $N = 4$ samples.

4 Decision Feedback Multiple-Symbol Differential Detection

Comparing Fig. 2 with Fig. 3, one sees that the MSDD phase correlator of (8) does not use all of the trellis memory for the computation of the detection index. A *full* MSDD implementation would considers instead the correlation of the differential detector outputs with all the possible phase differences over the $N-1$ trellis branches. The phase differences for $(N-1-j) \leq \ell \leq (N-2)$ can be obtained using the decisions from the previous detection instants. One possible decision feedback detector index is therefore given by

$$\Lambda_m^{\mathrm{FBACK}} = \Lambda_m^{\mathrm{MSDD}} + \mathrm{Re}\left[\sum_{j=1}^{N-2}\sum_{\ell=1}^{j} r(k-j)r^*(k-N-\ell+1)e^{-j\hat{\theta}^m_{(k-j)N-2-j+\ell}}\right] \tag{11}$$

$$m = 1, 2, \ldots M^{N-1}$$

where the previous phase difference decisions $\Delta\hat{\phi}_{k-N+1-i}$ are introduced in

$$\hat{\theta}^m_{(k-j)N-2-j+\ell} = \theta^m_{(k-j)N-2-j} + \sum_{i=0}^{\ell-1} \Delta\hat{\phi}_{k-N+1-i}\ . \tag{12}$$

Equations (11) and (12) show how to complete the trellis of Fig. 3 by using decision feedback. The new trellis is illustrated in Fig. 4, where it is assumed that the decision $\hat{\theta}_{(k-3)0} = \Delta\hat{\phi}_{k-3}$, at time $k-3$, is such that state "a" was the most likely one at this point. Then previous decisions $\Delta\hat{\phi}_{k-3}$ and $\Delta\hat{\phi}_{k-4}$ are used to limit the possible phase paths, in the *effective detector window* of $2N-2$ symbols, to those illustrated in Fig. 4. This diminution in the paths number over the effective window allows a reduction in the probability of choosing the wrong path, and therefore a decrease in the overall bit error probability compared to MSDD performed over an N-symbol window.

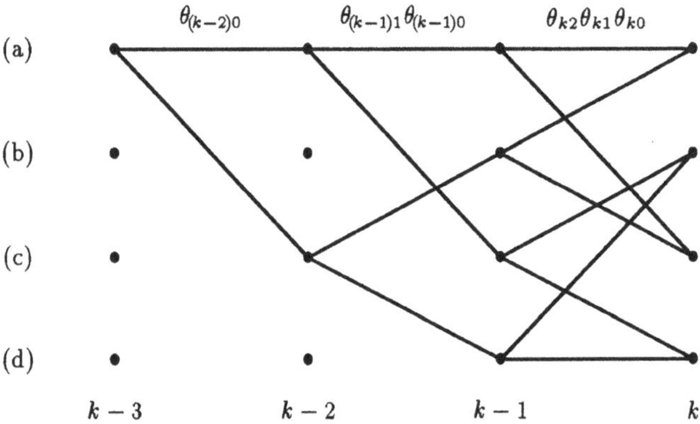

Fig. 3. The MSDD phase trellis for $M = 2$ and $N = 4$.

The index Λ_m^{FBACK} of (11) is used on a block-by-block basis, as in MSDD, by processing the received samples on a window of N symbols, in order to obtain $N-1$ new decisions, and by taking in N new symbols again. It can also be used on a branch-by-branch basis. In this case, the operations based on the phase trellis of Fig. 4 stay the same, except that only one new branch (or new received sample) is considered at a time. The index Λ_m^{FBACK} is computed for $m = 1, 2, \ldots M$ and a decision on one symbol $\theta_{(k-N+2)0}$ is taken.

Another form of decision feedback MSDD was recently proposed by Edbauer [11]. The detector index is given by

$$\Lambda_m^{\text{EDB}} = \text{Re}\left[r(k)r^*(k-1)e^{-j\Delta\phi_k^m}\right] + \text{Re}\left[\sum_{\ell=1}^{N-2} r(k)r^*(k-\ell-1)e^{-j\hat{\theta}_{k\ell}^m}\right] \quad (13)$$

$$m = 1, 2, \ldots, M$$

with

$$\hat{\theta}_{k\ell}^{m} = \Delta\phi_k^m + \sum_{i=1}^{\ell} \Delta\hat{\phi}_{k-i} \ . \tag{14}$$

The particularity of this index is that it relies on only one new received phase transition $\Delta\phi_k^m$ and $N-2$ previous decisions $\hat{\Delta}\phi_{k-1}, \hat{\Delta}\phi_{k-2}, \cdots$ to perform a single new decision. An example of the corresponding phase trellis is illustrated in Fig. 5. The number of possible phase paths over the effective detector window of length N is limited to a minimum by the decision feedback process, which again increases the performance of the system without feedback (the conventional two-symbol differential detector in this case).

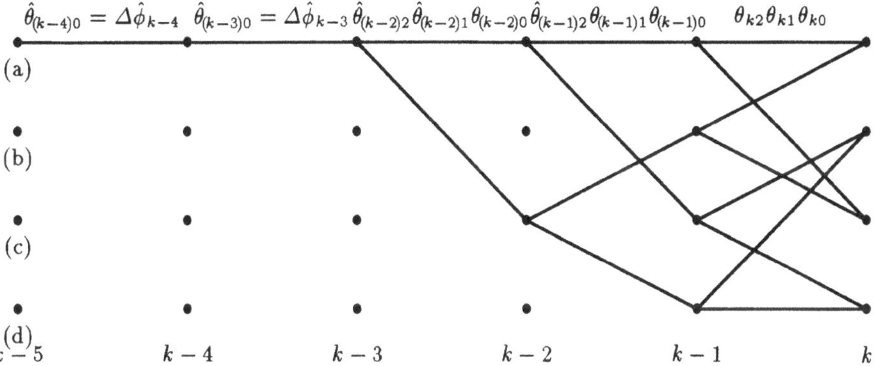

Fig. 4. The new decision feedback phase trellis for $M = 2$ and $N = 4$.

4.1 Hard Decision Feedback MSDD

The detection schemes presented so far make use of the unquantized outputs of the differential detectors to perform the correlation with the different hypotheses. These *soft* outputs, defined in (10), can be quantized (i.e. decisions on the closest possible phase transition are made) and correlated with the hypotheses, which gives a *hard* decision feedback detector. Every detection index of the previous sections can be defined in term of the quantized outputs

$$\mathcal{D}_{(k-j)\ell}^{Q} = [r(k-j)r^*(k-\ell-j-1)]_Q \ , \tag{15}$$

where Q refers to the decision-making process.

In particular, once $\mathcal{D}^{Q}_{(k-j)\ell}$ is used in the index $\Lambda_m^{\mathrm{FBACK}}$ of (11) *on a branch-by-branch basis*, one obtains the so-called *differential detection with nonredundant error correction* for MPSK and CPM signals. For a binary symbol set, this form of decoding of systematic convolutional codes has been called *feedback decoding* and is nothing but a sliding block decoder which implements the optimum decoding procedure for binary symmetric channels [12]. Feedback decoders can be efficiently implemented using the concept of syndromes and look-up tables. This constitutes the proposed implementation of the NEC schemes found in [6] and [7].

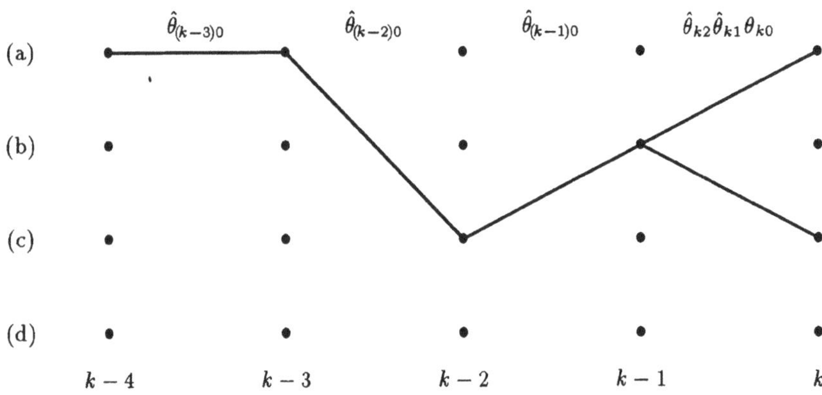

Fig. 5. The Edbauer decision feedback phase trellis for $M = 2$ and $N = 4$.

5 Performance and Computational Complexity

The performance of many of the detection schemes discussed above has already been published and the results will not be repeated here. It is interesting to compare these results to some of the variants introduced in this paper. For example, how does NEC, which uses hard decisions, compare with *soft-decision-*NEC? Such a comparison, using a $\pi/4$-QPSK modulation, is given in Fig. 6 for different detector lengths. It is noted that, at a BER of 10^{-4}, soft-NEC is better than hard-NEC by less than 0.5 dB for any given window length. This small difference between soft-decisions and hard-decisions schemes is common to all the other decision feedback methods, for window lengths of 3 to 5 symbols. Since soft schemes are in general more computationally intensive than hard ones, it appears that the latter should be chosen if the system power budget is not too tight.

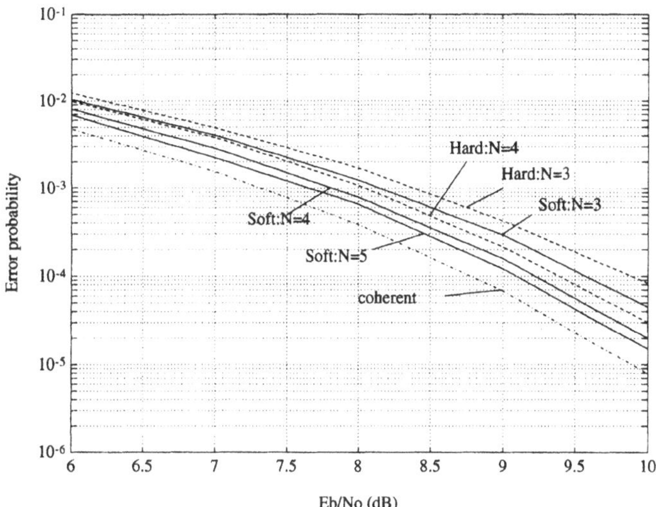

Fig. 6. The performance of soft (continuous curves) and hard (broken curves) decision feedback MSDD using $\Lambda_m^{\mathrm{FBACK}}$ on a branch-by-branch basis. The modulation scheme is $\pi/4$-QPSK.

Another interesting simulation result is the difference between the use of the index $\Lambda_m^{\mathrm{FBACK}}$ block-by-block and branch-by-branch. It is found that this difference is less than 0.2 dB at a BER of 10^{-4}, in favor of the branch-by-branch technique, for N=3,4,5 and soft decisions.

It is also worth comparing the different methods for equal effective window lengths. For $\Lambda_m^{\mathrm{MSDD}}$ with N=4, $\Lambda_m^{\mathrm{FBACK}}$ with N=3 and Λ_m^{EDB} with N=4, all using soft-decisions and an effective window length of 4, the performances at a BER of 10^{-4} are as indicated in Fig. 7. Also indicated are the number of computations per symbol decision required for that window length. This number refers to the sum of complex multiplications of the form

$$\mathcal{D}_{(k-j)\ell}e^{-j\theta^{m}_{(k-j)\ell}} \tag{16}$$

plus the number of complex additions. It is given using the formulas shown in Fig. 8, which are based on the different index expressions.

For an equal effective window length, all the detectors perform almost equally well. However, the feedback decision methods require at most half the number of computations of MSDD, due to their limited number of search trellis paths.

Method, window	Eb/No (dB)	no. operations per decision
MSDD, N=4	9.7	128
EDB, N=4	9.6	52
Branch-FBACK, N=3	9.7	64
Block-FBACK, N=3	9.55	32

Fig. 7. The performance and the computational complexity of the soft-decision methods, for $\pi/4$-QPSK (M=4) and for an effective window length of 4 symbols at a BER of 10^{-4}.

Since no error propagation behavior has been noted during the simulations, it appears that these methods should be favored over straight MSDD. The choice among the feedback decision methods would itself be dictated by implementation considerations.

6 Conclusion

The view of the ML index computation in terms of a differential phase trellis allows one to consider the MSDD methods of [1], [7] and [11] as special cases of decision feedback MSDD. It also unifies these different forms of detections to the theory of optimum noncoherent detection over an AWGN channel. On such a channel, decision feedback improves the performance due to the fact that it extends the effective noncoherent window, while keeping the number of possible trellis paths equal to that of the MSDD implementation. The simulation results show that decision feedback is a way to improve the MSDD performance, while reducing its implementation costs.

MSDD	Edbauer	Block-FBACK	Branch-FBACK
$(N/2)M^{N-1}$	$(4N-3)M$	$(N-1)M^{N-1}$	$(N-1)^2M^{N-1}$

Fig. 8. The computational complexity of the soft-decision schemes, in operations per decision.

References

1. D. Divsalar and M. K. Simon, "Multiple-symbol differential detection of MPSK," *IEEE Trans. on Commun.*, vol. 38, pp. 300–308, March 1990.
2. D. Makrakis, A. Yongaçoglu and K. Feher, "A sequential decoder for differential detection of trellis coded PSK signals," *Proc. Intl. Communications Conference*, Philadelphia, pp. 43.7.1–43.7.6, June 1988.
3. D. Makrakis and P. Mathiopoulos, "Trellis coded noncoherent QAM: a new bandwidth and power efficient scheme," *Proc. IEEE Veh. Technol. Conference*, San Francisco, pp. 95–100, May 1989.
4. H. Leib and S. Pasupathy, "Noncoherent block demodulation of PSK," *Proc. IEEE Veh. Technol. Conference*, Orlando, pp. 407–411, May 1990.
5. P. E. K. Chow and D. H. S. Ho, "Improving DCPSK transmission by means of error control," *IEEE Trans. on Commun. Tech.*, pp. 715–719, October 1971.
6. T. Masamura, S. Samejima, Y. Morihiro and H. Fuketa, "Differential detection of MSK with nonredundant error correction," *IEEE Transactions on Communications*, vol. Com-27, pp. 912–918, June 1979.
7. S. Samejima, K. Enomoto and Y. Watanabe, "Differential PSK system with nonredundant error correction," *IEEE Transactions on Communications*, vol. Sac-1, pp. 74–81, January 1983.
8. J. G. Proakis, *Digital Communications, 2nd Ed.*, McGraw-Hill, 1989.
9. C.-E. Sundberg, "Continuous phase modulation," *IEEE Communications Magazine*, vol. 24, pp. 25–38, April 1986.
10. J. K. Cavers and W. P. Leblanc, *A 2400 bps MSK Modem Based on Digital Signal Processing Techniques*, Faculty of Engineering Science, Simon Fraser University, 1985.
11. F. Edbauer, "Bit error rate of binary and quaternary DPSK signals with multiple differential feedback detection," *IEEE Transactions on Communications*, vol. 40, pp. 457–460, March 1992.
12. A. J. Viterbi and J. K. Omura, *Principles of Digital Communication and Coding*, McGraw-Hill, 1979.

Error Reduction of Coded DPSK over a Fading Channel

Michel Fattouche [1] and *Hatim Zaghloul* [2]

[1] Department of Electrical & Computer Engineering
The University of Calgary, Calgary, Alberta,
T2N 1N4, Phone: (403) 220-4877

[2] Research & Development, Wi-LAN Inc.
Calgary, Alberta, Phone: (403) 273-9133

Abstract: *The paper offers a theoretical justification for using the Hilbert transform to estimate: (1) the phase differential of a CW transmitted over a flat fading channel from its envelope, and (2) the group delay of a frequency-selective fading channel from its amplitude response. In the former case, the theoretical justification is based on expressing the observed CW signal transmitted over a fading channel as a product of z'-domain zeros while in the latter case the theoretical justification is based on expressing the observed frequency response of a selective fading channel as a product of z-domain zeros. The paper also justifies the existence of a sign ambiguity in the resulting estimates based on the fact that the Hilbert transform assumes that the z'-domain zeros representing the CW signal as well as the z-domain zeros representing the selective fading channel are necessarily inside the unit circle which is not always true. The resulting estimate is applied to a Differential Quadrature Phase Shift Keying (DQPSK) signal transmitted over a flat fading channel and to a Differential Multilevel Phase Shift Keying (DMPSK) signal with Orthogonal Frequency Division Multiplexing (OFDM) transmitted over a frequency-selective channel. In both cases, it is shown that the irreducible BER due to the fading channel is reduced substantially.*

1. Introduction

In [1], we show that it is possible to use the Hilbert transform to estimate the absolute value of the phase derivative of a CW signal transmitted over a fading channel from the envelope of the signal. The Hilbert transform is based on Voelcker's phase-envelope relationship [2]. In his analysis, Voelcker first transforms an analytic signal into the complex time domain, then determines the zeros of the signal in this domain and finally establishes whether an analytic signal is Minimum Phase or non-Minimum Phase based on the location of the zeros. In this paper we show that Voelcker's definition of the Minimum Phase (MP) property is different from the conventional MP definition [3-6]. This is due to the fact that the conventional MP definition is based on the location of the zeros in the z-domain while Voelcker's MP definition is based on the location of the zeros in the complex time domain.

Some minimum phase research has been carried out in the context of the location of the zeros in the complex time domain [7-9]. However, the bulk of the minimum phase research has been carried out in the context of the location of zeros in the z-domain [10,11]. For this reason, we replace in this paper, Voelcker's complex time domain by a new domain we refer to as the z'-domain.

The new z'-domain reduces to the time domain when evaluated over the unit circle, i.e. it is the time-domain counterpart of the z-domain. It can thus be used to establish whether an analytic signal is Minimum Phase or non-Minimum Phase in Voelcker's sense based on the location of the zeros in the z'-domain, as done to establish whether a causal system is Minimum Phase (MP) or non-Minimum Phase based on the location of the zeros in the z-domain. In this paper, we distinguish between Voelcker's MP property and the conventional MP property by referring to Voelcker's as the Time-domain MP (TMP) property.

The North American standard for digital cellular telephones (IS-54) [12] calls for a differential phase modulation scheme; namely, $\pi/4$-offset Differential Quadrature Phase Shift Keying (DQPSK). Also the Eureka project [13,14] which is the emerging standard for the European Digital Audio Broadcasting (DAB) project calls for a frequency domain differential phase modulation; namely DQPSK using Orthogonal Frequency Division Multiplexing (OFDM) [15-17]. In both cases, one can show that the effect of the phase differential due to the fading channel is to create an irreducible Bit Error Rate (BER). Traditionally, research aiming at removing such an irreducible BER was confined either to using diversity [18-20] or to sending a pilot tone to monitor the fading channel [21]. Diversity is complex and costly, while sending a pilot tone represents a waste of power and is effective only over narrowband signals. For this reason, we aim in this paper at reducing the irreducible BER by estimating the phase differential due to fading using the amplitude of the received signal. Research aiming at estimating the phase differential using the amplitude of the received signal is not new. Independently of [22] and of this paper, Poletti and Vaughan have estimated in [23] the phase differential of a fading channel from the amplitude of a narrowband signal transmitted over the channel.

In section 2, we reexamine Voelcker's phase-envelope relationship as well as the definition of the TMP property. In section 3, we introduce the z'-domain and define the TMP property in the context of the z'-domain, as well as review the MP definition in the context of the z-domain. In section 4 we use Voelcker's phase-envelope relationship to estimate the absolute value of the phase derivative of a CW signal transmitted over a flat fading channel from the envelope of the signal, and justify its use in the context of the z'-domain in section 5. In section 6, we use the resulting estimate of the absolute value of the phase derivative of a CW signal transmitted over a fading channel to reduce the irreducible BER for Differential Multilevel Phase Shift Keying (DMPSK) transmitted over such a channel. In section 7, we use a frequency-domain version of Voelcker's phase-envelope relationship to estimate the absolute value of the group delay of a selective fading channel from the amplitude response of the channel, and justify its use in the context of the z-domain. In section 8, we use the resulting estimate of the absolute value of the group delay of the fading channel to reduce the irreducible BER for DMPSK-OFDM transmitted over a frequency-selective fading channel. Section 9 concludes the paper with a summary of results.

2. Voelcker's Phase-Envelope Relationship

Voelcker in [2] introduces a time domain signal *m(t)* which is analytic, i.e. its real and imaginary parts form a Hilbert pair. The spectrum *M(f)* of *m(t)* is therefore single-sided in the sense that it is nonzero only for $t > 0$. Voelcker then obtains the inverse Fourier transform *m(s)* of *M(f)* into the complex time-domain s

$$\begin{aligned} m(s) &= \int_0^\infty M(f) e^{j2\pi f s} df \\ &= \int_0^\infty M(f) e^{-2\pi f \sigma} e^{j2\pi f \tau} df \end{aligned} \tag{1}$$

where $s = \tau + j\sigma$ and $j = \sqrt{-1}$. The existence of *m(s)* implies that (1) must converge for any $\sigma \geq 0$ or equivalently, *m(s)* must be free of singularities in the open upper half of the s-plane. Because *m(s)* is analytic, $\frac{d \ln m(s)}{ds}$ is also analytic if *m(s)* is zero-free in the upper half of the s-plane. When *m(s)* is both analytic and zero-free in the upper half of the s-plane we have

$$\frac{d\phi(t)}{dt} = H[\frac{d \ln |m(t)|}{dt}] \tag{2}$$

where $|m(t)|$ and $\phi(t)$ are the envelope and the phase of *m(t)* respectively and $H[\ ]$ denotes a Hilbert transform operation. Relation (2) is Voelcker's phase-envelope relationship. He refers to signals that satisfy (2) as Minimum Phase (MP) signals.

As mentioned in [2], Voelcker's use of the MP term is rather unconventional. The MP property is traditionally attributed to causal systems with no z-domain zeros inside the unit circle, while Voelcker attributes it to analytic signals with no s-domain zeros in the upper half plane. For this reason, we distinguish in this paper between the two definitions by referring to signals that satisfy Voelcker's MP definition as Time-domain MP (TMP) signals while referring to causal systems with no z-domain zeros in the unit circle as MP systems. Moreover, since the MP property has been extensively studied in the literature in the context of the location of z-domain zeros, we replace in the following section Voelcker's s-domain by a new z'-domain which is the time-domain counterpart of the z-domain. This allows us to attribute the TMP property to analytic signals with no z'-domain zeros in the unit circle.

3. Definitions of Systems and Signals

The first four definitions refer to systems in the context of the z-domain while the remaining definitions refer to signals in the context of the z'-domain.

3.1 System Definitions

Definition 1: A *Causal* system defined by its transfer function *G(f)* is one whose impulse response *g(t)* contains no temporal components at negative time t. In other words, a discrete time system defined by its z-domain transfer function *G(z)* is causal iff *G(z)* has all its poles inside the unit circle. Causality implies that the real part of *G(f)* together with its imaginary part form a Hilbert pair [3,4].

Definition 2: Systems with no poles nor zeros outside the unit circle in the z-plane are called *Minimum Phase* (MP) systems. The logarithm of the amplitude response $\ln|G(f)|$ and the phase response $\psi(f)$ of a MP system form a Hilbert pair, i.e.

$$\psi(f) = -H \ln|G(f)| \quad \text{and} \quad \ln|G(f)| = H[\psi(f)]$$

Definition 3: A *Maximum Phase* (MaxP) system is a causal system whose z-plane zeros are all outside the unit circle.

Definition 4: An *anti-Causal* system has all its poles outside the unit circle. An anti-causal system with no zeros outside the unit circle is referred to as an anti-MP system while an anti-causal system with no zeros inside the unit circle is referred to as an anti-MaxP system.

3.2 Signal Definitions

Definition 5: The *z'-transform* of a periodic time signal *m(t)* of period T, is $m(z')$ where

$$m(z') = \sum_{i=-\infty}^{\infty} C_i z'^{-i} \tag{3}$$

and the $\{C_i\}_{i=-\infty}^{\infty}$ are the Fourier coefficients of *m(t)*. By analogy, with the z-transform, evaluating the z' -transform on the unit circle in the z'-plane yields the time-domain signal, i.e. replacing z' by $e^{-j2\pi t/T}$ in $m(z')$ in (3) yields *m(t)*.

Definition 6: Similar to Voelcker's definition, an *analytic* signal is one whose Fourier Transform (FT) contains no spectral components at negative frequencies. In other words, the periodic time signal *m(t)* is analytic iff $m(z') = \sum_{i=0}^{\infty} C_i z'^{-i}$. It is possible to translate the analytic condition $m(z') = \sum_{i=0}^{\infty} C_i z'^{-i}$ to that all the poles in the z'-plane are confined within a circle of radius R_1 where R_1 is the distance between the origin and the furthest pole; i.e. the Region of Convergence (ROC) is the exterior of this circle. Since calculating the z'-transform on the unit circle yields the time domain signal, it is essential that the ROC include the unit circle.

The single-sidedness of the spectrum of analytic signals leads to a Hilbert relationship between the real and imaginary parts of the signal. In other words, if a signal, *m(t)* = $a(t)+jb(t)$, is analytic then its real part $a(t)$ and its imaginary part $b(t)$ form a Hilbert pair, i.e.

$$b(t) = H[a(t)] \quad \text{and} \quad a(t) = -H[b(t)]$$

If $m(t) = |m(t)|e^{j\phi(t)}$, is analytic, then $\ln(m(t)) = \ln|m(t)| + j\phi(t)$ is analytic if *m(t)* has no zeros in the ROC; i.e., if $m(z')$ has no poles nor zeros outside the unit circle in the z'-plane, where $|m(t)|$ is the envelope and $\phi(t)$ is the phase of $m(t)$. With $m(t)$ analytic, $m(z')$ has no poles outside the unit circle and $\ln(m(t))$ is analytic if $m(z')$ has no zeros outside the unit circle.

Definition 7: Signals with no poles nor zeros outside the unit circle in the z'-plane are called *Time-domain Minimum Phase* (TMP) signals. The logarithm of the envelope and the phase of TMP signals form a Hilbert pair, i.e.

$$\phi(t) = H[\ln|m(t)|] \quad \text{and} \quad \ln|m(t)| = -H[\phi(t)] \tag{4}$$

One should note that the phase $\phi(t)$ in (4) is arbitrary to within multiples of 2π; this arbitrariness, however, is minimized by requiring $\phi(t)$ to be a continuous function of time. One should also note that, if a signal is not TMP, then the Hilbert transform of its envelope gives a lower limit on its phase lag. The non uniqueness of the phase is the well known common envelope problem [10,11] where analytic signals can have the same envelope but different phases. All signals of the same envelope but different phases are related to each other by functions of unity amplitude called *allpass* functions. If an analytic signal contains L zeros, there are 2^{L-1} possible all pass functions. If $L = 1$ then there is only one all pass function which converts the TMP signal to a Time-domain Maximum-Phase (TMaxP) signal.

Definition 8: A *TMaxP* signal is an analytic signal whose z'-plane zeros are all outside the unit circle.

For completeness, we provide the following two definitions.

Definition 9: Signals whose FT contains no spectral components at positive frequencies will be called *anti-analytic* signals. In other words, the periodic time signal $m(t)$ is anti-analytic iff $m(z') = \sum_{i=-\infty}^{0} C_i z'^{-i}$.

It is possible to translate the anti-analytic condition $m(z') = \sum_{i=-\infty}^{0} C_i z'^{-i}$ to that all the poles in the z'-plane are confined outside a circle of radius R_1 where R_1 is the distance between the origin and the nearest pole; i.e., the ROC is the interior of this circle.

Definition 10: An *anti-TMP* signal is an anti-analytic signal whose z'-plane zeros are all outside the unit circle. An anti-TMaxP signal is an anti-analytic signal whose z'-plane zeros are all inside the unit circle.

Fig-1 illustrates the relationship between the different domains: the discrete time domain, the discrete frequency domain, the z-domain and the z'-domain. From Fig-1, one can conclude that the method to move between the z-domain and the z'-domain has not been defined yet.

4. Application of Voelcker's Relationship to a Fading CW

In [1] we use Voelcker's relationship to estimate the phase derivative $\frac{d\phi(t)}{dt}$ (with sign ambiguity) of a CW signal $m(t)$ transmitted over a fading channel from its envelope $|m(t)|$. The fading channel in [1] is first chosen as a mobile radio channel, then as an indoor radio channel. We include here only the results corresponding to the mobile radio channel which was characterized by Bultitude [24] measuring the impulse response of the channel as a function of distance. Scott [25] calculated the envelope and phase of a CW signal $m(t)$ transmitted over this channel as a function of distance. A block of the generated envelope $|m_k|_{k=1}^{N}$ is shown in Fig-2a versus the traveled distance where $m_k = m(kT_0)$, T_0 is the sampling interval and $N = 2047$ is the number of observed samples. Two curves are shown in Fig-2b. The first curve displays the phase differential $\{\delta\phi_k\}_{k=1}^{N}$ calculated by Scott [25] while the second displays the phase differential $\{\delta\hat{\phi}_k\}_{k=1}^{N}$ obtained through equation (2) where $\delta\phi_k = \phi_k - \phi_{k-1}$,

$\phi_k = \phi(kT_0)$ and $\delta\hat{\phi}_k = H[\ln|m_k| - \ln|m_{k-1}|]$. In Figs-2a and b, the wavelength λ of the traveling wave is equal to 0.28669m.

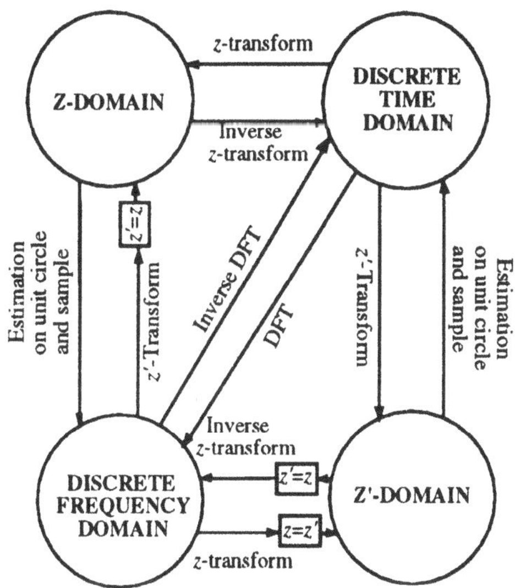

Fig-1 The relationship between the time domain, the frequency domain, the z-domain and the z'-domain.

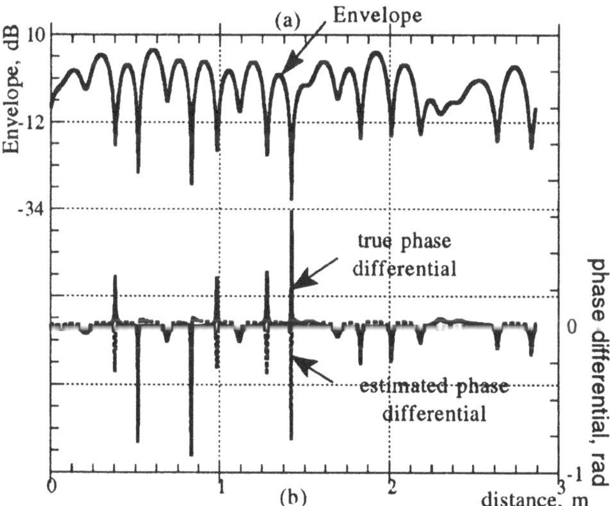

Fig-2 (a) The envelope |m(t)| of m(t). (b) The true phase differential of m(t) compared with the estimated phase differential (2).

A measure of the goodness of fit between a time series: $\{x_k\}_{k=1}^{N}$ and its estimated series $\{\hat{x}_k\}_{k=1}^{N}$ is the Relative Mean Square Error (RMSE) which is defined as

$$\text{RMSE} = \frac{\sum_{k=1}^{N}[x_k - \hat{x}_k]^2}{\sum_{k=1}^{N}[x_k]^2}$$

where N is the length of each series. Note that the RMSE is not necessarily less than one.

In Fig-2b, the RMSE between $|\delta\phi_k|_{k=1}^{N}$ and the unbiased estimate $|\delta\hat{\phi}_k + \mu|_{k=1}^{N}$ is 0.07 indicating a good fit, where $\mu = \frac{1}{N}\sum_{k=1}^{N}\delta\phi_k$. The mean μ of $\delta\phi_k$ is not necessarily zero. For this reason, it is added to $\delta\hat{\phi}_k$ in order to make $|\delta\hat{\phi}_k + \mu|$ an unbiased estimate of $|\delta\phi_k + \mu|$. It can be shown that for a CW signal $m(t)$ with a Rayleigh distributed envelope $|m(t)|$, the function $H[\frac{d\ln|m(t)|}{dt}]$ in (2) has a student-t distribution over $H[\frac{d\ln|m(t)|}{dt}] < 0$. This is consistent with the fact that the phase derivative $\frac{d\phi(t)}{dt}$ corresponding to a Rayleigh distributed $m(t)$ has also a student-t distribution [26,27]. Hence, expression (2) provides an estimate for $\frac{d\phi(t)}{dt}$ which preserves the distribution of the true phase derivative over the region of interest when $|m(t)|$ has a Rayleigh distribution.

5. Theoretical Justification of Voelcker's Relationship

In this section, we justify the use of (2) in the estimation (with sign ambiguity) of the phase derivative $\frac{d\phi(t)}{dt}$ of a CW signal $m(t)$ transmitted over a fading channel from its envelope $|m(t)|$.

Since $m(t)$ is bandlimited [26] to the frequency band $[-f_d, f_d]$ where f_d is the maximum Doppler frequency ($f_d = V/\lambda$, V is the velocity of the mobile and λ is the wavelength of the traveling wave), then by observing $m(t)$ over a finite interval $[0,T]$ where $T = \frac{n}{f_d}$ and n is a positive integer, one can obtain $M(f)$, the spectrum of $m(t)$, sampled with a $1/T$ frequency spacing. This is achieved by forcing $m(t)$ to be periodic with a period T and finding its Fourier coefficients $\{C_i\}_{i=-n}^{n}$ spaced by $1/T$ in the frequency domain. In this case, the sampled spectrum $M(i/T)$ is equal to

$$M(i/T) = \begin{cases} C_i & \text{for } i = -n,\cdots,n \\ 0 & \text{otherwise} \end{cases} \qquad (5)$$

and from (3) we have

$$m(z') = \sum_{i=-n}^{n} C_i z'^{-i}$$

From the fundamental theorem of algebra, the factorization of $m(z')$ into $2n$ zeros is guaranteed, i.e.

$$m(z') = C_{-n} z'^{n} \prod_{i=1}^{2n} (1-\alpha_i z'^{-1}) \qquad (6)$$

where α_i is the ith root of $m(z')$ in the z'-plane, assuming that $C_{-n} \neq 0$. If $C_{-n} = C_{-n+1} = \cdots = C_{-n+p-1} = 0$, then we factorize $m(z') = \sum_{i=-n+p}^{n} C_i z'^{-i}$ into $(2n\text{-}p)$ zeros, i.e. $m(z') = C_{-n+p} z'^{n-p} \prod_{i=1}^{2n-p} (1-\alpha_i z'^{-1})$. We will assume without loss of generality that $p = 0$ for the remainder of the paper.

Since the calculation of $m(z')$ on the unit circle reduces to $m(t)$ then the envelope $|m(t)|$ and the phase $\phi(t)$ of $m(t)$ can be written as

$$|m(t)| = |C_{-n}| \prod_{i=1}^{n} |m_i(t)| \qquad (7a)$$

$$\phi(t) = \prec C_{-n} - \frac{2\pi n t}{T} + \sum_{i=1}^{2n} \prec m_i(t) \qquad (7b)$$

where $m_i(t) = (1-\alpha_i e^{j2\pi t/T})$ and $\prec$ denotes phase. In order to justify using (2) to estimate $\frac{d\phi(t)}{dt}$ we have to compare $H[\frac{d\ln|m(t)|}{dt}]$ with $\frac{d\phi(t)}{dt}$. From (7a&b) we have

$$H[\frac{d\ln|m(t)|}{dt}] = \sum_{i=1}^{2n} H[\frac{d\ln|m_i(t)|}{dt}] \qquad (8a)$$

$$\frac{d\phi(t)}{dt} = \frac{-2\pi n}{T} + \sum_{i=1}^{2n} \frac{d \prec m_i(t)}{dt} \qquad (8b)$$

Let us investigate each $\prec m_i(t)$ in (8b) separately. When $|\alpha_i| < 1$, $m_i(t) = (1-\alpha_i e^{j2\pi t/T})$ is a TMP signal and

$$\frac{d \prec m_i(t)}{dt} = H[\frac{d\ln|m_i(t)|}{dt}] \qquad (9a)$$

When $|\alpha_i| > 1$, $m_i(t) = (1-\alpha_i e^{j2\pi t/T})$ is a TMaxP signal and can be written as $m_i(t) = -\alpha_i e^{j2\pi t/T}(1-\frac{1}{\alpha_i}e^{-j2\pi t/T})$. Hence,

$$\frac{d \prec m_i(t)}{dt} = -H[\frac{d\ln|m_i(t)|}{dt}] + \frac{2\pi}{T} \qquad (9b)$$

From the above, $m_i(t)$ can be either a TMP or a TMaxP depending on $|\alpha_i|$. By ordering $\{m_i(t)\}_{i=1}^{2n}$ such that $\{m_1(t), m_2(t), \cdots, m_{n_1}(t)\}$ are TMaxP and $\{m_{n_1+1}(t), \cdots, m_{2n}(t)\}$ are TMP (where n_1 is the number of TMaxP terms and $2n-n_1$ is the number of TMP terms in (6)), $\frac{d\phi(t)}{dt}$ in (8b) reduces to

$$\frac{d\phi(t)}{dt}=\frac{-2\pi n}{T}-\sum_{i=1}^{n_1}H[\frac{d\ln|m_i(t)|}{dt}] + \sum_{i=n_1+1}^{2n}H[\frac{d\ln|m_i(t)|}{dt}]+\frac{2\pi n_1}{T} \quad (10)$$

Comparing $\frac{d\phi(t)}{dt}$ in (10) with $H[\frac{d\ln|m(t)|}{dt}]$ in (8a), one can see that they are equal except for two discrepancies. The first discrepancy is the extra term $\frac{2\pi(n_1-n)}{T}$ in (10) which renders the mean of $\frac{d\phi(t)}{dt}$ not necessarily zero. The second discrepancy is the negative sign in $-\sum_{i=1}^{n_1}H[\frac{d\ln|m_i(t)|}{dt}]$ in (10) when $m_i(t)$ is TMaxP which causes a sign ambiguity in estimate (2). Both discrepancies are resolved in section 6 when (2) is used to estimate $\frac{d\phi(t)}{dt}$ in order to reduce the irreducible BER for a DQPSK signal transmitted over a flat fading channel.

One should note that the product $\prod_{i=1}^{2n}(1-\alpha_i z'^{-1})$ in (6) is analytic. Hence, each term $m_i(t)=(1-\alpha_i e^{j2\pi t/T})$ in it is either TMP or TMaxP. On the other hand, if $m(z')$ in (6) is written as

$$m(z')=C_n z'^{-n}\prod_{i=1}^{2n}(z'-\alpha_i) \quad (11)$$

then the product $\prod_{i=1}^{2n}(z'-\alpha_i)$ is anti-analytic and subsequently each term $m_i(t)=(e^{-j2\pi t/T}-\alpha_i)$ in it is either anti-TMP or anti-TMaxP depending on whether $|\alpha_i|$ is larger than 1 or smaller than 1. It can be shown that (10) is valid regardless whether $m(z')$ is written as (6) or as (11).

6. Reducing the Effects of Random FM due to Flat Fading

The North American standard for digital cellular telephones (IS-54) calls for a convolutionally-coded differential phase modulation scheme; namely, $\pi/4$-offset DQPSK partially coded with a rate 1/2 code of constraint length 5 (as illustrated in Fig-3). When transmitted over a flat fading channel, IS-54 possesses an irreducible BER. In this section we aim at reducing such an irreducible BER by estimating the phase differential associated with the channel using equation (2), estimating the mean μ of the phase differential, adding it to the estimate to obtain an unbiased estimate, resolving the sign ambiguity associated with the estimate and removing the resulting unbiased estimate from the phase differential of the received DMPSK signal.

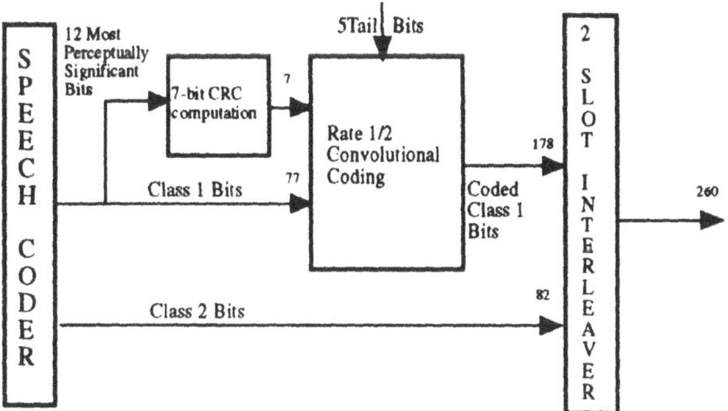

Fig. 3 Forward Error Correction for the Speech Coder. Class 1 bits are convolutionally coded and interleaved with Class 2 bits and transmitted over two time slots.

The complex envelope $\xi_o(t)$ of a DQPSK signal transmitted over a flat fading channel can be written as the complex envelope of the DQPSK signal *u(t)*, multiplied by the complex envelope *m(t)* of a CW signal transmitted over a flat fading channel and contaminated with AWGN. Mathematically, we have

$$\xi_o(t) = u(t)m(t) + n_o(t) \qquad (12)$$

where $u(t) = \sqrt{E_b / T_o} \sum_{i=-\infty}^{\infty} e^{j2\pi\alpha_i/M} p(t - iT_o)$,

$\alpha_i = \beta_i + \alpha_{i-1}$, β_i is the information digit transmitted once every T_o and takes one of the M values in the alphabet $\{0,1,2,3\}$, $p(t)$ is a shaping pulse, E_b is the average transmitted bit energy and $n_o(t)$ is the zero mean white Gaussian noise.

The proposed algorithm consists of five steps. The first step consists of obtaining an estimate $|\hat{m}_k|$ of the sampled envelope $|m_k|$ where $m_k = m(kT_o)$. This is accomplished by first obtaining the sampled envelope $|\xi_o(kT_o)|$ of $\xi_o(kT_o)$, then bandlimiting it to twice the Doppler bandwidth, i.e. to the frequency band $[-2f_d, 2f_d]$. The second step consists of obtaining an estimate for $\delta\phi_k$ for every sample time (or equivalently for every symbol interval) using (2), i.e. obtaining

$$\delta\hat{\phi}_k = H[\ln|\hat{m}_k| - \ln|\hat{m}_{k-1}|] \qquad (13)$$

The third step consists of estimating the mean μ of $\delta\phi_k$. This is achieved by obtaining the estimate $\hat{\mu}$ of μ where $\hat{\mu} = \frac{1}{N}\sum_{k=1}^{N} \prec\xi_{o,k} - \prec\xi_{o,k-1}$ and $\prec\xi_{o,k}$ is the phase of $\xi_o(kT_o)$. The fourth step consists of estimating the sign of $\delta\hat{\phi}_k + \hat{\mu}$ for each fade .This is achieved by computing the two values:

$$S_+ = \sum_{k=1}^{N} \min_{\alpha_k \in \{0,\cdots,3\}} \{(\prec\xi_{o,k} - \prec\xi_{o,k-1}) + \{\delta\hat{\phi}_k + \hat{\mu}\} - \alpha_k\}^2$$

$$S_- = \sum_{k=1}^{N} \min_{\alpha_k \in \{0,\cdots,3\}} \{(\prec\xi_{o,k} - \prec\xi_{o,k-1}) - \{\delta\hat{\phi}_k + \hat{\mu}\} - \alpha_k\}^2$$

If $S_+ < S_-$, $\delta\hat{\phi}_k + \hat{\mu}$ is added to $\prec\xi_{o,k} - \prec\xi_{o,k-1}$, otherwise, it is subtracted from $\prec\xi_{o,k} - \prec\xi_{o,k-1}$. One has to note that the third and fourth steps in the above algorithm attempt to remove the two discrepancies associated with estimate (2), namely the nonzero bias in $\delta\hat{\phi}_k$ and its sign ambiguity.

Fig-4 displays 6 BER curves for a simulation of IS-54 transmitted over a flat fading channel assuming that the mobile station is moving at 100Km/Hr, that the sampling error is 25% of one symbol duration and that hard decision detection is applied at the receiver. In Fig-4, the dashed curves correspond to IS-54 without phase correction, while the solid curves correspond to IS-54 with phase correction using (13).

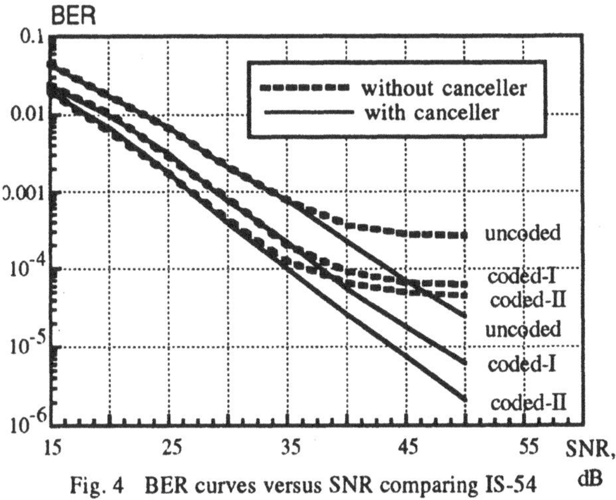

Fig. 4 BER curves versus SNR comparing IS-54 with canceller to IS-54 without canceller.

In Fig-4, coded-I corresponds to FEC coding as shown in Fig-3 without interleaving, while coded II corresponds to the same code with interleaving. Several conclusions can be drawn from Fig-4: (1) partial coding as described in Fig-3 does not prevent the irreducible BER to occur unless phase correction is applied, (2) interleaving adds a 4dB gain for coding to be effective at such high speeds (100Km/Hr), (3) the phase correction removes the irreducible BER regardless whether coding is applied with interleaving or not. The implication of the last conclusion is that interleaving can be avoided in some cases, (at the expense of losing 4dB) thus reducing the amount of delay affecting the transmission of data. When coding is applied on the full frame as opposed to a fraction of the frame, the irreducible BER is reduced so that to allow data transmission at BER less than 10^{-5}.

Real time implementation of the algorithm has been achieved on a floating-point Digital signal processor (AT&T's C30).

7. Application of Voelcker's Relationship to Selective Fading

In this section we derive estimates for the group delay associated with a selective fading channel and in the following section we use the derived estimates to reduce the irreducible BER for a DMPSK-OFDM signal transmitted over such a channel. In both sections we use a similar approach to the one used above for a flat fading channel after replacing the time domain by the frequency domain, the z'-domain by the z-domain and the flat fading channel by a selective fading channel.

When a system is MP, we have a frequency domain version of Voelcker's relationship (2)

$$\frac{d\psi(f)}{df} = -H[\frac{d\ln|G(f)|}{df}] \qquad (14)$$

between the amplitude response $|G(f)|$ and the phase response $\psi(f)$ of the complex frequency response $G(f)$ of the channel. Such a relationship is used in [1] to estimate $\frac{d\psi(f)}{df}$ for an indoor radio channel (with a carrier frequency centered at 1700MHz) [30] from the amplitude response $|G(f)|$.

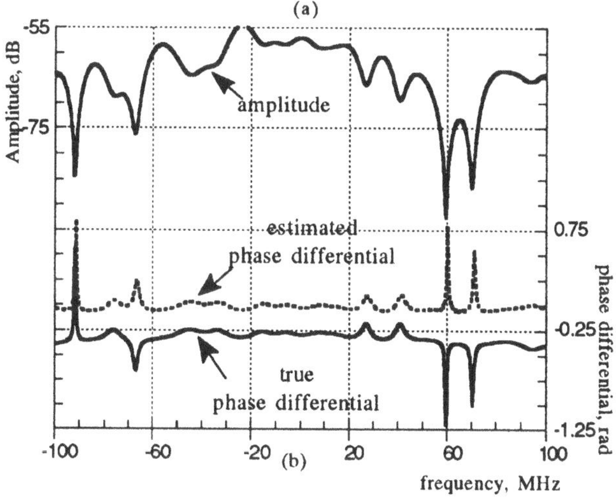

Fig-5 (a) The amplitude |G(f)| of G(f); (b) The true phase differential of G(f) compared with its estimated phase differential.

In Fig-5a, the amplitude response $|G(f)|$ is shown for -100MHz < f< 100MHz. Two curves are displayed in Fig-5b. The first curve displays the true phase differential $\{\delta\psi_k\}_{k=1}^{N}$ of $G(f)$ while the second curve displays its estimate $\delta\hat{\psi}_k = H[\ln|G_k| - \ln|G_{k-1}|]$ obtained using (14), where $\delta\psi_k = \psi_k - \psi_{k-1}$, $\psi_k = \psi(kB_o)$, B_o is the sampling interval and N = 511 is the number of observed samples. In Fig-5b, the RMSE between $|\delta\psi_k|$ and its unbiased estimate $|\delta\hat{\psi}_k + \mu|$ is 0.16, where $\mu = \frac{1}{N}\sum_{k=1}^{N}\delta\psi_k$. The fit is not as good as for the CW case probably due to the fact that the sampling rate is smaller than for the CW case.

In order to justify using (14) to estimate $\frac{d\psi(f)}{df}$ we first sample the complex impulse response $g(t)$ corresponding to the complex transfer function $G(f)$ then we z-transform the result. The sampling is achieved by observing $G(f)$ over a finite band

$[-B, B]$ then, by forcing $G(f)$ to be periodic with a period $2B$ one can find its Fourier coefficients $\{c_i\}$ spaced with a $1/2B$ time spacing. For a causal system, $g(t)$ is nonzero only over the interval $[T_1, T_2]$ i.e.

$$g(i/2B) = \begin{cases} c_i & for\ i = p, \cdots, p+n \\ 0 & otherwise \end{cases} \qquad (15)$$

where $p = 2B/T_1$ and $n = 2B/(T_2 - T_1)$. Taking the z-transform of $g(i/2B)$, we have

$$G(z) = \sum_{i=p}^{p+n} c_i z^{-i} \qquad (16)$$

By factorizing $G(z)$ into n zeros we obtain

$$G(z) = c_p z^{-p} \prod_{i=1}^{n} (1 - \alpha_i z^{-1}) \qquad (17)$$

where α_i is the ith root of $G(z)$ in the z-plane. Since the calculation of $G(z)$ on the unit circle reduces to $G(f)$ then the amplitude response $|G(f)|$ and the phase response $\psi(f)$ are

$$|G(f)| = c_p \prod_{i=1}^{n} G_i(f) \qquad (18a)$$

$$\psi(f) = \prec c_p - \frac{2\pi pf}{2B} + \sum_{i=1}^{n} \prec G_i(f) \qquad (18b)$$

where $G_i(f) = (1 - \alpha_i e^{-j2\pi f/2B})$. In order to justify using (14) to estimate $\frac{d\psi(f)}{df}$ we have to compare $-H[\frac{d \ln |G(f)|}{df}]$ with $\frac{d\psi(f)}{df}$. From (18a&b), we have

$$H[\frac{d \ln |G(f)|}{df}] = \sum_{i=1}^{n} H[\frac{d \ln |G_i(f)|}{df}] \qquad (19a)$$

$$\frac{d\psi(f)}{df} = \sum_{i=1}^{n} \frac{d \prec G_i(f)}{df} - \frac{2\pi p}{2B} \qquad (19b)$$

When $|\alpha_i| < 1$, $G_i(f) = (1 - \alpha_i e^{-j2\pi f/2B})$ is a MP function and

$$\frac{d \prec G_i(f)}{df} = -H[\frac{d \ln |G_i(f)|}{df}] \qquad (20a)$$

When $|\alpha_i| > 1$, $G_i(f) = (1 - \alpha_i e^{-j2\pi f/2B})$ is a MaxP function and can be written as $G_i(f) = -\alpha_i e^{-j2\pi f/2B}(1 - \frac{1}{\alpha_i} e^{j2\pi f/2B})$. Hence,

$$\frac{d \prec G_i(f)}{df} = H[\frac{d \ln |G_i(f)|}{df}] - \frac{2\pi}{2B} \qquad (20b)$$

From the above, $G_i(f)$ can be either MP or MaxP. By ordering $\{G_i(f)\}_{i=1}^{n}$ such that $\{G_1(f), G_2(f), \cdots, G_{n_1}(f)\}$ are MaxP and $\{G_{n_1+1}(f), \cdots, G_n(f)\}$ are MP (where n_1

is the number of MaxP terms and $n-n_1$ is the number of MP terms in (17)) $\frac{d\psi(f)}{df}$ in (19b) reduces to

$$\frac{d\psi(f)}{df} = +\sum_{i=1}^{n_1} H[\frac{d\ln|G_i(f)|}{df}] - \sum_{i=n_1+1}^{n} H[\frac{d\ln|G_i(f)|}{df}] - \frac{2\pi(n_1+p)}{2B} \tag{21}$$

Comparing $\frac{d\psi(f)}{df}$ in (21) with $H[\frac{d\ln|G(f)|}{df}]$ in (19), one can see that they are equal except for two discrepancies. The first discrepancy is the extra term $-\frac{2\pi(n_1+p)}{2B}$ in (21). The second discrepancy is the positive sign in $+\sum_{i=1}^{n_1} H[\frac{d\ln|G_i(f)|}{df}]$ in (21) when $G_i(f)$ is MaxP. Both discrepancies are removed in the following section.

8. Reducing the Effects of the Group Delay due to Selective Fading

The European emerging standard for Digital Audio Broadcasting (DAB) calls for a frequency domain differential phase modulation scheme; namely, DQPSK-OFDM. When transmitted over a selective fading channel, DQPSK-OFDM possesses an irreducible BER. In this section we aim at reducing the irreducible BER for DMPSK-OFDM transmitted over a selective fading channel by estimating the group delay of the channel using (14).

The complex envelope $\xi_o(f)$ of a DMPSK-OFDM signal transmitted over a selective fading channel over the frequency band $[-B,B]$ can be written as the complex envelope of the DMPSK-OFDM signal $U(f)$, multiplied by the complex frequency response $G(f)$ of the selective fading channel, contaminated with AWGN. Mathematically, we have

$$\xi_o = U(f)G(f) + N_o(f) \tag{22}$$

where $U(f) = \sqrt{E_b / N_o} \sum_{i=-\infty}^{\infty} e^{j2\pi\alpha_i/M} P(f - iB_o)$

$\alpha_i = \beta_i + \alpha_{i-1}$, β_i takes one of the M values in the alphabet $\{0,1,\cdots,M-1\}$, $P(f)$ is a shaping pulse, E_b is the average transmitted bit energy and $n_o(t)$ is the zero mean white Gaussian noise. β_i is the information digit between adjacent frequencies separated by B_o where $NB_o = 2B$. The proposed algorithm consists of five steps. The first step consists of obtaining an estimate $|\hat{G}_k|$ of the sampled amplitude $|G_k|$ where $|G_k|=|G(kB_o)|$ This is accomplished by first obtaining the sampled amplitude $|\xi_{o,k}|$ of $\xi_{o,k}$ where $\xi_{o,k} = \xi(kB_o)$, then timelimiting it to the interval $[(p-n/2)/2B,(p+3n/2)/2B]$. The second step consists of obtaining an estimate for $\delta\psi_k$ using (14) i.e. obtaining

$$\delta\hat{\psi}_k = H[\ln|\hat{G}_k| - \ln|\hat{G}_{k-1}|] \qquad (23)$$

The third step consists of estimating the mean μ of $\delta\psi_k$ as $\hat{\mu} = \frac{1}{N}\sum_{k=1}^{N} \prec\xi_{o,k} - \prec\xi_{o,k-1}$ and $\prec\xi_{o,k}$ is the phase of $\xi_{o,k}$ The fourth step consists of estimating the sign of $\hat{\psi}_k + \hat{\mu}$ for each fade. This is achieved by computing the two values

$$S_+ = \sum_{k=k_1}^{k_2} \min_{\alpha_k \in \{0,\cdots,M-1\}} \{(\prec\xi_{o,k} - \prec\xi_{o,k-1}) + \{\delta\hat{\psi}_k + \hat{\mu}\} - \alpha_k\}^2$$

$$S_- = \sum_{k=k_1}^{k_2} \min_{\alpha_k \in \{0,\cdots,M-1\}} \{(\prec\xi_{o,k} - \prec\xi_{o,k-1}) - \{\delta\hat{\psi}_k + \hat{\mu}\} - \alpha_k\}^2$$

If $S_+ < S_-$, $\delta\hat{\psi}_k + \hat{\mu}$ is added to $(\prec\xi_{o,k} - \prec\xi_{o,k-1})$, otherwise, it is subtracted from $(\prec\xi_{o,k} - \prec\xi_{o,k-1})$.

Assuming perfect match-filtering and perfect timing recovery at the DMPSK receiver, Fig-6 displays 6 BER curves for a simulated OFDM signal transmitted over the frequency selective fading channel illustrated in Figs-5, with M = 4, 8 and 16, for each of the following three cases. The first case (Case A) corresponds to no correction for the group delay due to the channel. The second case (Case B) corresponds to correction for the group delay due to the channel using (23). The third case (Case C) corresponds to perfect adjustment for the group delay due to the channel, i.e. by setting $\delta\psi_k$ equal to zero. From Fig-6 one can see that by adjusting for the delay using (23) one can approach the limiting case (Case C) where the delay distortion is entirely removed.

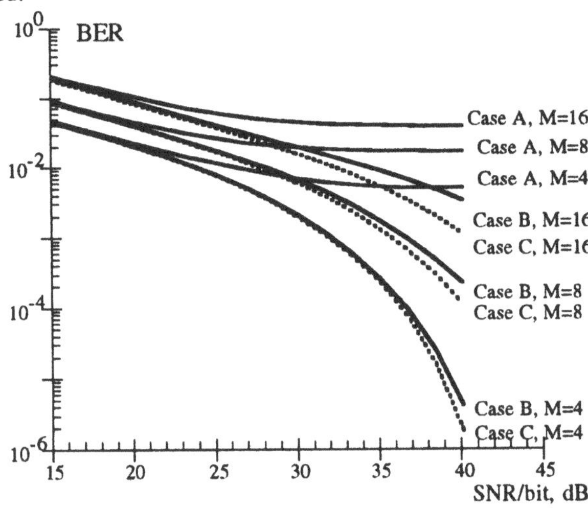

Fig-6 BER curves for M=4, 8 and 16.

9. Conclusions

This paper offered a theoretical justification for using the Hilbert transform to estimate: (1) the phase differential (also known as random FM or Doppler shift) of a CW signal transmitted over a flat fading channel from its envelope and (2) the group delay of a selective fading channel from its amplitude response. In the first case, the paper shows that at the proximity of a fade the signal can be considered either TMP or TMaxP, while in the second case, in the proximity of a frequency-domain null, the system can be considered either MP or MaxP. The first case is justified in the context of the z'-domain while the second case is justified in the context of the z-domain. This is based on the concept that a bandlimited signal observed over a finite duration $[0,T]$ can be regarded as a product of a finite number of zeros in the z'-domain. Similarly, a system whose impulse response is of finite duration and whose frequency response is observed over a finite band $[-B,B]$ can be regarded as a product of a finite number of zeros in the z-domain.

Also, the paper used the Hilbert-based estimate to successfully reduce the irreducible BER caused by the phase differential due to fading.

Acknowledgment:

This work was supported by a grant from the Natural Sciences and Engineering Research Council of Canada and AGT Limited.

REFERENCES

[1] M. Fattouche and H. Zaghloul, "Estimating the Phase Differential of Signals Transmitted over Fading Channels," *Electron. Lett.*, Vol. 27, no. 18, pp. 1823-1824, June 1991.

[2] H. Voelcker, "Toward a unified theory of modulation-part 1: phase-envelope relationships," *Proceedings of the IEEE*, vol. 54, pp. 340-353, March 1966.

[3] A. Papoulis, "The Fourier integral and its applications," McGraw-Hill, 1962, New York, pp. 205-217.

[4] H. W. Bode, "Network Analysis and Feedback Amplifier Design," Robert E. Krieger Publishing Co., 1975, New York, pp. 303-345.

[5] G. Longo and B. Pincinbono, "Time and Frequency Representation of Signals and Systems," Springer-Verlag, New-York, 1989, pp. 1-9.

[6] J. Dugundji, "Envelopes and Pre-Envelopes of Real Waveforms," *IRE Trans. on Information Theory*, Vol. IT-4, pp. 53-57, March 1958.

[7] A.A. Requicha, "The Zeros of Entire Functions: Theory and Engineering Applications," *Proc. IEEE*, Vol. 68, no. 3, pp. 308-328, March 1980.

[8] Y.O. Al-Jalili, "Analysis and Detection Algorithms for Complex Time Zeros of Bandlimited Signals," *IEE Proceedings-I*, Vol. 138, no. 3, pp. 189-200, June 1991.

[9] G. Lockhart, T. Bayes, A.A. Mehdi and Y.O. Al-Jalili, "Embedding Data in DSM-AM using Zero-Synchronous Modulation," *Electron. Lett.*, Vol. 23, no. 14, pp. 745-746, 2nd July 1987.

[10] A.V. Oppenheim and R.W. Shafer, "Discrete-Time Signal Processing," Prentice-Hall, Englewood Cliffs, 1989.

[11] J.G. Proakis and D.G. Manolakis, "Introduction to Digital Signal Processing," McMillan Publishing Company, New York, 1991.

[12] Cellular System, "Dual-Mode mobile station-base station compatibility standard," EIA/TIA, Project Number 2215, Electronic Industries Association, Jan. 1990.

[13] P. Hoeher, J. Hagenauer, E. Offer, Ch. Rapp and H. Schulze, "Performance of an RCPC-Coded OFDM-Based Digital Audio Broadcasting (DAB) System," Globecom'91, pp. 2.1.1-2.1.7, Phoenix, Arizona, Dec. 2-5 1991.

[14] J.F. Helard and B. LeFloch, "Trellis Coded Orthogonal Frequency Division Multiplexing for Digital Video Transmission," Globecom'91, pp. 23.5.1-23.5.7, Phoenix, Arizona, Dec. 2-5 1991.

[15] B. Hirosaki, S. Hasegawa and A. Sabato, "Advanced Groupband data modem using orthogonally multiplexed QAM technique," *IEEE Trans. Commun.*, vol. COM-34, pp. 587-592, June 1986.

[16] L.J. Cimini, Jr., "Analysis and simulation of a digital mobile channel using orthogonal frequency division multiplexing," *IEEE Trans. Commun.*, vol. COM-33, pp. 665-675, July 1985.

[17] E.F. Casas and C. Leung, "OFDM for Data Communications over Mobile Radio FM Channels-Part I: Analysis and Experimental Results," *IEEE Trans. Commun.*, vol. COM-39, pp. 794-807, May 1991.

[18] F. Adachi, "Postdetection Selection Diversity Effects on Digital FM Land Mobile Radio," *IEEE Trans. Veh. Technol.*, Vol. VT-31, pp. 166-172, Nov. 1982.

[19] T. Miki and M. Hata, "Performance of 16Kbit/s GMSK Transmission with Postdetection Selection Diversity in Land Mobile Radio," *IEEE Trans. Veh. Technol.*, Vol. VT-33, pp. 128-133, Aug. 1984.

[20] F. Adachi and J.D. Parsons, "Random FM Noise with Selection Combining," *IEEE Trans. Comm.*, Vol. COM-36, pp. 752-755, June 1988.

[21] M. Yokoyama, "BPSK System with Sounder to Combat Rayleigh Fading in Mobile RadioCommunication," *IEEE Trans. Veh. Technol.*, Vol. VT-35, pp. 35-40, Feb. 1985.

[22] M. Fattouche and H. Zaghloul, "Equalization of IS-54 Transmitted over a Flat Fading Channels," to appear in the proceedings of ICC'92, Chicago, Ill., June 14-18 1992.

[23] M.A. Poletti and R.G. Vaughan, "Reduction of Multipath Fading Effects in Single Variable Modulations," ISSPA'90, pp. 672-676, Gold Coast, Australia, 27-31 August 1990.

[24] R.J.C. Bultitude and G.K. Bedal, "Propagation characteristics on microcellular urban mobile radio channels at 910 MHz," *IEEE Journal on Selec. Areas on Comm.*, Vol. SAC-7, No.1, pp. 31-40, Jan. 1989.

[25] K.E. Scott, "Antenna Diversity and Multichannel Adaptive Equalization in Digital Radio," Ph.D. Dissertation, Department of Electrical and Computer Engineering, University of Calgary, 1991.

[26] W.C. Jakes, "Microwave Mobile Communications," New York: Wiley, 1974.

[27] J. Andersen, S.L. Lauritzen and C. Thommesen, "Statistics of phase derivatives in mobile communications," 36th IEEE Vehicular Tech. Conf., Dallas, Texas, May 1986, pp. 228-231.

[28] W.D. Rummler, "A New Selective Fading Model: Application to Propagation Data," *B.S.T.J.*, Vol. 58(5), pp. 1037-1071, March-June 1979.

[29] Y.O. Al-Jalili, A.A. Al-Obaidi and M.K. Habib, "Characteristics of Reduced Bandwidth Envelope Detected Minimum-Phase Signals," *IEE Proceedings-I*, Vol. 137, no. 6, pp. 345-354, Dec. 1990.

[30] H. Zhaghloul, G. Morrison and M. Fattouche, "Frequency response and path loss measurements of the indoor channel," Electron. Lett., vol. 27, pp. 1021-1022, 6th June 1991,

[31] R.F. Pawula, S.O. Rice and J.H. Roberts, "Distribution of the phase angle between two vectors perturbed by Gaussian noise," *IEEE Trans. Commun.*, vol. COM-30, pp. 1828-1841, Aug. 1982.

Performance of BCH Codes with DES Encryption in a Digital Mobile Channel

Guy Ferland and Jean-Yves Chouinard

Department of Electrical Engineering, University of Ottawa
161 Louis-Pasteur, Ottawa, Ontario, Canada, K1N 6N5

Abstract. In a digital mobile communication channel, the use of an encryption algorithm is essential to prevent unwanted eavesdropping of private conversations and to protect sensitive data transfers against tampering. This paper presents a performance analysis of different modes of operation of the Data Encryption Standard (DES) encryption in a digital mobile radio channel environment. The objective is to determine the degradation effects of data enciphering on the channel reliability and to analyze the error correction capability of Bose-Chaudhuri-Hocquenghem (BCH) codes for the enciphered data when channel errors occur. The digital mobile communication channel is simulated using Fritchman's channel model, which can represent a large variety of error burst channels. Computer simulations are used to evaluate the bit error performance of BCH codes in the error burst channels, when the information data is first encrypted with DES.

1 Introduction

To protect sensitive data transfers against passive attacks, such as eavesdropping, or active attacks (e.g., information tampering), it is necessary to encrypt the data before transmission over the communication channel. The DES can be considered as the algorithm of choice for applications requiring low to medium security because of the high throughput rates achievable and because of the availability of inexpensive hardware to implement it. In a digital channel, the Data Encryption Standard algorithm can be operated in block cipher modes as well as in stream cipher modes; each mode of operation of DES provides a different level of cryptographic protection.

However, enciphering of data often results in a significant increase in the number of bit decoding errors at the receiver, due to error propagation which is inherent in most encryption modes. This additional cryptographic degradation of the channel reliability is usually unacceptable in a mobile communication channel where the signal already suffers from a relatively high channel bit error rate (BER) because of fading effects caused by multipath propagation. To reduce the BER at the receiver to acceptable levels, it is often necessary to utilize error control coding techniques such as Forward Error Correction (FEC) to enable the receiver to detect and correct some of these errors.

The objective here is to determine the impact of DES encipherment on the data bit error rate at the receiver and to analyze the ability of BCH block codes

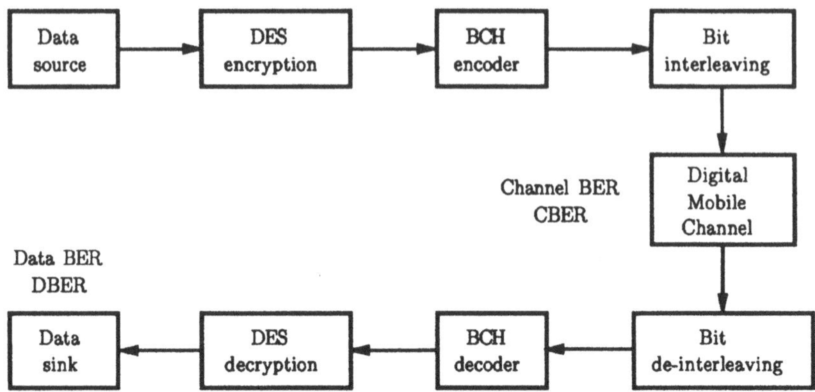

Fig. 1. Block diagram of DES secure communication link

to correct errors in the received vectors. The influence of bit interleaving on the bit error rate is also examined.

Figure 1 illustrates the block diagram of a communication link in which BCH channel coding and DES encryption/decryption are used, and shows the order in which the various data processing steps are carried out. To avoid confusion concerning the BER, the following definitions are used hereafter: $DBER$ is defined as the (information) data bit error rate at the receiver after FEC decoding and decryption. $CBER$ represents the channel bit error rate and refers to the number of bits that are corrupted by the channel.

2 Digital Mobile Channel Model

The mobile communication channel is characterized by slow and fast fading of the signal power envelope at the receiver [13]. The slow fading is generally caused by the topology of the terrain and distant scatterers. Fast fading is the result of multipath propagation due to scatterers near the mobile unit. As the mobile unit moves, the signal strength fluctuates rapidly and fading occurs. When the receiver is in a fade trough, the drop in signal power can be in the order of 20 to 30 dB [12]. This periodic variation in the signal strength results in time intervals during which the channel bit error rate ($CBER$) is low, followed by usually shorter intervals when the $CBER$ can be higher by many orders of magnitude. Thus, errors occur in bursts and the channel is said to have memory.

Memoryless digital channels, such as the binary symmetric channel model (BSC) which generates error sequences with randomly distributed errors are not representative of a real digital mobile channel. However, the BSC channel will be used in simulations to compare the performance of error control codes with channels having memory.

To represent the behavior of a digital channel with memory, many models have been proposed over the years [8, 10, 11]. The Gilbert model [10] consists of a two-state Markov chain. When the chain is in the "good" (G) state, no errors occur. However, when the channel goes into the "bad" (B) state, an error may occur with probability ϵ, where $0 < \epsilon < 0.5$. The transition probabilities between states, $P_{G,B}$ and $P_{B,G}$, control the persistence of the states; by modifying them, the statistical distribution of error-free intervals and error bursts can be modified. The crossover probability ϵ controls the density of error bursts when the Markov chain is in the state B.

The channel model proposed by Fritchman [8] is a general case of the Gilbert model [10] but is more flexible and can represent a wider variety of bursty channel conditions. The model consists of a Markov chain with N states $\{S_1, \ldots, S_N\}$ which is partitioned into two groups of states. The states S_1 to S_k represent error-free states while the remaining states, S_{k+1} to S_N, are error states. In its general formulation, transitions are allowed between all states. This general Fritchman model is a non-renewal model (which takes into account the interdependence between the different error bursts), but is mathematically complicated to analyze and the determination of its parameters [11], such as transition probabilities between the states, is difficult.

Fritchman has proposed a simplified version of the channel model. For this simplified model, the Markov chain consists of N states, of which the first $(N-1)$ are error-free states and the Nth one, i.e. S_N, is the only error state. Furthermore, transitions are only allowed to the same state (e.g., from S_i to S_i, for $1 \leq i \leq N$), from the good states to the bad one, and vice versa (i.e., from S_i to S_N and S_N to S_i). The simplified Fritchman channel model becomes a renewal model [11] but can still accurately describe the error distributions observed in most digital channels with memory. The error distribution is determined by the channel transition probability's matrix P:

$$P = \begin{pmatrix} P_{1,1} & 0 & \ldots & 0 & \ldots & 0 & P_{1,N} \\ 0 & P_{2,2} & \ldots & 0 & \ldots & 0 & P_{2,N} \\ \vdots & \vdots & \ddots & \vdots & \ddots & \vdots & \vdots \\ 0 & 0 & \ldots & P_{i,j} & \ldots & 0 & P_{i,N} \\ \vdots & \vdots & \ddots & \vdots & \ddots & \vdots & \vdots \\ 0 & 0 & \ldots & 0 & \ldots & P_{N-1,N-1} & P_{N-1,N} \\ P_{N,1} & P_{N,2} & \ldots & P_{N,j} & \ldots & P_{N,N-1} & P_{N,N} \end{pmatrix} \tag{1}$$

where $P_{i,j}$ indicates the transition probability from the state S_i to the state S_j.

For the Fritchman's channel model, the error-free interval length distribution $P(0^m|1)$ and the error burst length distribution $P(1^m|0)$ can be expressed as sums of exponential functions:

$$P(0^m|1) = \sum_{i=1}^{N-1} \alpha_i \beta_i^m \quad \text{and} \quad P(1^m|0) = \alpha_N \beta_N^m \quad \text{for } m \geq 1 \tag{2}$$

where the model's parameters $\{\alpha_i\}$ and $\{\beta_i\}$ are related to the transition probabilities by:

$$\alpha_i = \frac{P_{N,i}}{P_{i,i}} \quad \text{and} \quad \beta_i = P_{i,i} \tag{3}$$

The error-free interval length distribution $P(0^m|1)$ gives the probability of having m or more consecutive error-free bits, provided that an error had occurred. As for the Gilbert model, the transition probabilities control the persistence of the states, and therefore, determine the distribution of errors.

3 Data Encryption Standard (DES)

3.1 Data Encryption Algorithm

The Data Encryption Standard constitutes an interesting encryption approach for applications requiring low to medium levels of security, such as secure speech communications, because of the high throughput rates achievable (up to 100 $Mbits/s$) and the availability of economical hardware logic to implement it.

DES is a substitution-permutation encryption algorithm [1, 5] which encrypts blocks of 64 plaintext bits with a 56-bit key. The data is cycled 16 times through a set of substitution and permutation transformations to yield a highly nonlinear and complicated relationship between the input plaintext data and the 64-bit output ciphertext (see Figure 2), thereby ensuring that computer-aided attempts to break the code will be prohibitively long [1]. The cryptographic properties of secure cryptosystems have been stated by Shannon in 1949 [17] as being *i)* confusion, which ensures that the relationship between the ciphertext and the key is highly complex, and; *ii)* diffusion, which dissipates the changes in the plaintext or the key all over the resulting ciphertext.

3.2 Modes of Operation of DES and Effects of Channel Errors

In a digital communication channel, the DES algorithm can be implemented in either block cipher modes, where the plaintext data bits are encrypted as 64-bit data blocks; or in stream cipher modes for which the data bits are enciphered individually. The block cipher mode [5] of operation is further subdivided into the Electronic Codebook (ECB) and the Cipher Block Chaining (CBC) modes.

For the Electronic Codebook case, a message M is broken into blocks of 64 data bits (see Figure 3). Each block is encrypted as a unit and transmitted over the channel. The receiver simply decrypts the received cipher block with the same key as is used at the transmitter and recovers the original data. Note that because of the intricate relationship of all data bits within the encrypted block, a single bit in error within a block will invariably corrupt the entire 64 bit block (i.e., about half of the 64 message bits will be deciphered incorrectly). Thus, bit errors propagate within the block in which they occur. However, since each group of 64 bits is encrypted independently from previous or subsequent ones, errors do not propagate outside the block of 64 bits in which they initially

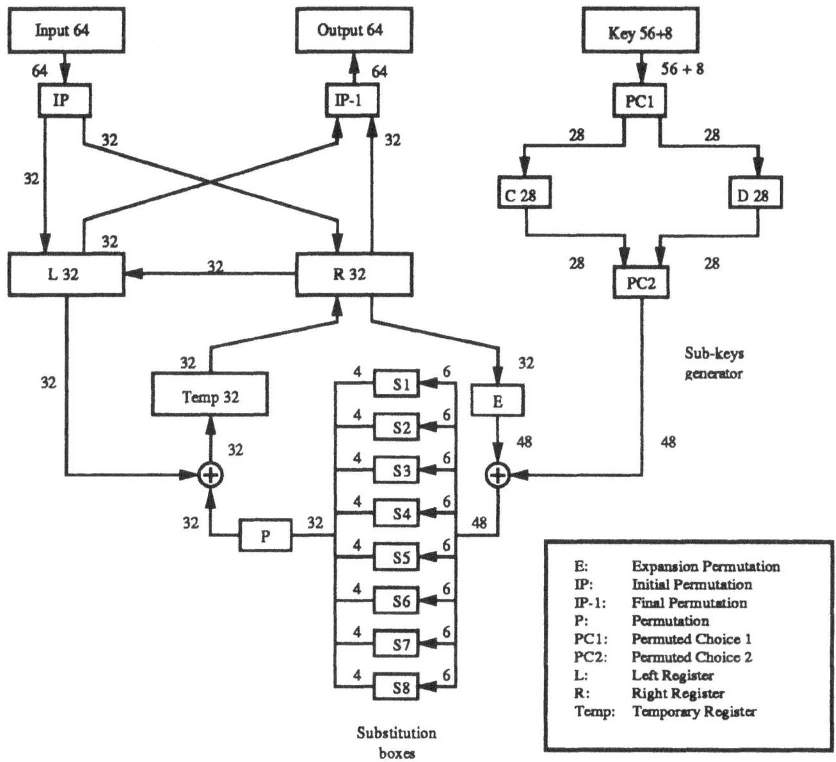

Fig. 2. DES encryption/decryption algorithm

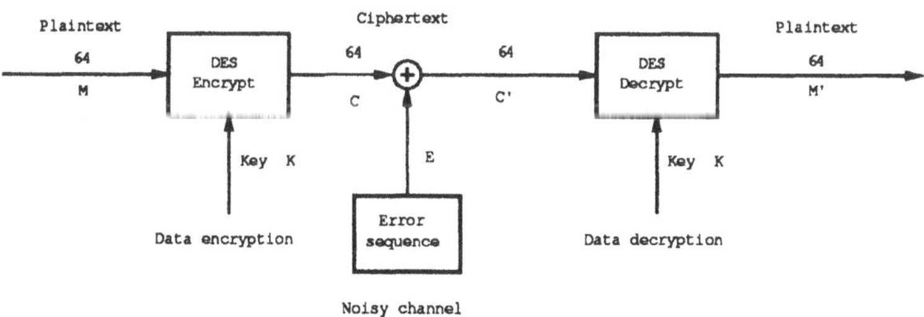

Fig. 3. Electronic Codebook (ECB) mode of operation of DES

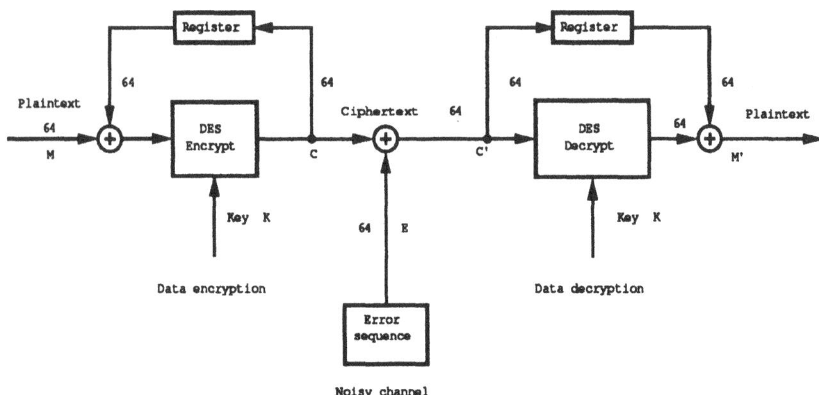

Fig. 4. Cipher Block Chaining (CBC) mode of operation of DES

occur. Another important property of the ECB mode is that an external system is required to maintain the receiver and transmitter in block synchronization.

In CBC mode, the encryption/decryption of a block of data is made dependent on the previously transmitted block, as shown in Figure 4. The plaintext block M_i to be encrypted is added (modulo 2) to the cipher block from the previous round of encryption C_{i-1}. For the first data block, a predetermined initial vector (IV) is added to the data block to produce C_i. At the receiver, C_i' is acquired, where $C_i' = C_i \oplus E_i$ (the element e_i of the 64-bit error vector E_i is 1 when the channel introduces a bit error, or 0 otherwise). C_i' is decrypted by the DES algorithm and then added to the cipher block received in the previous step (the same initial vector IV used for the transmission). The encryption transformation is given by:

$$C_i = DES_K(M_i \oplus C_{i-1}) \tag{4}$$

The decipherment is done by applying the decryption transformation $DES_K^{-1}(\bullet)$ to the present ciphertext block C_i', and then, by adding (modulo-2) the previous decrypted cipher block C_{i-1}':

$$DES_K^{-1}(C_i') \oplus C_{i-1}' = DES_K^{-1}(C_i \oplus E_i) \oplus (C_{i-1} \oplus E_{i-1}) \tag{5}$$

if $C_i' = C_i$ and $C_{i-1}' = C_{i-1}$ (i.e, no transmission errors), then

$$\begin{aligned} DES_K^{-1}(C_i') \oplus C_{i-1}' &= DES_K^{-1}(C_i) \oplus (C_{i-1}) \\ &= DES_K^{-1}[DES_K(M_i \oplus C_{i-1})] \oplus C_{i-1} \\ &= M_i \oplus C_{i-1} \oplus C_{i-1} \\ DES_K^{-1}(C_i') \oplus C_{i-1}' &= M_i \end{aligned} \tag{6}$$

A single bit received in error in block C_i' corrupts 65 bits since the decoded block $DES_K^{-1}(C_i') \oplus C_{i-1}$ will not be decoded correctly and the subsequent decoding of $DES_K^{-1}(C_{i+1}) \oplus C_i'$ will have one bit in error. In general, j bits received in error within a block of 64 bits will corrupt $(64 + j)$ bits. Thus, in CBC mode, errors propagate within the block in which they occur and in the next 64-bit block.

Furthermore, for the CBC mode, as for the ECB mode, block synchronization is required between the receiver and transmitter. An important advantage of CBC is that the chaining of contiguous data blocks prevents the insertion, deletion or replacement of some cipher blocks during transmission over the channel [15], whereas ECB is vulnerable to these attacks.

For stream ciphers [1, 5], the data bits are encrypted individually or in blocks of s bits where $1 \leq s \leq 64$. Figures 5 and 6 illustrate the encryption and decryption processes for two stream cipher modes: Output Feedback (OFB) and Cipher Feedback (CFB) modes.

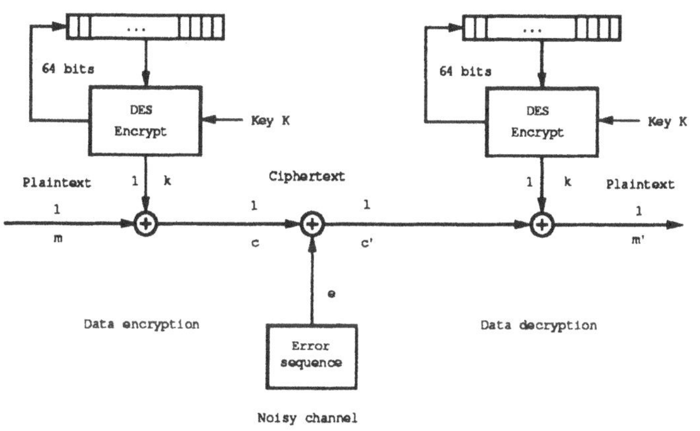

Fig. 5. Output Feedback (OFB) mode of operation of DES

For OFB, note that the generated bit stream added to the data bits (to form the cipher) is independent of the data bits being encrypted. Because of this, external bit synchronization is required between the transmitter and receiver. However, OFB is the only cipher mode without any inherent error propagation: a transmitted bit corrupted by the channel will not affect any other bit at the receiver.

In CFB mode, bits are still encrypted individually, but this time, not independently of each other. At the transmitter, bit $c_i = m_i \oplus k_i$ where k_i is a function of the 64 previously transmitted bits (or the initial vector in the first

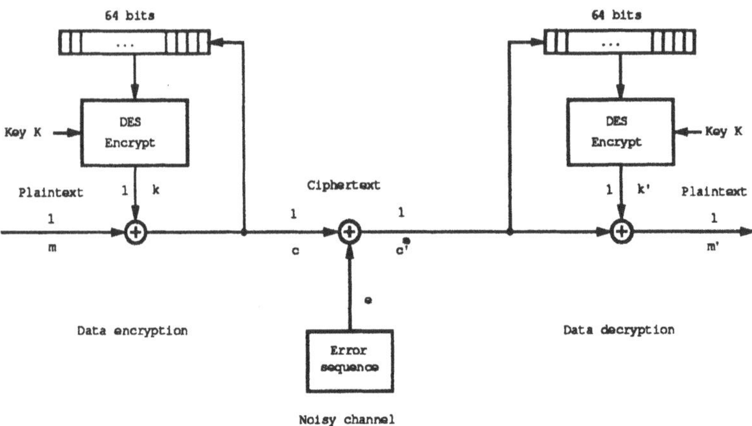

Fig. 6. Cipher Feedback (CFB) mode of operation of DES

encryption round):

$$k_i, \ldots, k_{i+s-1} = f_s \left[DES_K(c_{i-64}, c_{i-63}, \ldots, c_{i-2}, c_{i-1}) \right] \tag{7}$$

where $f_s(\bullet)$ indicates that only a subset s of the 64 bits generated by DES, that is k_i to k_{i+s-1}, is used while the other bits are discarded. In this paper, we assume that $s = 1$. A bit received in error $c_i' = c_i \oplus e_i$ is decrypted incorrectly at the receiver since:

$$m_i' = c_i' \oplus k_i \neq c_i \oplus k_i \tag{8}$$

Furthermore, since the key stream bits

$$\begin{aligned} k_{i+1} &= f_1 \left[DES_K(c_{i-63}, c_{i-62}, \ldots, c_{i-1}, c_i') \right] \qquad \text{up to} \\ k_{i+64} &= f_1 \left[DES_K(c_i', c_{i+1}, \ldots, c_{i+62}, c_{i+63}) \right] \end{aligned} \tag{9}$$

at the receiver are functions of the corrupted bit c_i', the next 64 consecutive bits (i.e., m_{i+1} to m_{i+64}) will be affected by that one error. Because of the diffusion property of DES, on average, about half of those 64 bits will be erroneous.

The cipher feedback mode is a self-synchronous stream cipher. If bit synchronization is lost, this is equivalent to a bit error. Once the erroneous digit is cleared from the receiver memory, correct deciphering operation resumes. This is evident from the above example where the next key stream bit:

$$k_{i+65} = f_1 \left[DES_K(c_{i+1}, c_{i+2}, \ldots, c_{i+63}, c_{i+64}) \right] \tag{10}$$

is the same as the one generated at the transmitter, resulting in correct deciphering (at least until the next error). Finally, note that CFB is resistant to bit deletion, modification or replacement (like ECB and CBC for block ciphers) while OFB is vulnerable to these attacks.

4 Simulation Results

A series of computer simulations has been done to assess the performances, in terms of data bit error rate (i.e., $DBER$) versus the channel bit error rate ($CBER$), for digital channels when DES encryption is used to protect the information. Figure 1 shows the components of the digital mobile communication link under consideration. For the simulations, both binary symmetric channel and Fritchman' s channel model are studied to determine the cryptographic degradation of the block and stream cipher modes of DES over memoryless channels as well as channels with memory.

Table 1. Fritchman's model parameters for fading channels

Channel label		Fritchman's model parameters					
	$CBER$	α_1	β_1	α_2	β_2	α_3	β_3
HIG	1.605×10^{-1}	0.729324	0.592677	0.119140	0.915145	0.014270	0.994939
SUH	8.920×10^{-2}	0.640335	0.743248	0.129841	0.950203	0.013075	0.997700
URB	2.588×10^{-2}	0.566301	0.815557	0.107606	0.997096		
SU2	1.457×10^{-2}	0.629742	0.643552	0.161549	0.997519		
OP1	8.970×10^{-3}	0.556226	0.857084	0.096573	0.998904		
SU1	3.988×10^{-3}	0.551425	0.685453	0.314832	0.998711		

For the simulations of the Fritchman channel model, six different sets of channel parameters are used. These are based on actual measurements carried out in Quebec city [16]. Table 1 lists the Fritchman's model parameters with their corresponding channel bit error rate $CBER$. Each $CBER$ listed in Table 1 represents the average error rate measured over the duration of the recording for each channel. Since each recording provides only one value of $CBER$, six sets of Fritchman's model parameters were selected to obtain $CBER$ values. The simulation results obtained are for specific mobile channel conditions only (urban (**URB**, **HIG**), suburban (**SU1**, **SU2**, and **SUH**) and open (**OP1**)), and therefore care is needed in the interpretation of the results. Nevertheless, cryptographic degradation and information bit error rate $DBER$ are estimated over the range $10^{-1} \leq CBER \leq 10^{-3}$.

4.1 Cryptographic degradation in DES block and stream ciphers modes

Figures 7 and 8 show that, under identical channel and coding conditions, the ranking in terms of the data bit error rate (i.e., $DBER$), obtained with the four DES modes of operation is, from the lowest to highest $DBER$: OFB, ECB, CFB and CBC. These figures provide the measured data bit error rate as a function of the channel bit error rate ($CBER$) for both the BSC and Fritchman's channels.

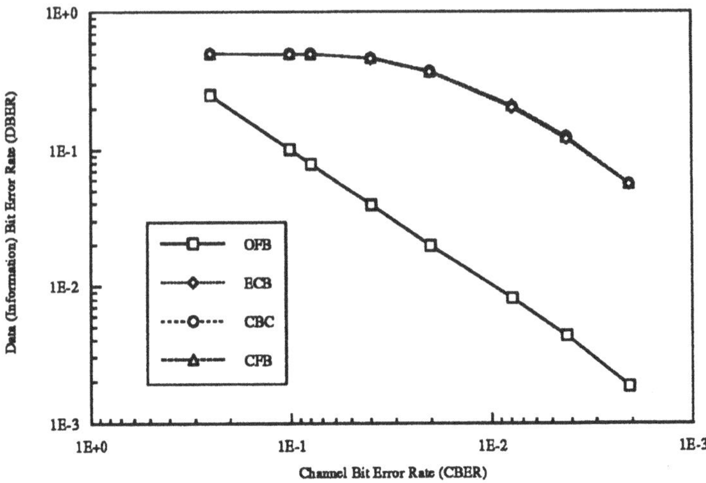

Fig. 7. BER performance over a BSC channel for all four DES modes without channel coding

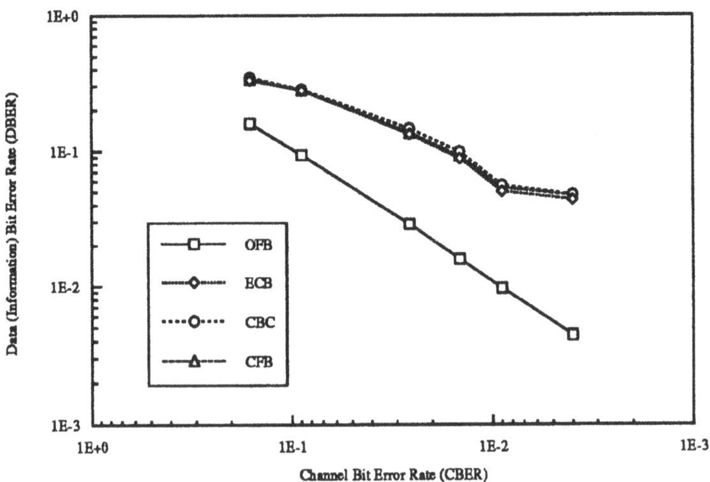

Fig. 8. BER performance over a Fritchman channel for all four DES modes without channel coding

For the particular case of OFB mode, the $DBER$ is identical to that of unciphered data since the cipher bits are decrypted independently from each other. The three other DES modes suffer from cryptographic degradation. Because of the chaining of the adjacent encrypted blocks, $DBER$ for the CFB and CBC modes are consistently higher (simulations have shown a $DBER$ increase of up to a 50% $DBER$ in certain cases) than for the ECB mode under the same channel and coding conditions. Even if the difference is small, CFB mode has consistently shown a slightly lower data bit error rate than CBC (in some cases, a $DBER$ decrease of up to 15% has been noticed); the only exception being for the case where the actual information block length $k = 64$ (before coding).

For digital channels with memory, such as Fritchman's channel, the clustering of errors leads to a lower $DBER$ degradation for ECB, CBC and CFB modes, termed D_{ECB}, D_{CBC} and D_{CFB}, than in the case of a memoryless BSC with similar channel bit error rate ($CBER$).

4.2 BCH error control coding

To reduce the data bit error rate $DBER$ caused by DES cryptographic degradation, Forward Error Correction codes can be used, where k bits of enciphered data are encoded in a vector of n bits ($n > k$) prior to transmission over the channel, allowing the receiver to correct up to t errors within the n bit block. The FEC codes considered here are binary BCH block codes [2, 14]. The error correcting capability t increases as the number of parity bits $(n - k)$ increases, but this results in a decrease in the actual code rate, $R \equiv k/n$, hence reducing the real throughput of information across the digital communication channel. Table 2 lists the BCH codes used along with their code rate R and blocklength n. The $BCH(n,k,t)$ codes are chosen such that the code rate $R \approx 75\%, 50\%$ or 33% for different codeword blocklenghts.

Table 2. Code rates R and blocklengths n of the BCH codes used in the simulations

Code rate	Blocklength			
	$n = 15$	$n = 31$	$n = 63$	$n = 127$
$k/n \approx 75\%$	73%	84%	71%	72%
$BCH(n,k,t)$:	$BCH(15,11,1)$	$BCH(31,26,1)$	$BCH(63,45,3)$	$BCH(127,92,5)$
$k/n \approx 50\%$	47%	52%	48%	50%
$BCH(n,k,t)$:	$BCH(15,7,2)$	$BCH(31,16,3)$	$BCH(63,30,6)$	$BCH(127,64,10)$
$k/n \approx 33\%$	33%	35%	38%	34%
$BCH(n,k,t)$:	$BCH(15,5,3)$	$BCH(31,11,5)$	$BCH(63,24,7)$	$BCH(127,43,14)$

For a given channel bit error rate $CBER$, the error control coding should reduce the data bit error rate $DBER$. As for the uncoded case, the ranking in terms of decreasing $DBER$ for the four DES modes is: OFB, ECB, CFB and

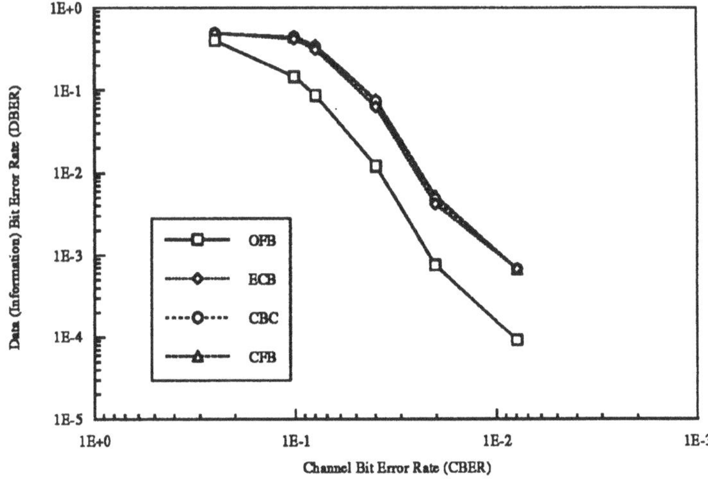

Fig. 9. BER performance over a BSC channel for the four DES modes with a $BCH(31, 16, 3)$ block code

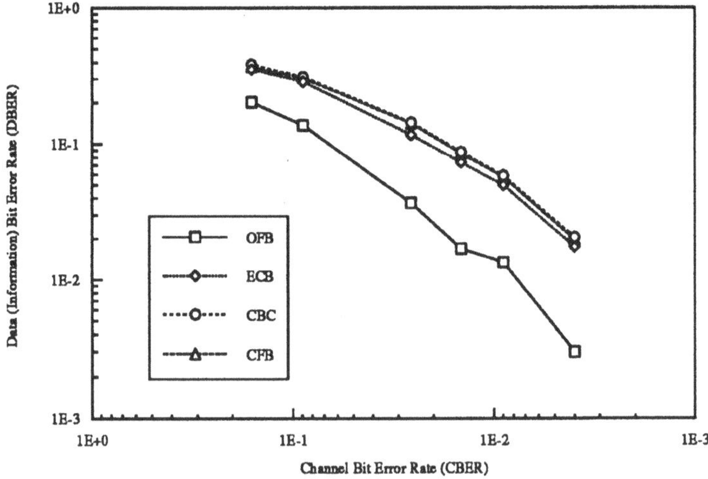

Fig. 10. BER performance over a Fritchman channel for the four DES modes with a $BCH(31, 16, 3)$ block code

CBC; this is illustrated on Figures 9 and 10 for a $BCH(31,16,3)$ triple error correcting code. Figure 9 shows also that the $BCH(31,16,3)$ code is efficient for correcting random errors in a memoryless channel while, as evidenced by Figure 10, there is almost no improvement when this same triple error correcting code is used over the Fritchman's channel; this is caused by the grouping of errors into high error density clusters.

Figures 11 and 12 show the $DBER$, as a function of $CBER$, when DES is used in the ECB mode, for BCH codes having the same code rate $R \approx 50\%$ (as per Table 2), for blocklengths n = 15, 31, 63, and 127. For both BSC and Fritchman's channels, an increase in n gives a smaller data bit error rate; this performance improvement is more significant for the memoryless channel.

As shown for the CBC mode of operation (see Figures 13 and 14), given a fixed blocklength $n = 63$, an increase in the error correcting capability t of the BCH code (and therefore a reduction in its actual code rate $R = k/n$), decreases the $DBER$ as expected. Computer simulations have also shown that BCH codes are more effective in reducing the $DBER$ for OFB mode than for ECB, CBC or CFB modes; the latter having a higher degree of error propagation.

It should be noted that the use of BCH codes may result in a $DBER$ which is higher than if no coding had been used (avalanche effect). In a BSC channel, this is only observed for the OFB mode. However, in a fading channel this may occur in any cipher mode. The average number of codewords N_c required [7] to encode a complete block of N encrypted bits (as a first approximation, $N_c \approx \lceil N/k \rceil$) has a significant impact on the performance of the BCH code ($N_c = 1$ for OFB mode): given two BCH codes (n, k_1, t_1) and (n, k_2, t_2) of the same length n, where $R_1 > R_2$, it is possible for the code with highest rate to outperform the other one when the former results in a lower value of N_c than the latter. For instance, the number of codewords required to encode a whole block of 64 information bits is $N_{c_1} = 1$ for a $BCH(127,64,10)$ code, while a $BCH(127,57,11)$ code needs $N_{c_2} = 2$ codewords. Even if the $BCH(127,64,10)$ code is less powerful in terms of error correcting capability t than the $BCH(127,57,11)$ code, an incorrectly decoded 127-bit codeword will affect only a single 64-bit ciphertext block for the former $BCH(127,64,10)$ code, while affecting the decipherment of one or two 64-bit ciphertext blocks for the $BCH(127,57,11)$ code. This may lead to a higher $DBER$ for the $BCH(127,57,11)$ code. Simulations have indicated that this effect is most apparent for ECB mode. It is less pronounced for the CBC mode and appears to be nonexistent for the CFB mode.

4.3 Bit interleaving

BCH channel coding provides better results when the channel is corrupted by random errors. To correct error bursts, bit interleaving may be used in conjunction with the BCH codes. By applying bit interleaving on I consecutive $BCH(n,k,t)$ codewords, the errors in bursts are redistributed more evenly over the $(I \times n)$ bits [14]. For channels with memory where error correction codes are used, bit interleaving is effective in decreasing the data bit error rate if; *i)* the error correcting capability t of the error control code is high enough to correct

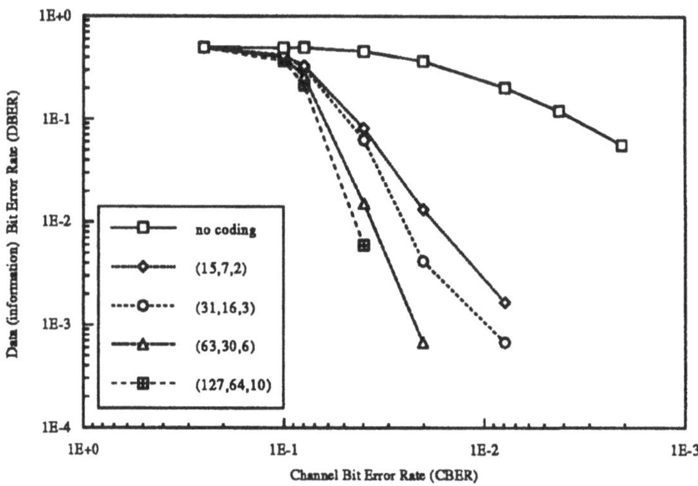

Fig. 11. BER performance with ECB mode over a BSC channel ($R \approx 50\%$)

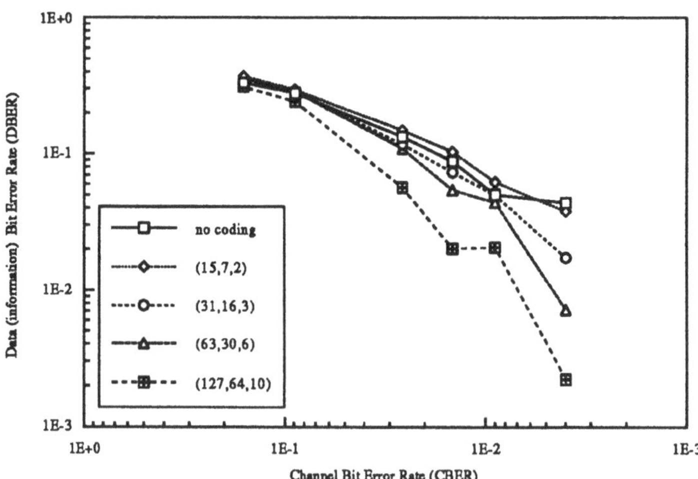

Fig. 12. BER performance with ECB mode over a Fritchman channel ($R \approx 50\%$)

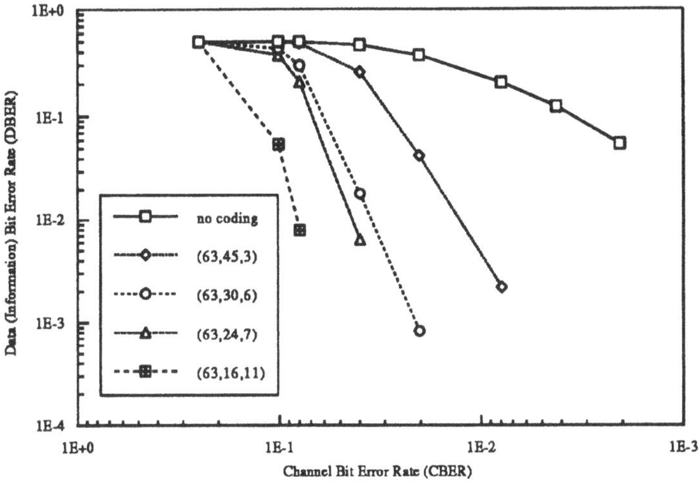

Fig. 13. Bit Error Rate performance with CBC mode over a BSC channel

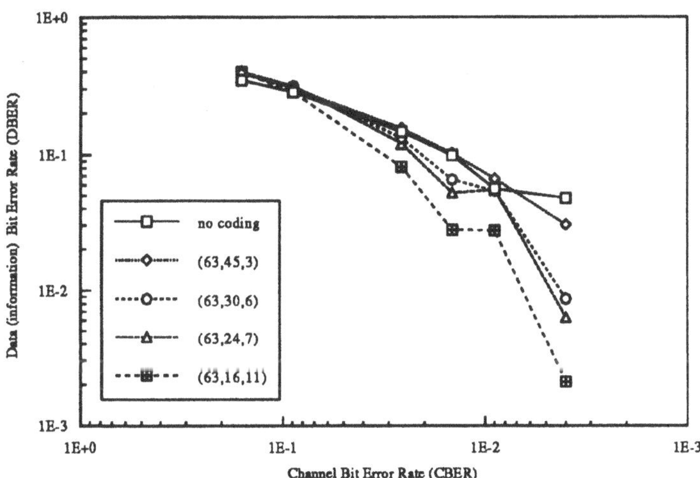

Fig. 14. Bit Error Rate performance with CBC mode over a Fritchman channel

most channel errors after deinterleaving, and if; *ii)* the interleaving degree I is large enough to reduce the memory effect of the channel by effectively breaking up the error clusters in the error sequence. However, as the interleaver at the transmitter, as well as the deinterleaver at the receiving end, needs I codewords before interleaving (or deinterleaving) can be done, a systematic delay of $(I \times n)$ bits is introduced. For large values of interleaving degrees I and blocklengths n, this delay may be unacceptable if real-time processing is required (e.g., for encrypted speech applications). The interleaving degrees I employed for the simulations are $I = 1$ (i.e., no interleaving), 25, 100, and 300.

Figures 15 and 16 provide the BER performance for the output feedback mode obtained by using bit interleaving, along with channel coding, over the Fritchman's channel for different interleaving degrees ranging from $I = 1$ (i.e., no interleaving) to $I = 300$. In the OFB mode of operation, a single channel error affects only a single decrypted bit. BCH coding with the same blocklength ($n = 31$) is used to correct the data stream. As can be seen from Figure 15 for a $BCH(31, 26, 1)$ code, error control coding can, in fact, lead to an increase in the data bit error rate if the error correcting capability t is less than the density of transmission errors in the channel. Here the $BCH(31, 26, 1)$ single error correction code is not sufficiently powerful to correct the corrupted codewords. However, as shown in Figure 16 the actual data bit error can be significantly reduced if the error correcting capability t if sufficient to cope with most error patterns within a block of n received bits. The best $DBER$ performances are obtained when the largest interleaving degree, i.e., $I = 300$, is chosen, since it spreads the channel errors more uniformly.

The same conclusions can be drawn for the other operation modes of DES. Figure 17 shows the results obtained for the CBC block mode with the $BCH(31, 11, 5)$ code over the Fritchman channel. Again, the largest interleaving degree provides the lowest data bit error rate, since t is greater than the channel error density. As noted previously, the tradeoff for this data bit error rate is a longer systematic interleaving delay which may prove unacceptable for real-time mobile applications such as digital speech transmission.

4.4 Dependence on the key and message strings

Finally, it should be noted that for the ECB, CBC, and CFB modes, there is a slight dependence of the $DBER$ obtained on the actual DES key K used for encryption, or on the specific data sequence M transmitted. This dependency has been observed during the simulations. This is caused by diffusion property of DES which translates as about, but not exactly, half of the encrypted bits being modified when a single key bit out of 56 bits, or a message bit out of 64 bits, is changed. The data bit error rate in OFB stream cipher mode is independent of these two factors since it depends only on the error sequence from the digital mobile channel.

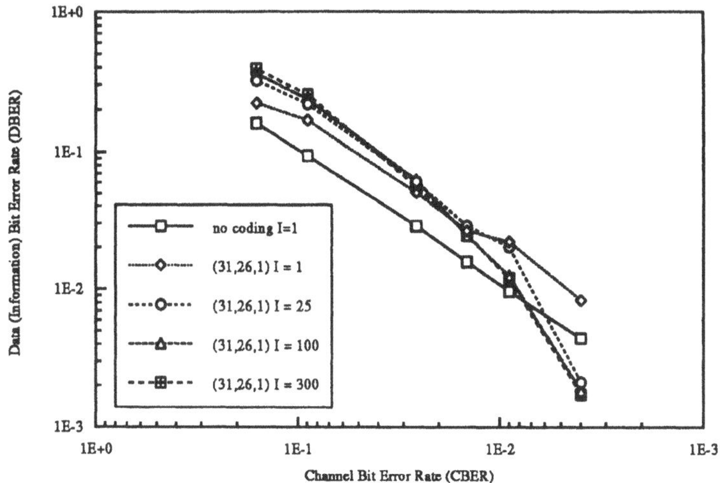

Fig. 15. BER performance with OFB mode over a Fritchman channel using a $BCH(31, 26, 1)$ code with bit interleaving

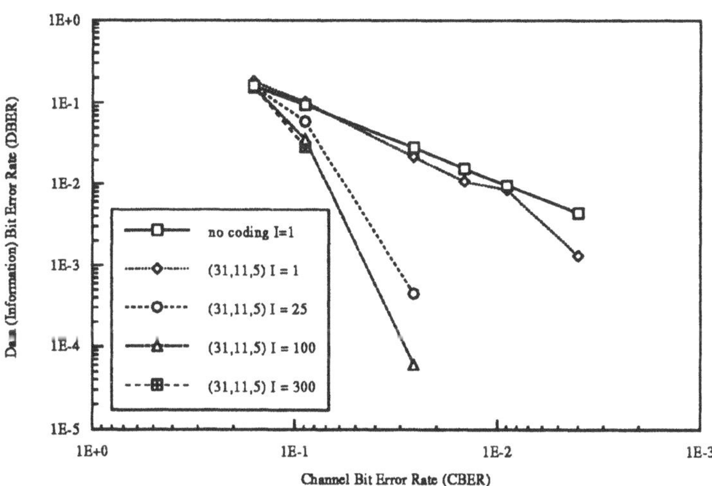

Fig. 16. BER performance with OFB mode over a Fritchman channel using a $BCH(31, 11, 5)$ code with bit interleaving

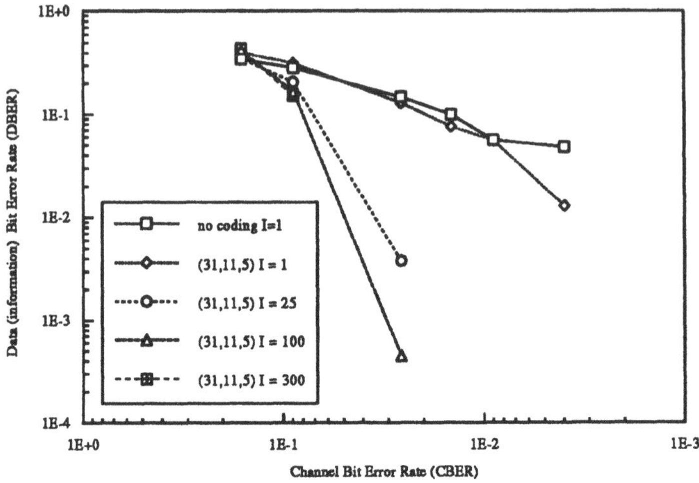

Fig. 17. BER performance with CBC mode over a Fritchman channel ($BCH(31,11,5)$ code with bit interleaving)

5 Conclusion

Simple encipherment methods such as the Data Encryption Standard can provide medium level security at high data rates for digital mobile communication applications. The four different modes of operation of DES: Electronic Codebook (ECB), Cipher Block Chaining (CBC), Output Feedback (OFB) and Cipher Feedback (CFB), provide different levels of protection of information. CBC offers the strongest protection among these four modes. However, the use of DES encryption in the ECB, CBC and CFB modes leads to a cryptographic degradation which translates into a significant increase in the data bit error rate, for memoryless channels as well as for channels with memory. No error propagation occurs in the OFB mode of operation of DES. On the other hand, OFB requires bit synchronization whereas ECB and CBC block cipher modes require only block synchronization, and CFB requires neither.

Error control coding techniques, such as BCH codes, reduce the actual information error rate provided that their error correcting capability is larger than the error density. It has been shown that the performance of BCH codes is strongly dependent on the burst error length distribution. BCH channel coding is less efficient in channels with memory, such as the Fritchman's digital channel model, than for discrete memoryless channels. For channels with memory, channel errors can be spread more evenly in time by interleaving the encoded bits before transmission and then deinterleaving the corrupted received bit stream, leading to a more efficient use of the error control code. The bit error rate performance with bit interleaving depends, however, on the actual error correcting capability of the error correction code and, thus, its code rate. In this paper, only BCH codes with bit interleaving are considered: it would be interesting to compare the performances of burst error correcting codes, in particular Reed-Solomon codes, under the same channel and cryptographic degradation conditions.

Acknowledgements

This research work is supported in part by the Natural Sciences and Engineering Research Council of Canada (NSERC) and the Telecommunications Research Institute of Ontario (TRIO). The authors would like to thank Dr. Michel Lecours and his group at Laval University (Québec, Canada) for the measurement data of the digital mobile channel. The authors also wish to thank Dr. Stafford Tavares from Queen's University (Kingston, Canada) for his valuable suggestions.

References

1. H. Beker, F. Piper, *"Cipher Systems: The Protection of Communications"*, Northwood Books, London, 1982.
2. R.E. Blahut, *"Theory and Practice of Error Control Codes"*, Addison-Wesley, Reading, Massachusetts, 1984.
3. R.E. Blahut, *"Principles and Practice of Information Theory"*, Addison-Wesley, Reading, Massachusetts, 1987.
4. J.-Y. Chouinard, M. Lecours, G.Y. Delisle, *"Estimation of Gilbert's and Fritchman's Models Parameters Using the Gradient Method for Digital Mobile Radio Channels"*, IEEE Transactions on Vehicular Technology, Vol. 37, No. 3, August 1988, pp. 158-166.
5. D.E.R. Denning, *"Cryptography and Data Security"*, Addison-Wesley, Reading, Massachusetts, 1983.
6. G. Ferland, J.-Y. Chouinard, *"Error Rate Performance Analysis of Stream and Block Ciphers in a Digital Mobile Communication Channel"*, Third Annual Conference on Vehicular Navigation and Information Systems (VNIS 92, Oslo, Norway), September 1992, pp. 426-433.
7. G. Ferland, *"Error Rate Performance Analysis of Stream and Block Ciphers in a Digital Mobile Communication Channel"*, M.A.Sc. thesis, University of Ottawa, Ottawa, Canada, July 1993.
8. B.D. Fritchman, *"A Binary Channel Characterization Using Partitioned Markov Chains"*, IEEE Transactions on Information Theory, Vol. IT-13, No. 2, April 1967, pp. 221-227.
9. R.G. Gallager, *"Information Theory and Reliable Communication"*, John Wiley and Sons, 1968.
10. E.N. Gilbert, *"Capacity of a Burst-Noise Channel"*, Bell System Technical Journal, September 1960, pp. 1253-1265.

11. L.N. Kanal, A.R.K. Sastry, *"Models for Channels with Memory and Their Applications to Error Control"*, Proceedings of the IEEE, Vol. 66, No. 7, July 1978, pp. 724-744.
12. M. Lecours, J.-Y. Chouinard, G.Y.Delisle, J. Roy. *"Statistical Modelling of the Received Signal Envelope in a Mobile Radio Channel"*, IEEE Transactions on Vehicular Technology, Vol 37, No. 4, November 1988, pp. 204-212.
13. W.C. Lee *"Mobile Communications Engineering"*, McGraw-Hill, 1982.
14. S. Lin, D.J. Costello, *"Error Control Coding: Fundamentals and Applications"*, Prentice-Hall, New Jersey, 1983.
15. C.P. Pfleeger, *"Security in Computing"*, Prentice-Hall, 1989.
16. A. Semmar, M. Lecours, J.-Y. Chouinard, J. Ahern, *"Characterization of Error Sequences in UHF Digital Mobile Radio Channels"*, IEEE Transactions on Vehicular Technology, Vol. 40, No. 4, November 1991, pp. 769-775.
17. C.E. Shannon, *"Communication Theory of Secrecy Systems"*, Bell System Technical Journal, Vol. 28, October 1949, pp. 656-715.
18. D.J. Torrieri, *"Principles of Secure Communication Systems"*, first edition, Artech House.

Reduced-Complexity Channel estimation and Equalization For TDMA Mobile Radio Systems

Armelle Wautier, Jean-Claude Dany
Ecole Supérieure d'Electricité, Gif sur Yvette, France
Christophe Mourot
Alcatel Radiotéléphone, Colombes, France

Abstract : A non-iterative block processing algorithm for channel estimation is presented. This is suitable for time-division multiple access (TDMA) digital radio systems with a signal structure based on time-slots. Such channel estimator can be applied in conjunction with decision feedback equalizer, maximum likelihood sequence estimator (MLSE) or near-MLSE. A theoretical analysis of the estimator performances for slow and fast time-varying channels is presented. A procedure for determining the required training sequence length and the number of estimated coefficients in the channel impulse response is developed. Depending on channel characteristics, there is a particular length which yields optimum performances at the beginning of the data block. This length is analytically evaluated for constant amplitude zero-autocorrelation (CAZAC) sequences. Finally, the effects of channel estimation optimization are assessed by computer simulations.

1. Introduction

Land mobile radio channels are generally characterized as fading multipath channels with time dispersion. When multipath delay spread is greater than symbol duration, it results in intersymbol interference (ISI), and necessitates the use of an equalizer [1-3]. In time-division multiple access (TDMA) digital radio systems, the signal structure is based on time-slots. A time-slot carries information data symbols and an equalizer training symbol sequence *a priori* known to the receiver (see Fig.1). Usually, the training sequence is placed in the middle of the time slot in order to minimize the effect of channel fluctuation which the receiver has to cope with. In such a system, the received time-slot is completely stored so that block processing algorithms can be used. The choice of such an algorithm depends on the time slot structure which in turn has to be designed to optimize the training sequence overhead with respect to the information data symbols.

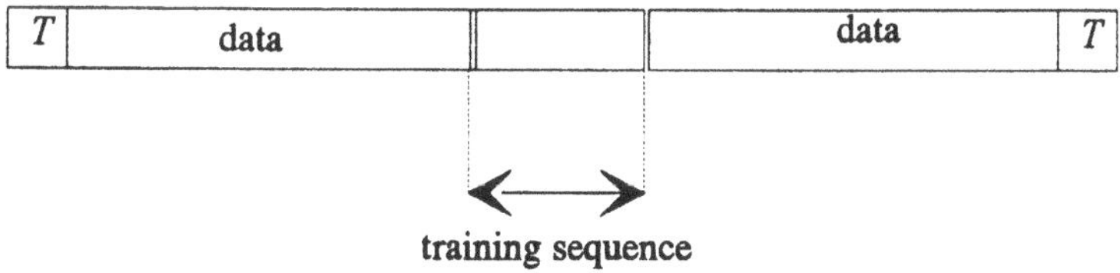

Fig.1. Time slot structure

We present a non-iterative block processing algorithm for channel estimation based on Least Sum of Squared Errors (LSSE) (see Fig.2). This work is related to, but independent from [4-5]. This is useful for designing very low complexity channel estimators for slowly time-varying channels. Such channel estimators can be applied in conjunction with decision feedback equalizer (DFE), maximum likelihood sequence estimator (MLSE) using the Viterbi algorithm or near-MLSE [1-3, 6]. The quality and the length of the estimated channel impulse response heavily influence the performance and the complexity of the equalizer.

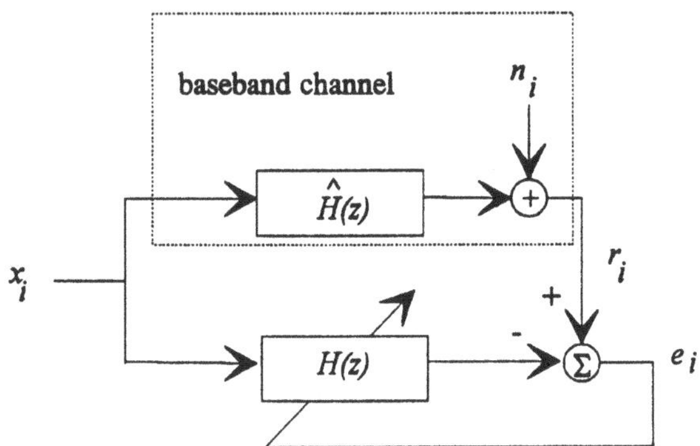

Fig. 2. Channel estimator block diagram.

We first describe the channel estimation principle in section II. Then, we give a theoretical analysis of the estimator performance over slow and fast time-varying channels in section III. Usefulness of particular types of training sequence such as constant amplitude zero-autocorrelation (CAZAC) sequences is demonstrated. The channel estimator performance is evaluated in conjunction with time-slot structure for mobile Rayleigh fading channels. Simple analytical expressions of bounds on performance are derived for CAZAC training sequences. Optimized channel estimation is discussed in section IV. We conclude like in [1] that a certain training sequence length optimizes the estimator performances. The conclusion in [1] is based on computer simulations for DFE, we give an analytical expression of the optimal training sequence length related to the proposed channel estimator for CAZAC sequences. The expression of the optimal length brings to the fore an exhaustive parameter relating channel characteristics (Doppler spread, noise variance, delay power profile), symbol rate and estimated channel impulse response length. We also develop procedures to select the most significant coefficients in the estimated channel impulse response. Based on simple comparisons, these procedures reduce the channel estimation error by cancelling noisy coefficients. Computer simulation results (for a TDMA receiver including equalizer) are presented to clearly depict the gain in receiver performance achieved by optimized channel estimation in section V.

2. Channel estimator principle

A coherent detection receiver operating on a band-limited linearly modulated signal is assumed. In spite of the band-limited assumption, the tapped delay line model for the channel is truncated at $K + 1$ tap coefficients. The sampled baseband equalizer input can be expressed in the form :

$$r_i = \sum_{k=0}^{K} s_{i-k} h_k(i) + n_i \tag{1}$$

where $r_i = r(iT_S + t_0)$; $n_i = n(iT_S + t_0)$; T_S denotes symbol interval ; $0 < t_0 < T_S$. $n(t)$ represents an equivalent complex noise of the additive white Gaussian noise passed through the receceiver filter. $h_k(i)$ represents the complex time-varying lowpass equivalent impulse response of the whole transmission system including the pulse shape modulation filter, the channel and the receiver filter. Assuming that the channel impulse response is constant during the training period of a time-slot, the sampled baseband equalizer input at time $iT_S + t_0$ is given by

$$r_i = \sum_{k=0}^{K} h_k x_{i-k} + n_i \tag{2}$$

$\{h_k\}$ are the channel tap coefficients ; $\{x_i\}$ are the transmitted training symbols *a priori* known to the receiver. In the proposed method, the estimated channel coefficients, denoted by $\hat{h}_k, k = 0, \ldots, K$, are adjusted to minimize the sum of squared error samples J defined as :

$$J = \sum_{i=i_1}^{i_2} |e_i|^2 \text{ where } e_i = r_i - \sum_{k=0}^{K} \hat{h}_k x_{i-k} \text{ (see Fig.2).} \tag{3}$$

i_1 and i_2 define the index limits for which the above sum J is minimized with respect to the $\hat{h}_k$'s. Letting L be the training sequence length, i_1 must be taken equal to $K + 1$ and i_2 to L in order to know all the variables involved in (3). The channel tap coefficient vector minimizing J satisfies the classical least-squares equation [7]:

$$\left(\mathbf{A}^{\mathrm{H}}\mathbf{A}\right)\hat{\mathbf{H}} = \mathbf{A}^{\mathrm{H}}\mathbf{R} \quad \text{where}$$

$$\hat{\mathbf{H}}^{H} = \left[\hat{h}_0^* \ldots \hat{h}_K^*\right], \quad \mathbf{R}^{H} = \left[r_{K+1}^* \ldots r_L^*\right], \mathbf{A}^{H} = \begin{bmatrix} x_{K+1}^* & \cdot & \cdot & \cdot & x_L^* \\ x_K^* & \cdot & \cdot & \cdot & \cdot \\ \cdot & \cdot & \cdot & \cdot & \cdot \\ x_1^* & \cdot & \cdot & \cdot & x_{L-K}^* \end{bmatrix} . \quad (4)$$

The operator $(.)^{H}$ represents the Hermitian transpose operator, $(.)^{*}$ represents the conjugate operator. The matrix $\mathbf{A}^{H}$ is completely defined by the set of the training sequence symbols $\{x_j\}$ and the vector $\mathbf{R}$ by the received samples $\{r_i\}$. A unique solution for $\hat{\mathbf{H}}$ may be expected if the number of columns in the matrix $\mathbf{A}^{H}$ is greater than or equal to the number of rows. This implies that the training sequence must be composed of at least $2K + 1$ symbols :

$$L \geq 2K + 1 \quad (5)$$

If $(\mathbf{A}^{H}\,\mathbf{A})^{-1}$ is non-singular, the estimated tap coefficient vector is expressed as

$$\hat{\mathbf{H}} = \left(\mathbf{A}^{H}\mathbf{A}\right)^{-1}\mathbf{A}^{H}\mathbf{R} \quad (6)$$

When the matrix $\mathbf{A}$ is a square matrix, i.e. for $L = 2K + 1$, this method becomes equivalent to the one proposed by Clark [8]. The least squares method presented here has the advantage of not fixing the training sequence length to $2K + 1$ only.

3. Channel Estimator Performance and Complexity

3.1. Performance of the estimator on a time-invariant channel

When the noise process n_i has zero-mean, the estimate in (6) is unbiased for a time-invariant channel. When the noise process is white with zero mean and variance σ_n^2, the covariance matrix of the estimate vector $\hat{\mathbf{H}}$ equals $\sigma_n^2\,(\mathbf{A}^{H}\,\mathbf{A})^{-1}$. Performance quality of the estimator can be evaluated by the trace of this matrix :

$$\mathrm{trace}\left(\mathrm{cov}\left[\hat{\mathbf{H}}\right]\right) = \sigma_n^2\,\mathrm{trace}\left(\left(\mathbf{A}^{H}\mathbf{A}\right)^{-1}\right), \text{ or}$$

$$\sum_{i=0}^{K} \mathrm{E}\left[\left|\hat{h}_i - h_i\right|^2\right] = \sigma_n^2\,\mathrm{trace}\left(\left(\mathbf{A}^{H}\mathbf{A}\right)^{-1}\right). \quad (7)$$

From (7), it can be seen that the mean squared error between the estimated channel tap coefficients and the actual ones depends as usual on the correlation properties of the training sequence. With the constraint of constant amplitude modulation, and for a given training sequence length L, the algorithm yields optimum performance for sequences satisfying

$$\mathbf{A}^{\mathrm{H}}\mathbf{A} = (L-K)|s|^2\,\mathbf{I}, \tag{8}$$

where $\mathbf{I}$ is the identity matrix, $|s|$ denotes the amplitude of the training sequence symbols. The minimum mean-square-error is then

$$\frac{K+1}{L-K}\frac{\sigma_n^2}{|s|^2}. \tag{9}$$

Sequences having constant amplitude zero-autocorrelation (CAZAC), i.e. no sidelobe in the autocorrelation function, satisfy the above stated property [8, 9]. Such sequences can be found in [8] for several modulations. The training sequence is then composed as described in Fig.3 of $P + K$ symbols ; P is the periodicity of the CAZAC training sequence. According to (5), P has to satisfy

$$P \geq K+1. \tag{10}$$

Unfortunatly, such training sequences exist for a limited number of P values. Pseudo-CAZAC sequences having zero-autocorrelation function only near the origin are however sufficient as $P \geq K+1$. uch sequences can be designed for that purpose. It should also be noted that the so-called polyphase sequences are designed to approximate the above stated property. Let us also notice that the training sequence needs K precursors only, instead of both K precursors and K postcursors as a training sequence for a correlation estimator does [10].

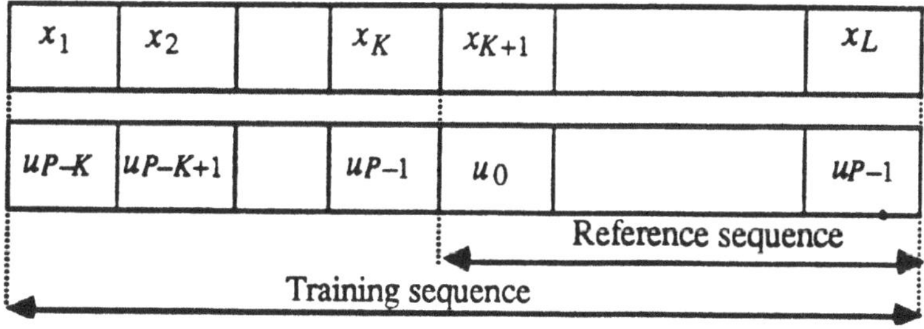

Fig. 3. Training sequence composition.

For CAZAC and pseudo-CAZAC sequences, the channel estimator yields the following performance :

$$\mathrm{E}\left|\hat{h}_i - h_i\right|^2 = \frac{1}{P}\frac{\sigma_n^2}{|s|^2}, \tag{11}$$

$$\mathrm{E}\left|\left(\hat{h}_i - h_i\right)^*\left(\hat{h}_j - h_j\right)\right| = 0 \qquad \text{for } i \neq j, \tag{12}$$

$$\sum_{i=0}^{K}\mathrm{E}\left|\hat{h}_i - h_i\right|^2 = \frac{K+1}{P}\frac{\sigma_n^2}{|s|^2} \tag{13}$$

For a static channel, the parameter $(K + 1) / P$ in (13) shows that the longer the training sequence is, the better the channel is estimated.

3.2. Performance of the estimator on a multipath mobile radio channel

The response of the channel may vary according to the fading due to the vehicle movement. The fading rate is determined from the Doppler spectrum. For discrete-time Gaussian wide sense stationary uncorrelated scattering channels (GWSSUS), the normalized Doppler spectra are specified for different delay ranges of the received paths [11] ; $R_i(\Delta t)$ denote their related normalized correlation functions. For a time-varying channel, a general formula like (13) is impossible. Indeed, besides the correlation properties of the training sequence and the statistical parameters of the channel, the particular symbol values of the considered training sequence affect the estimate. This dependency causes difficulty to use the estimator performance evaluation to design a system.

We avoid taking into account the choice of a particular CAZAC sequence by determining upper and lower bounds on the mean squared error between the estimated channel coefficients and the actual ones, independent of this factor [12]. Indeed, assuming the different taps of the channel to be statistically uncorrelated, the mean square error is bounded [12] as

$$\sum_{i=0}^{K}\mathrm{E}\left|\hat{h}_i - h_i(n_S)\right|^2 \leq \frac{K+1}{P}\frac{\sigma_n^2}{|s|^2} + \sum_{i=0}^{K}2\sigma_i^2\left[1 - \frac{1}{P}\sum_{\lambda=1}^{P}R_i\left((\lambda+n_S)T_S\right)\right], \tag{14}$$

$$\sum_{i=0}^{K}\mathrm{E}\left|\hat{h}_i - h_i(n_S)\right|^2 \geq \frac{K+1}{P}\frac{\sigma_n^2}{|s|^2}$$

$$+\sum_{i=0}^{K}\sigma_i^2\left[1 - \frac{2}{P}\sum_{\lambda=1}^{P}R_i\left((\lambda+n_S)T_S\right) + \frac{1}{P^2}\sum_{\lambda=1}^{P}\sum_{\gamma=1}^{P}R_i\left((\lambda-\gamma)T_S\right)\right] \tag{15}$$

n_s corresponds to the data symbol sample number within the data block ($n_s = 0$ corresponds to the first data symbol following the training sequence) (cf Fig.4). Analytical expressions of the bounds involve TDMA system parameters and channel characteristics : symbol period T_s, training sequence parameters (P, K and $|s|$), data symbol number within the data block n_s, noise variance σ_n^2, channel profile (path variances σ_i^2), correlation functions of the channel coefficients $R_i(\Delta t)$. The bounds provide an interesting general behavior of the estimator for any CAZAC sequence. In addition, it should be noted that the upper bound is reached for P equal to $K + 1$.

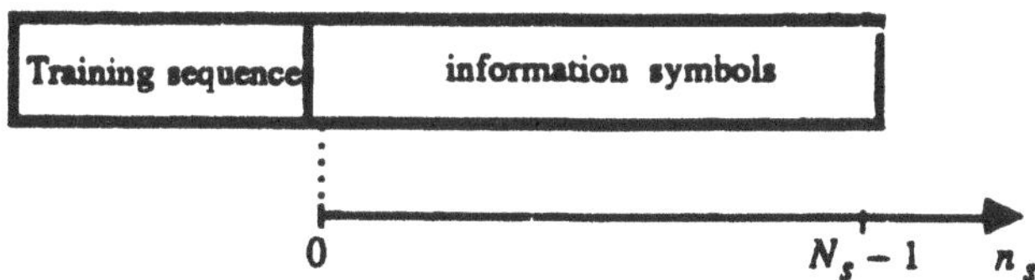

Fig.4.. Symbol position numbering.

3.3. Complexity

The algorithm for obtaining the estimate of the channel coefficients is given by (6). There is no matrix inversion to perform in real time ; the receiver is assumed to have prior knowledge of the matrix $(\mathbf{A}^H\mathbf{A})^{-1}$ stored with its corresponding training sequence. In the general case, the channel estimator requires $4(K + 1)(L + 1)$ multiplications between real numbers and $4(K + 1)L$ real additions. By making use of the correlation properties of the considered sequence and for certain modulation techniques, this complexity can be highly reduced. For constant amplitude sequences making the matrix $(\mathbf{A}^H\mathbf{A})^{-1}$ equal to a multiple of the identity matrix, it only requires $2(K + 1)$ multiplications and $4(K + 1)(L - K - 1 / 2)$ additions between real numbers. This complexity is very low compared to recursive least squares adaptation algorithms [3].

4. Optimized Channel Estimation

The normalized power spectra are now assumed to be identical, in a time-dispersive channel. Assuming that the vehicle is moving with a constant speed v and that multipath signal due to a large number of scatterers has uniform distribution with angle, the normalized correlation function $R(\Delta t)$ (Fourier transform of the classical normalized Doppler spectrum) for each channel path is given by

$$R(\Delta t) = J_0(2\pi f_d \Delta t). \tag{16}$$

$J_0(.)$ denotes the zeroth-order Bessel function of the first kind ; f_d is the maximum Doppler frequency shift given by $f_d = f_0\ V / c$ where f_0 denotes the carrier frequency, c the propagation velocity of the electromagnetic wave, V the speed of the mobile. This approach is a worst case of the time variation of the mobile radio channel [11]. Actually, the significant parameter characterizing the channel fluctuation is the normalized Doppler rate f_dT_S. When $2\pi f_dT_Sn_S$ is small, the following approximation of (16) is valid and can be used in (14) and (15)

$$R(n_ST_S) \approx 1 - (\pi f_dT_S)^2 n_S^2 \tag{17}$$

Let $\varepsilon(n_S)$ denote the normalized error estimation defined by

$$\varepsilon(n_S) = \frac{1}{\sigma^2}\sum_{i=0}^{K} \mathrm{E}\left|\hat{h}_i - h_i(n_S)\right|^2, \tag{18}$$

where σ^2 is defined by :

$$\sigma^2 = \sum_{i=0}^{K} \sigma_i^2 \tag{19}$$

In the following, quaternary phase-shift-keying (QPSK) modulation is considered ; raised cosine spectral shaping, evenly distributed between the transmitter and the receiver, is assumed. This consideration lets us substitute $2E_b / N_0$ for $|s|^2\sigma^2 / \sigma_n^2$; E_b denotes the energy per bit, and $N_0 / 2$ the double-sided spectral density of the additive white Gaussian noise.

4.1. Reference sequence length

We now consider the initial degradation of the channel estimator performance at the symbol position $n_S = 0$ (cf Fig.4) :

$$\varepsilon(0) \le \frac{K+1}{P}\frac{N_0}{2E_b} + 2\left[1 - \frac{1}{P}\sum_{\lambda=1}^{P} R(\lambda T_S)\right], \tag{20}$$

$$\varepsilon(0) \ge \frac{K+1}{P}\frac{N_0}{2E_b} + 1 - \frac{2}{P}\sum_{\lambda=1}^{P} R(\lambda T_S) + \frac{1}{P^2}\sum_{\lambda=1}^{P}\sum_{\gamma=1}^{P} R((\lambda-\gamma)T_S) \tag{21}$$

Although the channel is time-dispersive, the error bounds do not depend anymore on its power delay profile of the channel. Assuming that (17) is valid, (20) and (21) become :

$$
\begin{aligned}
\varepsilon(0) &\geq \frac{K+1}{P}\frac{N_0}{2E_b}+2(\pi f_d T_S)^2\frac{(P+1)^2}{4} \\
\varepsilon(0) &\leq \frac{K+1}{P}\frac{N_0}{2E_b}+2(\pi f_d T_S)^2\frac{(P+1)(2P+1)}{6}
\end{aligned} \tag{22}
$$

These expressions of lower and upper bounds highlight the influence of the reference sequence length on the estimate of the channel impulse response. The sequence must be long enough to estimate the channel coefficients ($P \geq K+1$), but short enough to avoid significant variation of the actual channel during the estimation period. An oversized training sequence can degrade receiver performance and this maximum length can be determined from the analytical expressions of the bounds (22). The initial square error bounds of 10 $\log_{10}(\varepsilon(0))$ are plotted in Fig.5 for E_b/N_0=20dB as a function of the reference sequence length for different normalized fade rates and for two values of K.

Firstly, recall that the upper bound is reached for P equal to $K+1$. Secondly, the two bound minima are rather tight (less than 0.5 dB in Fig.5). The upper bound is minimum for

$\sqrt[3]{\frac{3}{4}}\sqrt[3]{\frac{(K+1)}{(\pi f_d T_s)^2}\frac{N_0}{2E_b}}$ and the lower bound is minimum for

$$
\sqrt[3]{\frac{(K+1)}{(\pi f_d T_s)^2}\frac{N_0}{2E_b}}
$$

We can say that the optimal length minimizing the mean-square-error is close to

$$
0.78\sqrt[3]{\frac{(K+1)}{(\pi f_d T_s)^2}\frac{N_0}{2E_b}} \tag{23}
$$

Figure 6 plots the optimal reference sequence length P_{opt} as a function of E_b / N_0 for different fade rates.

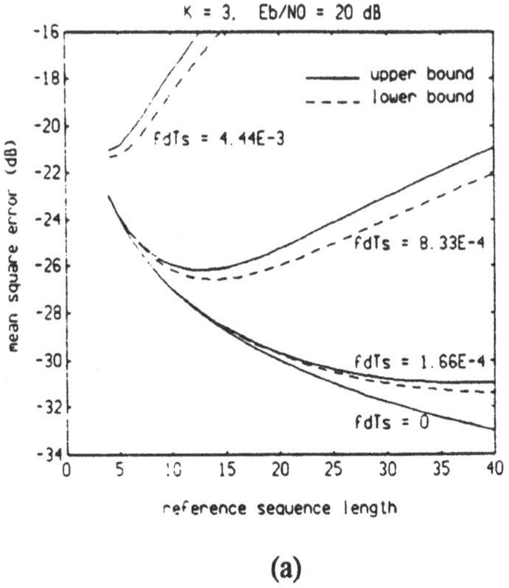

(a)

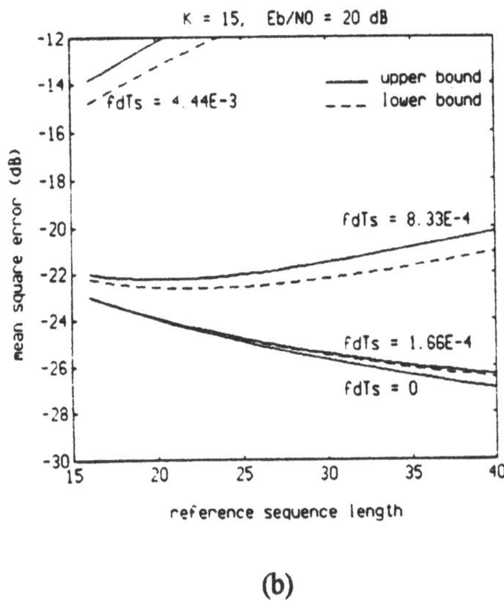

(b)

Fig.5. Squared error of the initial channel estimate $\varepsilon(0)$ versus reference sequence length P.
(——) upper bound, (- - -) lower bound
(a) E_b / N_0 = 20 dB ; K = 3, (b) E_b / N_0 = 20 dB ; K = 15.

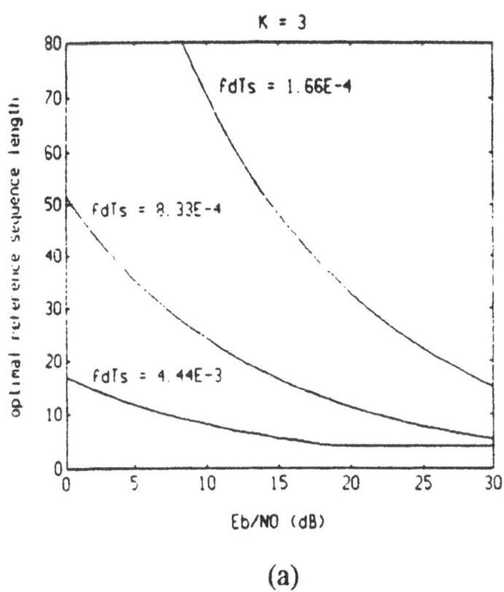

(a)

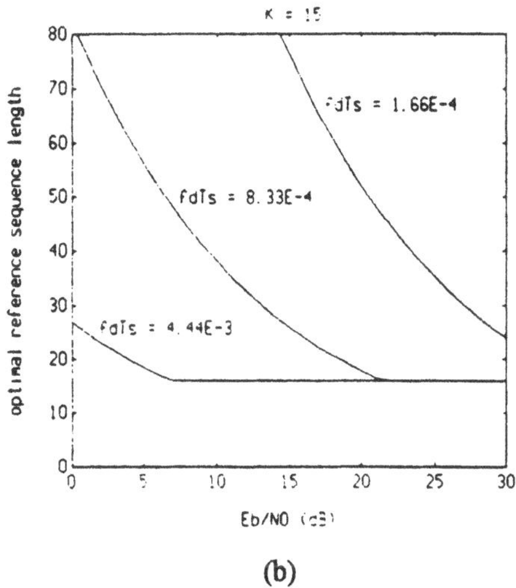

(b)

Fig.6. Optimal reference sequence length versus E_b/N_0 for a fixed value of K and for different values of f_dT_s (a) K = 3, (b) K = 15.

Figure 7 shows the effect of the selection of a particular reference sequence length. The curves plotting the initial square error bounds versus E_b / N_0 cross each other as the optimal length depends on the normalized fade rate and on the signal-to-noise ratio as well.

4.2. Number of taps in the channel impulse response

From equations (11) and (13), it can be seen that the mean squared estimation error is proportional to the number of non-zero coefficients in the estimated channel impulse response (CIR). To minimize this error, it could be useful to minimize the number of these coefficients. The channel estimator described above was modified to do so.

First, the estimator looks for the coefficient having the smallest contribution to the total CIR energy. It then substracts this contribution from the total energy and compares the result to a pre-defined threshold. If the result stays above, the coefficient is forced to zero, the total energy is modified accordingly and the operation is repeated for another coefficient. If the result of comparison falls below the threshold, the coefficient and the total energy are kept unchanged and the operation is stopped.

In addition to minimizing the estimation noise, this operation will also minimize branch metric calculation complexity in a near-MLSE or the number of filter coefficients in a DFE.

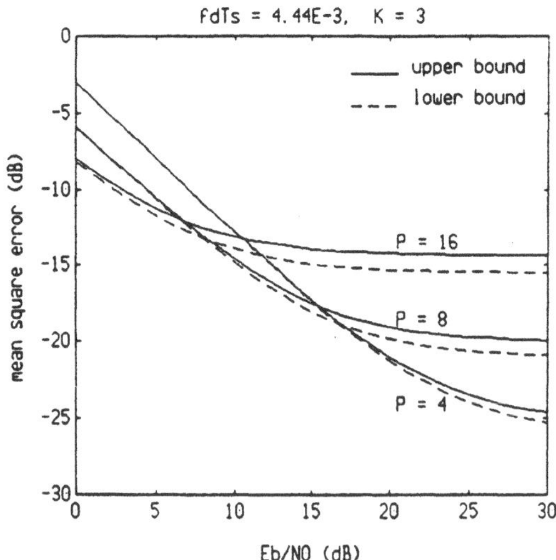

Fig.7. Influence of the choice of a particular reference sequence length on $\varepsilon(0)$: $\varepsilon(0)$ as a function of E_b/N_0 ; $K = 3$; $f_dT_s = 4.44\ 10^{-4}$.
(–) upper bound, (- - -) lower bound

5. Simulation Results

The performances of the above described techniques have been evaluated by means of computer simulations for a TDMA mobile communication system with no interferer and a GSM-like frame structure. The time-slots are composed of known start and stop symbols, information data symbols with the training sequence embedded in the middle. The modulation is raised cosine QPSK and the carrier frequency is 1.8 GHz. Used QPSK CAZAC training sequences can be found in [8]. The simulation program uses COST 207 (*European Cooperation in the field of Scientific and Technical research*) channel models [11]. The equalizer is a decision feedback sequence estimator (DFSE) which is a hybrid near-MLSE with feedback. The memory of the channel is truncated, which is used to build the near-MLSE trellis. The CIR coefficients corresponding to the truncated part are used to weight the previously estimated symbols found along survivors and feed the result back in the decision process. To minimize error propagation, the DFSE is preceded by a prefilter which turns the estimated CIR into a minimum phase response and shifts signal energy towards its earliest coefficients. More details on the simulation chain, DFSE and prefiltering can be found in [12- 14].

5.1. Comparison of the proposed estimator with a correlator estimator

The channel estimation proposed here is compared to the classical correlation method [9] in figure 8 for hilly terrain environment (HT) at 100 km/h and 1 Mbit/s (72 symbols per time slot ; $P = 16$; $K = 13$; $f_dT_s = 3.33\ 10^{-4}$) and typical urban environment (TU) at 5 km/h and 4 Mbit/s (186 symbols per time slot ; $P = 16$; $K = 13$; $f_dT_s = 4.16\ 10^{-6}$). For the same bit error rate after detection, the proposed method allows to reduce the overhead in a TDMA block. The overhead[1] is 67.5%

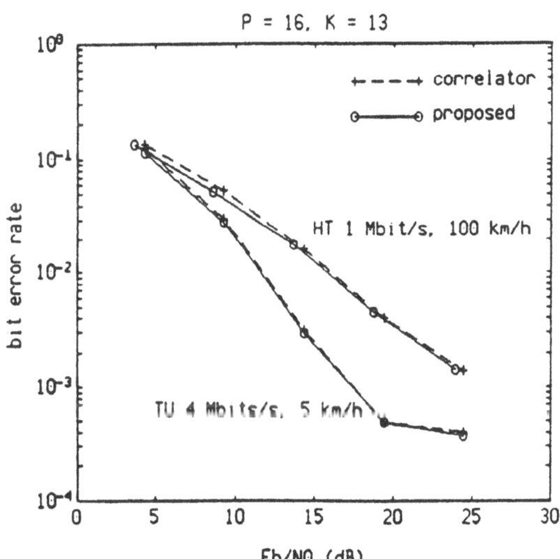

Fig.8. Proposed channel estimation vs. correlation channel estimation.
Typical urban, 4Mbit/s, 5km/h : 186 symbols per time slot; $P = 16$; $K = 13$; $f_dT_S = 4.16\ 10^{-6}$.
Hilly terrain, 1Mbit/s, 100km/h : 72 symbols per time slot; $P = 16$; $K = 13$; $f_dT_S = 3.33\ 10^{-4}$.

---

[1]The overhead is the ratio of the number of non information symbols in the slot over the number of information symbols.

instead of 140% for HT test case, 18.5% instead of 29% for TU test case. In the following, the proposed block least square method for channel estimation is always used.

5.2. Receiver bit-error-rate performance and block structure design

The effect of time-slot structure and channel estimation on bit-error-rate performance is illustrated here by means of computer simulation, this point is further developed in [12]. We consider a rural area environment (RA). The bit rate equals 300 kbit/s and the number of information symbols per time-slot is 40. Figure 9(a) illustrates the case of a high normalized fade rate (f_dT_S = 4.44 10^{-3}), the considered speed of 400 km/h corresponds to a high speed train), and figure 9(b) illustrates a medium normalized fade rate (f_dT_S = 1.0 10^{-3}, considered mobile speed is 90 km/h).

Curves plotting bit-error-rate versus bit-energy-to-noise ratio for different values of P cross each other as curves plotting channel estimate error versus bit-energy-to-noise ratio do (see figure 7). For large values of f_dT_S, decreasing P yields better performances as E_b/N_0 increases, in particular the irreducible bit error rate decreases (see Fig.9(a)). For low f_dT_S, the curves cross each other for greater values of E_b/N_0, increasing P can then improve receiver performances (see Fig.9(b)). This improvement depends on the number of information symbols per time-slot N_S.

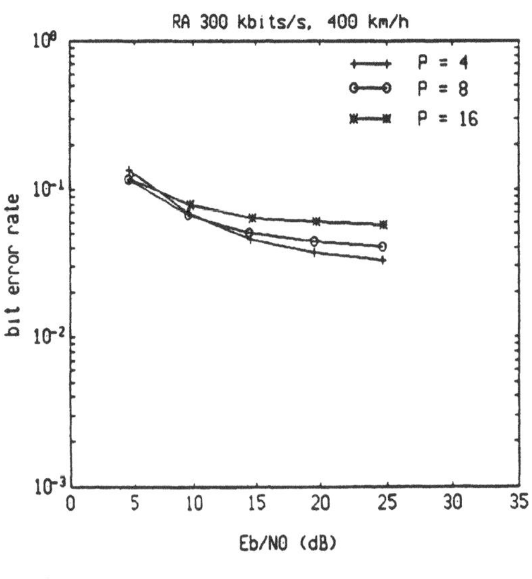

(a)

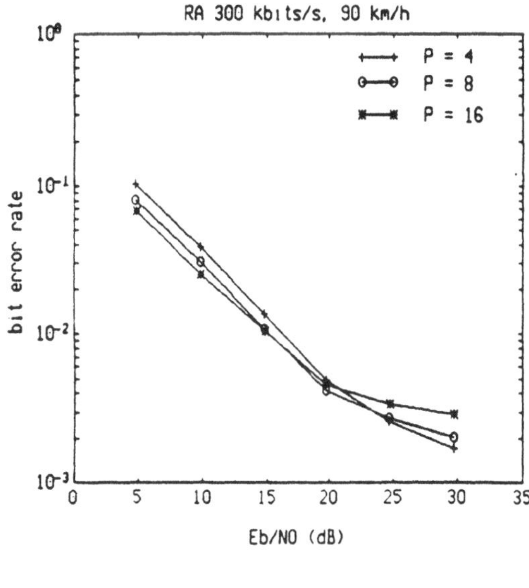

(b)

Fig. 9. Receiver performance for high and medium normalized fade rates parameterized by P. (a) $f_dT_s = 4.44\ 10^{-3}$; $K = 3$; $N_s = 40$. (b) $f_dT_s = 1.0\ 10^{-3}$; $K = 3$; $N_s = 40$.

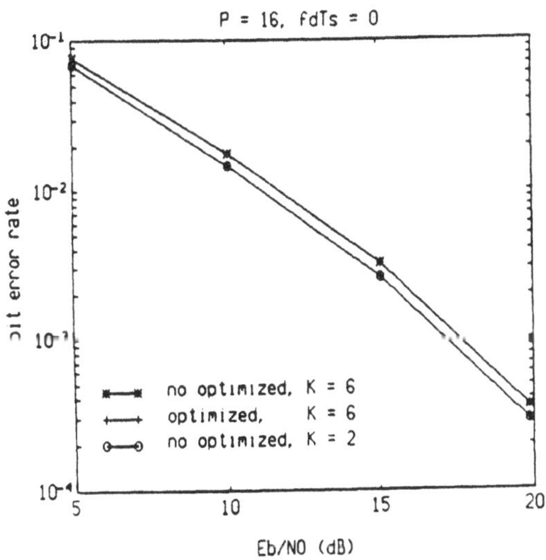

(a)

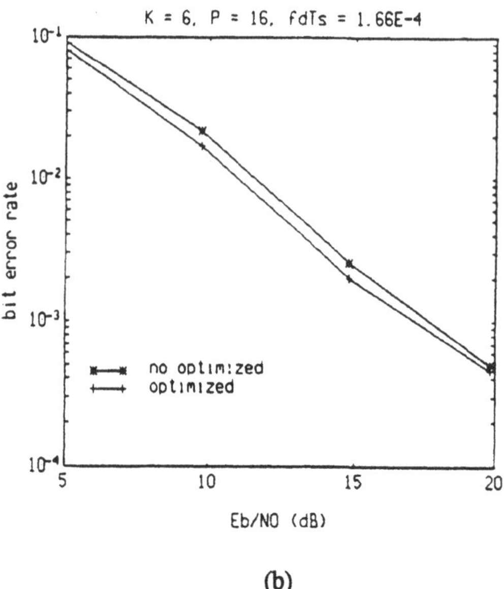

(b)

Fig. 10. Effect of CIR optimization for time-invariant and time-varying channels.
(a) 2 Rayleigh fading unequal paths with $2T_s$ delay dispersion, f_dT_s=0, P=16, N_s=150.
(b) 3 Rayleigh fading equal paths with $6T_s$ dispersion, f_dT_s=1.66 10^{-4}, P=16, K=6, N_s=150.

5.3. Effect of channel impulse response optimization on the bit-error-rate

To demonstrate the effect of CIR optimization, simulations have been carried out. We first consider a time-invariant two-ray Rayleigh fading channel having a delay power profile given by $R_h(\tau) = \delta(\tau) + 0.5\delta(\tau - 2T_s)$. The channel estimator is forced to estimate the CIR over an arbitrary too large window of $6T_s$. One can see (Fig. 10(a)) that CIR optimization with a 95% threshold improves receiver performances for all values of E_b/N_0 just as if the estimator were told to estimate the CIR over an appropriate window of $2T_s$. Again, note that the branch metric calculation is reduced. For a time-varying three-ray channel with its delay power profile given by $R_h(\tau) = \delta(\tau) + \delta(\tau - 2T_s) + \delta(\tau - 6T_s)$, there is again an improvement for all values of E_b / N_0 (Fig. 10(b)).

6. Conclusion

Fast start-up channel estimators have been analyzed in [8, 4] for time-invariant channels. The start-up behavior of block processing algorithms assuming a time-invariant channel has been evaluated by means of computer simulations [1]. We studied into detail a block processing algorithm for channel estimation suitable for TDMA transmission and analyzed its performance for both time-invariant and varying multipath-fading channels. Upper and lower bounds of the channel estimation mean square error have been determined in the case of CAZAC reference sequences (which are optimal if the channel is time-invariant). The analytical expressions of the bounds involve TDMA system parameters and channel characteristics, which quantify their related effects on channel estimate quality. For given channel conditions, the optimal length of the training sequence can be derived from the bounds of the initial channel estimation mean square error. Exceeding this length degrades receiver performances. This investigation shows that the estimator and the training sequence have to be jointly designed for best efficiency. Also, when channel tracking is not envisaged during data detection, the training sequence should be designed in conjunction with the number of symbols per time-slot. The number of coefficients in the estimate corresponds to the longest encountered time dispersion of the channel. However, when the time dispersion is lower, receiver performances can be improved to some extent by reducing the number of non-zero channel taps. An algorithm optimizing this number has been proposed and assessed by computer simulations.

REFERENCES

1. G.W. Davidson, D.D. Falconer and A.U.H. Sheikh, "An investigation of block-adaptive decision feedback equalisation for frequency selective fading channels", *Can. J. Elect. & Comp. Eng.*, vol. 13, pp. 106-112, Mar. 1988.
2. G. d'Aria, V. Zingarelli, "Fast adaptive equalizers for narrow-band TDMA mobile radio", *IEEE Trans. on veh. tech.*, vol. 40, pp. 392-404, May 1991.
3. J.G. Proakis, "Adaptive equalization for TDMA digital mobile radio", *IEEE Trans. on veh. tech.*, vol. 40, pp. 333-341, May 1991.
4. S.N. Crozier, D.D. Falconer and S.A. Mahmoud, "Least sum of squared errors (LSSE) channel estimation", *IEE Proc. Pt. F*, vol. 138, pp. 371-378, Aug. 1991.
5. R.A. Ziegler and J.M. Cioffi, "Estimation of time-varying digital channels", *IEEE Trans. on veh. tech.*, vol. 41, pp. 134-151, May 1992..
6. S.N. Crozier, "Short-block data detection techniques employing channel estimation for fading, time-dispersive channels", *Ph.D. dissertation, Carleton univ. Ottawa, ont., Canada*, Apr. 1990.
7. S. Haykin, *Adaptive filter theory.* Prentice-Hall, 1986.
8. A.P Clark, Z.C. Zhu and J.K. Joshi, "Fast start-up channel estimation", *IEE Proc. Pt. F*, vol. 131, pp. 375-382, July 1984.
9. A. Milewski, "Periodic sequences with optimal properties for channel estimation and fast start-up equalization", *IBM J. Res. Develop.*, vol. 27, pp. 426-431, Sept. 1983.
10. D. Poppen, "Design of training sequences for channel impulse response measurement", *COST 231, Vienna, Austria*, pp. 2-10, Jan. 1992.
11. M. Failli, "Digital land mobile radio communications" *COST 207 final report, CIC Inf. Techno. and Sciences, Brussels*, pp. 135-166, 1989.
12. A. Wautier, "Influence de l'estimation du canal sur les performances d'un égaliseur dans le cadre des radiocommunications avec le mobiles ", *Thèse de doctorat, univ. Paris XI, France*, Dec. 1992.

13. C. Mourot, "A high bit rate transmission technique in mobile radio cellular environment", *Proc. of the 42nd IEEE Veh. Tech. Conf., Denver, USA*, pp. 740-743, May 1992.
14. A. Wautier, J-C. Dany, C. Mourot, "Filtre correcteur de phase pour égaliseurs sous-optimaux", *Ann. Telecommun.*, vol. 47, pp. 359-369, Sep. 1992.

Probability of Packet Success for Asynchronous DS/CDMA with Block and Convolutional Codes

C. Trabelsi and A. Yongaçoğlu

Department of Electrical Engineering
University of Ottawa, Ontario, Canada K1N 6N5
Tel: +1 613 564 8251, Fax: +1 613 564 6882

Abstract. In this paper an approximation for the probability of packet success for asynchronous direct sequence code division multiple access (DS/CDMA) system is developed. In this approximation, an improved Gaussian model for the multiple access interference (MAI) is used and the effect of bit-to-bit error dependence caused by the delays and phases of the relative interfering signals is taken into account. Both block and convolutional codes are considered. A comparison is made with the results obtained by other methods which employ Gaussian model for the MAI and ignore the effect of bit-to-bit error dependence.

1 Introduction

There have been many publications on the computation of the probability of packet success for an asynchronous DS/CDMA system [1-5]. The exact calculation is computationally difficult, therefore the emphasis has been on approximations and bounds. One particularly attractive approximation is based on the assumptions that MAI is Gaussian and bit errors are independent. Such approximations do not generally give sufficiently accurate results [1].

An accurate but complicated approximation for the probability of packet success which takes into account the effect of bit-to-bit error dependence and uses an improved Gaussian model for the MAI has been recently presented [1-2]. However, in these papers [1-2] only block coding was considered.

Pursley and Taipale [3] analyzed the performance of a Viterbi decoder and derived an approximation for the probability of packet success in a DS/CDMA system using convolutional coding. However, the derivation of this bound assumes that the relative delays and phases of the interfering signals are zero (synchronous DS/CDMA). For asynchronous DS/CDMA such an assumption is no longer valid.

In this paper, we derive an accurate but also computationally simple approximation for the probability of packet success in an asynchronous DS/CDMA system. Results are presented for packets which employ convolutional or block coding. Initially we assume that unsuccessful transmissions are caused entirely by MAI. Later we also investigate the effect of thermal noise on the probability of packet success.

The organization of the paper is as follows: Section 2 presents the DS/CDMA system model considered in our study. In Section 3, we derive the approximation for the probability of packet success using block codes. The derivation of the approximation for convolutional codes is presented in Section 4. Section 5 gives some numerical results. Finally, in Section 6, we draw the conclusions of this work.

2 DS/CDMA System Model

The DS/CDMA system model used in our analysis is shown in Figure 1. If BPSK is assumed, the k'th user's transmitted signal has the form of:

$$s_k(t) = \sqrt{2P}\, b_k(t)\, a_k(t) \cos\left(w_c t + \theta_k\right) \tag{1}$$

where $b_k(t)$ is the data sequence of unit amplitude, positive and negative, rectangular pulses of duration T_b, $a_k(t)$ is the spreading signal consists of a periodic sequence of unit amplitude, positive and negative, rectangular pulses of duration T_c, P is the signal power, w_c represents the common center frequency and θ_k is the phase of the k'th carrier.

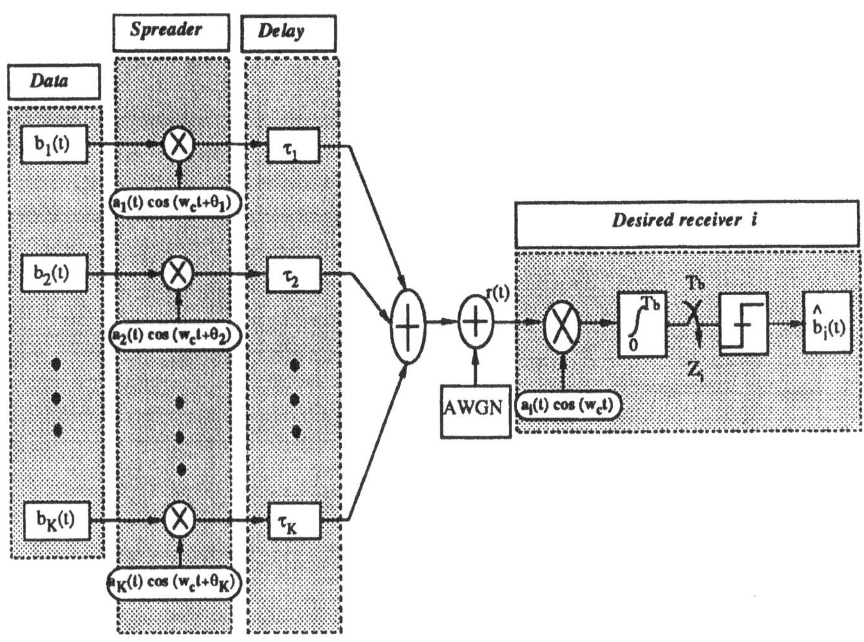

Figure 1: *DS/CDMA system model.*

For asynchronous DS/CDMA system, The received signal $r(t)$ could be expressed as:

$$
\begin{aligned}
r(t) &= \sum_{k=1}^{K} s_k(t-\tau_k) + n(t) \\
&= \sum_{k=1}^{K} \sqrt{2P}\, b_k(t-\tau_k)\, a_k(t-\tau_k) \cos(w_c t + \phi_k) + n(t)
\end{aligned} \tag{2}
$$

where τ_k and $\phi_k = (\theta_k - w_c\tau_k)$ are the time and phase shifts respectively, K is the number of simultaneous users and $n(t)$ is the thermal noise which is modeled as additive white Gaussian noise with zero mean and two-sided spectral density of $N_o/2$. If the received signal $r(t)$ is the input to a correlation receiver i matched to $s_i(t)$, the output is:

$$
Z_i = \int_0^{T_b} r(t)\, a_i(t-\hat{\tau}_i) \cos(w_c t + \hat{\phi}_i)\, dt \tag{3}
$$

where T_b is the data bit duration, $\hat{\tau}_i$ and $\hat{\phi}_i$ are the estimates of the time and phase shift of signal i. If the effect of thermal noise is neglected then, it has been shown [1] that the decision statistic Z_i of the desired receiver i, normalized with respect to the chip duration T_c, with all the signals at received power $P = 2$ and given $K-1$ interfering transmitters is given by:

$$
Z_i = N + \gamma \tag{4}
$$

The first term in Z_i represents the desired signal where N is the sequence length which is also the spreading factor ($N = T_b/T_c$) and the second term denoted by γ represents the MAI. If it is assumed that the MAI is a Gaussian process with variance ψ, then the probability of bit error for DS/CDMA is given by:

$$
p = Q(\frac{N}{\sqrt{\psi}}) \tag{5}
$$

3 Probability of Packet Success with Block Codes

In this section we present different methods of computing the probability of packet success for asynchronous DS/CDMA with block codes.

3.1 Using the Assumptions of Gaussian Distributed Interference and Independent Bit Errors

If the MAI is modeled as a Gaussian process with the interfering signals having random chip delays and phases, then it has been shown that the variance ψ is given by [1]:

$$
\psi = \frac{(K-1)\,N}{3} \tag{6}
$$

Substituting (6) into (5) the probability of channel bit error for asynchronous DS/CDMA is given by:

$$p = Q\left(\sqrt{\frac{3N}{K-1}}\right) \tag{7}$$

Now if we assume that the bit errors are independent, the probability of packet success for asynchronous DS/CDMA using block coding that can correct t or fewer errors is given by:

$$P_c(K) = \sum_{i=0}^{t} \binom{L_c}{i} p^i (1-p)^{L_c - i} \tag{8}$$

where p is given by (7) and $L_c = L/r$ is the number of coded bits per packet (r is the code rate and L is the number of information bits per packet).

3.2 Using the Assumptions of Improved Gaussian Distributed Interference and Dependent Bit Errors

It is well known that the computation of the probability of bit error for asynchronous DS/CDMA based on Gaussian assumption is not accurate especially for small number (e.g. less than 10) of simultaneous users [1]. Furthermore, for asynchronous DS/CDMA, the assumption of independent bit errors is no longer valid, because they are related through the phase and time delay of their transmission [1].

It has been shown in [1] that the variance of the MAI (ψ) is a function of the delays and phases of the interfering signals. Now if we treat ψ as a random variable which depends essentially on the delays and phases of the interfering signals, then the probability of channel bit error for asynchronous DS/CDMA becomes a function of Ψ and is given by:

$$p(\Psi) = Q\left(\frac{N}{\sqrt{\Psi}}\right) \tag{9}$$

The probability of packet success with block codes that can correct t or fewer errors is given by:

$$P_c(\Psi) = \sum_{i=0}^{t} \binom{L_c}{i} \left[Q\left(\frac{N}{\sqrt{\Psi}}\right)\right]^i \left[1 - Q\left(\frac{N}{\sqrt{\Psi}}\right)\right]^{L_c - i} \tag{10}$$

One can view (9) as channel conditional bit error probability and (10) as conditional probability of packet success. To obtain the probability of packet success when Ψ is random, we must average $P_c(\Psi)$ over the probability density function of Ψ. That is, we have to evaluate

$$P_c = E\{P_c(\Psi)\} = \int_0^{\infty} P_c(\psi)\, f_\Psi(\psi)\, d\psi \tag{11}$$

where $E\{.\}$ represents the expectation operation and $f_\Psi(\psi)$ is the probability density function of Ψ. Morrow and Lehnert [1-2] have evaluated equation (11) by finding the probability density function of Ψ. It has been shown that this way significant accuracy improvement over the standard Gaussian approximation is achieved. Furthermore, the bit-to-bit error dependencies within the packet is taken into consideration. However, the computations required to evaluate the probability density function of Ψ are very cumbersome. Holtzman [6] used a simple approximation for the computation of probability of bit error which does not require the probability density function of Ψ directly. By using a similar approach, we have computed the probability of packet success. The objective here is to compute (11) without carrying out the integration. This can be done by expanding $P_c(\Psi)$ using the Taylor series, i.e.,

$$P_c(\Psi) = P_c(\mu_\Psi)+(\Psi-\mu_\Psi)\; P_c^{'}(\mu_\Psi)+\frac{(\Psi-\mu_\Psi)^2}{2!}\; P_c^{''}(\mu_\Psi)+...+\frac{(\Psi-\mu_\Psi)^n}{n!}\; P_c^{n}(\mu_\Psi) \tag{12}$$

where μ_Ψ is the mean of Ψ and $P_c^n(\mu_\Psi)$ is the n'th derivative of $P_c(\Psi)$ evaluated at $\Psi = \mu_\Psi$. Taking the expectation of (12), one obtains

$$E\{P_c(\Psi)\} = P_c(\mu_\Psi) + \frac{\sigma_\Psi^2}{2!}\; P_c^{''}(\mu_\Psi) + ... + \frac{\mu_{\Psi,n}}{n!}\; P_c^{n}(\mu_\Psi) \tag{13}$$

where σ_Ψ^2 is the variance of Ψ and $\mu_{\Psi,n}$ is the n'th moment given by $E\{(\Psi - \mu_\Psi)^n\}$. We know that the first derivative of a function $P_c(\Psi)$ in which Ψ is a random variable could be approximated by:

$$P_c^{'}(\Psi) = \frac{P_c(\Psi+h) - P_c(\Psi-h)}{2h} \tag{14}$$

and the second derivative by:

$$P_c^{''}(\Psi) = \frac{P_c(\Psi+h) - 2P_c(\Psi) + P_c(\Psi-h)}{h^2} \tag{15}$$

where h is the step size, $h = \zeta\,\sigma_\Psi$, and ζ is a constant (a good value is found to be approximately 1.7). In the following analysis, we assume that only the first two terms in equation (13) are important. Substituting (15) into (13) one obtains:

$$E\{P_c(\Psi)\} = (1-\frac{1}{\zeta^2})\; P_c(\mu_\Psi) + \frac{1}{2\zeta^2}\; P_c(\mu_\Psi + \zeta\,\sigma_\Psi) + \frac{1}{2\zeta^2}\; P_c(\mu_\Psi - \zeta\,\sigma_\Psi) \tag{16}$$

From [1] we can deduce that the mean of Ψ is given by:

$$\mu_\Psi = \frac{(K-1)\,N}{3} \tag{17}$$

and the variance of Ψ is given by:

$$\sigma_\Psi^2 = \frac{K-1}{360}\;[23\,N^2 - (2-10K)\,N + (2-10K)] \tag{18}$$

Substituting (17), (18) into (16) one obtains a new approximation for the probability of packet success for asynchronous DS/CDMA using block codes:

$$\begin{aligned} P_c(K) &= (1-\frac{1}{\zeta^2}) \sum_{i=0}^{t} \binom{L_c}{i} p_1^i (1-p_1)^{L_c-i} \\ &+\frac{1}{2\zeta^2} \sum_{i=0}^{t} \binom{L_c}{i} p_2^i (1-p_2)^{L_c-i} \\ &+\frac{1}{2\zeta^2} \sum_{i=0}^{t} \binom{L_c}{i} p_3^i (1-p_3)^{L_c-i} \end{aligned} \tag{19}$$

where p_1, p_2 and p_3 are given by

$$p_1 = Q\left[\left(\frac{K-1}{3N}\right)^{-1/2}\right] \tag{20}$$

$$p_2 = Q\left[\left(\frac{(K-1)(N/3)+\zeta\,\sigma_\Psi}{N^2}\right)^{-1/2}\right] \tag{21}$$

$$p_3 = Q\left[\left(\frac{(K-1)(N/3)-\zeta\,\sigma_\Psi}{N^2}\right)^{-1/2}\right] \tag{22}$$

Although many terms appear in (19), it is much easier to compute than (11).

4 Probability of Packet Success with Convolutional Codes

If we consider uncoded packet transmission in packets of length L bits, then the probability of packet success in a binary symmetric channel is given by:

$$P_c = (1-p)^L \tag{23}$$

where p in this case is the probability of bit error. When convolutional codes are used, we can no longer assume that the bit errors are independent, because errors at the output of the Viterbi decoder are dependent even if the errors at the decoder input may be independent. Furthermore, in an asynchronous DS-CDMA system the bit errors on the channel(at the input of the decoder) are dependent.

In the analysis of convolutional codes, it has been assumed that bit errors at the output of Viterbi decoder cluster within a short segment of decoded data [3,8]. These clusters are known as error events. These error events are statistically independent, so that the probability of an event starting at a given time is independent of an event starting at any other time. By assuming an independent error event, the probability of packet success could be approximated by:

$$P_c \simeq (1 - P_u(p))^L \tag{24}$$

where $P_u(p)$ is the union bound on the first event error probability and p is the channel bit error probability for DS/CDMA. In [3] Pursley and Taipale considered two values of p in their analysis. The first one is computed by using the characteristic function of the decision statistic with the assumption that the relative phases and delays are zero (synchronous DS/CDMA) and is given by:

$$p = Q(\sqrt{\frac{E_b}{N_o}}) + \frac{1}{\pi} \int_0^\infty u^{-1} \sin(u)\, \Phi_2(u)\, (1 - \Phi_1(u))\, du \tag{25}$$

where E_b is the energy per information bit, $\Phi_2(u)$ and $\Phi_1(u)$ are given by:

$$\Phi_1(u) = \left[\cos(\frac{u}{N})\right]^{N(K-1)} \tag{26}$$

$$\Phi_2(u) = \exp\left(\frac{-N_o}{2E_b}\, u^2\right) \tag{27}$$

The second p value used is based on the assumption that MAI is Gaussian with the interfering signals are chips and phases aligned with the desired signal (a rate 1/2 convolutional code was considered for both cases). Thus p can be written as:

$$p = Q\left\{\left(\frac{K-1}{N} + \frac{1}{E_b/N_o}\right)^{-1/2}\right\} \tag{28}$$

As mentioned before, due to different delays and phases of the interfering signals, the errors at the decoder input could not be assumed independent. To take into account the effect of bit-to-bit error dependence, a new approximation for the probability of packet success using convolutional codes is developed.

Now if we treat the variance of the MAI, Ψ as a random variable as in the previous section, then, the conditional probability of packet success for asynchronous DS/CDMA with convolutional codes is given by:

$$P_c(\Psi) = [1 - P_u(p(\Psi))]^L \tag{29}$$

where $p(\Psi)$ is the conditional channel bit error probability. To obtain the probability of packet success when Ψ is random we have to evaluate (11) with $P_c(\Psi)$ given by (29). Following the same procedure as in the previous section, we can develop a new approximation for the probability of packet success for asynchronous DS/CDMA using convolutional codes:

$$P_c(K) = (1 - \frac{1}{\zeta^2})\,[1 - P_u(p_1)]^L + \frac{1}{2\zeta^2}\,[1 - P_u(p_2)]^L + \frac{1}{2\zeta^2}\,[1 - P_u(p_3)]^L \tag{30}$$

where p_1, p_2 and p_3 are given by (20), (21) and (22) respectively.

There exists many approximations for $P_u(p)$. In this paper we consider the bound taken in [3] because it represents a tight bound with respect to many other bounds. It is given by:

$$P_u(p) = \binom{2n_o - 1}{n_o} 2^{-2n_o-1} \{ [T(D) + T(-D)] + D\ [T(D) - T(-D)] \} |_{D=2\sqrt{p}} \quad (31)$$

where $T(D)$ is the transfer function of the convolutional code and n_o is half the free distance of the code (or half the "free distance plus one", if the distance is odd). In our study convolutional codes with rate 1/2 and constraint length (CL) 3 and 7 were considered. For the convolutional code with $r = 1/2$ and $CL = 3$, $T(D)$ was taken as:

$$T(D) = \frac{D^5}{1 - 2D} \quad (32)$$

and for $CL = 7$, $T(D)$ is developed in [7]. The first four terms are given by:

$$T(D) = 11D^{10} + 38D^{12} + 193D^{14} + 1331D^{16} + \ldots \quad (33)$$

In our numerical evaluation, the first 64 terms developed in [7] were considered.

The previous computation of the probability of packet success for asynchronous DS/CDMA using block and convolutional codes assumes that MAI is the only source of errors. The inclusion of the thermal noise is straightforward. If the additive white Gaussian noise is added to the sum of the K spread-spectrum signals then p_1, p_2 and p_3 are modified as:

$$p_1 = Q\left[\left(\frac{K-1}{3N} + \frac{1}{2rE_b/N_o}\right)^{-1/2}\right] \quad (34)$$

$$p_2 = Q\left[\left(\frac{(K-1)(N/3) + \zeta\,\sigma_\psi}{N^2} + \frac{1}{2rE_b/N_o}\right)^{-1/2}\right] \quad (35)$$

$$p_3 = Q\left[\left(\frac{(K-1)(N/3) - \zeta\,\sigma_\psi}{N^2} + \frac{1}{2rE_b/N_o}\right)^{-1/2}\right] \quad (36)$$

5 Numerical Results

Figure 2 shows the probability of packet success for asynchronous DS/CDMA using block codes versus the number of simultaneous users for the following three cases ($N = 31$, $t = 0$, 10, $L = 1000$ and $E_b/N_o = \infty$):

Case 1: using Gaussian model for MAI and independent bit errors assumption.

Case 2: using equation (11) evaluated in paper [1].

Case 3: using equation (19) developed in this paper.

Comparing case 2 and case 1, we can see that using Gaussian model for MAI and independent bit errors assumption results in very optimistic values for small

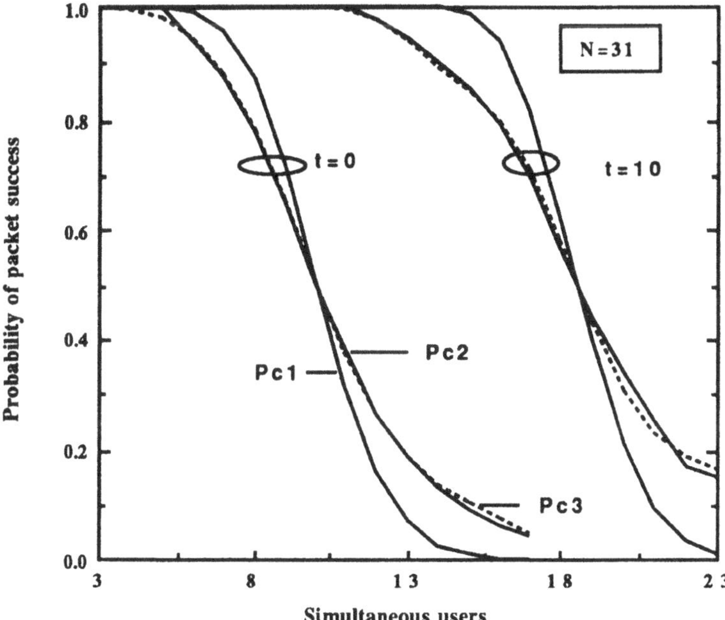

Figure 2: *Probability of packet success for asynchronous DS/CDMA with block codes (Pc1: based on Gaussian and independent bit errors, Pc2: paper [1], Pc3: this paper); spreading factor* $N = 31$.

number of users and very pessimistic values for large number of users. From this figure we can say that our equation for the probability of packet success for asynchronous DS/CDMA is in good agreement with the one given in [1]. Note that our equation is much simpler to compute.

Figure 3 shows the probability of packet success for DS/CDMA using convolutional codes versus the number of simultaneous users for the following cases ($N = 31$, $L = 1000$, $r = 1/2$, $CL = 3$, 7 and $E_b/N_o = \infty$).
Case 1: $P_c(K)$ considered in paper [3] based on equations (24), (28) and (33) where bit errors are independent and MAI is modeled as Gaussian with the interfering signals are chips and phases aligned with the desired signal (synchronous case).
Case 2: $P_c(K)$ for asynchronous DS/CDMA based on equation (24), (7) and (33) where bit errors are independent and MAI is modeled as Gaussian with random chip delays and phases.
Case 3: $P_c(K)$ for asynchronous DS/CDMA based on equations (30), (20-22) and (31) which are developed in this paper. Here the bit errors are dependent and MAI is modeled as an improved Gaussian process.
As we have deduced with block codes, the probability of packet success using convolutional codes based on Gaussian model for the MAI and independent bit errors is optimistic when the number of simultaneous users is small and pes-

simistic for large number of users. The probability of packet success based on the Gaussian model for the MAI with the interfering signals are chips and phases aligned with the desired signal is a very loose lower bound.

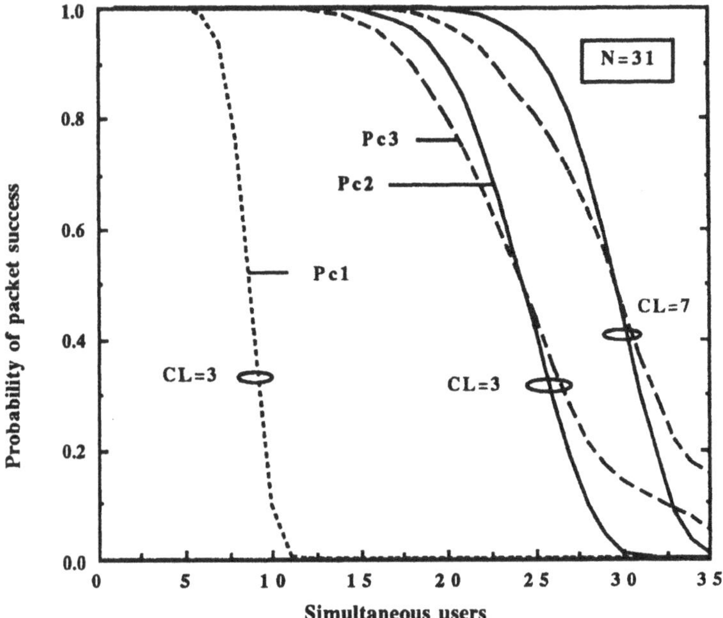

Figure 3: *Probability of packet success for DS/CDMA with* $CL = 3$, *7, rate 1/2 convolutional code (Pc1: case 1, Pc2: case 2, Pc3: case 3); spreading factor* $N = 31$.

Figure 4 shows the probability of packet success for asynchronous DS/CDMA for the uncoded case with different values of the spreading factor, N =63, 127 ($L = 1000$ and $E_b/N_o = \infty$). From these figures we can see that as N increases the difference on the probability of packet success becomes less significant between the models based on the Gaussian and independent bit errors and the improved Gaussian and dependent bit errors.

All previous figures that were shown for the probability of packet success did not take into account the effect of thermal noise. Figure 5 shows the probability of packet success for asynchronous DS/CDMA for the uncoded case with spreading factor, $N = 255$ and different E_b/N_o. We can see that the effect of thermal noise can be significant and in that case its effect should be taken into account in the analysis of the DS/CDMA packet network.

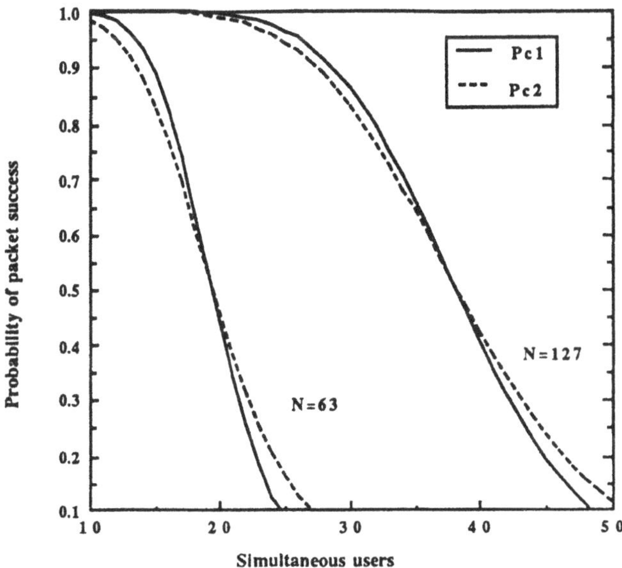

Figure 4: *Probability of packet success for uncoded asynchronous DS/CDMA with spreading factor $N = 63$ and 127. (Pc1: based on Gaussian and independent bit errors, Pc2: this paper (improved Gaussian and dependent bit errors)).*

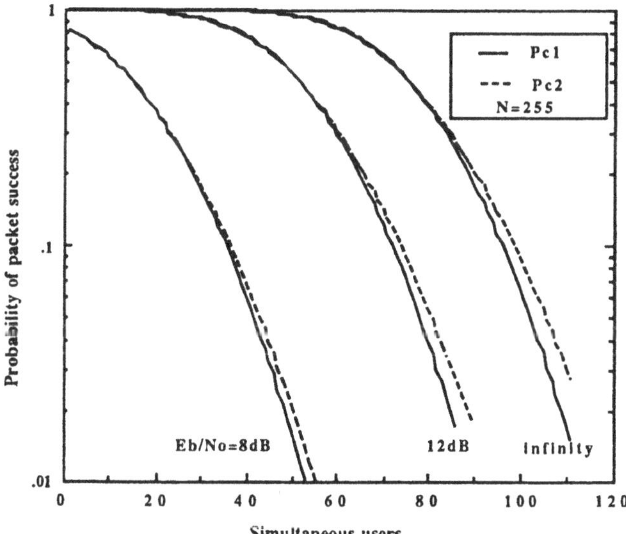

Figure 5: *Probability of packet success for uncoded asynchronous DS/CDMA with spreading factor $N = 255$ and different E_b/N_o. (Pc1: based on Gaussian and independent bit errors, Pc2: this paper (improved Gaussian and dependent bit errors)).*

6 Conclusions

A simple approximation for the probability of packet success in an asynchronous DS/CDMA system with block and convolutional codes has been developed. This approximation is more accurate than those based on the standard Gaussian assumption, and also takes into account the effect of bit-to-bit error dependence caused by relative delays and phases of the interfering signals. As demonstrated, for some special cases, this approximation for the probability of packet success can be used to easily analyze the throughput-delay performance of slotted asynchronous DS/CDMA packet networks [9-10].

References

[1] R. Morrow and J. Lehnert, "Bit-to-Bit Error Dependence in Slotted DS/SSMA Packet Systems with Random Signature Sequences", IEEE Trans. on Commun., Vol. Com-37, No.10, pp. 1052-1061, October 1989.

[2] R. Morrow and J. Lehnert, "Packet Throughput in Slotted ALOHA DS/SSMA Radio Systems with Random Signature Sequences", IEEE Trans. on Commun., Vol. Com-40, No.7, pp. 1223-1230, July 1992.

[3] M. B. Pursley, and D. J. Taipale, "Error Probabilities for Spread-Spectrum Packet Radio with Convolutional Codes and Viterbi Decoding", IEEE Trans. on Commun., Vol. Com-35, No.1, pp. 1-12, January 1987.

[4] J. S. Lehnert and M. B. Pursley, "Error Probabilities for Binary Direct Sequence Communications with Random Signature Sequences", IEEE Trans. on Commun., Vol. Com-35, No.1, pp. 87-98, January 1987.

[5] K. Yao, "Error Probability of Asynchronous Spread Spectrum Multiple-Access Communication Systems", IEEE Trans. on Commun., Vol. Com-25, No.8, pp. 803-809, August 1977.

[6] J. M. Holtzman, "A simple, Accurate Method to Calculate Spread-Spectrum Multiple-Access Error Probabilities", IEEE Trans. on Commun., Vol. Com-40, No.8, pp. 461-465, March 1992.

[7] J. S. Storey and F. Tobagi, "Throughput Performance of a Direct Sequence CDMA Packet Radio Network", SEL Technical Report No.85-277, Stanford University, June 1985.

[8] D. P. Taylor and M. Grossman, "The Effect of FEC on Packet Error performance in a VSAT Network", Proc. of IEEE Globecom, pp. 42.1.1-42.1.4, December 1986.

[9] C. Trabelsi and A. Yongaçoğlu, "Throughput Analysis of Slotted Asynchronous DS/CDMA System", Proc. of IEEE Globecom, pp. 47.5.1-47.5.5, December 1993.

[10] C. Trabelsi and A. Yongaçoğlu, "Performance of Asynchronous DS-CDMA for Personal Access Satellite System", Proc. of IEEE Int. Conf. on Universal Personal Commun., pp. 819-823, October 1993.

Sequential Decoding on Intersymbol Interference Channels under the Pe-Criterion

Ivonete Markman[1*] and John B. Anderson[2]

[1] ECE Dept., Rose-Hulman Institute of Technology, Terre Haute, IN 47803-3999, USA. E-mail: Markman@EE.Rose-Hulman.Edu

[2] ECSE Dept., Rensselaer Polytechnic Institute, Troy, NY 12180-3590, USA. E-mail: anderson@ecse.rpi.edu

Abstract. The Pe-criterion is a recent analysis of sequential decoding of tree codes based on the design condition that the decoder must achieve a set error probability *Pe*. This work extends the criterion to channels with intersymbol interference (ISI), assuming binary random coding, and both hard and soft decision decoding. The result is a definite boundary for the tree search region and a well defined estimate of path numbers searched. Comparisons among the channels show the influence of ISI in the decoding process. We show how ISI increases the search region and number of paths searched as compared to memoryless channels. Furthermore, we show that the use of soft decisions may not only result in large savings as compared to hard decisions, but indeed be the key to assure feasibility of the limited search decoder.

1 Introduction

The Pe-criterion is a recent analysis [1] of limited search sequential decoding of tree codes that diverges from the standard approach [4], by assuming that the decoder may fail with a probability $Pe > 0$. It therefore represents a simpler analysis, which more closely resembles the behavior of real decoders. The criterion was initially applied to the BSC [1] and later extended to the binary-input memoryless soft decision channels resulting from the output quantization of an additive white Gaussian noise (AWGN) channel [5].

This work extends the Pe-criterion to the sequential decoding of binary tree codes on a class of channels with memory which arises when the waveform channel is linear and produces intersymbol interference in the received signal. Intersymbol interference (ISI) is the effect by which an input symbol to the channel influences the channel output in more than one interval. Some causes of ISI are bandwidth constraints in the system introduced by filters, and multipath propagation in radio channels. In order to analyze the ISI effect, we assume a discrete-time channel model of a PAM system [1, 3], as shown in Fig. 1. For this

* Supported by CNPq - Conselho Nacional de Desenvolvimento Científico e Tecnológico, Brazil.

model, the unquantized channel output samples are given by the expression

$$y'_j = v_j + w_j = \sum_{k=0}^{m} H_k x_{j-k} + w_j \tag{1}$$

where $\mathbf{H} = (H_0, H_1, \ldots, H_m)$ represents the unit-sample response of a finite length channel filter with m memory elements, and w_j, for $j > 0$, are i.i.d. Gaussian random variables with mean zero and variance $N_0/2$. In addition, we restrict our attention to binary PAM signaling, i.e., $x_j \in \{1, -1\}$ (corresponding to bits 0 and 1, respectively), and real, time-invariant, unit-sample response $\mathbf{H}$ of the discrete-time channel filter. The model in (1) can be shown [3] to be the discrete-time equivalent to many continuous-time (waveform) digital communication systems with ISI.

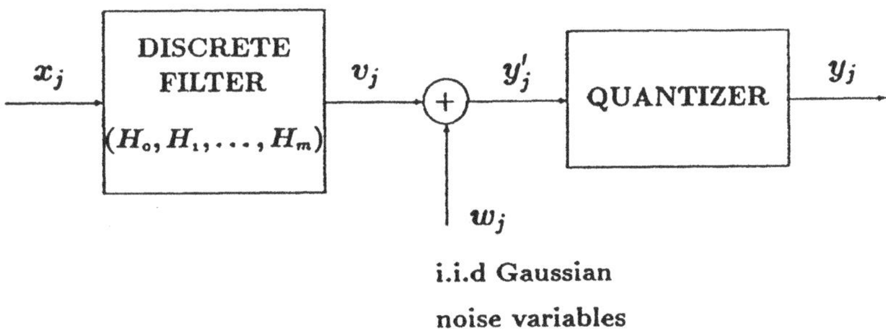

Fig. 1. Discrete-time channel model of the PAM system.

For a discrete channel filter with m memory elements, the last state s_{j-1}, at the interval j, is given by the contents of the memory elements of the filter at that interval, so that there are $M = 2^m$ distinct states. Therefore, as we can see from Fig. 1, the current state at the same interval, s_j, is given by the current channel input symbol x_j and the contents of the first $m-1$ memory elements of the channel filter. This implies that the state transition probability $P(s_j/x_j, s_{j-1})$ assumes only the values 0 or 1, which results in the following property:

Proposition 1. *For the discrete ISI channel, knowledge of the couple* (x_j, s_{j-1}) *completely defines* s_j.

We are then interested in applying the Pe-criterion for sequentially decoding binary tree codes that are transmitted through the discrete ISI channel model. Therefore, the input sequence to the channel is a code word of a tree code, $\mathbf{x_i}$, for $i = 1, 2, \ldots, 2^{nR}$, where R is the code rate and n is the code word length. We assume a first symbol decoder[3] in this analysis, and a random code in which the symbols are i.i.d. equiprobable random variables. These codes do not necessarily represent the capacity-maximizing code ensemble, but they resemble the pattern of practical tree codes.

This paper is organized as follows. Sect. 2 discusses the optimality criterion and metric. Sect. 3 applies the Pe-criterion to binary-input binary-output ISI channels, which correspond to hard decision decoding. Sect. 4 applies the criterion to binary-input continuous-output ISI channels, which correspond to the ultimate case of soft decision. In both Sects. 3 and 4, the shape of the search region is sketched in a generalized distance versus depth diagram, the number of paths searched is calculated, and simple examples are numerically analyzed. Sect. 5 concludes by showing the main results of this work.

2 The Metric Definition

In order to apply the Pe-criterion to the discrete ISI channel, we have to consider the optimality criterion and definition of a metric. Two possible cases may be considered, depending on the information available at the decoder.

2.1 No State Information Available

For this type of decoder, when the channel output sequence $\mathbf{y}$ is received, there is no knowledge of the channel state sequence. The maximum a posteriori (MAP) decision rule would choose the code word $\mathbf{x_i}$ which maximizes

$$P(\mathbf{x_i}/\mathbf{y}) = \frac{P(\mathbf{x_i})P(\mathbf{y}/\mathbf{x_i})}{P(\mathbf{y})} \tag{2}$$

where $\mathbf{x_i} = (x_{i1}, x_{i2}, \ldots, x_{in})$, $\mathbf{y} = (y_1, y_2, \ldots, y_n)$ and n is the sequence length.

A metric choice could then be the same used for the sequential decoding on memoryless channels [1, 4, 5], given by

$$z(\mathbf{y}, \mathbf{x_i}) = \log_2 \left[\frac{P(\mathbf{y}/\mathbf{x_i})}{P(\mathbf{y})}\right] - nB \tag{3}$$

where B is a generic bias term.

However, it can be easily shown that for the discrete ISI channel, the probabilities on the right-hand side of (3) are not suitable for sequential decoding, since they turn out not to be recursive.

---

[3] A first symbol decoder is a machine whose job is the decoding and release of the first code tree branch only. Practical decoders are adaptations of this one.

2.2 Starting State (s_0) Information Available

In this case, there is knowledge of the starting state s_0 before the channel output sequence $\mathbf{y}$ is received. The MAP decision rule would choose the code word $\mathbf{x_i}$ which maximizes

$$P(\mathbf{x_i}/\mathbf{y}, s_0) = \frac{P(\mathbf{x_i})P(\mathbf{y}/\mathbf{x_i}, s_0)}{P(\mathbf{y}/s_0)} \tag{4}$$

A metric would then be defined by the expression

$$z(\mathbf{y}, \mathbf{x_i}/s_0) = \log_2 \left[\frac{P(\mathbf{y}/\mathbf{x_i}, s_0)}{P(\mathbf{y}/s_0)}\right] - nB \tag{5}$$

again with B being a generic bias term.

For the discrete ISI channel, the probabilities on the right-hand side of (5) become

$$\begin{aligned} P(\mathbf{y}/\mathbf{x_i}, s_0) &= \prod_{j=1}^{n} P(y_j/x_{ij}, s_{j-1}) = \prod_{j=1}^{n} P(y_j/s_{j-1}, s_j) \\ P(\mathbf{y}/s_0) &= \sum_{\mathbf{x_i}} P(\mathbf{x_i})P(\mathbf{y}/\mathbf{x_i}, s_0) \end{aligned} \tag{6}$$

where Property 1 was applied in solving $P(\mathbf{y}/\mathbf{x_i}, s_0)$.

We observe in (6) that $P(\mathbf{y}/\mathbf{x_i}, s_0)$ is a recursive expression, however, $P(\mathbf{y}/s_0)$ is not. Therefore, we still cannot derive a sequential decoder for the optimum metric in (5). Instead, let us define a suboptimum decoder that substitutes for $P(\mathbf{y}/s_0)$ the expression

$$\prod_{j=1}^{n} P(y_j/s_{j-1}) = \prod_{j=1}^{n} \left[\sum_{x_{ij}} P(x_{ij})P(y_j/x_{ij}, s_{j-1})\right] \tag{7}$$

which is a recursive expression, therefore suitable for sequential decoding.

Hence, from (6) and (7), we can define the suboptimum metric to be

$$z(\mathbf{y}, \mathbf{x_i}/s_0) = \sum_{j=1}^{n} z_i = \sum_{j=1}^{n} \left\{\log_2 \left[\frac{P(y_j/s_{j-1}, s_j)}{P(y_j/s_{j-1})}\right] - B\right\} \tag{8}$$

In this study, our analysis of the discrete ISI channel is based on the suboptimum metric defined above. The particular choice in (8) was empirical, and other choices are possible. One question that arises at this point is how good our choice is, compared to the optimum metric. We discuss this in the next section.

3 The Binary-Input Binary-Output ISI Channel

The binary-input, binary-output ISI channel, which corresponds to hard decision decoding, is obtained in Figure 1 when the quantizer delivers binary outputs to the decoder.

3.1 Evaluation of the Drop Line

The drop line represents an important mode of failure for a Pe-criterion decoder, i.e., that of dropping the correct path during the decoding process. It is defined as the line outside of which the correct path wanders with probability Pe or less, and its plot in a generalized distance versus depth diagram characterizes the shape of the search region.

The derivation of the drop line for discrete ISI channels utilizes the general drop line bound [6], a special case of the Chernoff bound [4], which is stated as follows:

Theorem 2 (General Drop Line Bound). *The probability that the metric z of the correct path ever wanders below $-\eta$ satisfies*

$$P\{z \leq -\eta\} \;\leq\; 2^{h\eta}\,, \qquad \eta > 0 \tag{9}$$

with h being the negative solution value of

$$1 = G(z) = E\left[2^{hz}\right] = \sum_{z} P(z) 2^{hz} \tag{10}$$

where $z = \sum_{j=1}^{n} z_j$, z_j are the symbol metric increments, n is the sequence length and $P(z)$ is the probability of the metric of the correct path.

Proof. Given in reference [6]. □

To make sure the Pe-criterion is satisfied, we set

$$2^{h\eta} = Pe \tag{11}$$

By assuming starting state information available, the discrete ISI channel metric was defined in (8), as a derivation of the optimum metric in (5). The general drop line bound may be used to state a property of the suboptimum metric.

Theorem 3. *For the discrete ISI channel, the application of the Pe-criterion to the suboptimum decoder (whose metric is defined in (8)) with a certain probability of error, Pe, is equivalent to applying the Pe-criterion to the optimum decoder[4] (whose metric is defined in (5)), but with a possibly smaller probability of error ($\leq Pe$).*

Proof. Given in reference [6]. □

The importance of Theorem 3 resides in the fact that the Pe-criterion is still satisfied for the optimum decoder, when it is satisfied for the suboptimum one. But for a fixed probability of error, Pe, the equivalent bound for the optimum decoder will be looser. Although looser, we expect that it is still relatively tight.

An important consequence of Theorem 3 is stated as a corollary below.

[4] The Pe-criterion may still be applied to the optimum metric, however, the decoder is not sequential.

Corollary 4. *For the discrete ISI channel, the number of paths searched, when applying the Pe-criterion to the suboptimum decoder (whose metric is defined in (8)) with a certain probability of error, is an upper bound to the number of paths searched by the optimum decoder (whose metric is defined in (5)) with the same probability of error.*

Proof. Given in reference [6]. □

Altogether, the results in Theorem 3 and Corollary 4 mean that, when utilizing the suboptimum decoder, we are being conservative. We are upper bounding the number of paths needed to estimate the correct path. Since the probabilities of error for the optimum and suboptimum equivalent decoders in Theorem 3 are expected to be close, then so should be the number of paths searched for the suboptimum and optimum decoders in Corollary 4.

For the ISI channel, since the metric in (8) is a function of the starting state, and the input and output sequences, we may observe that

$$G(z) = \sum_{z} P(z)\, 2^{hz} = \sum_{\mathbf{s}} \sum_{\mathbf{y}} P(\mathbf{s}, \mathbf{y})\, 2^{h\left(\log_2 \left[\prod_{j=1}^{n} \frac{P(y_j/s_{j-1}, s_j)}{P(y_j/s_{j-1})}\right] - nB\right)} \tag{12}$$

where we used Property 1, in the second equality.

By applying (12) into (10), and reordering terms, we derive the expression

$$2^{nhB} = \sum_{s_0} P(s_0) \prod_{j=1}^{n} \left\{ \sum_{s_j} P(s_j/s_{j-1}) \sum_{y_j} \frac{P(y_j/s_{j-1}, s_j)^{(h+1)}}{P(y_j/s_{j-1})^{h}} \right\} \tag{13}$$

However, the form of (13) can be simplified if we define the modified state transition matrix

$$\boldsymbol{\Lambda}_\ell = \left[\Lambda_{s_{\ell-1}, s_\ell}\right] = \left[P(s_\ell/s_{\ell-1}) \sum_{y_\ell} \frac{P(y_\ell/s_{\ell-1}, s_\ell)^{(h+1)}}{P(y_\ell/s_{\ell-1})^{h}} \right] \tag{14}$$

where $s_{\ell-1}, s_\ell \in \{0, 1, \ldots, M-1\}$ and $\ell = 1, 2, \ldots, n$.

If we then substitute (14) into (13), we get

$$2^{nhB} = \sum_{s_0} P(s_0) \sum_{s_1} \Lambda_{s_0, s_1} \sum_{s_2} \Lambda_{s_1, s_2} \;\cdots\; \sum_{s_n} \Lambda_{s_{n-1}, s_n} \tag{15}$$

It can be easily verified from (14) that $\boldsymbol{\Lambda} = \boldsymbol{\Lambda}_1 = \boldsymbol{\Lambda}_2 = \ldots = \boldsymbol{\Lambda}_n$. In addition, we observe that

$$\sum_{s_\ell} \Lambda_{s_{\ell-1}, s_\ell} \sum_{s_{\ell+1}} \Lambda_{s_\ell, s_{\ell+1}} = \sum_{s_{\ell+1}} \boldsymbol{\Lambda}^2_{\ell+1} = \sum_{s_\ell} \boldsymbol{\Lambda}^2_\ell \tag{16}$$

where $\boldsymbol{\Lambda}^2_\ell$ represents the matrix product $\boldsymbol{\Lambda}_\ell \times \boldsymbol{\Lambda}_\ell$.

Hence, by applying (16) into (15), we derive

$$2^{nhB} = \sum_{s_0} P(s_0) \sum_{s_1} \boldsymbol{\Lambda}^n_1 \tag{17}$$

From the theory of matrices, it can be proven [4, 6] that, as n grows

$$\frac{1}{n}\log\left[\sum_{s_0} P(s_0)\sum_{s_1} \boldsymbol{\Lambda}_1^n\right] \longrightarrow \frac{1}{n}\log\left[\lambda_{\max}^n\right] = \log\left[\lambda_{\max}\right] \tag{18}$$

where $\lambda_{\max}$ is the largest eigenvalue of the matrix $\boldsymbol{\Lambda}$ defined in (14). Then, by substituting (18) into (17), we get

$$B = \frac{1}{h}\log_2\left[\lambda_{\max}\right] \tag{19}$$

For values of h in the range $-1 \leq h \leq 0$, the bias B in (19) assumes corresponding values in the range $0 \leq B \leq 1$.

Now, by assuming binary equiprobable random coding, we have that

$$P(s_j/s_{j-1}) = \begin{cases} P(x_{ij}) = \frac{1}{2} & \text{if } P(y_j/s_{j-1}, s_j) \neq 0 \\ 0 & \text{otherwise} \end{cases} \tag{20}$$

Hence, from (20) above, we can express $P(y_j/s_{j-1})$ as

$$P(y_j/s_{j-1}) = \sum_{s_j} P(s_j/s_{j-1})P(y_j/s_{j-1}, s_j) = \frac{1}{2}\sum_{s_j} P(y_j/s_{j-1}, s_j) \tag{21}$$

And, from (8) and (21), we derive

$$z(\mathbf{y}, \mathbf{x_i}/s_0) = \sum_{j=1}^{n}\log_2\left[\frac{P(y_j/s_{j-1}, s_j)}{\sum_{s_j} P(y_j/s_{j-1}, s_j)}\right] + (1-B)n \tag{22}$$

By rewriting (22) above as a linear function of the code word depth n, we get the form

$$z = s(h)n - D \tag{23}$$

where $s(h) = 1 - B$ is a function of h and the generalized distance D is defined as

$$D = -\sum_{j=1}^{n}\log_2\left[\frac{P(y_j/s_{j-1}, s_j)}{\sum_{s_j} P(y_j/s_{j-1}, s_j)}\right] \tag{24}$$

The constraint introduced in (11) implies that the region identified by the expression $z \geq -\eta$ contains the search region, since each path will be dropped by the decoder whenever its metric falls below $-\eta$. From (23), this region is equivalent to

$$D \leq s(h)n + \eta(h) \tag{25}$$

with the equality above defining the boundary of the region, for a given value of h.

Equation (25) shows that, by varying the value of h, or equivalently the bias B, different values are generated for the slope $s(h)$. Identically, different values of the generalized distance D-axis intercept, $\eta(h)$, are also generated through (11). Figure 2 shows a typical plot of the generalized distance D versus n, according to the equality in (25), for different values of the bias B. The figure is composed of a series of straight lines with varying slopes $s(h)$ and intercepts $\eta(h)$. Now consider a depth value, n. If, for all possible values of h, we choose the smallest among all the generated values of D, and call it $g(n)$, we will derive the curve

$$g(n) = s(\hat{h})n + \eta(\hat{h}) \tag{26}$$

which is the loci of the tangency points to the straight lines defined by the equality in (25). In other words, for each n, there is a value of h (named $\hat{h}$), for which (25) holds with equality and is tangent to the curve in (26).

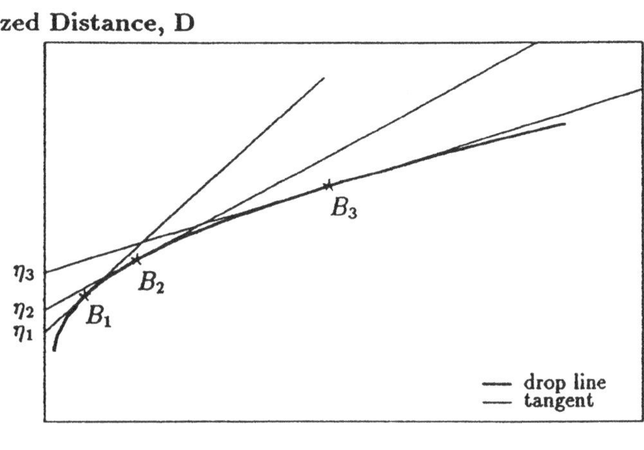

Fig. 2. Illustration of the drop line construction, showing three of the tangents to the line, corresponding to three different values of the bias B (defined by the corresponding value $\hat{h}$ in (26)).

It is desirable to find an analytical expression for $g(n)$. By calculus, it can be shown that [1, 6]

$$\begin{aligned} n &= -\frac{\log_2(Pe)}{\hat{h}^2 \left.\frac{dB}{dh}\right|_{h=\hat{h}}} = -\frac{\eta(\hat{h})}{\hat{h} \left.\frac{dB}{dh}\right|_{h=\hat{h}}} \\ g(n) &= \frac{\log_2(Pe)}{\hat{h}} \left[1 - \frac{s(\hat{h})}{\hat{h} \left.\frac{dB}{dh}\right|_{h=\hat{h}}}\right] \end{aligned} \tag{27}$$

where

$$\frac{dB}{dh} = -\frac{B}{h} + \frac{1}{h\lambda_{\max}\ln(2)}\frac{d\lambda_{\max}}{dh} \tag{28}$$

The numerical solution of (27) shows that, in order to obtain non-negative values for n and $g(n)$, which define our region of interest, $\hat{h}$ must be in the range $-1 \leq \hat{h} \leq 0$. This is equivalent to the bias being in the range $0 \leq B \leq 1$, in (19).

The curve in (27) is the drop line. We observe that it depends on the channel parameters and on the probability of error Pe, but not on the coding rate, R. Of course, this curve is an approximation to the real drop line, which is a staircase type of function. In the real problem, both the drop line $g(n)$ and the depth n can only assume discrete values, because of the discrete nature of the channel. By assuming those variables to be continuous, we derived a drop line that has, however, the same trend as the real one.

3.2 Number of Paths Searched

The calculation of the number of paths searched for non-backtracking searches is based on estimating the number of incorrect subset (ISS) paths that lie below the drop line. Every drop line tangent is a bound to the drop line. By calculating the number of ISS paths below each tangent, and optimizing this number over all the tangents, it is possible to get a tight bound. The method is derived from the one used for memoryless channels, based on difference equations [1, 6]. The new method takes into account the dependence among channel symbols introduced by the channel memory, by considering a length τ in branches. This can be accomplished by a recursive use of the difference equation τ times. We then derive

$$\mathrm{EN}\eta(z) = b^{\tau}\sum_{\mu_\tau} P(\mu_\tau)\mathrm{EN}\eta(z+\mu_\tau) \tag{29}$$

where $\mathrm{EN}\eta(z)$ is the expected number of ISS paths that hit the metric limit $-\eta$, in front of an original node with metric z, $\mu_\tau = \sum_{j=1}^{a\tau}\mu_j$ is the accumulated metric over τ branches ($a\tau$ symbols), given by $\mu_\tau = \sum_{j=1}^{a\tau}\mu_j$, μ_j is the j-th symbol metric increment, and $P(\mu_\tau)$ is the probability of the accumulated metric increment μ_τ.

Since (29) is a difference equation, its solution can be found by replacing $\mathrm{EN}\eta(z)$ with a general term $K\sigma^z$, and solving the resulting characteristic equation for the values of σ. The substitution in (29) results in

$$1 = b^{\tau}\sum_{\mu_\tau} P(\mu_\tau)\sigma^{\mu_\tau} \tag{30}$$

Because of random coding, the ISS paths are independent of the correct path after the node from which they initially diverge. And since the channel output sequence $\mathbf{y}$ depends only on the correct path and the added noise, it is independent of the ISS path from the diverging node on, if we assume knowledge of the node state. If we then assume that the starting node (with starting state

s_0) is also the diverging node, the entire ISS path is independent of the entire channel output sequence. Since, for an ISS path, the metric is a function of the starting state s_0, the ISS code word $\mathbf{x_i}$ and the channel output sequence $\mathbf{y}$, then the average in (30) may be equivalently taken over the set $(s_0, \mathbf{x_i}, \mathbf{y})$. However, for the ISS paths that diverge from the correct path at the root node, we have

$$P(s_0, \mathbf{x_i}, \mathbf{y}) = P(s_0, \mathbf{x_i}) P(\mathbf{y}/s_0, \mathbf{x_i}) = P(\mathbf{s}) P(\mathbf{y}/s_0) \tag{31}$$

where the last equality used Property 1 and the independence between $\mathbf{y}$ and $\mathbf{x_i}$, for the ISS paths being considered.

Hence, by substituting the suboptimal metric defined in (8) with $n = a\tau$ and (31) into (30), we derive

$$1 = b^{\tau} \sigma^{-a\tau B} \sum_{\mathbf{s}} P(\mathbf{s}) \sum_{\mathbf{y}} P(\mathbf{y}/s_0) \prod_{j=1}^{a\tau} \left[\frac{P(y_j/s_{j-1}, s_j)}{P(y_j/s_{j-1})} \right]^{(\log_2 \sigma)} \tag{32}$$

In order to simplify the solution of (32), we substitute (7) for $P(\mathbf{y}/s_0)$, as we did for the derivation of the suboptimum metric, so that we get

$$\sigma^{a\tau B} = b^{\tau} \sum_{s_0} P(s_0) \prod_{j=1}^{a\tau} \left\{ \sum_{s_j} P(s_j/s_{j-1}) \sum_{y_j} \frac{P(y_j/s_{j-1}, s_j)^{(\log_2 \sigma)}}{P(y_j/s_{j-1})^{(\log_2 \sigma)-1}} \right\} \tag{33}$$

The substitution implied by (33), with respect to (32), will affect the calculation of the number of paths searched for the suboptimum decoder. However, the values are expected to be close, since the values of $P(\mathbf{y}/s_0)$ and $\prod_{j=1}^{n} P(y_j/s_{j-1})$ are expected to remain close, even for the ISS paths [6].

The form of (33) can be simplified, if we define the modified state transition matrix

$$\boldsymbol{\Lambda}'_{\ell} = \left[\Lambda'_{s_{\ell-1}, s_{\ell}} \right] = \left[P(s_{\ell}/s_{\ell-1}) \sum_{y_{\ell}} \frac{P(y_{\ell}/s_{\ell-1}, s_{\ell})^{(\log_2 \sigma)}}{P(y_{\ell}/s_{\ell-1})^{(\log_2 \sigma)-1}} \right] \tag{34}$$

where $s_{\ell-1}, s_{\ell} \in \{0, 1, \ldots, M-1\}$ and $\ell = 1, 2, \ldots, a\tau$. We can then apply the same derivation as in (15)–(18), when we substitute (33) for (13), $\boldsymbol{\Lambda}'_{\ell}$ for $\boldsymbol{\Lambda}_{\ell}$, Λ' for Λ, and $\lambda'_{\max}$ for $\lambda_{\max}$. As a result we obtain the characteristic equation

$$\log \left[\sigma^{B} \right] = \log \left[2^{R} \lambda'_{\max} \right] \tag{35}$$

where $R = \log_2(b)/a$ is the code rate, and $\lambda'_{\max}$ is a function of σ and the channel parameters.

A numerical solution of the characteristic equation shows that it has two real roots σ, for $-1 \leq h \leq 0$. Hence, $\mathrm{EN}\eta(z)$ is a combination of the two exponential solutions, given by

$$\mathrm{EN}\eta(z) = K_1 \sigma_1^{z} + K_2 \sigma_2^{z} \tag{36}$$

and $0 < \sigma_1 < \sigma_2$ are the real roots.

In order to determine K_1 and K_2 in (36), it is necessary to define the boundary conditions. Whenever the metric $z + \mu_{\tau}$ hits the $-\eta$ line, one count is made for the node z. On the other hand, when $z + \mu_\tau$ hits a metric limit A ($A > 0$), a zero count is made for the node z. The boundary conditions may thus be expressed as

$$\mathrm{EN}\eta(z) = \begin{cases} 1/b & -\eta + a\tau\beta \leq z \leq -\eta \\ 0 & A \leq z \leq A + a\tau\alpha \end{cases} \tag{37}$$

where α and β are the highest and lowest symbol metric possible, respectively.

It can be shown [2] that if a combination of these two exponential solutions is a lower bound of $\mathrm{EN}\eta(z)$ in its boundary regions, it will be a lower bound throughout its domain. Hence, by applying the two innermost boundary conditions in (37) to the solution in (36), a lower bound to the real solution is obtained. In addition, since the boundary A is artificial, it can be eliminated by making it arbitrarily large. In that case, by taking the limit of K_1 and K_2 as A tends to infinity, the solution in (36) will satisfy

$$\mathrm{EN}\eta(z) \geq \left[\frac{\sigma^\eta}{b}\right] \sigma^z \tag{38}$$

per incorrect subset, with σ being the smallest real root of (35).

From above, we may have a lower bound to the solution of (33) on a per incorrect subset basis. There are $b - 1$ such subsets. By assuming the root node ($z = 0$), and using (11) to replace η, we derive the expression

$$\mathrm{EN}\eta(0) \geq \frac{b-1}{b} \sigma^{\left(\frac{\log_2(Pe)}{h}\right)} \tag{39}$$

as a function of the bias B (or h). Similarly, an upper bound can be derived by applying the two outermost boundary conditions in (37) to the solution in (36):

$$\mathrm{EN}\eta(0) \leq \frac{b-1}{b} \sigma^{\left(\frac{\log_2(Pe)}{h} - a\tau\beta\right)} \tag{40}$$

By comparing (39) and (40), and observing that σ and the term $a\tau\beta$ are not functions of Pe, we realize that (39) tends in exponent to the actual solution in (36), as Pe tends to zero. Therefore, we will assume (39) as an estimate of the number of paths searched. Additionally, it can be shown [6] that this estimate has minimum value for σ and B satisfying

$$B = R \qquad\qquad \sigma = 2^{(h'+1)} \tag{41}$$

with $h = h'$ solving (19), for $B = R$. When it is substituted into (39), the result is

$$\mathrm{EN}\eta(0) \approx \frac{b-1}{b} 2^{\left[\frac{(h'+1)\log_2(Pe)}{h'}\right]} \tag{42}$$

It is interesting to find that $B = R$ minimizes the path number estimate for the suboptimum Pe-criterion sequential decoder over the ISI channel. The same result was previously derived for the optimum analysis over the memoryless channels [1, 6], where the estimate was shown to be an asymptotic upper bound to the actual number of paths searched as Pe tends to zero.

3.3 Example

We will analyze the effect of hard decision decoding on the ISI channel through the study of a simple case, when the channel filter has only one memory element ($m = 1$) and two states. Without loss of generality, it is assumed that the filter taps satisfy the expression $H_0^2 + H_1^2 = 1$, so that the filter does not affect the input signal energy. More complex cases with a finite number of memory elements can be seen as straightforward generalizations of this simple case. However, the eigenvalues need to be calculated through numerical methods, instead of the closed form solutions that appear in this section. Algorithms for this can be found, e.g., in [7].

For the two-state hard decision ISI channel, an equivalent channel model may be seen in Fig. 3, where each input symbol, output symbol or state appears in its binary representation (0 or 1). Note that by forcing hard decisions at the channel output in Fig. 1, much of the state information in the noisy filter output will be destroyed. We will be interested to measure the effect of this action, and later compare it with the use of soft decisions.

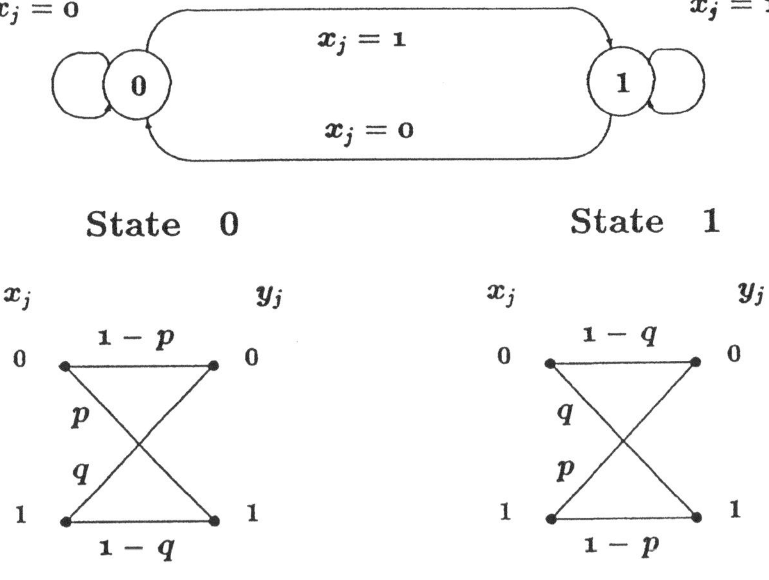

Fig. 3. Equivalent FSC channel model of the hard decision ISI channel, where the input symbol, output symbol and state appear in their binary representation.

In Fig. 3, the channel parameters p and q are defined as

$$p = Q\left(-\sqrt{\frac{2}{N_0}}(H_0 + H_1)\right) \qquad q = Q\left(-\sqrt{\frac{2}{N_0}}(H_0 - H_1)\right) \tag{43}$$

where $Q(.)$ is the Q-function [6], given by

$$Q(x) = \int_{-\infty}^{x} \frac{1}{\sqrt{2\pi}} \exp\left[-\frac{t^2}{2}\right] dt \tag{44}$$

If we evaluate (14) for this example, we have that

$$\begin{aligned} \Lambda_{00} = \Lambda_{11} &= 2^{(h-1)} \left\{ \frac{(1-p)^{(h+1)}}{(1-p+q)^h} + \frac{p^{(h+1)}}{(1-q+p)^h} \right\} \\ \Lambda_{01} = \Lambda_{10} &= 2^{(h-1)} \left\{ \frac{(1-q)^{(h+1)}}{(1-q+p)^h} + \frac{q^{(h+1)}}{(1-p+q)^h} \right\} \end{aligned} \tag{45}$$

In order to find the eigenvalues λ of Λ, we need to determine the roots of the determinant of the matrix $[\lambda I - \Lambda]$, where I is the identity matrix of order 2, in this case. The maximum eigenvalue is given by

$$\lambda_{\max} = 2^{(h-1)} \left\{ \frac{(1-p)^{(h+1)}}{(1-p+q)^h} + \frac{p^{(h+1)}}{(1-q+p)^h} + \frac{(1-q)^{(h+1)}}{(1-q+p)^h} + \frac{q^{h+1}}{(1-p+q)^h} \right\} \tag{46}$$

From (46), we may easily derive $d\lambda_{max}/dh$. Then, from (19) and (46), we are able to calculate the bias B, and from (28), we can calculate dB/dh. These give the drop line.

Figure 4 shows plots of drop lines for different channel parameters, transmitted bit energy-to-noise ratio $Eb_t/N_0 = 4.323$ dB and $Pe = 10^{-4}$. The channel parameters are represented by the angle $\theta = \arctan(H_1/H_0)$. We observe that the search regions increase with increased angle θ, for $0° \leq \theta < 90°$. The special case of $\theta = 90°$ can be shown to be equivalent to $\theta = 0°$, *i.e.*, a memoryless channel case, except for a one-period delay (The model fails to recognize delays, and does not work in that case).

We can also observe that the drop line is invariant with the sign of the angle θ. This happens because a change in the sign of θ results in an exchange of the values of p and q, which does not affect the overall performance on this channel.

The drop lines in Fig. 4 show how the presence of intersymbol interference increases the search region, compared to the memoryless case ($\theta = 0°$), corresponding to the equivalent BSC. In order to observe the influence of ISI in terms of the number of paths searched, we have to evaluate (34), that is,

$$\begin{aligned} \Lambda'_{00} = \Lambda'_{11} &= 2^{[(\log_2 \sigma)-2]} \left\{ \frac{(1-p)^{(\log_2 \sigma)}}{(1-p+q)^{(\log_2 \sigma)-1}} + \frac{p^{(\log_2 \sigma)}}{(1-q+p)^{(\log_2 \sigma)-1}} \right\} \\ \Lambda'_{01} = \Lambda'_{10} &= 2^{[(\log_2 \sigma)-2]} \left\{ \frac{(1-q)^{(\log_2 \sigma)}}{(1-q+p)^{(\log_2 \sigma)-1}} + \frac{q^{(\log_2 \sigma)}}{(1-p+q)^{(\log_2 \sigma)-1}} \right\} \end{aligned} \tag{47}$$

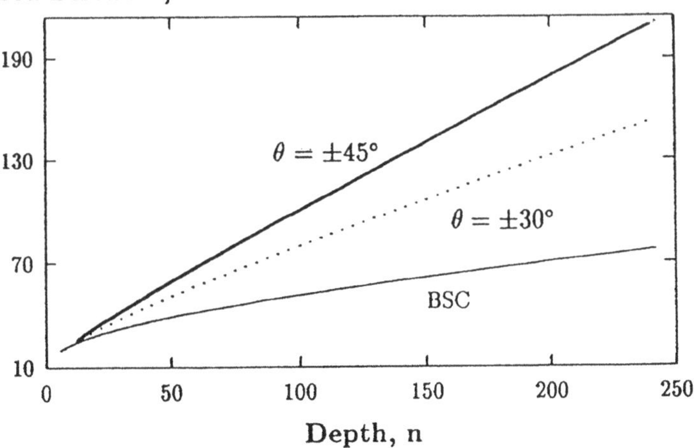

Fig. 4. Drop lines for hard decision ISI channels, for $Eb_t/N_0 = 4.323$ dB and $Pe = 10^{-4}$.

From the roots of the determinant of the matrix $[\lambda' I - \boldsymbol{\Lambda}']$ (where I is the identity matrix of order 2), we may find the maximum eigenvalue of $\boldsymbol{\Lambda}'$, which is

$$\lambda'_{\max} = 2^{(\nu-2)} \left\{ \frac{(1-p)^{\nu}}{(1-p+q)^{(\nu-1)}} + \frac{p^{\nu}}{(1-q+p)^{(\nu-1)}} + \right. \\ \left. + \frac{(1-q)^{\nu}}{(1-q+p)^{(\nu-1)}} + \frac{q^{\nu}}{(1-p+q)^{(\nu-1)}} \right\} \tag{48}$$

where $\nu = \log_2 \sigma$.

Then, from (35) and (48), we may derive the characteristic equation. By solving the characteristic equation for the values of σ, we are able to calculate the number of paths searched as given by (42).

Table 1 shows these values for $Eb_t/N_0 = 6.0$ dB, different channel parameters (represented by the angle θ), and different values of Pe and R. They show the growth in the number of paths searched with θ, for $0° \leq \theta < 90°$, and its invariance with the sign of θ, in accordance with the shape of their drop lines. They also show that the number of paths searched increases with increasing R, and decreasing Pe and Eb_t/N_0, that is, whenever the use of the channel becomes more demanding.

Finally, we observe that, as θ and R increase and Eb_t/N_0 decreases, the number of paths searched may grow unbounded (as shown in the tables by a dash). We should note, however, that we are using a suboptimum analysis which overbounds the optimum one. In an optimum analysis, unbounded growth would suggest that $R > C$. In our case, we cannot come to the same conclusion,

Table 1. Number of paths searched for the hard decision ISI channel with one memory element and $Eb_t/N_0 = 6.0$ dB.

R	θ	Pe 10^{-4}	10^{-6}	10^{-8}
1/3	$0°$	1.6	2.9	5.2
	$\pm30°$	3.8	10.3	28.2
	$\pm45°$	—	—	—
1/2	$0°$	4.1	11.6	33.0
	$\pm30°$	274.9	$6.4E+3$	$1.5E+5$
	$\pm45°$	—	—	—
2/3	$0°$	28.6	176.4	$1.1E+3$
	$\pm30°$	$1.1E+71$	$1.2E+107$	$1.3E+142$
	$\pm45°$	—	—	—
3/4	$0°$	128.6	$1.56E+3$	$1.90E+4$
	$\pm30°$	—	—	—
	$\pm45°$	—	—	—

although we can conclude that C is underbounded by any rate R at which the number of paths searched is finite.

4 The Binary-Input Continuous-Output Channel

We now consider the binary-input continuous-output ISI channel, the ultimate case of soft decision, when the channel outputs are delivered unquantized to the Pe-criterion decoder. The application of the Pe-criterion to this channel follows closely the one developed in Sect. 3, for the hard decision ISI channel. The changes are due to the semicontinuous nature of this channel. The results allow us to estimate the maximum savings in path searching attainable through the use of soft decision, for a binary-input ISI channel.

4.1 Evaluation of the Drop Line

For the semicontinuous ISI channel, the derivation of the drop line follows the one in Sect. 3.1, if we substitute the probabilities and summations involving the received sequence $\mathbf{y}$ with probability densities and integrals, respectively. With that in mind, the modified state transition matrix will be

$$\boldsymbol{\Lambda}_\ell = \left[\Lambda_{s_{\ell-1},s_\ell}\right] = \left[P(s_\ell/s_{\ell-1})\int_{y_\ell} \frac{p(y_\ell/s_{\ell-1},s_\ell)^{(h+1)}}{p(y_\ell/s_{\ell-1})^h}\,\mathrm{d}y_\ell\right] \tag{49}$$

where $s_{\ell-1}, s_\ell \in \{0,1,\ldots,M-1\}$ and $\ell = 1,2,\ldots,n$. Here $p(y_\ell/s_{\ell-1},s_\ell)$ is given by

$$p(y_\ell/s_{\ell-1},s_\ell) = \frac{1}{\sqrt{\pi N_0}}\exp\left[-\frac{(y_\ell - v_\ell)^2}{N_0}\right] \tag{50}$$

where $N_0/2$ is the variance of the i.i.d. additive Gaussian noise samples and v_ℓ is a function of $(s_{\ell-1}, s_\ell)$ or, equivalently, $(s_{\ell-1}, x_\ell)$, as in (1). Also, $p(y_\ell/s_{\ell-1})$ is given by

$$p(y_\ell/s_{\ell-1}) = \sum_{s_\ell} P(s_\ell/s_{\ell-1})p(y_\ell/s_{\ell-1}, s_\ell) \tag{51}$$

From then on, the derivation is exactly the same expressed by (15)–(28), when the probability densities in (50) and (51) are substituted accordingly, and the modified state transition matrix is given by (49). Thus, the drop line points may be calculated as in Sect. 3.1.

4.2 Number of Paths Searched

In estimating the number of paths searched by a Pe-criterion decoder for the semicontinuous ISI channel, a variation of the method used in Sect. 3.2 will be applied. The method is based on integral equations, as was done for the memoryless semicontinuous channel [6].

In Sect. 3.2, the expected number of paths searched by a non-backtracking decoder was estimated by the difference equation (29). For the semicontinuous ISI channel, μ_j is a continuous variable, so that (29) will be approximately

$$\mathrm{EN}\eta(z) \approx b^\tau \sum_{\mu_\tau} p(\mu_\tau)\, \Delta\mu_\tau\, \mathrm{EN}\eta(z+\mu_\tau) \tag{52}$$

where $p(.)$ represents the probability density function.

If we then take the limit of (52) as $\Delta\mu_\tau$ tends to zero, we derive

$$\mathrm{EN}\eta(z) = b^\tau \int_{\mu_\tau} p(\mu_\tau)\, \mathrm{EN}\eta(z+\mu_\tau)\, \mathrm{d}\mu_\tau \tag{53}$$

By assuming a solution of the type $K\sigma^z$, and placing it into (53), we derive

$$1 = b^\tau \int_{\mu_\tau} p(\mu_\tau)\, \sigma^{\mu_\tau}\, \mathrm{d}\mu_\tau \tag{54}$$

Let us assume s_0 to be the state for the root node. With a being the number of coded symbols per branch and μ_τ being the accumulated suboptimal metric along τ branches ((8) with $n = a\tau$), we may rewrite (54) as

$$1 = \frac{b^\tau}{\sigma^{a\tau B}} \sum_{\mathbf{s}} P(\mathbf{s}) \int_{\mathbf{y}} p(\mathbf{y}/s_0) \left[\frac{p(\mathbf{y}/\mathbf{s})}{\prod_{j=1}^{a\tau} p(y_j/s_{j-1})}\right]^{(\log_2 \sigma)} \mathbf{dy} \tag{55}$$

in the same way (32) was derived for the binary-input binary-output ISI channel.

By substituting (7) in (55), we derive a similar expression to (33), given by

$$\sigma^{a\tau B} = b^\tau \sum_{s_0} P(s_0) \prod_{j=1}^{a\tau} \left\{ \sum_{s_j} P(s_j/s_{j-1}) \int_{y_j} \frac{p(y_j/s_{j-1}, s_j)^{(\log_2 \sigma)}}{p(y_j/s_{j-1})^{(\log_2 \sigma)-1}} \mathrm{d}y_j \right\} \tag{56}$$

We can then define the state transition matrix as

$$\boldsymbol{\Lambda}'_\ell = \left[\Lambda'_{s_{\ell-1},s_\ell}\right] = \left[P(s_\ell/s_{\ell-1}) \int_{y_\ell} \frac{p(y_\ell/s_{\ell-1},s_\ell)^{(\log_2\sigma)}}{p(y_\ell/s_{\ell-1})^{(\log_2\sigma)-1}} \mathrm{d}y_\ell\right] \tag{57}$$

where $s_{\ell-1}, s_\ell \in \{0,1,\ldots,M-1\}$ and $\ell = 1,2,\ldots,a\tau$.

From this point on, the derivation will be the same one applied in Sect. 3.2 for the binary-input binary-output ISI channel.

4.3 Example

We now analyze the influence of the use of soft decisions in the application of the Pe-criterion on the ISI channel. We consider an equivalent unquantized soft decision ISI channel to the hard decision one described in Sect. 3.3. As in Sect. 3.3, more complex ISI channels with a finite number of memory elements, can be seen as generalizations of this simple case. The equivalent FSC channel model for this example is still characterized by the state transitions in the upper part of Fig. 3. Due to the semicontinuous nature of the channel, its description for each state is given by the transition probability densities described in (50).

If we evaluate (49), from (50) and (51), we derive

$$\Lambda_{00} = \Lambda_{11} = \frac{2^{(h-1)}}{\sqrt{\pi N_0}} \int_{-\infty}^{\infty} \frac{\exp\left[-(h+1)\frac{(y-H_0)^2}{N_0}\right]}{\left\{\exp\left[-\frac{(y-H_0)^2}{N_0}\right] + \exp\left[-\frac{(y+H_1)^2}{N_0}\right]\right\}^h} \mathrm{d}y \tag{58}$$

$$\Lambda_{01} = \Lambda_{10} = \frac{2^{(h-1)}}{\sqrt{\pi N_0}} \int_{-\infty}^{\infty} \frac{\exp\left[-(h+1)\frac{(y-H_1)^2}{N_0}\right]}{\left\{\exp\left[-\frac{(y-H_1)^2}{N_0}\right] + \exp\left[-\frac{(y+H_0)^2}{N_0}\right]\right\}^h} \mathrm{d}y$$

The maximum eigenvalue of $\boldsymbol{\Lambda}$ is given by

$$\lambda_{\max} = \frac{2^{(h-1)}}{\sqrt{\pi N_0}} \left\{ \int_{-\infty}^{\infty} \frac{\exp\left[-(h+1)\frac{(y-H_0)^2}{N_0}\right]}{\left\{\exp\left[-\frac{(y-H_0)^2}{N_0}\right] + \exp\left[-\frac{(y+H_1)^2}{N_0}\right]\right\}^h} \mathrm{d}y \right. + \left. + \int_{-\infty}^{\infty} \frac{\exp\left[-(h+1)\frac{(y-H_1)^2}{N_0}\right]}{\left\{\exp\left[-\frac{(y-H_1)^2}{N_0}\right] + \exp\left[-\frac{(y+H_0)^2}{N_0}\right]\right\}^h} \mathrm{d}y \right\} \tag{59}$$

From (59) above, we may derive $d\lambda_{\max}/dh$. From (19) and (59), we are able to calculate the bias B and, from (28), we may calculate dB/dh. Finally, we can plot the drop line.

Figure 5 shows plots of drop lines for unquantized soft decision channels and the corresponding hard decision ones, for $Eb_t/N_0 = 4.323$ dB, $Pe = 10^{-4}$ and different channel parameters, represented by the angle θ. The plots show the decreased search regions resulting from the use of soft decisions, as compared to

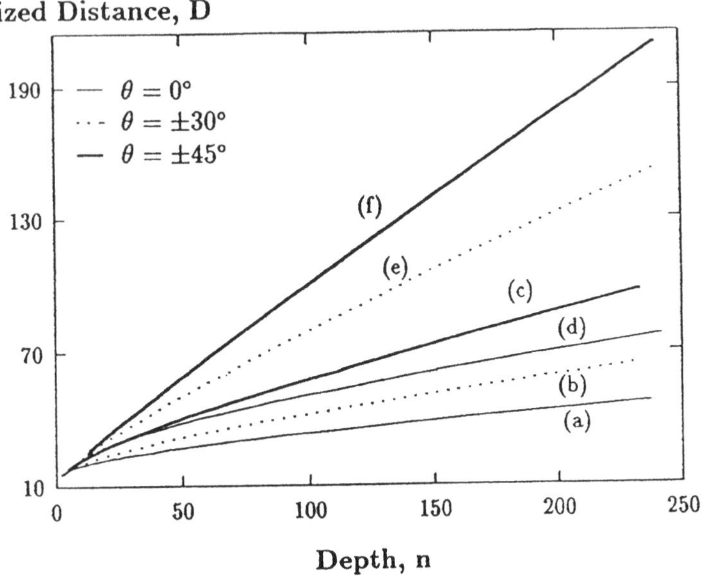

Fig. 5. Drop lines for the unquantized soft decision ISI channels (a,b,c) and the corresponding hard decision ones (d,e,f), for $Eb_t/N_0 = 4.323$ dB and $Pe = 10^{-4}$.

hard decisions. It also shows that the soft decision channels are less sensitive to an increase in the angle θ than the hard decision ones.

In order to observe the effect of soft decision decoding on the number of paths searched, we have to evaluate (34), that is

$$\Lambda'_{00} = \Lambda'_{11} = \frac{2^{(\log_2 \sigma - 2)}}{\sqrt{\pi N_0}} \int_{-\infty}^{\infty} \frac{\exp\left[-(\log_2 \sigma)\frac{(y-H_0)^2}{N_0}\right]}{\left\{\exp\left[-\frac{(y-H_0)^2}{N_0}\right] + \exp\left[-\frac{(y+H_1)^2}{N_0}\right]\right\}^{(\log_2 \sigma - 1)}} dy \tag{60}$$

$$\Lambda'_{01} = \Lambda'_{10} = \frac{2^{(\log_2 \sigma - 2)}}{\sqrt{\pi N_0}} \int_{-\infty}^{\infty} \frac{\exp\left[-(\log_2 \sigma)\frac{(y-H_1)^2}{N_0}\right]}{\left\{\exp\left[-\frac{(y-H_1)^2}{N_0}\right] + \exp\left[-\frac{(y+H_0)^2}{N_0}\right]\right\}^{(\log_2 \sigma - 1)}} dy$$

The maximum eigenvalue of $\boldsymbol{\Lambda}'$ is given by

$$\lambda'_{\max} = \frac{2^{(\log_2 \sigma - 2)}}{\sqrt{\pi N_0}} \left\{ \int_{-\infty}^{\infty} \frac{\exp\left[-(\log_2 \sigma)\frac{(y-H_0)^2}{N_0}\right]}{\left\{\exp\left[-\frac{(y-H_0)^2}{N_0}\right] + \exp\left[-\frac{(y+H_1)^2}{N_0}\right]\right\}^{(\log_2 \sigma - 1)}} dy \; + \right. \tag{61}$$

$$+ \int_{-\infty}^{\infty} \frac{\exp\left[-(\log_2 \sigma)\frac{(y-H_1)^2}{N_0}\right]}{\left\{\exp\left[-\frac{(y-H_1)^2}{N_0}\right]+\exp\left[-\frac{(y+H_0)^2}{N_0}\right]\right\}^{(\log_2 \sigma - 1)}} dy \Bigg\}$$

By substituting (61) into (35), we derive the characteristic equation. And after solving the characteristic equation for the values of σ, we calculate the number of paths searched given by (42).

Table 2 shows values of the number of paths searched for $Eb_t/N_0 = 6.0$ dB, different channel parameters (represented by the angle θ), and different values of Pe and R. If we compare these results with those in Table 1, we are able to verify the savings in the number of paths searched resulting from the use of soft decisions as compared to hard decisions, for the ISI channel. These results agree with the drop lines in Fig. 5, in the sense that higher drop lines can reasonably be expected to enclose more tree paths.

Finally, we also observe that as θ and R increase and Eb_t/N_0 decreases, the use of soft decisions may be the key to feasibility because, eventually, there is a bounded number of paths searched for the soft decision case, even though it has grown unbounded for the hard decision case. As in the hard decision case, out of bound growth does not necessarily represent that R is greater than C, but any rate R for which the number of paths is bounded constitutes an underbound to C.

Table 2. Number of paths searched for the binary-input continuous-output ISI channel with one memory element and $Eb_t/N_0 = 6.0$ dB.

		Pe		
R	θ	10^{-4}	10^{-6}	10^{-8}
1/3	0°	0.7	0.9	1.0
	±30°	0.8	1.1	1.4
	±45°	1.2	1.8	2.8
	±60°	5.6	18.5	61.6
1/2	0°	0.0	1.3	1.8
	±30°	1.3	2.0	3.1
	±45°	2.6	5.9	13.3
	±60°	193.4	$3.8E+3$	$7.5E+4$
2/3	0°	2.1	3.6	6.1
	±30°	3.7	8.1	17.9
	±45°	18.2	89.4	439.8
	±60°	$1.0E+13$	$3.9E+19$	$1.4E+26$
3/4	0°	3.5	6.9	13.9
	±30°	7.7	22.8	67.5
	±45°	120.5	$1.4E+3$	$1.7E+4$
	±60°	—	—	—

5 Conclusion

This work has extended the application of the Pe-criterion to channels with intersymbol interference. We now review the most important results.

1. The derivation of the drop line demands calculation of the maximum eigenvalue of a modified state transition matrix which is a function of the state transition probabilities and the transmitted bit energy-to-noise ratio, Eb_t/N_0.
2. The calculation of the number of paths searched demands the calculation of the maximum eigenvalue of a modified state transition matrix which is a function of the state transition probabilities, Eb_t/N_0 and R. The estimate of the number of paths searched is minimized when the bias B is set to R.
3. Considerable savings may be achieved through the use of soft decisions, especially for medium to high rates. Soft decisions may be the key for the decoder feasibility, more so than for memoryless channels.
4. A lower bound to the capacity C of the channel is given by the highest code rate R for which the number of paths searched is finite.
5. For a given Eb_t/N_0, the tap values of the equivalent discrete channel filter dictate the decoder performance. For a two-state ISI channel, the angle θ established by the two filter taps is the parameter to consider. Higher values of θ represent greater numbers of paths searched, for $0° < \theta < 90°$.

References

1. Anderson, J. B.: Sequential decoding based on an error criterion. IEEE Trans. on Info. Theory, **IT-38** (1992), 987–1001.
2. Anderson J. B., Jelinek, F.: A 2-cycle algorithm for source coding with a fidelity criterion. IEEE Trans. on Info. Theory, **IT-19** (1973), 77–92.
3. Forney, G. D.: Maximum-likelihood sequence estimation of digital sequences in the presence of intersymbol interference. IEEE Trans. on Info. Theory, **IT-18** (1972), 363–78.
4. Gallager, R. G.: Information theory and reliable communication (1968), John Wiley & Sons Inc., New York.
5. Markman, I., Anderson, J. B.: Sequential decoding on memoryless soft decision channels under the Pe-criterion. Proceedings of the 1993 IEEE International Symposium on Information Theory (1993), San Antonio, TX.
6. Markman, I.: Sequential decoding on channels with and without memory under a probability of error criterion". Ph.D. Thesis (1993), ECSE Dept., Rensselaer Polytechnic Institute, Troy, NY.
7. Smith, B. T. *et al.*: Matrix eigensystem routines - EISPACK GUIDE. Lecture Notes in Computer Science, **6** (1976), 2nd edition, Springer-Verlag, New York.

SEPARABLE CONCATENATED CODES WITH ITERATIVE MAP FILTERING

J. Lodge, R. Young, and P. Guinand

Communications Research Centre
3701 Carling Avenue, Ottawa Canada K2H 8S2
(tel) 613-998-2284, (fax) 613-990-6339

Abstract. In practice, very efficient signalling over radio channels requires more than designing very powerful codes. It requires designing very powerful codes that have special structure so that practical decoding schemes can be used with excellent, but not necessarily optimal, results. Examples of two such approaches include the concatenation of convolutional and Reed-Solomon coding, and the use of very large constraint-length convolutional codes with reduced-state decoding. In this paper, powerful codes are obtained by using simple block codes to construct multidimensional product codes. The decoding of multidimensional product codes, using separable symbol-by-symbol maximum *a posteriori* (MAP) "filters", is described. Simulation results are presented for three-dimensional product codes constructed with the (16,11) extended Hamming code. The extension of the concept to concatenated convolutional codes is given. The relationship between the free distance and the interleaving factors is examined, and then exemplified with computer simulation. Potential applications are briefly discussed.

1 Background

The work discussed in this paper was motivated by concepts introduced in [1] for the decoding of concatenated convolutional codes. In that paper it is shown that symbol-by-symbol MAP decoding for the inner code allows soft decisions to be passed to the outer decoder, resulting in impressive performance. The inner decoding algorithm can be thought of as a type of nonlinear filter that accepts as its input a noisy signal. Then it makes use of the structure inherent in the inner code to produce a noisy output "decoded" signal (that is hopefully less corrupted in some sense than the original input signal). Here we apply and extend the same philosophy for multidimensional product codes. A logical extension to the case of concatenated convolutional codes is then described and computer simulation results are given.

Here, the system model shown in Figure 1 is used, and codes that can be represented by a trellis of finite duration are considered. It is well known that decoding trellises exist for convolutional codes [2] as well as for linear block codes [3]. Furthermore, it has been known for some time that well-structured MAP decoding algorithms exist for codes that can be represented by trellises of finite

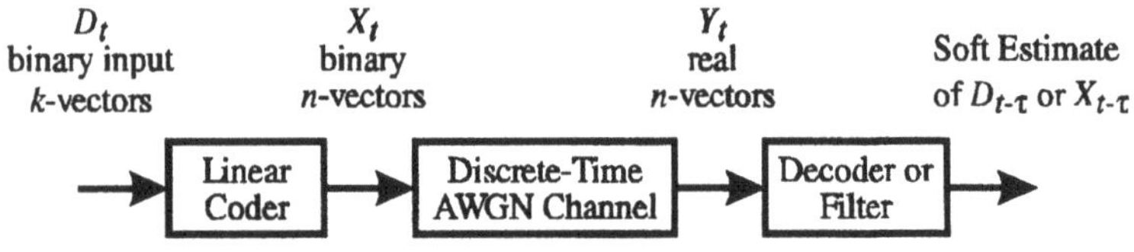

Figure 1. A block diagram of the system model.

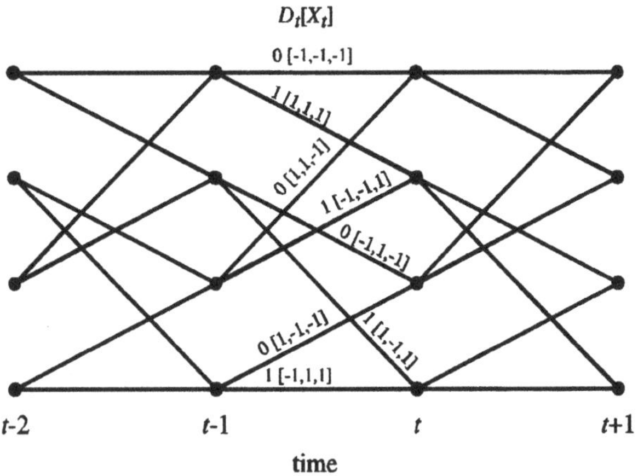

Figure 2. A trellis corresponding to a rate 1/3, 4 state convolutional code.

duration [2][4]. The simple time-invariant 4-state trellis, shown in Figure 2, is used to illustrate the concepts. This trellis corresponds to a rate-1/3 convolutional code. In general (e.g., for block codes) the trellis may be time-varying with the number of states, M_t, being a function of the time index t. It is assumed that at the start and the end of the time interval of interest, the coder is in the zero state. Any given input sequence[1] ${}_iD_{i'}$ of binary (e.g., 0 or 1) k-vectors, that satisfies the above end conditions, will correspond to a particular path through the trellis that is described by a sequence of states

$$ {}_{i-1}S_{i'} = \{S_{i-1} = 0, \ldots, S_t = m, \ldots, S_{i'} = 0\} \tag{1} $$

where $S_t \in \{0,\ldots,M_t\text{-}1\}$.

[1]Here, the following notation is used. If Q is a long sequence (e.g., a discrete-time signal), then ${}_iQ_{i'}$ denotes the segment of the sequence extending from time index i to time index i', inclusive. Of course the segment of the original sequence is itself a sequence.

For the kth path through the trellis, the coder produces a particular channel input sequence (i.e., code word) denoted by

$$ {}_{i}X_{i'}(k)=\{X_i(k),\ldots,X_t(k),\ldots,X_{i'}(k)\}, \tag{2}$$

where $X_t(k)$ is an n-vector denoted by

$$X_t(k)=[x_{t_1}(k),\ldots,x_{t_n}(k)] \tag{3}$$

of binary (e.g., -1 or +1) elements. It is assumed here that all valid code words are equally likely. For most commonly used codes, this assumption is consistent with the assumption that the data source generates a sequence of independent and identically distributed binary random variables, with the two values being equally likely. In the example trellis of Figure 2, k=1, n=3 and M=4 for all t. For notational convenience, the functional dependence of $X_t(k)$ on S_{t-1} and S_t is only shown when required. The corresponding channel output sequence is given by

$$ {}_{i}Y_{i'}=\{Y_i,\ldots,Y_t,\ldots,Y_{i'}\}, \tag{4}$$

where Y_t is an n-vector denoted by

$$Y_t=[y_{t_1},\ldots,y_{t_n}] \tag{5}$$

with the real-valued elements having conditional probability density functions given by

$$p(y_{t_j}\,|\,x_{t_j})=\left(2\pi\sigma^2\right)^{-1/2}\exp\left[-(y_{t_j}-x_{t_j})^2/2\sigma^2\right]. \tag{6}$$

Clearly this model is appropriate for antipodal signalling in an additive thermal noise environment typical of satellite communications.

Begin by considering the case where the coder in Figure 1 is unknown. In this case, one can do little better than to assume that the input to the additive white Gaussian noise channel is a sequence of independent and identically distributed binary (i.e., -1 or +1) random variables. If the two values for the input bits are equally likely

$$\Pr\left\{x_{t_j}=1\,|\,y_{t_j}\right\}=R_{t_j}^{-1}\cdot p\left(y_{t_j}\,|\,x_{t_j}=1\right), \tag{7}$$

where

$$R_{t_j}=p\left(y_{t_j}\,|\,x_{t_j}=1\right)+p\left(y_{t_j}\,|\,x_{t_j}=-1\right) \tag{8}$$

is not dependent upon the hypothesized value of x_{t_j}.

Next we take into account the knowledge of the structure of the code C. The conditional probability that the kth code word was the transmitted one is given by

$$\Pr\{X={}_iX_{i'}(k)|{}_iY_{i'};C\} = \frac{p({}_iY_{i'}|X={}_iX_{i'}(k))\cdot\Pr\{X={}_iX_{i'}(k)\}}{\sum_{j=1}^{J} p({}_iY_{i'}|X={}_iX_{i'}(j))\cdot\Pr\{X={}_iX_{i'}(j)\}}, \tag{9}$$

where J is the total number of possible codewords and the conditioning on C refers to the knowledge of the coding structure. According to the independence assumption inherent in the channel model,

$$\begin{aligned} p({}_iY_{i'}|X={}_iX_{i'}(k)) &= \prod_{t,m} p\left(y_{t_m}|x_{t_m}=x_{t_m}(k)\right) \\ &= \prod_{t,m} R_{t_m}\cdot\Pr\left\{x_{t_m}=x_{t_m}(k)|y_{t_m}\right\} \\ &= \left(\prod_{t,m} R_{t_m}\right)\left(\prod_{t,m}\Pr\left\{x_{t_m}=x_{t_m}(k)|y_{t_m}\right\}\right) \end{aligned}, \tag{10}$$

where the second line follows from equation (7). Note that the first product on the third line is not dependent upon the hypothesized code word. Combining equations (9) and (10), along with the assumption that all code words are equally likely, with *a priori* probability $1/J$, yields

$$\begin{aligned} \Pr\{X={}_iX_{i'}(k)|{}_iY_{i'};C\} &= \frac{(1/J)\left(\prod_{t,m} R_{t_m}\right)\left(\prod_{t,m}\Pr\left\{x_{t_m}=x_{t_m}(k)|y_{t_m}\right\}\right)}{\sum_{j=1}^{J}(1/J)\left(\prod_{t,m} R_{t_m}\right)\left(\prod_{t,m}\Pr\left\{x_{t_m}=x_{t_m}(j)|y_{t_m}\right\}\right)} \\ &= \frac{\prod_{t,m}\Pr\left\{x_{t_m}=x_{t_m}(k)|y_{t_m}\right\}}{\sum_{j=1}^{J}\prod_{t,m}\Pr\left\{x_{t_m}=x_{t_m}(j)|y_{t_m}\right\}} \end{aligned} \tag{11}$$

In order to facilitate the computation of the probability that any given branch in the trellis is traversed by the transmitted code word, the set $B(m',m)$ is defined as the index set of all code words that traverse the trellis branch between $S_{t-1}=m'$ and $S_t=m$. The state transition probability is given by

$$\Pr\{S_{t-1}=m';S_t=m|{}_iY_i\} = \frac{\sum_{k\in B(m',m)}\Pr\{X={}_iX_i(k)|{}_iY_i;C\}}{\sum_{(q',q)}\left(\sum_{k\in B(q',q)}\Pr\{X={}_iX_i(k)|{}_iY_i;C\}\right)} \tag{12}$$

The computations required for equations (11) and (12) can be performed efficiently using the recursive approach given by

$$\sigma_t(m',m) = \sum_{k \in B(m',m)} \Pr\{X =_i X_i(k) |_i Y_i ; C\} = \alpha_{t-1}(m') \cdot \gamma_t(m',m) \cdot \beta_t(m) \ , \qquad (13)$$

where

$$\begin{aligned} \alpha_t(m) &= \sum_{m'=0}^{M_{t-1}-1} \alpha_{t-1}(m') \cdot \gamma_t(m',m) \\ \beta_t(m) &= \sum_{m'=0}^{M_{t+1}-1} \beta_{t+1}(m') \cdot \gamma_{t+1}(m',m) \quad . \\ \gamma_t(m',m) &= \prod_{j=1}^{n} \Pr\{\boldsymbol{x}_{t_j} = x_{t_j}(m',m) | y_{t_j}\} \end{aligned} \qquad (14)$$

Clearly, $\alpha_t(m)$ is computed in a recursive fashion in the forward direction, while $\beta_t(m)$ is computed by the analogous computation in the backward direction. This recursive approach is similar to that introduced by Bahl et al. [4], except here the recursions are based upon conditional probabilities instead of the conditional probability density functions used in [4]. Here, we have formulated the processing using only probabilities in order to facilitate iterative processing without the need of converting between probabilities and density functions.

We would like to use the above MAP processing to determine the probabilities $\Pr\{\boldsymbol{x}_{t_j} = -1 |_i Y_{i'} ; C\}$. This can easily be done by defining the set of all transitions for which $x_{t_j} = -1$;

$$A = \{(m',m) \colon x_{t_j}(m',m) = -1\}, \qquad (15)$$

and then computing

$$\Pr\{x_{t_j} = -1 |_i Y_{i'} ; C\} = \frac{\sum_{(m',m) \in A} \sigma_t(m',m)}{\sum_{(m',m)} \sigma_t(m',m)} \ . \qquad (16)$$

The noisy codeword enters the MAP "filter" as a vector of independent probabilities, and then is output from the filter with the probabilities (which are no longer independent) being refined according to the structure of the code. A similar procedure can be used for determining the probability that the information bit d_{t_j} is zero by replacing the set A by

$$A' = \{(m',m)\colon d_{t_j}(m',m) = 0\}. \tag{17}$$

In this paper, we distinguish between the terms "MAP filter" and "MAP decoder", with the former computing the *a posteriori* probabilities of the coded bits and the latter the *a posteriori* probabilities of the decoded bits. (Clearly for systematic codes, the *a posteriori* probabilities of the information bits are a subset of the probabilities for the coded bits.) If hard decisions are performed on the output of the MAP filter, the minimum average probability of coded bit error is achieved assuming that the assumptions made about the transmission model are accurate. However, in general the "code word" that minimizes the average probability of bit error may not be a valid code word! A good choice for a valid code word can be obtained by iterating the filtering operation until a valid code word is "captured". Of course, the assumption of independent probabilities by the MAP algorithm is erroneous when the algorithm is used iteratively. Nevertheless, the MAP algorithm demonstrates an ability to "capture" valid code words when it is iterated. However, valid code words selected in this way are not truly optimal by any commonly-used criterion.

2 Iterative MAP Filtering for Block Product Codes

Two linear block codes can be used to construct a more powerful code by constructing a product code with them. In this case the new code word can be viewed as a rectangular matrix (or equivalently a two-dimensional array) with each row being a code word of one of the linear block codes, while each column is a code word of the other code. The rate, Hamming distance, and asymptotic coding gain of the product code are equal to the products of those quantities for the individual codes. While the following condition is not necessary, identical systematic codes will be assumed in both dimensions here. Therefore, the overall code word is a square array, with parity bits along two of its four sides, and the distance and rate of the overall code are the squares of those values for the original code. Clearly this concept can be extended to higher dimensions. For example, in three dimensions the overall code word is a cubic array with parity bits along three of the six sides, while in four dimensions the overall code word is a four-dimensional hypercube with parity bits along four of the eight sides.

In multidimensional signal processing, digital filtering is often performed using "separable" filters. That is, in order to avoid excessive computational requirements, one-dimensional filtering is performed sequentially in each of the N dimensions, rather than performing a single massive N-dimensional digital filter. In this paper, we investigate the analogous approach for the decoding of multidimensional product codes. That is, one-dimensional MAP filters will be used sequentially in each dimension. Of course this approach is no longer optimal in the sense of minimizing average probability of bit error, but it is more likely to be computationally feasible than performing true MAP decoding for the entire composite code. Consider the two-dimensional case first. One-dimensional MAP filtering can be done across the rows giving a new set of refined probabilities, taking into account only the horizontal structure of the overall code. These new probabilities (for which the elements in a column are still independent) are then further refined by one-dimensional MAP filtering down the columns to complete a single filtering cycle. This process can be iterated any number of times. The extension to the cases with more than two dimensions is obvious, with a single filtering cycle consisting of sequentially performing one-dimensional filtering in each of the dimensions. In the multidimensional signal processing case, iterating the filtering does not make sense because the filters are linear. However, in the product coding case, the filters are highly nonlinear and additional filtering cycles may significantly improve the performance. Note that the independence assumption for each MAP filter operation is valid for the entire first cycle. Thus, up to this point the probabilities computed by each MAP filter operation are the "real probabilities" of symbol error given the MAP processing that has been performed up to that point. Therefore, an accurate estimate of the average probability of bit error can be computed at the receiver after each filtering pass in the first cycle. For subsequent iterations the independence assumption is erroneous and consequently the significance of the resulting "probabilities" is less clear.

Straightforward application of the iterative MAP filtering algorithm for the decoding of multidimensional product codes can produce very good results [5]. However, in many cases one of a number of variations of the MAP filtering approach can yield improved performance. Here, we will outline two such variations. Note that in all cases the overall decoding is suboptimal. At this point in time, it is unknown why a given variation of the approach performs better for one code, while a different variation performs better for another code.

The first variation makes use of the following factorization. From equations (13) through (16), it can be shown that

$$\Pr\left\{x_{t_j} = -1 |_i Y_{i'} ; C\right\} = \frac{\Pr\left\{x_{t_j} = -1 | y_{t_j}\right\} \sum_{(m',m)\in A} \sigma_t'(m',m)}{\sum_{(m',m)} \sigma_t(m',m)} \tag{18}$$

where

$$\sigma_t(m',m) = \Pr\left\{x_{t_j} = -1 | y_{t_j}\right\} \cdot \sigma_t'(m',m).$$

Note that the first term in the numerator of equation (18) is dependent only upon the jth element at time t and the summation is independent of this element and therefore represents the "refinement factor" due to the structure of the coder. In general such a factorization is not possible for the denominator of equation (18). However, a similar expression exists for $\Pr\left\{x_{t_j} = 1 |_i Y_{i'} ; C\right\}$ and the denominator terms cancel when the likelihood ratio, given by,

$$L = \frac{\Pr\{x_{t_j} = -1|_i Y_{i'};C\}}{\Pr\{x_{t_j} = 1|_i Y_{i'};C\}} = \frac{\Pr\left\{x_{t_j} = -1|y_{t_j}\right\} \cdot \left(\sum_{(m',m)\in A} \sigma_t'(m',m) \right)}{\Pr\left\{x_{t_j} = 1|y_{t_j}\right\} \cdot \left(\sum_{(m',m)\notin A} \sigma_t'(m',m) \right)} \tag{19}$$

is computed. Therefore the likelihood ratio factors into the form

$$L_{t_j}(1) = L_{t_j}(0) \cdot f_{t_j}(1), \tag{20}$$

for one iteration and

$$L_{t_j}(K) = \left[\prod_{q=1}^{K} f_{t_j}(q) \right] L_{t_j}(0) \tag{21}$$

for K iterations, where $f_{tj}(q)$ represents the refinement factor for the qth iteration. Note that the above factorization is only applicable to MAP filtering and the MAP decoding of systematic codes, and is not generally applicable to MAP decoding. "**Partial factor MAP filtering**" is a variation for which the input to the MAP filters, for component codes in a given dimension, includes only those refinement factors that correspond to the filtering passes in the other dimensions. This variation reduces the error introduced by the independence assumption in the MAP processing.

A second variation, referred to here as the "**real probabilities**" approach, takes advantage of the fact that the probabilities resulting from one complete filtering cycle are the last truly valid probabilities (because the independence assumption for each MAP filter operation is valid up to this point of the processing). The resulting "real probabilities" are saved, and for subsequent cycles equation (13) is replaced by

$$\sigma_t(m',m) = \alpha_{t-1}(m')\gamma_t'(m',m)\beta_t(m) \tag{22}$$

where the modified branch probability γ_t' is computed using the "real probabilities". The forward and backward recursions are still computed using the filtered probabilities.

3 Block Coding Example

In this section a three-dimensional product code constructed using the (16,11) extended Hamming code is considered. The one-dimensional MAP filter for this code has 32 states, where the parity bits are used to define the state vector. The three-dimensional product code has a rate of 0.325 and an asymptotic coding gain of 13.18 dB. Monte Carlo simulations were performed for which 100 code words, each with 1331 information bits, were processed for a range of signal-to-noise ratios. Figure 3 shows the number of code words containing errors for one-cycle, three-cycle, and six-

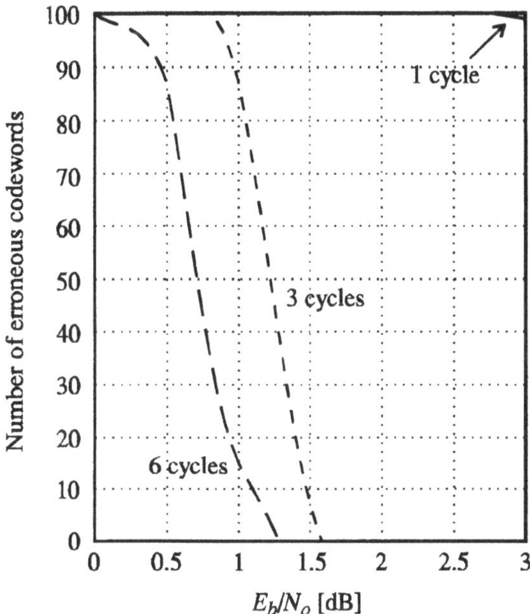

Figure 3. The number of code words containing errors for 1 cycle, 3 cycles and 6 cycles of MAP filtering. A total of 100 code words was processed for each value of E_b/N_o, with each code word containing 1331 information bits.

cycle MAP filtering. A couple of points are evident. Firstly, impressive performance can be achieved by constructing multidimensional product codes using comparatively simple component codes. Partial factor MAP filtering was used here. The results are notably better than those presented in [5] for full factor MAP filtering. Secondly, the improvement due to iterative filtering can be very dramatic. Note that the results for one cycle of filtering correspond to the more conventional approach to the decoding of concatenated codes, where soft-in soft-out decoding algorithms are used. The capability of iterative filtering to capture the correct code word is quite striking.

This example is intriguing from another point of view. Consider the performance at an E_b/N_o of 1 dB. After 6 cycles (i.e., 18 filtering passes) of MAP filtering, over 85 percent of the 1331-bit blocks are being received error-free. For these successfully processed blocks, each filtering pass must improve the signal in some sense. Consequently, the first filtering pass must deal with the worst quality of signal. The input symbol-energy-to-noise-spectral-density ratio experienced by this filter is -3.88 dB. Since this first filter is a filter for a rate-11/16 code, it must operate at an equivalent E_b/N_o of -2.25 dB. Figure 4 shows the operating point of this filter relative to the Shannon limit. Interestingly, the filter is improving the

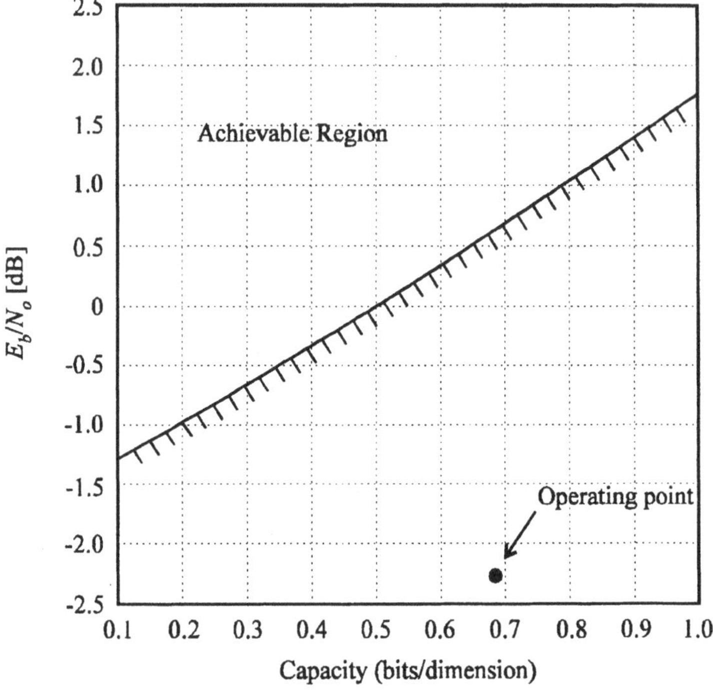

Figure 4. The operating point of the first filter relative to the region of achievable rates for error-free transmission.

signal quality even though it is operating at almost 3 dB below the Shannon limit for a code of rate 11/16! Of course, this does not violate any fundamental law because error-free performance is not achieved by the filter.

4 Extension to Convolutional Codes

In this paper, very good performance has been achieved by constructing multidimensional product codes, using fairly simple block codes. For many radio applications convolutional coding has proven to be more attractive than block coding. Therefore it is natural to seek similar techniques for convolutional codes.

A product code can be thought of as a special case of concatenated coding with interleaving between the coding stages. For example, a two-dimensional product code can be viewed as concatenated coding with a block interleaver that has a number of columns equal to the word length of the first component code, and a number of rows equal to the number of information bits for the second component code. The special property of this choice of interleaver/coding arrangement is that the composite

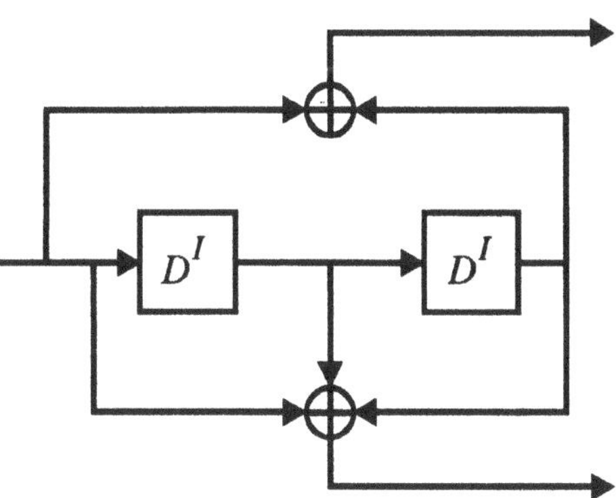

Figure 5. An example convolutional encoder including I-fold time division interleaving. Here, D is the delay operator.

code word can be viewed as being composed of a number of code words corresponding to any given component code, by appropriately grouping the output code symbols. It is this special property, referred to here as "separability", that is exploited above for efficient decoding. Here, we discuss a class of convolutional codes that is directly analogous to block product codes. These codes are constructed from smaller component convolutional codes, and they possess the separability property mentioned above.

Convolutional encoders are linear and shift-invariant [6], and consequently a sum of valid code words, each with a different delay, is still a valid code word. An interleaving scheme that does not destroy the shift-invariant property is desirable. One such scheme is time-division interleaving, which can be implemented as is illustrated in Figure 5. Therefore this type of combined encoder/interleaver can be used as a building block for the type of composite code that is desired. This concept is illustrated in Figure 6 for a two-tier example code. Each tier contains a number of identical coders with inputs interconnected to the coder outputs of the previous tier. The interconnection must be done so that the code words arriving from the previous tier are linearly combined, through the current tier, in such a way that the outputs can be subdivided into valid code words for the previous tier. For example, in Figure 6, c_{11}, c_{12} and c_{13} are three valid code words for code CE1. In general, these three code words may not be identical to the two code words generated by the first tier.

Here, we develop such an interconnection using a recursive approach. Starting with a rate k_1/n_1 convolutional coder at the first tier, we wish to add a second tier consisting of rate k_2/n_2 coders. In our interconnection there will be k_2 coders at tier 1 and n_1 coders at tier 2. The concatenation of tier 1 and tier 2 is treated as a supercoder of rate $k'_2/n'_2 = k_1k_2/n_1n_2$. To connect a third tier of rate k_3/n_3 coders we repeat the above process. There will be k_3 supercoders at tier 2 and n'_2 coders at tier 3 and after the interconnection this will produce a supercoder of rate $k'_3/n'_3 = k'_2k_3/n'_2n_3$. In general, interconnecting tier i to tier i+1 requires k_{i+1}

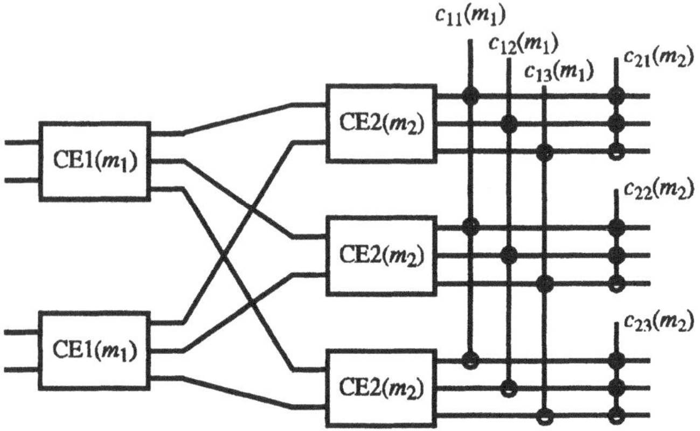

Figure 6. Two-tier coding with rate 2/3 component codes. CE1(I_1) is an encoder with I_1-fold interleaving for code 1. CE2(I_2) is an encoder with I_2-fold interleaving for code 2. c_{1q} is the qth valid codeword for code 1, with I_1-fold interleaving. c_{2q} is the qth valid codeword for code 2, with I_2-fold interleaving.

supercoders at tier i and n'_i coders at tier i+1. This concatenation is treated as a supercoder of rate $k'_i k_{i+1}/n'_i n_{i+1}$ for subsequent interconnections. The final supercoder resulting from concatenating N tiers of convolutional coders has a rate

$$k'_N/n'_N = \prod_{i=1}^{N}\left(k_i/n_i\right) . \tag{23}$$

The actual interconnection of tier i to tier i+1 is straightforward. If we denote the jth coder at stage i as $c_{i,j}$, then our interconnection strategy is to connect the mth output of supercoder $c_{i,j}$ to the jth input of coder $c_{i+1,m}$. The recursive nature of the code construction has the consequence that it is sufficient to consider the properties of this class of code by studying two-tier codes. Generally, a given property of two-tier codes can be extended to multi-tier codes by induction.

The individual code words from the convolutional coders are dispersed as they propagate through subsequent tiers. In order to facilitate MAP filtering, we must be able to construct valid code words from each tier. Let us denote the output sequence of n'_N bits as

$$\{b(0), b(1), b(2), \ldots, b(n'_N - 1)\}$$

Then, the mth code symbol from the ith tier is

$$\{b(q), b(q+p), b(q+2p), \ldots, b(q+(n_i - 1)p)\}$$

where

$$p = \begin{cases} \prod_{j=i+1}^{N} n_j, & \text{for } i < N \\ 1, & \text{for } i = N \end{cases}, \tag{24}$$

$$q = \prod_{j=1}^{N} n_j \bullet \left\lfloor \frac{m}{p} \right\rfloor + m \bmod p, \tag{25}$$

$$m = \left\{ 0, 1, 2, \ldots, \frac{n'_N}{n_i} - 1 \right\}, \tag{26}$$

where $\lfloor x \rfloor$ denotes the largest integer less than or equal to x. Note that each of the component code words (appropriately interleaved) is present at the output.

Separable concatenated convolutional codes can be compactly represented making use of the notational approach presented in [6]. Here we will discuss the representation for two-tier codes. Let code CE1 (n_1 outputs, k_1 inputs), with a k_1 x n_1 transfer function matrix $G_1(D)$, be the component code for the first tier. Here, D denotes the delay operator. Similarly let code CE2 (n_2 outputs, k_2 inputs), with a k_2 x n_2 transfer function matrix $G_2(D)$, be the component code for the second tier. The input data is represented by the k_2 x k_1 matrix $X(D)$, with elements $x_{qp}(D)$ denoting pth input to the qth coder in the first tier. Clearly, each row of the input data vector represents the input to one of the k_2 coders in the first tier. The output of the first tier is the k_2 x n_1 matrix given by

$$W(D) = X(D)G_1(D^{I_1}), \tag{27}$$

with elements $w_{qp}(D)$ denoting pth output from the qth coder in the first tier. Referring to Figure 6, the next operation to consider is the rearranging of the outputs of the first tier, to form the inputs to the n_1 coders in the second tier. The appropriate rearranging of the outputs corresponds to matrix transposition. Thus, the input to the second tier is $W^T(D)$, and the output of the second tier is given by the $n_1 \times n_2$ matrix

$$Y(D) = W^T(D)G_2(D^{I_2}) = G_1^T(D^{I_1})X^T(D)G_2(D^{I_2}), \tag{28}$$

with elements $y_{qp}(D)$ denoting pth output from the qth coder in the second tier. The transpose of the output matrix is given by

$$Y^T(D) = G_2^T(D^{I_2})X(D)G_1(D^{I_1}). \tag{29}$$

Comparing equations (28) and (29), it is clear that the code construction is commutative in the sense that if code CE2 was used in the first tier and code CE1 was used in the second tier, the resulting composite code would be the same as the

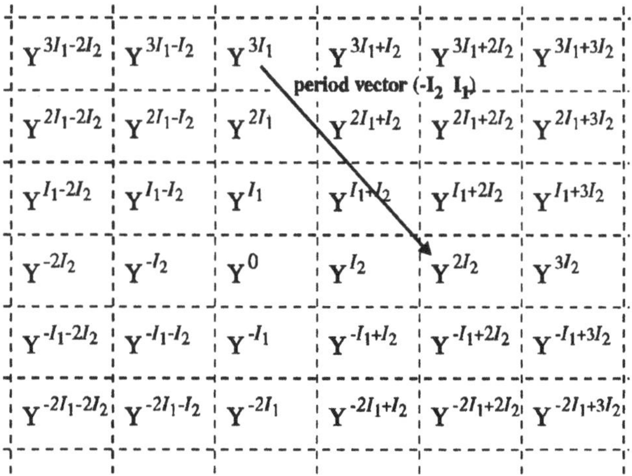

Figure 7. The product diagram for a two-tier code. The period vector illustrated assumes that $I_1=2$ and $I_2=3$.

original one, with the exception that the input and output matrices would be replaced by their transposes.

For product codes, the Hamming distance of the composite code is equal to the product of the Hamming distances of the component codes. For separable concatenated convolutional codes, we would like the minimum free distance of the composite code to be equal to the product of the minimum free distances of the component codes. However, the minimum free distance of the composite code is a function of the interleaving factors. Furthermore, it is desirable to choose interleaving factors that are reasonably small in order to achieve low decoding delays and memory requirements. These observations motivate a study of the relationship between the interleaving factors and the minimum free distance of the composite code. Here, we will consider the two-tier case.

Recall that the output of the composite code is a matrix of the form

$$Y(D)=\sum_{p} Y^p D^p, \tag{30}$$

where Y^p is an n_1 x n_2 matrix with the elements being the outputs at time p. Consider the product diagram shown in Figure 7. Each square in the diagram represents the $n_1 x n_2$ matrix of outputs at a particular instant in time. Therefore, each square in the diagram is intersected by n_1 rows and n_2 columns. Each row in the diagram represents a code word of code CE2, and each column represents a code word of code CE1. Note that the time indices are of the form

$$p=m\cdot I_1+q\cdot I_2, \tag{31}$$

where m and q are integers. An equation of the form of (31) can be solved for all p if and only if I_1 and I_2 are relatively prime [7]. If they are not relatively prime, then the situation can be represented by several distinct product diagrams, but the effective interleaving factors are reduced by the greatest common divisor of the actual interleaving factors. Therefore, we will only consider the case where the factors are relatively prime. In this case, not only can equation (31) be solved for any integer value of p, but each value of p has an infinite number of solutions because if the vector $(m' \; q')$ is a solution so are all pairs of integers of the form

$$\begin{pmatrix} m & q \end{pmatrix} = \begin{pmatrix} m' & q' \end{pmatrix} + l \cdot \begin{pmatrix} -I_2 & I_1 \end{pmatrix}, \tag{32}$$

where l is any integer. Thus, each output symbol is present on the product diagram an infinite number of times, with the spacing between occurrences of the same symbol being separated by $(-I_2 \;\; I_1)$ symbol intervals.

Assume that transmission has been error free up to time p=0, and that an error event begins at this time. Since the columns are valid code words for code CE1, each column with an erroneous bit must have at least d_{f1} erroneous bits, where d_{f1} is the minimum free distance of CE1. Similarly, the rows are valid code words for code CE2 and consequently each row with an erroneous bit must have at least d_{f2} erroneous bits. For sufficiently large interleaving factors, it is clear the minimum free distance for the composite code will be equal to the product of the component distances because it is possible that all erroneous rows and columns will have a number of erroneous bits equal to the minimum free distance of their respective component codes. Furthermore, if the interleaving factors are sufficiently large the periodic images of the error events will not overlap on the product diagram. However, if the interleaving factors are too small, then images of minimum distance error events will overlap and the minimum free distance of the composite code can be less than the desired product. For the example illustrated in Figure 7, consider the case where an error event in one of the columns intersecting Y^0 has d_{f1} erroneous bits, some of which occur at time $3I_1$. Also, all the corresponding error events in the rows have d_{f2} erroneous bits, including some at time $2I_2$. But $3I_1$ and $2I_2$ are the same time (i.e., 6), and therefore the composite error event may have less erroneous bits than the desired product. This situation can be avoided if I_1 is chosen to be greater than the required number of symbol intervals to guarantee that code CE2 has accumulated at least d_{f2} erroneous bits since the start of an error event. Alternatively, I_2 can be chosen to be greater than the required number of symbol intervals to guarantee that code CE1 has accumulated at least d_{f1} erroneous bits since the start of an error event. The required number of bits can easily be determined by iterating the state equations [8] while applying the appropriate constraints.

As an example, consider the two-tier code with both component codes being the 8-state rate-3/4 code given in [8]. Numerical analysis, using the state equations, shows that this code is guaranteed to accumulate four (i.e., the free distance for the code) erroneous bits within eight symbol intervals. Here, four cycles of the "real probabilities" approach to separable MAP filtering were used to decode the composite code. From the above discussion it is expected that the performance should significantly improve as the interleaving factor in one tier is increased up to 8, while the other interleaving factor is kept at 1. Further increases in the interleaving factors should result in only incremental performance improvement, because although the overall weight structure of the code will continue to improve the maximum free distance has already been achieved. Simulation results for a variety of interleaving factors are shown in Figure 8. These results confirm our expectations.

Other examples of the performance of separable concatenated convolutional codes, in a variety of channel types, can be seen in [9],[10] and [11].

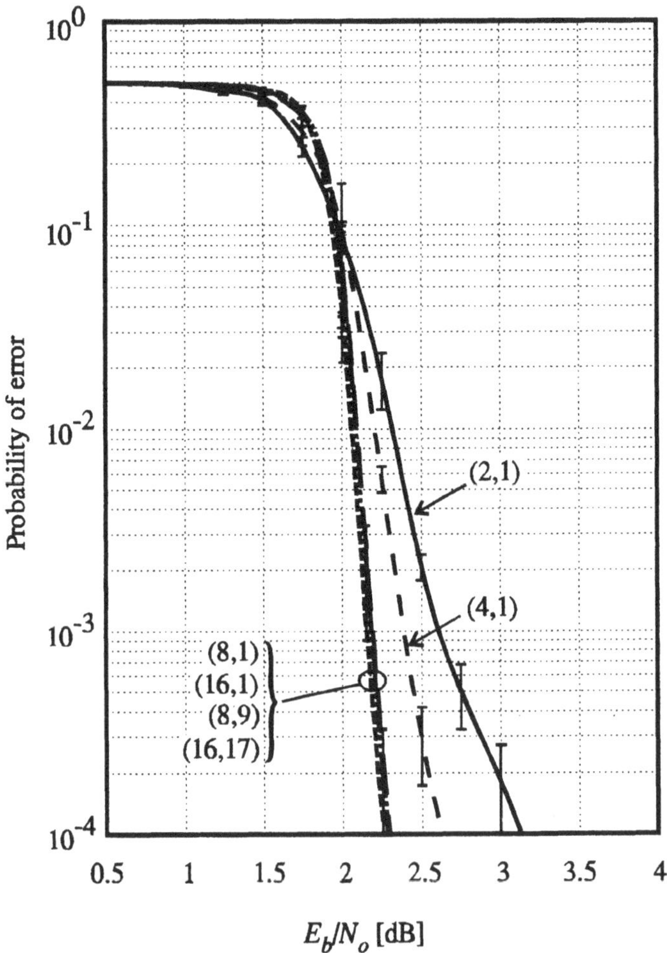

Figure 8. Performance as a function of the interleaving factors (I_1,I_2) for a two-tier code constructed using an 8-state rate-3/4 composite codes. Four cycles of MAP filtering were used to decode the composite code. The error bars identify a 99% confidence interval based upon 8 independent trials.

5 Conclusions

The concept of decoding multidimensional product codes using iterative MAP filtering was described. Impressive simulation results were presented for the example of a three-dimensional product code constructed using the (16,11) extended Hamming code. Significant reductions in the computational requirements may still be possible. To date trellises based upon state vectors defined by the parity bits have been used. In general, such trellises are not the simplest trellises that can be used to represent a given block code [12]. The use of minimal trellises for reducing the computational requirements for iterative MAP filtering is an area of current study.

The concept has been extended to convolutional codes, for which component convolutional codes are combined to form a large convolutional code. A compact matrix description for the code construction was presented, and used to show that the construction is "commutative". The relationship between the free distance and the interleaving factors was examined and it was shown to be straightforward to select appropriate interleaving factors to guarantee that the free distance of the composite code is equal to the product of the free distances of the component codes. The utility of this approach to the choice of interleaving factors was illustrated by an example using computer simulation results.

As would be expected with such powerful coding techniques, the decoding process is quite computationally intensive. Therefore, the development of efficient implementation techniques is an important area for future work. For some codes, it is possible that simpler algorithms (e.g., [1]) can replace the MAP processing without severely degrading the performance.

A number of potential application areas are evident. Firstly, powerful codes with a high coding rate can be implemented. For example, the four-dimensional code, constructed with (25,24) single parity bit component codes, has a rate of 0.85 and an asymptotic coding gain of 11.3 dB. Secondly powerful codes with relatively short block lengths can be implemented for packet data applications. Perhaps the ultimate example is the two-dimensional code constructed with the (24,12) extended Golay code, which provides an asymptotic coding gain of 12 dB with a word length of only 144 information bits. The third application is for extremely power efficient coding (e.g., deep-space applications).

Acknowledgment

The authors would like to acknowledge the many useful discussions with Dr. Peter Hoeher of the German Aerospace Research Establishment (DLR) and Prof. Joachim Hagenauer of the Technical University of Munich.

References

[1] J. Hagenauer and P. Hoeher, "A Viterbi algorithm with soft-decision outputs and its applications," Proc. GLOBECOM'89, Dallas, Texas, pp. 47.1.1-47.1.7, Nov. 1989.

[2] G.D. Forney, "The Viterbi algorithm," Proc. IEEE, vol. 61, pp. 268-278, Mar. 1973.

[3] J.K. Wolf, "Efficient maximum likelihood decoding of linear block codes using a trellis," IEEE Trans. Inform. Theory, vol. IT-24, pp. 76-81, January 1978.

[4] L. Bahl, J. Cocke, F. Jelinek, and J. Raviv, "Optimal decoding of linear codes for minimizing symbol error rate," IEEE Trans. Inform. Theory, vol. IT-20, pp. 284-287, Mar. 1974.

[5] J.H. Lodge, P. Hoeher and J. Hagenauer, "The decoding of multidimensional codes using separable MAP 'filters'," Proc. Queen's University 16th Biennial Symp. on Communications, pp.343-346, May 1992.

[6] G.D. Forney, "Convoluational codes I: Algebraic structure," IEEE Trans. Inform. Theory, vol. IT-16, pp. 720-738, Nov. 1970.

[7] J.H. McClellan and C.M. Rader, *Number Theory in Digital Signal Processing*, Englewood Cliffs N. J., Prentice-Hall Inc., 1979, Chapter III.

[8] J.G. Proakis, *Digital Communications*, New York NY, McGraw-Hill Book Company, 1983, Chapter 5.

[9] J. Lodge, R. Young, P. Hoeher, and J. Hagenauer, "Separable MAP 'filters' for the decoding of product and concatenated codes," Proceedings of the IEEE International Conference on Communications, pp. 1740-1745, Geneva, Switerzerland, May 1993.

[10] J.H. Lodge and R.J. Young, "Separable concatenated codes with iterative MAP decoding for Rician fading channels," Proceedings of the Third International Mobile Satellite Conference IMSC '93, pp. 467-472, Pasadena, CA, June 1993.

[11] J.H. Lodge and R.J. Young, "Separable concatenated codes with iterative MAP decoding for Rayleigh fading channels," Proceedings of the Fifth Annual Conference on Wireless Communications, pp. 703-712, Calgary, Canada, July 1993.

[12] G.D. Forney, "Coset codes II: Binary lattices and related codes," IEEE Trans. on Inform. Theory, vol. 34, pp. 1152-1187, Sept. 1988.

Adapting the Strongly-Connected Trellis Concept for Use with Trellis-Coded Modulation*

Sébastien Roy[1] and Paul Fortier[2]

[1] Systèmes Tertius Oculus, RR 2, Site 17, Box 9, Beresford, NB, E0B 1H0
[2] Département de génie Électrique, Université Laval, Sainte-Foy, QC, G1K 7P4
fortier@gel.ulaval.ca

Abstract. The strongly-connected trellis was originally proposed by Chang and Yao as a way of increasing the efficiency of Viterbi decoders implemented on locally-connected parallel processor arrays. Originally, this concept was derived with binary convolutional codes in mind. Its adaptation to TCM is non-trivial due to the presence of parallel branches in the trellis. The method we propose utilizes a modified strongly-connected trellis concept to increase the efficiency of TCM decoders on locally-connected processor arrays. A TCM decoder based on these principles was implemented on a MasPar massively parallel computer for experimental purposes.

1 Introduction

The complexity of the Viterbi algorithm is proportional to the coding gain desired. In fact, obtaining high coding gains at high symbol rates requires enormous computational power. This has lead to increased interest in parallel implementations of the Viterbi algorithm, parallel processing being the last possible way to increase processing speed once the physical barrier has been reached in VLSI. Modern VLSI Viterbi decoders are made up of parallel processing elements and can provide high throughput for codes up to length $K = 7$.

Unfortunately, it would appear that the Viterbi algorithm is difficult to implement in parallel [1]. Indeed, the algorithm cannot be divided into independent sub-tasks and calls for high connectivity between processing elements. In VLSI, links occupy more space than transistors and high degrees of connectivity can be very costly. Indeed, VLSI design criterions call for regular, repetitive circuit patterns with local links only (systolic arrays). Physical limits are quickly reached since the number of links necessary for efficient Viterbi decoding grows exponentially with the constraint length. For this reason, it is interesting to consider locally-connected methods even if it implies a loss of efficiency.

Chang and Yao [2] have proposed the strongly-connected trellis in 1989 as a way to increase the efficiency of Viterbi decoding on locally-connected processor

* This work was supported in part by grants from NSERC and FCAR.

arrays. This concept was originally intended for binary convolutional codes. We propose a method to adapt the strongly-connected trellis to the more powerful trellis-coded modulation schemes [3, 4]. Using these principles, a software decoder for TCM has been constructed on a MasPar MP1 massively parallel computer.

While the application of the strongly-connected trellis to parallel Viterbi decoding leads to increased efficiency for binary convolutional codes, it is shown herein that similar benefits can be derived for parallel decoding of trellis-coded modulation schemes.

This paper is organized as follows. Section 2 discusses the architecture of the MasPar computer. In Section 3, the Viterbi algorithm is presented as a matrix operation. Section 4 introduces the concept of the strongly-connected trellis, while its application to TCM is presented in Section 5. In Section 6, an implementation of the parallel Viterbi algorithm to decode TCM on the MasPar is shown. Finally, the paper concludes in Section 7.

2 Parallel Architectures and the MasPar Computer

Many types of parallel architectures exist and each is programmed differently. The MasPar MP1 [5, 6] is made up of 2048 processor elements (PEs) arranged in a rectangular matrix of 64 × 32. Each PE is connected to its eight nearest neighbors by highspeed links (see Fig. 1). Furthermore, processors on the edges of the array are linked to the opposite edge by wraparound links. The processor array is in fact toroidal in shape if we consider the presence of the wraparound links. These local connections provide a bandwidth of 24 Gigabytes per second. In addition, a global router mechanism is provided to permit random point-to-point linking. Such global links are much slower at 1.5 Gigabyte per second and important setup delays are incurred whenever a new global link is established. Since our work is concerned with locally-connected processor arrays, we have chosen not to use the global router.

The MasPar is a SIMD machine (Single Instruction Multiple Data). This means that the active portion of the processor array executes the same instructions at the same time. However, each processor operates on its own local dataset. This type of parallel architecture is not as powerful or as flexible as the MIMD class (Multiple Instruction Multiple Data) where each PE is an independent computer with its own memory and program (see Fig. 2). On the other hand, SIMD machines are cheaper and much easier to program since the software retains a sequential flow.

With any parallel computer, maximum efficiency is almost impossible to achieve because parallelism itself and interprocessor communications constitute an additional overhead. In fact, the efficiency tends to decrease as the number of processors increases [7]. Certain problems are better suited to certain architectures. It is important to point out that the SIMD architecture of the MasPar is very specialized and is best suited to problems involving a large number of independent sub-tasks such as image processing.

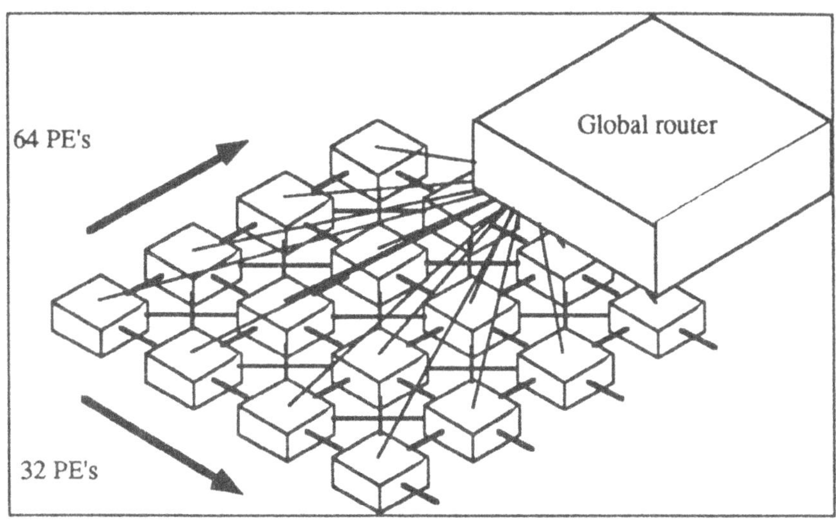

Fig. 1. The structure and interconnections of the MasPar processor array

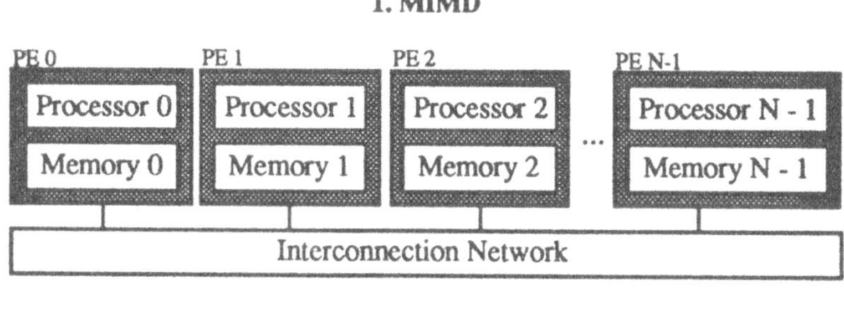

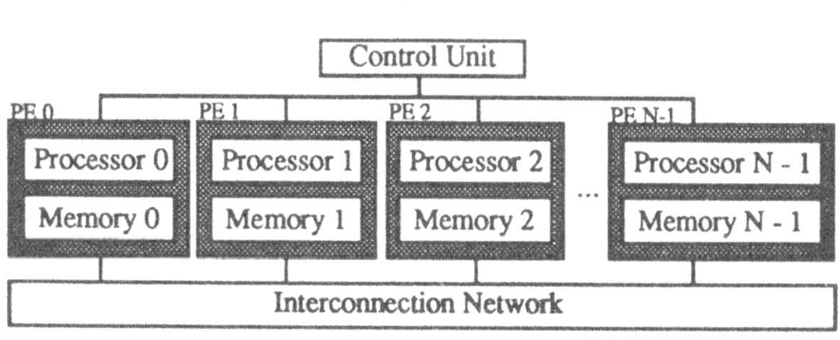

Fig. 2. The SIMD and MIMD paradigms of parallelism

Let us take a look at Fast Fourier Transforms as a problem for SIMD computers. The FFT algorithm is characterized by the same type of interdependence pattern as the Viterbi algorithm. Because of this, it is not a good algorithm for parallelisation on any architecture, but especially SIMD. While running a single FFT problem on the MasPar's 2048 processors would be much faster than on most sequential machines, it would probably not be very efficient. However, if you happen to have a large number of such problems to resolve, it is much more efficient to run 2048 FFTs simultaneously. This approach is better suited to SIMD since each processor runs its own independent yet identical problem.

Since our goal is to accelerate a single TCM decoder and not 2048 such decoders, our problem is a poor candidate for SIMD parallelisation. The strongly-connected trellis concept helps to attain the maximum efficiency possible on a locally-connected architecture (SIMD or MIMD).

Our parallel implementation is unidimensional; in other words, all states in a single stage are treated simultaneously but each stage is treated sequentially. In this manner, we are making a better use of the number of processors and the architecture of the MasPar computer [8].

3 Parallel Viterbi Decoding Using Matrix Operations

Let us define an adjacency matrix that represents the trellis stage corresponding to the kth transition. The element holds the metric of the branch form state i to state j. Elements for which no branch exists in the trellis are given an infinite (or very large) value (see Fig. 3).

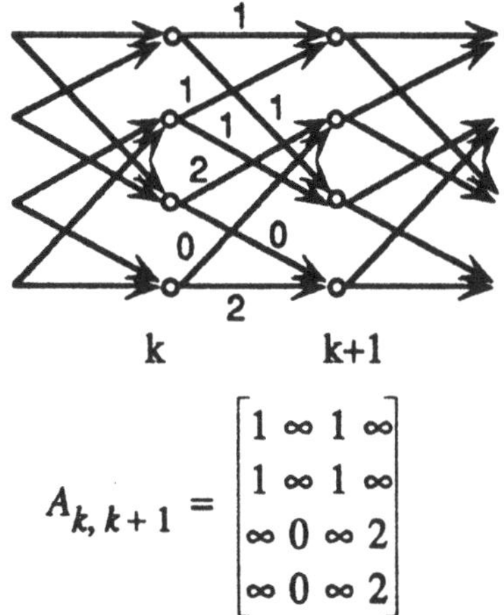

$$A_{k,k+1} = \begin{bmatrix} 1 & \infty & 1 & \infty \\ 1 & \infty & 1 & \infty \\ \infty & 0 & \infty & 2 \\ \infty & 0 & \infty & 2 \end{bmatrix}$$

Fig. 3. A trellis stage and its adjacency matrix

The add-compare-select step of the Viterbi algorithm can be redefined with the help of a matrix-vector operation similar in structure to matrix-vector multiplication. This approach has been helpful in the process of parallelizing the Viterbi algorithm since matrix-vector multiplication is well adapted to parallel architectures and extensively covered in the literature [7].

Given a vector $\mathbf{P}_k$ of N elements containing the lengths of the survivors after the kth transition, the matrix-vector form of the survivor update operation can be written as:

$$\mathbf{P}_{k+1} = \mathbf{A}_{k+1} \otimes \mathbf{P}_k \tag{1}$$

where the operator $\otimes$ is defined as follows:

$$p_{k+1}[i] = \min(a_{k+1}[i,j] + p_k[j]) \qquad j = 0, 1, \ldots, N-1 \ . \tag{2}$$

Let us compare (2) with the definition of matrix-vector multiplication:

$$p_{k+1}[i] = \sum_j (a_{k+1}[i,j] \times p_k[j]) \qquad j = 0, 1, \ldots, N-1 \ . \tag{3}$$

It can be verified that the operator $\otimes$ is structurally equivalent to matrix-vector multiplication. The sum and multiplication operations in the definition of matrix-vector multiplication have merely been replaced by the minimum operator and addition, respectively.

We need also know from which state the new survivors come so we must compute for each survivor i:

$$\hat{j} = \min{}^{-1}(a_{k+1}[i,j] + p_k[j]) \qquad j = 0, 1, \ldots, N-1 \ . \tag{4}$$

Now that the core of the Viterbi algorithm has been formally defined as a matrix-vector operation, we need to define parallel algorithms based on (2). Given a linear array of N processors with wraparound (ring array), the vector-matrix operation can be executed in $o(N)$ steps for an N-state convolutional code. Each processor initially contains an element of vector $\mathbf{P}_k$. Elements of matrix $\mathbf{A}_{k+1}$ are entered form the top as illustrated in Fig. 4. Each processor adds its element of $\mathbf{P}_k$ to its element of $\mathbf{A}_{k+1}$ and stores the result in a local accumulator. As the next set of values of $\mathbf{A}_{k+1}$ are entered, the values of $\mathbf{P}_k$ are rotated one processor to the right. Those values are again summed individually by each processor and the result is compared with the content of the local accumulator. The smallest value is stored in the accumulator. After N steps, the accumulators contain the elements of vector $\mathbf{P}_{k+1}$.

While the method described above is simple and elegant, it is desirable to exploit the power of more than N processors. A second method, called the division and fusion algorithm, can be scaled to fit processor arrays of various sizes. The available processor array is initially divide in M groups of N processors (see Fig. 5). Each group performs only a fraction of the matrix-vector operation. The first group deals exclusively with the first $\frac{N}{M}$ steps of the matrix-vector operation and so needs only generate the corresponding fraction of the branch metrics. The final result is merely the minimum of the M partial results for each element of $\mathbf{P}_{k+1}$.

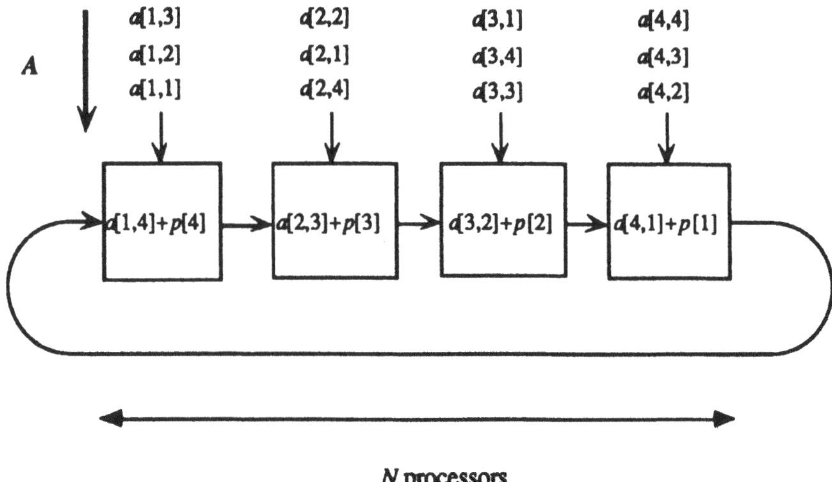

Fig. 4. The linear parallel algorithm

4 The Strongly-Connected Trellis

The performance of Viterbi decoders implemented on systolic arrays or similar parallel local connectivity structures suffers from inherent inefficiency. This is due to the fact that most convolutional codes result in a trellis with low connectivity. This in turn leads to sparse-matrix operations by the parallel Viterbi decoder and, therefore, inefficient use of processing resources.

To overcome this problem, Chang and Yao [2] have proposed to combine a number r of stages into one to obtain a strongly-connected trellis where all state-pairs are linked by one branch. There is a one-to-one correspondence between each branch of the strongly-connected trellis and each r-branch path in the primitive trellis (see Fig. 6). It has also been shown that the generation of metrics for these composite branches was straightforward and involved little additional overhead. The composite metric is merely the sequential concatenation of the individual branches' metrics.

Let us define the compression ratio $r_{\max}$ as the maximum number of stages of the primitive trellis that can be "compressed" into one without creating ambiguity, i.e. losing the one-to-one correspondence between the primitive trellis and the strongly-connected trellis:

$$r_{\max} = L(\log_b N) \tag{5}$$

where b stands for the number of branches entering or leaving any state (a measure of the connectivity of the primitive trellis) and N stands for the number of state. The $L()$ function rounds downward to the nearest integer. For codes with a bit rate of $\frac{1}{n}$, b is always equal to 2 since each state transition results of a single bit entering the encoder. Such codes have low-connectivity trellises which

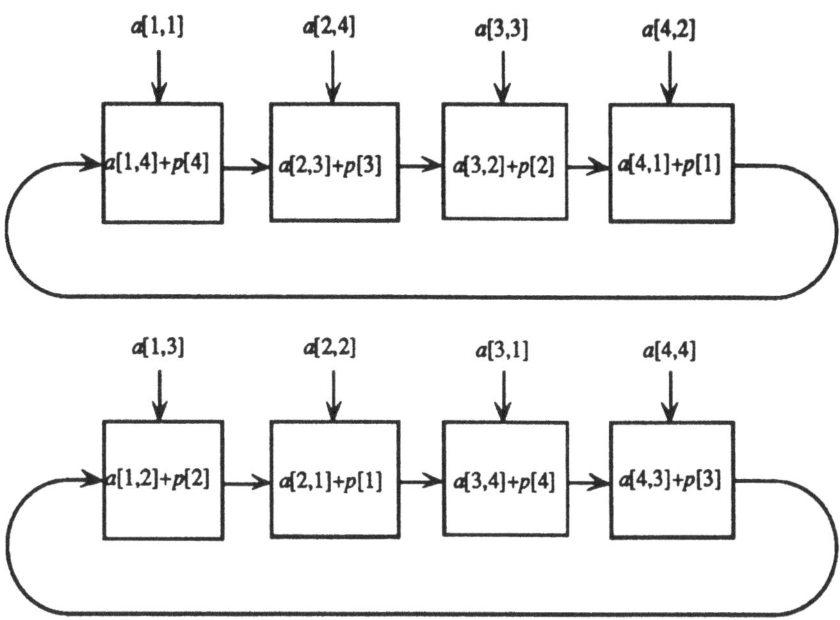

Fig. 5. The division and fusion parallel algorithm

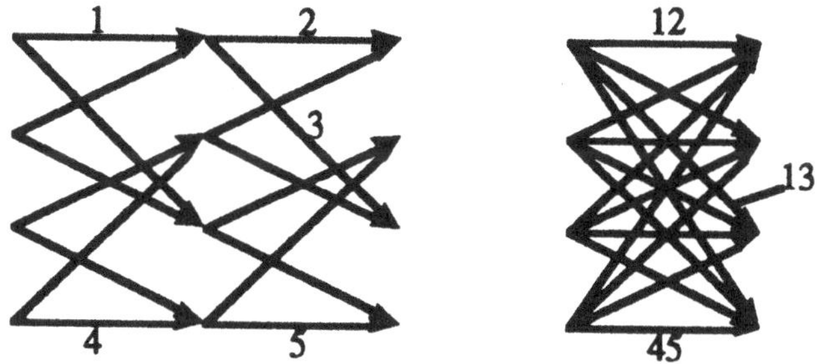

Fig. 6. Two stages of a 4-state trellis and the corresponding strongly-connected trellis stage. Branch labels illustrate the concatenation

implies high compression ratios. For example, a 256 states rate $\frac{1}{n}$ convolutional code will allow a compression of eight stages, resulting in a fully-connected trellis.

On the other hand, TCM codes are usually built around a rate $\frac{n}{n+1}$ in which case b will be equal to n^2 in (5). Indeed, TCM is characterized by high trellis connectivity. Accordingly, the compression ratio $r_{\max}$ will typically be smaller for TCM codes than binary convolutional codes with the same number of states N.

5 Trellis-Coded Modulation

Consider a 256-state 16-QASK TCM encoder as depicted in Fig. 7. The signal set is partitioned in eight subsets, each of which contains two maximally distant symbols [3, 4] The output of a rate $\frac{2}{3}$ convolutional encoder is used to select one of the eight subsets while a single message bit selects the symbol to be transmitted within the subset. The corresponding trellis is characterized by pairs of parallel branches while the overall connectivity remains low.

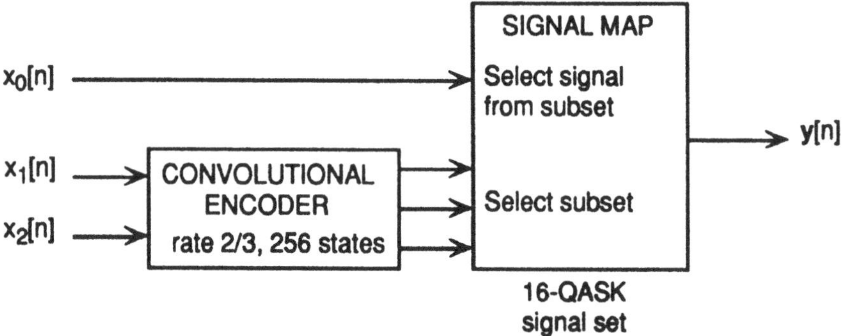

Fig. 7. A typical TCM encoder

The standard decoding procedure for TCM comprises three steps:

1. Metrics are generated for all branches in a trellis stage based on unquantized channel output and Euclidean distance (soft decisions).
2. The metrics for each group of parallel branches are compared and all but the shortest branches for each state-pair are eliminated.
3. The trellis being reduced to single branches, conventional Viterbi decoding can now be applied.

Incorporating the strongly-connected trellis concept in the TCM decoding process is non-trivial due to the presence of parallel branches in the original trellis. This obstacle can be avoided by eliminating parallel branches (steps 1 and 2 described above) at the primitive trellis level. Therefore, a reduced trellis, ridden of its parallel branches can be compressed instead of the original one (see Fig. 8). The reduced trellis is actually a representation of the subset part of the encoder where no information remains about which signal was selected within the subset.

From this point, decoding can proceed efficiently using the Viterbi algorithm at the strongly-connected trellis level.

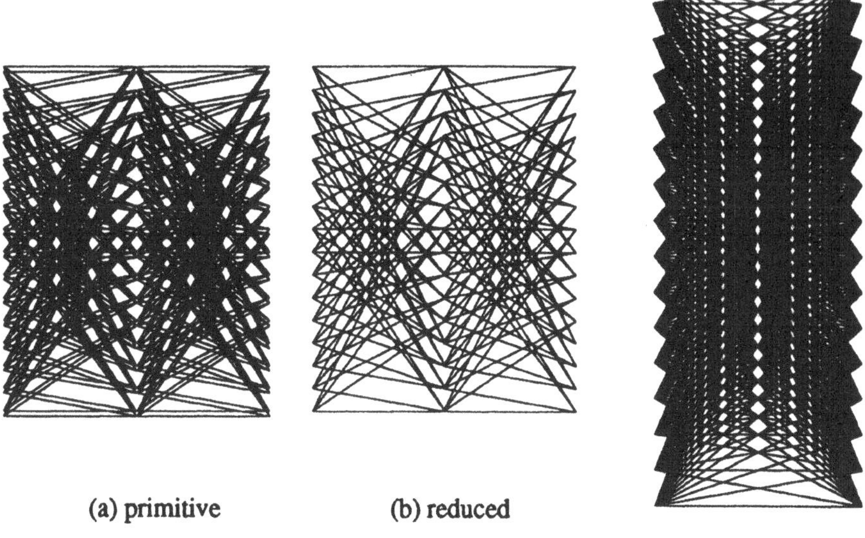

Fig. 8. The three types of trellises used in strongly-connected trellis decoding of TCM

6 Implementation

Initially, a table must be constructed where for each branch of the strongly-connected trellis, the corresponding path or transition sequence in the original trellis can be looked up. Since the reduced trellis is considered at this point, the transition sequence is identified by its subsets and is thus unique. The table can take the form of an $N \times N$ matrix with column and row indices corresponding to starting and ending states while the value itself reveals the in-between transition sequence. While such a data structure is rather large, it should be distributed to a number of local processor memories. In fact, each processor will only need to access its own local portion of the branch-to-path table. The purpose of this table is to provide a link between the primitive trellis and the strongly-connected trellis, allowing work to be conducted at both levels.

The decoding algorithm can be mapped to a parallel processing array in a number of ways. We have used a partitioning scheme based on ending states, i.e. processor 0 is assigned all branches which end at state 0. Local memories need only contain the portions of the branch-to-path table corresponding to their assigned branches.

For the TCM code described above, there are 4 branches entering or leaving any state in the reduced trellis. This results in a compression ratio of $r_{\max} = \log_4 256 = 4$. Consequently, the decoding is conducted on groups of 4 symbols. For each composite branch, the corresponding subset sequence is looked up in the branch-to-path table and eight branch metrics are then computed since there are 4 transitions and 2 parallel branches per transition. The smallest metrics for all 4 transitions are added up to form the composite branch metric.

During the reduction phase, note must be taken of which branch was selected within each group of parallel branches. For a given strongly-connected trellis branch, a transitional code gives this information for each branch of the corresponding path in the primitive trellis. In our example, the transitional code would consist of a 4-bit sequence in memory where each bit identifies the branch retained for each of the four transitions. This transitional code is illustrated in Fig. 9. Once the metrics and transitional codes for all branches of the strongly-connected trellis have been generated, Viterbi decoding can be performed using matrix operations.

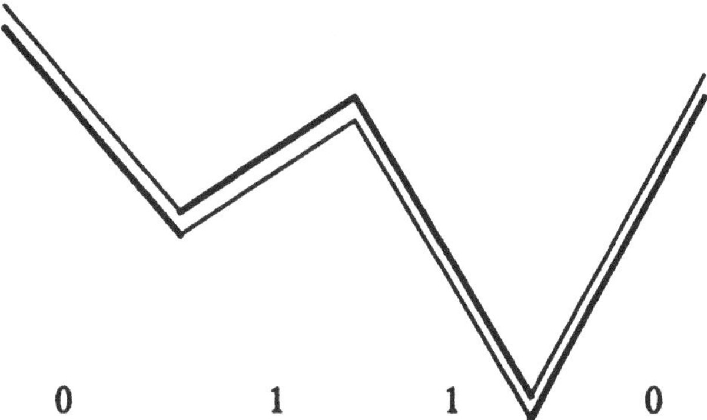

Fig. 9. Transitional code for a single composite branch. Within each pair of parallel branches, the bold branch has the shortest metric. A "1" indicates that the upper branch was chosen

When the end of a data frame is reached, the Viterbi decoder yields a state sequence. The survivor memory containing this state sequence also contains an associated transitional code sequence. The state sequence provides information at the subset level while the transitional code sequence identifies the signal within each subset. These two sequences are combined to obtain a maximum-likelihood signal sequence from which the original message can be derived.

Figure 10 gives a general block diagram of the parallel TCM decoding algorithm based on the strongly-connected trellis concept. One may observe that metric generation is performed at the primitive trellis level while Viterbi decoding itself (the add-compare-select step, the bulk of the work) occurs at the strongly-connected trellis level.

A simulation was conducted on the MasPar computer using a 16-QASK 256 states TCM code. Using all 2048 processors, a decoding rate of 190 message bits per second was obtained. We should point out that this is the performance of a software decoder, written in a high level language, and it is therefore much slower than comparable custom VLSI decoders. Previously, a similar simulation

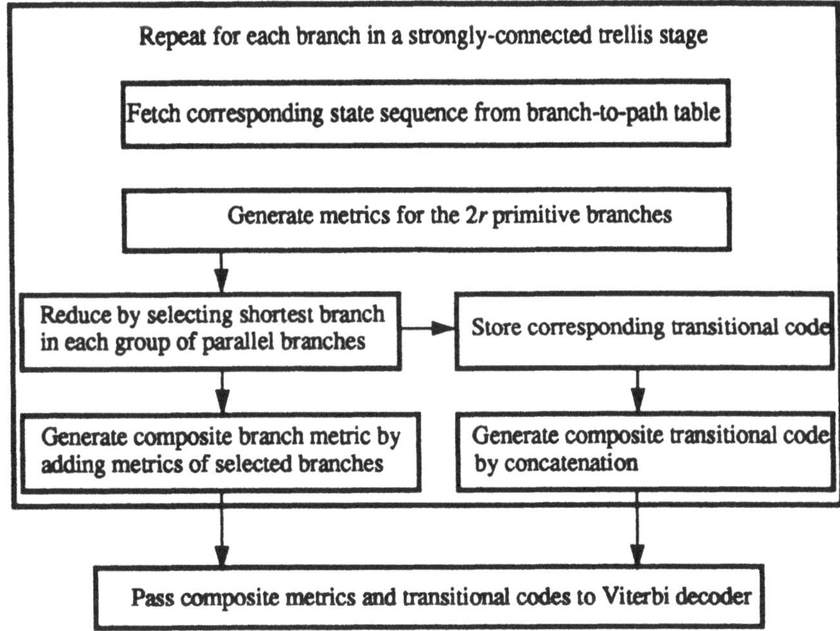

Fig. 10. Block diagram of the parallel TCM decoding algorithm

based on the strongly-connected trellis had been performed using a standard 256-states binary convolutional code and a decoding rate of 1320 bits per second was measured [9]. This performance is 69.5 times better than a sequential decoder running on a Sun SPARCstation IPC (19 bits per second).

On the other hand, the TCM simulation on the MasPar was 6.95 times slower than the binary convolutional code simulation with the same number of states. This slowdown is imputable to the higher computational complexity of the TCM code. Specifically:

- Floating point arithmetics are used for metrics instead of integer arithmetics.
- The Euclidean distance criterion is used instead of Hamming distance.
- Mapping message bits to modulated signals and vice-versa constitutes an additional overhead.
- The metric generation step has to be performed at the primitive trellis level.

In theory, for a binary convolutional code with 256 states and rate $\frac{1}{2}$, the use of the strongly-connected trellis speeds execution by a factor of 8 ($\log_2 256$). On the other hand, the theoretical speedup factor for a 16-QASK TCM code with 256 states is only 4. This is an additional justification for the performance gap between the two types of codes.

7 Conclusion

In this paper, we have shown that the strongly-connected trellis concept can be successfully applied to TCM codes and yield efficiency gains comparable to those obtained with binary convolutional codes. Even though the actual implementation was done on a particular computer, namely the MasPar MP1, its systolic-like architecture serves as a good model for possible VLSI implementations.

References

1. Sparso, J., Pederson, S., Paaske, E.: Design of a Fully Parallel Viterbi Decoder. Proceedings of VLSI 91. 2.2.1–2.2.10, 1991
2. Chang, C. Y., Yao, K.: Systolic Array Processing of the Viterbi Algorithm. IEEE Trans. Inform. Theory **35** (1989) 76–86
3. Ungerboeck, G.: Channel coding with multilevel/phase signals. IEEE Trans. Inform. Theory **28** (1982) 55–67
4. Ungerboeck, G.: Trellis-coded modulation with redundant signal sets - Part I: Introduction. IEEE Communications Magazine **25** (1987) 5–11
5. MasPar Parallel Application Language (MPL) User Guide. Sunnyvale: MasPar Computer Corporation, 1991.
6. MasPar Parallel Application Language (MPL) Reference Manual. Sunnyvale: MasPar Computer Corporation, 1991.
7. Leighton, F. T.: Introduction to Parallel Algorithms and Architectures. San Mateo, CA: Morgan-Kaufmann, 1992.
8. Roy, S.: Méthodes de réalisation de l'algorithme de Viterbi sur un ordinateur massivement paralèle de type SIMD. Masters' thesis, Univ. Laval, Sainte-Foy, QC, 1993.
9. Roy, S., Fortier, P.: An implementation of the Viterbi algorithm on a massively parallel computer. Proceedings of CCECE 1992, Toronto, Canada, MM10.2.1–MM10.2.4

Variable-Rate Punctured Trellis-Coded Modulation and Applications*

François Chan and David Haccoun

Département de génie électrique et de génie informatique
Ecole Polytechnique de Montréal
C.P. 6079, succ. "Centre-Ville"
Montréal, Canada, H3C 3A7

Abstract. The puncturing technique, which is widely used for binary convolutional codes, is applied to Trellis-Coded Modulation (TCM). Extensive computer searches have provided the best punctured TCM codes. The free Euclidean distance of most of these codes matches the distance of Ungerboeck's optimum codes. Since punctured codes are obtained from a low-rate code, decoding can be simplified at the expense of a small reduction in the coding gain. In addition to the decoding advantages, the rate of a punctured code can be easily changed and thus, variable bandwidth efficiency systems can be implemented with a single encoder/decoder.

1 Introduction

Trellis-Coded Modulation (TCM) can achieve a significant coding gain in the range of 3 to 6 dB over uncoded modulation without requiring more bandwidth [1]–[3]. A rate R=m/m+1 convolutional code is used and redundancy is introduced in the signal set, i.e., compared to uncoded modulation, the constellation set is expanded. Ungerboeck has found optimum codes with maximum free Euclidean distances for several constellations and number of states [3]. Unfortunately, the optimum code for a given signal constellation is different from the optimum code for another constellation. Hence, several encoders and decoders would be required to implement a system with various spectral efficiencies (e.g., 1, 2 or 3 b/s/Hz) or constellations. Furthermore, decoding a rate R=m/m+1 convolutional or TCM code with the Viterbi Algorithm [4, 5] requires (2^m—1) binary comparisons per state at each trellis level, in order to select the survivor, i.e., the most likely path, among the 2^m merging paths.

* This research has been supported in part by the Natural Sciences and Engineering Research Council of Canada, the Fonds pour la formation des Chercheurs et l'Aide à la Recherche (FCAR) du Québec and by a grant from the Canadian Institute for Telecommunications Research under the National Centers of Excellence program of the Government of Canada. This paper was presented in part at the IEEE International Symposium on Information Theory, San Antonio, Texas, Jan. 17–22, 1993.

Hence, if the number of states is large and if the coding rate is high, the decoding complexity may become prohibitive.

A so-called pragmatic approach to TCM has been proposed by Viterbi et al. [6] to avoid these drawbacks: a conventional 64–state rate-1/2 binary convolutional encoder is employed for QPSK, while for 8-PSK, one uncoded bit is added to the two output symbols of the same convolutional encoder to select a signal from the 8-PSK constellation. Similarly, for 16-PSK, two uncoded bits and the two coded output symbols are used. Hence, a single encoder/decoder is sufficient to implement various modulation schemes and achieve different spectral efficiencies. However, there is a limitation to this pragmatic approach since only a rate-1/2 convolutional code is used. Some of the optimum one-, two-, four-, and eight-dimensional TCM schemes use codes with rate R=2/3, 3/4 and 4/5 [3] and hence, limiting the pragmatic code to a rate-1/2 will entail sub-optimality. As an example, the optimum 8-PSK code is a rate-2/3 code and consequently, the pragmatic 8-PSK code yields an inferior performance with a 0.4 dB degradation at $P(B) = 1 \times 10^{-5}$ and an asymptotic degradation of 2 dB [6]. Our objective is to consider punctured codes in order to obtain the same flexibility (variable spectral efficiency) as Viterbi et al.'s pragmatic approach without being limited to using a rate-1/2 code in conjunction with parallel branches [7].

The practical advantages of puncturing for convolutional codes [8]–[12], that is, simplified decoding and variable rate, are well known. This technique is widely implemented in commercial convolutional encoders/decoders [13], [14]. In this paper, this technique is applied to Trellis-Coded Modulation. The paper is organized as follows. In Section 2, the punctured coding approach to Trellis-Coded Modulation is described and some of the best found codes are listed. Section 3 discusses decoding techniques and gives performance results of optimum and suboptimum decoding, obtained from computer simulations. In Section 4, applications are considered.

2 Puncturing technique for Trellis-Coded Modulation

The objective is to obtain an error performance identical, or only slightly inferior, to Ungerboeck's optimum code by replacing the high-rate convolutional code used in Ungerboeck's TCM scheme [1]–[3] by a punctured code. This technique is referred to as *Punctured Trellis-Coded Modulation* (PTCM).

2.1 Punctured Trellis-Coded Modulation (PTCM) codes

An example is used to illustrate how the puncturing technique can be applied to Trellis-Coded Modulation. A rate–1/2, 8–state encoder, depicted in Fig. 1, is considered. Just like in the construction of a rate $R = b/v$ punctured code derived from a low-rate $R = 1/v_0$ code, a puncturing matrix with b columns, v_0 rows and v 1's is applied at the output of the encoder [8]–[12]. The i-th column of the matrix corresponds to the i-th branch of the original low-rate trellis and a 1 (or a 0) in the i-th column means that the corresponding code symbol of the i-th branch is kept (or deleted, respectively). Here, the puncturing matrix $\mathbf{P} = \begin{bmatrix} 1 & 1 \\ 0 & 1 \end{bmatrix}$ is used, as illustrated in Fig. 2. Hence, our matrix yields a rate-2/3 code and its trellis, which is obtained by combining two levels of the original trellis is represented in Fig. 3. The code symbols are mapped onto the 8-PSK signal constellation of Fig. 4, which is the same as in Ungerboeck's scheme [1]. The shortest error-event path is shown in Fig. 3 and corresponds to the path with minimum free distance

$$d^2_{free} = 2.0 + 0.58 + 2.0 = 4.58$$

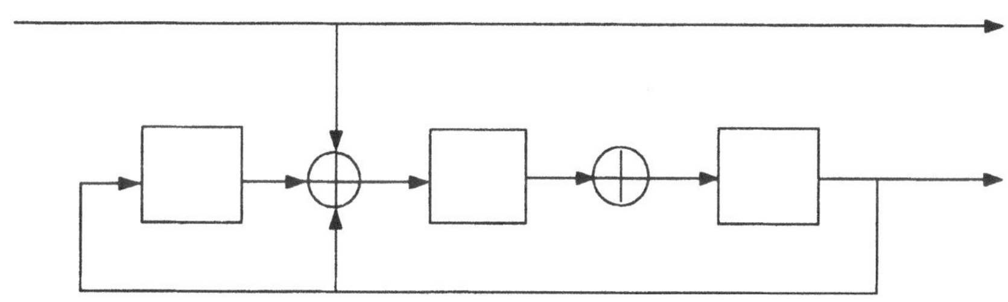

Parity Check Coefficients: 04, 15

Fig. 1. Rate-1/2, 8–state encoder

In the above example, a feedback code has been selected as the original code and punctured to yield a rate R=2/3 code. Usually, punctured binary codes are obtained from feedforward codes [8]–[10]. However, it has been noticed that for 8-PSK, a feedback code provides a better bit error performance than a feedforward code with the same Euclidean distance, even though their error-event performances may be almost the same [15]. Consequently, in the remaining of our search, feedback codes are considered. As for their decoding, the procedure will be described in detail in the next section.

One of Ungerboeck's heuristic rules which are conjectured to provide the best TCM schemes states that transitions originating from or merging into the same state receive signals with the second largest distance [1]. From the trellis of a rate-1/2 code (see e.g. Fig. 2) and our mapping of Fig. 4, we can deduce that in order to satisfy the above rule, the last column of the perforation matrix should contain

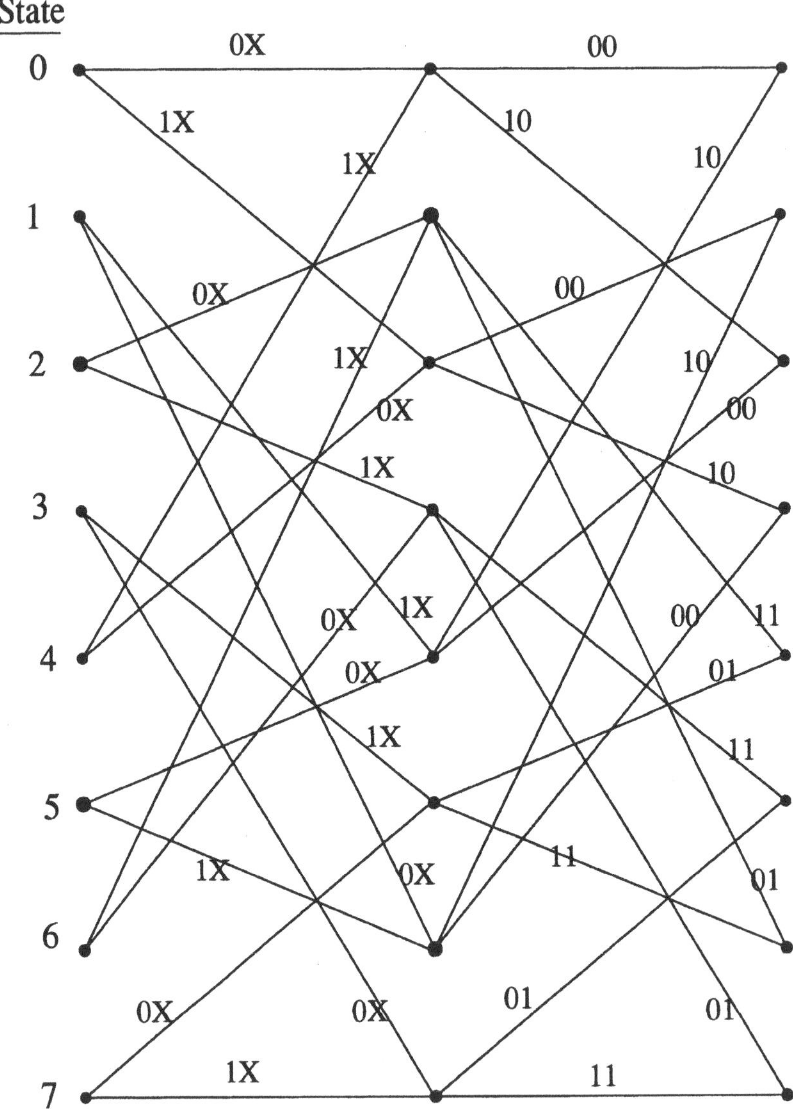

Fig. 2. Trellis for R=1/2, 8–state code

only 1's, i.e., the last branch symbols are not punctured. Otherwise, in our example, transitions originating from or merging into any state do not receive signals from one of the two QPSK subsets, {0,2,4,6} and {1,3,5,7}, but from a combination of both and hence, the distance between signals is reduced. Therefore, for a rate-2/3, 8-PSK PTCM code and the mapping of Fig. 4, the perforation matrix must be

$$\mathbf{P} = \begin{bmatrix} 1 & 1 \\ 0 & 1 \end{bmatrix}$$

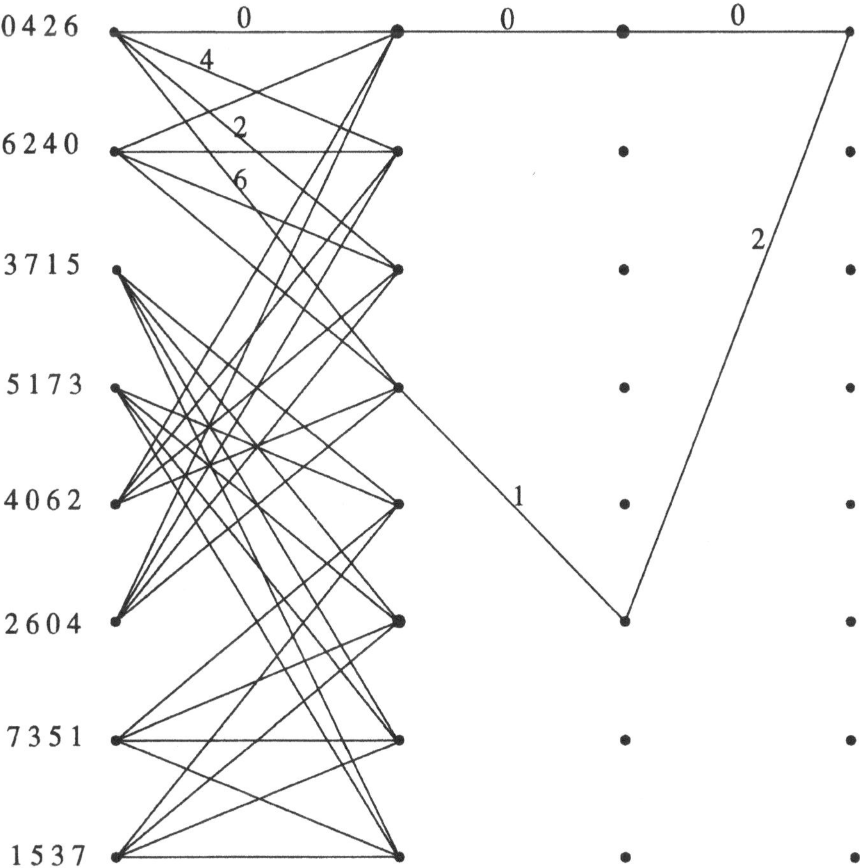

Fig. 3. Trellis for punctured R=2/3, 8–state code

The best 8-PSK PTCM codes obtained from a computer search are listed in Table 1, where $\tilde{m}$ represents the number of information bits that are actually used by the convolutional encoder and where the parity-check coefficients h^2, h^1, h^0 are expressed in octal notation. When several codes have the same free Euclidean distance, the code yielding the lowest error probability as obtained from computer simulations is selected. When the number of codes with the same distance is very large (10 or more codes), we select one of them at random. If the distance spectrum of the codes is known, we could also make the selection based on the spectrum. Compared to Ungerboeck codes, PTCM codes offer the same, or only slightly smaller, free Euclidean distance.

The optimum codes for 8-PSK with 8 or more states [3] are "true" rate-2/3 codes, i.e., there are no uncoded bits and therefore, no parallel transitions in the trellis. With parallel transitions, the maximum achievable free Euclidean distance is limited by the distance between signals assigned to parallel transitions since the free

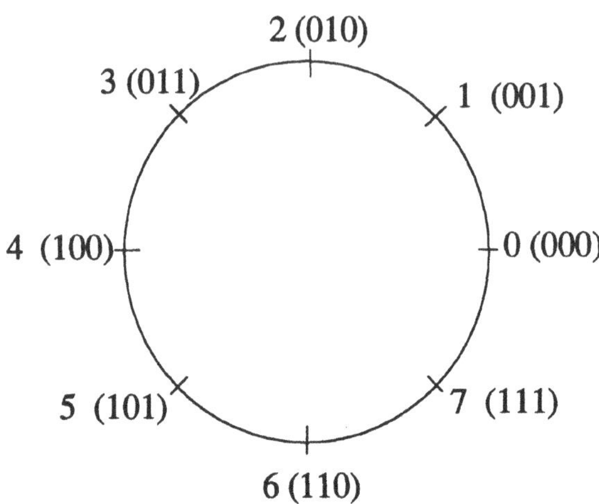

Fig. 4. 8-PSK constellation

Number of states	$\tilde{m}$	Punctured		Ungerboeck (1987)	
		Parity-check coefficients h^1, h^0	d^2_{free}	Parity-check coefficients h^2, h^1, h^0	d^2_{free}
8	2	04, 15	4.58	04, 02, 11	4.58
16	2	04, 23	5.17	16, 04, 23	5.17
32	2	34, 47	5.17	34, 16, 45	5.75
64	2	074, 163	6.34	066, 030,103	6.34
128	2	174, 311	6.34	122, 054, 277	6.58
256	2	240, 705	7.17	130, 072, 435	7.51

Table 1 Punctured and Ungerboeck TCM codes with largest d_{free} for 8-PSK

squared Euclidean distance of a code is given by

$$d^2_{free} = Min\left[d^2_{par}, d^2_{mul}\right]$$

where d^2_{par} is the distance between parallel branches and d^2_{mul} is the minimum squared Euclidean distance between multi-branch paths. However, parallel transitions reduce the connectivity in the trellis and hence, multi-branch error-events are longer. For some codes, especially short-memory codes, d^2_{par} is larger than d^2_{mul} and hence, it is preferable to have parallel transitions, which increase the minimum length and the free distance of multi-branch error-events. As a result, a larger free distance is achieved with parallel transitions for 4–state 8-PSK, 16-PSK with up to 128 states and for 16-QAM. Therefore, to achieve the same distance as the optimum codes, the puncturing technique should be able to implement codes with parallel transitions. This can be easily done by using a systematic rate-1/2 encoder and activating the clock of the shift-register only $\tilde{m}$ times every m cycles (or branches) if an encoder with a rate $R = m/m + 1$ and $(m - \tilde{m})$ uncoded bits is to be implemented [15]. Therefore, PTCM codes are quite general: they can include conventional high-rate codes or codes with parallel transitions.

A computer search for the largest Euclidean distance 16-PSK and 16-QAM PTCM codes has been performed; the best PTCM codes achieve the same free Euclidean distances as Ungerboeck's best known codes [7], [15]. In both cases, using the same reasoning as with the 8-PSK code given above, the perforation matrix is

$$P = \begin{bmatrix} 1 & 1 & 1 \\ 0 & 0 & 1 \end{bmatrix}$$

The rate of 16-PSK or 16-QAM codes should be R=3/4, but the true rate is $R = \tilde{m}/\tilde{m} + 1$, where $\tilde{m}$, the number of bits actually encoded in the convolutional encoder, is smaller than 3: $\tilde{m}$ is equal to 1 for 16–PSK with up to 128 states and to 2 for 16–QAM with a number of states between 8 and 512. This means that there are $(3 - \tilde{m})$ uncoded bits and $2^{(3-\tilde{m})}$ parallel branches in the trellis. The shift-register clock is activated $\tilde{m}$ times every m cycles (or branches) and these $\tilde{m}$ times occur at the end of the m cycles.

We have found quasi-optimum PTCM codes for 8-PSK, 16-PSK and 16-QAM. Following Ungerboeck's approach [3], codes for 32–QAM and 64–QAM could be obtained by using the 16–QAM codes and adding uncoded bits. The search could be extended to larger constellations (e.g., 256-QAM, 512-QAM, etc.), multidimensional constellations and coset codes. Codes with a larger number of states may also be investigated. However, this is an extremely time-consuming process since the search time increases with the rate of the code or constellation size and the number of states.

2.2 Variable-rate PTCM codes

The best original rate-1/2 codes provided by our computer search for 8–PSK are different from the best rate-1/2 codes for 16–PSK. Hence, two encoders and two decoders would be required to achieve 2 and 3 b/s/Hz bandwidth efficiencies. The advantage of punctured convolutional codes is the rate variability: the rate of a convolutional code can be easily changed by modifying only the puncturing pattern. The obtained codes are said to be rate-compatible. To maintain this advantage with PTCM codes, rate-1/2 codes which, when punctured, provide the largest free Euclidean distances for QPSK, 8–PSK and 16–PSK and not only for one of these constellations, must be found. Table 2 lists families of such codes, generated from rate-1/2 codes and compares their free distances with those of the best known codes, i.e., Larsen's codes [16] for QPSK and Ungerboeck's codes [3] for 8–PSK and 16–PSK. Although each PTCM code in the family may not necessarily be the best for that rate, its free Euclidean distance is equal, or only slightly smaller than the distance of the optimum code, which is not rate-compatible. Therefore, with PTCM, a single encoder/decoder is sufficient to achieve variable bandwidth efficiencies and hence, variable throughputs.

Number of states	Variable-rate PTCM codes				Best known codes		
	h^1, h^0	QPSK	8PSK	16PSK	QPSK	8PSK	16PSK
16	4, 23	7	5.17	1.628	7	5.17	1.628
32	10, 45	7	5.17	1.910	8	5.76	1.910
64	74, 163	8	6.34	2.0	10	6.34	2.0
128	154, 325	10	6.34	2.0	10	6.58	2.0

Table 2 Free distances of variable-rate PTCM codes

The 64–state code listed in Table 2 has the same distance as the best known codes for 8-PSK and 16-PSK. Its QPSK distance is, however, slightly inferior. This distance can be improved without affecting the trellis connectivity and hence, rate variability. This is achieved by using an original code with a rate lower than R=1/2. An encoder providing a free Hamming distance $d_{free} = 10$ has been obtained by computer search [15]. It is the same encoder that the one in Table 2, except that a third output has been added. State transitions are not modified, the trellis connectivity remains the same. The perforation matrix for QPSK is

$$\mathbf{P} = \begin{bmatrix} 1 \\ 0 \\ 1 \end{bmatrix}$$

yielding a rate R=1/2 code. For 8-PSK (rate–2/3), the matrix is

$$\mathbf{P} = \begin{bmatrix} 1 & 1 \\ 0 & 1 \\ 0 & 0 \end{bmatrix}$$

This means that the third output symbol is simply ignored or punctured and we have the same encoder as in Table 4. The same procedure is repeated for 16-PSK, i.e., the third symbol is always punctured.

3 Approximate decoding of PTCM codes

In Section 2, it has been mentioned that some PTCM codes have the same free Euclidean distance as Ungerboeck codes. Therefore, using a bound on the error-event probability P(E), they should provide a similar error performance asymptotically, i.e., at high signal-to-noise ratios. In this section, we verify that even at moderate signal-to-noise ratios, for a bit error probability in the range of 10^{-5} to 10^{-3}, performances are also similar. Only Viterbi decoding, which is a widely used optimum decoding algorithm for convolutional codes [4] and TCM [5], [17], is considered in this paper. The error performances of PTCM codes over an Additive White Gaussian Noise (AWGN) channel, as obtained from computer simulations, are presented and compared to the performance of Ungerboeck codes. A punctured code can be considered as a true high-rate code and optimally decoded using its high-rate trellis. For the 8–state, 8-PSK code in the example of section 2, this corresponds to the trellis of Fig. 3. This is referred to as "usual decoding" since this is the same type of decoding that is usually performed for TCM codes and it is optimum. On the other hand, a punctured code can also be viewed as a low-rate code with symbols periodically deleted, as illustrated in Fig. 2. Using this low-rate trellis simplifies the decoding but unlike binary convolutional codes, for TCM codes, it results in some ambiguity which may be resolved by suboptimum decoding as shown next.

We first consider optimum Viterbi decoding and hence, treat the punctured TCM code as a true high-rate code with a rate $R = m/(m+1)$. Let the received symbol at trellis level l be r_l and let the coded symbol of a branch, i.e., the signal, at the same trellis level be a^i, $i = 0, N-1$, where N is the cardinality of the signal set. Then, the branch metric $m(r_l, a^i)$ is

$$d^2(r_l, a^i) = \left\|r_l - a^i\right\|^2 \tag{7}$$

where $||.||^2$ denotes the squared Euclidean distance.

The error-event probability of a TCM code is lower bounded by [17]

$$P(E) \geq Q(d_{free}/2\sigma) \tag{8}$$

where d_{free} is the free Euclidean distance and σ^2 is the Gaussian noise variance. The error performance approaches this bound asymptotically at high signal-to-noise ratios. Since the PTCM code is considered as a true high-rate code and decoding proceeds in the same fashion as an Ungerboeck code, bound (8) on the error-event probability is also valid for PTCM: the decoder simply views the PTCM code as a code equivalent to Ungerboeck code. The bit error probability obtained by computer simulations of 8-PSK using Ungerboeck and our punctured codes with 64 states is illustrated in Fig. 5. Both codes have the same free Euclidean distance as indicated in the previous section and it can be seen that their error performances are also very close when they are both decoded optimally. Therefore, the free Euclidean distance not only defines the asymptotic coding gain, but is also an appropriate criterion for our code selection, even for a bit error probability in the range of $[10^{-3} - 10^{-5}]$. Similarly, for 16-PSK and 16-QAM, computer simulations show that the error performance of PTCM codes is very close to that of Ungerboeck codes when decoded in the same manner since the free distances are identical [15].

When a PTCM code is considered as a true high-rate code with a rate R=m/m+1 and decoded optimally, its performance is similar to that of Ungerboeck code with the same free distance but decoding is also similar: $2^m - 1$ binary comparisons per state are required at each trellis level. Clearly, if the code has a large number of states and if m is large, then decoding may become prohibitively complex and one must resort to suboptimum decoding. Following the punctured binary convolutional codes, a PTCM code is now viewed as a low-rate code with symbols periodically deleted. It should be noted that at the transmitter side, the PTCM code is always viewed as a true high-rate code, that is, a channel signal is a mapping of the symbols on two or more branches of the low-rate code. Fig. 6 depicts a small part of the trellis for the 8-PSK punctured code of Fig. 2 and illustrates the problem: at intermediate node[2] c, metrics of the converging branches, i.e., branches at the first level of the trellis, are unknown in PTCM, as opposed to punctured binary convolutional codes.

For the PTCM code of Fig. 6, a channel signal is defined as the symbol on a branch at the first level in combination with the symbols on the following second level branch. In our example, 0 is the first symbol of the channel signal 000 or 010. As indicated previously, while the metric $m(r_l, 000)$ of 000 can be computed as $d^2(r_l, 000)$, where r_l is the received channel signal, the metric of symbol 0 on the first level branch, is not known. However, in order to simplify decoding, a final decision must be made at the intermediate node c. Therefore, the decision rule at state c is to select branch 0 if

$$M(a) + m(r_l, 0) < M(b) + m(r_l, 1) \tag{9}$$

where $M(i)$ is the accumulated metric at state i and where $m(r_l, s)$ is the metric of the branch with symbol s to be defined later. Without loss of generality let s be 0. Then, $m(r_l, 0)$ can assume two values: $m(r_l, 000)$ or $m(r_l, 010)$ since, as already mentioned 0 is the first symbol of signals 000 or 010 (see Fig. 6). A solution is to use as $m(r_l, 0)$ an approximate metric and to base the decision at the intermediate node on that metric. This is accomplished as follows.

First, note that only four distinct signals can pass through an intermediate state. In our example, the four signals that can pass through state c are 0, 2, 4 and 6 (in octal representation). Signals 0 and 2 are grouped together since they both come

---

[2] When decoding a rate R=b/v punctured code using the trellis of the low-rate original code, a state at level b (and any multiple of b), corresponding to b (or multiple of b) information bits is denoted a "true" state since it corresponds to a state in the trellis of a conventional high-rate code. A state at any other trellis level is denoted an "intermediate" state.

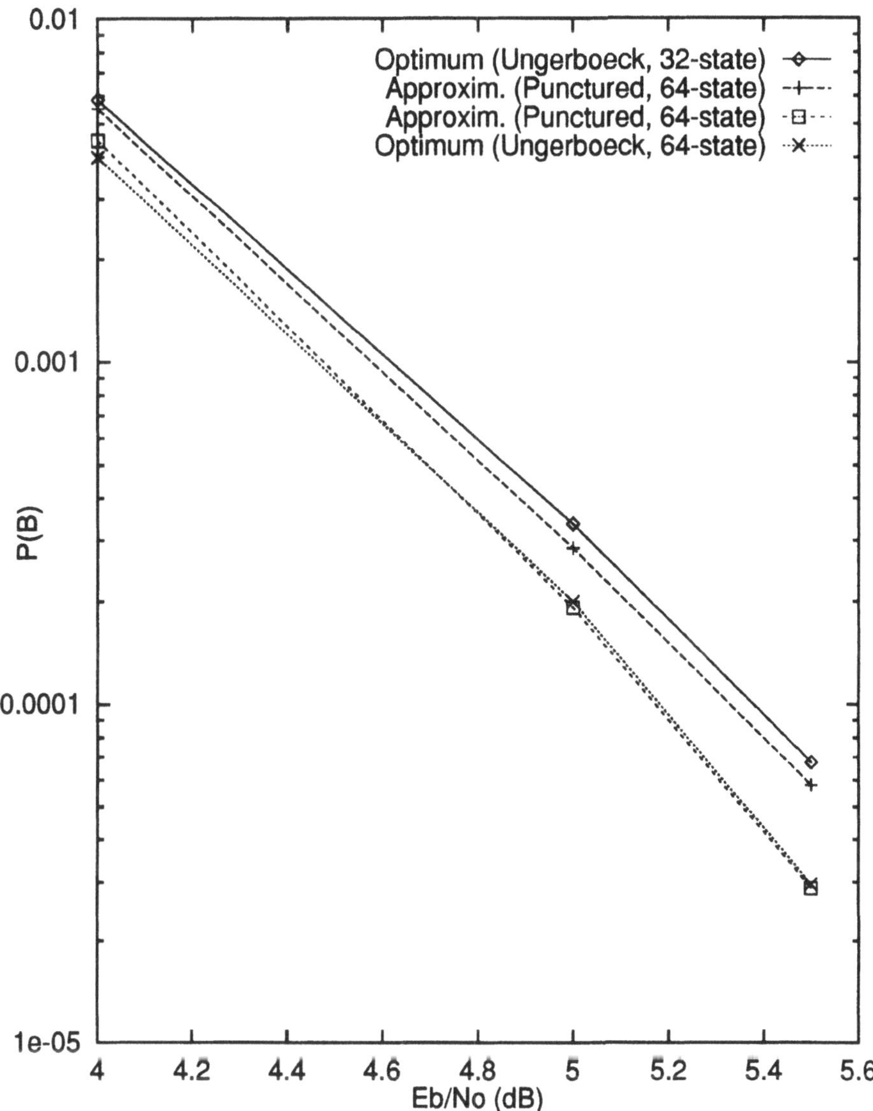

Fig. 5. Comparison of the bit error performance of optimum and approximate decoding techniques with Ungerboeck and PTCM codes — 8–PSK

from state a and have the same leading symbol 0 (branch from state a to state c). Similarly, signals 4 and 6 are grouped together. Hence, at state c, the choice between branch 0 and branch 1 is, in fact, a choice between group(0,2) and group(4,6). The

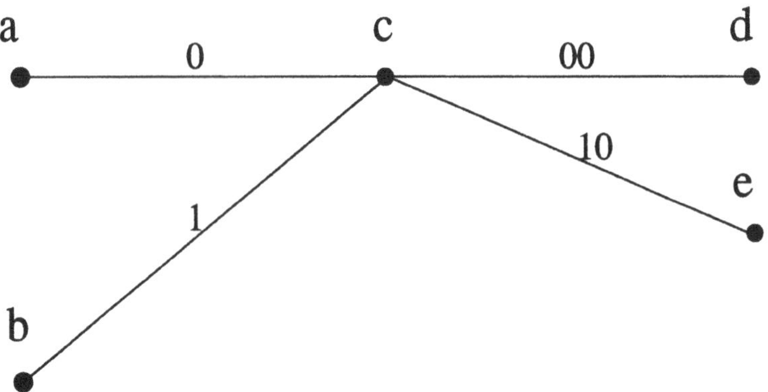

Fig. 6. Part of the trellis for 8-PSK punctured code

metric of branch 0, i.e., of group(0,2) is defined as the Euclidean distance between the received signal r_l and a point equidistant from signals 0 and 2. (It is assumed, of course, that all input signals are equiprobable.) As illustrated in Fig. 7, we have selected signal 1 as this equidistant point. Similarly, for the group(4,6), point 5 is chosen. Thus, at node c, the metric of branch 0 is

$$m(r_l, 0) = d^2(r_l, 001) \tag{10}$$

and the metric of branch 1 is

$$m(r_l, 1) = d^2(r_l, 101) \tag{11}$$

An optimum decoder uses two metrics for $m(r_l, 0)$, $m(r_l, 000)$ and $m(r_l, 010)$, since in a punctured high-rate trellis obtained by combining two levels of the low-rate trellis (see Figs. 2 and 3), branch 0 at the first level of the low-rate trellis corresponds to two branches of the high-rate trellis: branch 000 and branch 010. Therefore, $m(r_l, 0)$ is just an approximation to the two exact metrics and is used to simplify decoding by allowing a final decision at the intermediate state c. At the next trellis level, using as leading symbol the surviving branch at the intermediate state, the exact metrics can be computed.

In summary, the approximate decoding technique consists in merging constellation signals that have the same leading symbol in a group and use as approximate branch metric the Euclidean distance between the received signal and a point equidistant from the signals in the group. This technique is clearly suboptimum since it does not use the exact metric but an approximation which reduces the metric

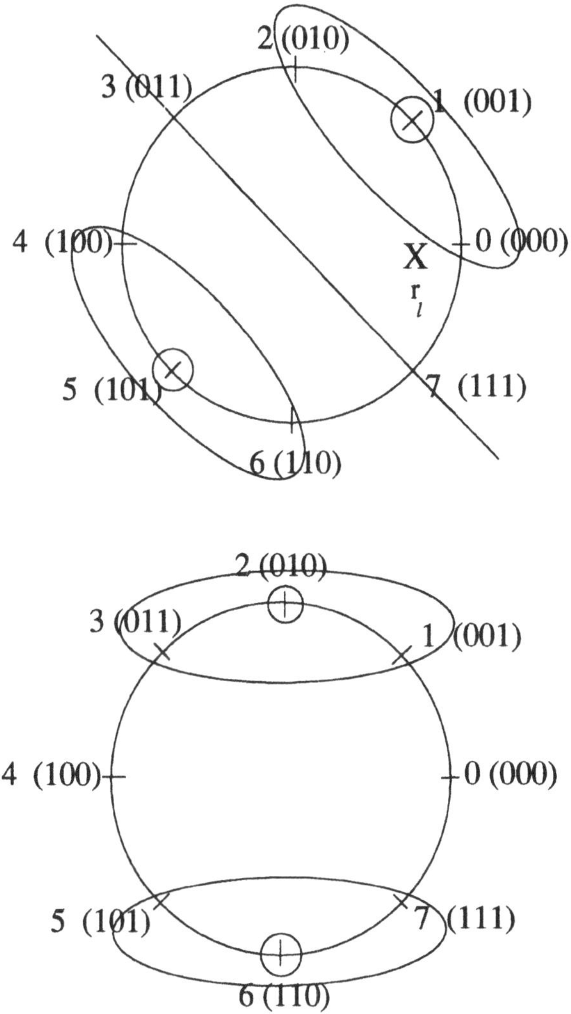

Fig. 7. 8-PSK constellation with merged points for approximate decoding

separation [15], hence leading to an error performance degradation. A more detailed analysis of this decoding technique can be found in [15]. .

Bit error performance obtained by simulation of PTCM 8-PSK codes with approximate decoding and with optimum decoding is shown in Fig. 5. It can be seen that approximate decoding leads to a slight degradation of approximately 0.15 dB at $P(B) = 1 \times 10^{-4}$ for a 64–state code. However, its complexity is reduced: it requires 2 comparisons per state, while optimum decoding necessitates 3 comparisons. In addition, the performance of this PTCM code is still superior to that of a 32–state Ungerboeck code. Compared to the 32–state and 64–state Ungerboeck codes, both the performance and the complexity of our approximate decoding lie between those of the two codes. Hence, PTCM codes allow a complexity/performance trade-off. The complexity reductions provided by the puncturing technique increase significantly with the coding rate R=m/m+1 since decoding the low-rate trellis requires only m comparisons per state instead of 2^m-1 for the high-rate trellis.

Our approximate decoding technique for PTCM codes described above provides good performance for 8–PSK, allowing a reduction of the decoding complexity at a cost of a very small performance degradation. With rate–2/3, 16-QAM PTCM codes, however, this technique does not provide satisfactory results. For example, the error performance of the 64–state code with approximate decoding is inferior to the performance of the 32–state Ungerboeck code with optimum decoding, while requiring a larger decoding complexity [15]. Hence, the approximate metrics lead to an unacceptable degradation. Another technique, called staged decoding, has been developed [15]. It is shown that with this technique, the error performance of the 64–state PTCM is now superior to that of the 32–state Ungerboeck code [15].

As a conclusion, the error performance of PTCM codes with suboptimum decoding is slightly inferior to that of Ungerboeck codes but decoding complexity is reduced. The puncturing technique yields the same complexity savings with TCM codes as with binary convolutional codes. It should be noted that the small performance degradation is not caused by the code but by the decoding technique. Hence, if the need arises, the performance can be improved by changing the decoder, while no modification is required at the transmitter side.

4 Applications

It has been shown that the best PTCM codes found using an extensive computer search can be as good as the best known codes discovered by Ungerboeck [3]. Furthermore, the proposed decoding techniques provide the same complexity savings as with punctured convolutional codes and these savings entail only a very slight degradation of the error performance. Families of rate-compatible PTCM codes have also been obtained and hence, the other advantage of punctured convolutional codes, rate variability, is maintained for PTCM.

Hence, PTCM can be used to replace the convolutional codes in Trellis-Coded Modulation when these codes are true high-rate codes. This would significantly reduce the decoding complexity while error performance is only slightly degraded and consequently, would allow the use of codes with a large number of states. For a given level of complexity, the puncturing technique and a code with a larger number of states may yield a better performance than optimum decoding of a conventional code with fewer states if the coding rate is high.

The second advantage is that a single encoder/decoder is sufficient to achieve variable bandwidth efficiencies and hence, variable throughputs. Therefore, PTCM is an attractive alternative to Viterbi et al.'s pragmatic codes [6], [18]. The puncturing technique is more flexible than Viterbi et al's pragmatic approach because it allows either a true high-rate code or a code with parallel branches. Since many good codes for the gaussian channel [3] and especially for fading channels (see e.g. [19], [5]) do not have parallel branches, the puncturing technique may be able to reproduce such codes whereas the pragmatic approach cannot. The free squared Euclidean distance of our 64–state code is $d^2_{free} = 6.34$ for 8-PSK and thus, the asymptotic coding gain is 5.0 dB, while the distance of the pragmatic code is $d^2_{free} = 4.0$, resulting in an asymptotic coding gain of 3.0 dB [6]. It can be seen that when parallel branches become a limitation, the puncturing technique can provide codes without parallel branches which have a larger free distance than pragmatic codes. Parallel branches become a limitation when the number of states increases

A variable-rate communications system may be useful when several types of information have to be transmitted: for example, QPSK can be used for data, 8-PSK for digitized voice and 16-PSK for video, which in general consists of a large number of bits. Furthermore, this technique can be used for the same applications as punctured convolutional codes [20]. In type I hybrid ARQ/FEC transmission schemes [16], 16-PSK, which provides the highest throughput, can be used first and, should an error be detected, 8-PSK, or even QPSK may be used in subsequent retransmissions.

PTCM may also be well suited for unequal error protection (UEP), assuming some information about the susceptibility to errors of the information bits is supplied by the source: the most sensitive data may be transmitted using QPSK, the less sensitive ones may be transmitted using 8-PSK or 16-PSK. Hence, by changing the rate of the code, three levels of error protection are available; the bit error probability is improved approximately by more than three orders of magnitude from one level to the next one, e.g., from 8-PSK to QPSK.

5 Conclusion

In this paper, we have applied the puncturing technique to Trellis-Coded Modulation and showed that the essential advantages puncturing provides to binary convolutional codes are maintained: simplified decoding and easy rate varying coding/decoding. Hence, variable bandwidth efficiency systems can be easily implemented with this technique. Finally, PTCM is very flexible and the code is not limited to rate R=1/2 with parallel branches; it can be either a true high-rate code or a code with parallel branches. PTCM appears to provide an interesting alternative to Viterbi et al's pragmatic approach to TCM [6].

Acknowledgement

Stimulating discussions at the beginning of this work with Professor H. Leib of McGill University are gratefully acknowledged.

References

[1] G. Ungerboeck, "Channel coding with multilevel/phase signals," *IEEE Trans. Inform. Theory*, vol. IT-28, pp. 55–67, Jan. 1982.

[2] G. Ungerboeck, "Trellis-coded modulation with redundant signal sets — part I: Introduction," *IEEE Commun. Mag.*, pp. 5–11, Feb. 1987.

[3] G. Ungerboeck, "Trellis-coded modulation with redundant signal sets — part II: State of the art," *IEEE Commun Mag.*, pp. 12–21, Feb. 1987.

[4] A. J. Viterbi, "Convolutional codes and their performance in communication systems," *IEEE Trans. Commun. Technol.*, vol. COM-19, pp. 751–772, Oct. 1971.

[5] E. Biglieri, D. Divsalar, P. J. Mclane, and M. K. Simon, *Introduction to Trellis-Coded Modulation with Applications*. Macmillan Publishing Company, New York, 1991.

[6] A. J. Viterbi, J. Wolf, E. Zehavi, and R. Padovani, "A pragmatic approach to trellis-coded modulation," *IEEE Commun. Mag.*, pp. 11–19, Jul. 1989.

[7] F. Chan and D. Haccoun, "High-rate punctured convolutional codes for trellis-coded modulation," in *Proc. IEEE International Symposium on Information Theory*, p. 414, San Antonio, Texas, 1993.

[8] J. B. Cain, G. C. Clark, and J. M. Geist, "Punctured convolutional codes of rate (n-1)/n and simplified maximum likelihood decoding," *IEEE Trans. Inform. Theory*, vol. IT-25, pp. 97–100, Jan. 1979.

[9] Y. Yasuda, K. Kashiki, and Y. Hirata, "High-rate punctured convolutional codes for soft decision Viterbi decoding," *IEEE Trans. Commun.*, vol. COM-32, pp. 315–319, Mar. 1984.

[10] D. Haccoun and G. Bégin, "High-rate punctured convolutional codes for Viterbi and sequential decoding," *IEEE Trans. Commun.*, vol. 37, pp.1113–1125, Nov. 1989.

[11] G. Bégin and D. Haccoun, "High-rate punctured convolutional codes: Structure properties and construction technique," *IEEE Trans. Commun.*, vol. 37, pp. 1381–1385, Dec. 1989.

[12] G. Bégin, D. Haccoun, and C. Paquin, "Further results on high-rate punctured convolutional codes for Viterbi and sequential decoding," *IEEE Trans. Commun.*, vol. 38, pp. 1922–1928, Nov. 1990.

[13] "Q0256 — k=7 multi-code rate Viterbi decoder technical data sheet," Qualcomm, Incorporated, San Diego, CA 92121, 1990.

[14] "Custom ASIC products short form catalog," Stanford Telecommunications, Incorporated, Santa Clara, CA, 1989.

[15] F. Chan and D. Haccoun, "A punctured coding approach to trellis-coded modulation," Tech. Rep. EPM/RT-93/22, Ecole Polytechnique de Montréal, Montréal, Canada, 1993.

[16] S. Lin and D. J. Costello, *Error Control Coding: Fundamentals and Applications*. Englewood Cliffs: Prentice-Hall, 1983.

[17] G. C. Clark and J. B. Cain, *Error-Correction Coding for Digital Communications*. Plenum Press, 1981.

[18] "Q1875 pragmatic trellis decoder technical data sheet," Qualcomm, Incorporated, San Diego, CA 92121, 1992.

[19] D. Divsalar and M. K. Simon, "The design of trellis coded MPSK for fading channels: Performance criteria," *IEEE Trans. Commun.*, vol. 36, No. 9, pp. 1004–1012, Sep. 1988.

[20] J. Hagenauer, "Rate-compatible punctured convolutional codes (RCPC codes) and their applications," *IEEE Trans. Commun.*, vol. 36, pp. 389–400, Apr. 1988.

Trellis-Based Decoding of Binary Linear Block Codes

Vladislav Sorokine, Frank R. Kschischang and Véronique Durand

Department of Electrical and Computer Engineering
University of Toronto, Toronto, Ontario, Canada M5S 1A4

Abstract. In this paper, we present an efficient algorithm for the iterative construction of the minimal trellis and corresponding tree for a binary linear block code. State-extension and branch-labeling is based only on the state labels at the current position and on the set of atomic generators, making it unnecessary to store the whole code tree. Using these results, we have devised a sequential stack algorithm for decoding binary linear block codes. By modifying the metric bias term according to received data, maximum-likelihood soft-decision decoding is guaranteed in the absence of stack overflow. At practical signal-to-noise ratios, the computational complexity of this sequential decoding scheme is much smaller on average than that of the Viterbi algorithm. Several modifications to the basic algorithm, including stack overflow handling, sorting of the received data, and "fast push-through" can be used to improve the characteristics of the variable-bias term sequential decoding algorithm. The result is a practical and general soft-decision decoding method for linear block codes.

1 Introduction

Like convolutional codes, block codes can be described as the set of paths through a trellis. The primary motivation for such a representation is that the maximum-likelihood decoding problem for memoryless channels can be translated into the problem of finding a minimum-cost path through the corresponding trellis [1]. The complexity of decoding (using, e.g., the Viterbi algorithm) can be measured in terms of various parameters of the minimal trellis [2], an important one being the maximum number of states in the trellis.

Another useful representation of a block code is a tree. A binary $[n, k]$ block code can be described as a collection of 2^k leaf nodes in a binary tree of depth n. Some soft-decision decoding algorithms, most notably sequential decoding algorithms like the stack algorithm, are suited for dealing with codes represented as trees, rather than as trellises.

An important practical issue in dealing with block code trellises and trees is that their large size precludes storage of the entire trellis or tree in computer

* This work was supported in part by the Natural Sciences and Engineering Research Council, Canada.

memory. For example, the minimal trellis for a [127, 64] cyclic code—a BCH code, say—contains $2\sum_{i=0}^{63} 2^i = 2^{65} - 2$ nodes (vertices) [3]. In this paper, we describe an algorithm for the iterative construction of a trellis (or a tree), that avoids storage of the whole trellis or tree for the code. The novelty of our approach does not lie in the characterization of the minimal trellis, as this problem is reasonably well understood [1, 2, 4, 5]; rather, the novelty of this work is in an algorithm to compute efficiently the set of successor (or predecessor) states, with the associated branch labels, from a given state.

The large trellis size associated with block codes will usually preclude use of the Viterbi algorithm to search for the maximum-likelihood codeword. In this paper, using the iterative tree construction algorithm, we devise a sequential stack algorithm for soft-decision decoding an arbitrary binary linear block code. The algorithm differs from the usual Zigangirov-Jelinek (ZJ) algorithm described, e.g., in [6], by using a variable, data-dependent bias term. The new bias term guarantees that the coder will output the maximum-likelihood codeword (in the absence of stack overflows). Even in the presence of stack overflow, the maximum-likelihood path is often retained, particularly when the stack overflow results in the loss of only high cost paths.

We study several modifications of the basic algorithm that can improve performance in practice. For example, we find that sorting the received data in order of decreasing reliability can yield a reduction in the number of nodes searched, over some signal-to-noise ratio regions. We also study a "fast push-through" decoder that quickly determines candidates for the maximum-likelihood path, and then iteratively updates these estimates after searching the code tree more thoroughly. Such a decoder would be useful in the practical situation in which an upper limit is placed on the total decoding time (or number of nodes searched) for a given received block. If time runs out before a decoding decision is made, the decoder can simply output its best candidate codeword.

This paper is structured as follows. Section 2 covers the algorithm used for iterative construction of trellises and trees for binary linear block codes. In Section 3, we present a sequential maximum-likelihood soft-decision algorithm for binary linear block codes with variable bias term. A number of performance enhancements are also presented. In Section 4, simulation results that illustrate the performance of decoders are presented and discussed. Finally, Section 5 summarizes the algorithms presented in the paper.

2 Trellises and Trees for Binary Linear Block Codes

2.1 Trellises

We begin with block code trellises. Graph-theoretic definitions for block code trellises have been given by Massey [2] and Muder [4]. For the purposes of this paper, we find it convenient to combine the two definitions.

Briefly, a length n block code trellis is an edge-labeled directed graph. The set of vertices of the graph is partitioned into a collection of disjoint subsets

$V_0, V_1, \ldots, V_n$. The set V_i is referred to as the set of *states* at time index i. All edges (or *branches*) in the graph are of the type that connect a state at time index i with a state at time index $i+1$. If a state $v_{i-1} \in V_{i-1}$ is connected by a branch to a state $v_i \in V_i$, then v_i is said to be a successor of v_{i-1}; equivalently, v_{i-1} is a predecessor of v_i. Each branch $v_{i-1} \xrightarrow{\alpha} v_i$, $v_{i-1} \in V_{i-1}, v_i \in V_i$, is labeled with an element α from a nonempty label set A_i. For codes over $GF(2)$, we will usually take $A_i = GF(2)$ for all i. We will insist that a trellis may not have redundant branches, i.e., parallel edges $v_{i-1} \xrightarrow{\alpha} v_i$, $v_{i-1} \xrightarrow{\alpha'} v_i$ with the same branch label ($\alpha = \alpha'$) are disallowed.

A block code trellis also has the following properties: (1) V_0 contains a unique state v_0 (the root); V_n contains a unique state v_n (the goal); (2) each state can be reached by at least one directed path from the root; (3) the goal can be reached by at least one directed path from each state. This definition differs slightly from that of [2] (for example, all branches in the graph must be labeled); it also differs from that of [4] (for example, the tree associated with a code is not a trellis). A path from the root to the goal will be referred to as a path *through* the trellis.

Each path $v_0 \xrightarrow{\alpha_1} v_1 \xrightarrow{\alpha_2} \cdots \xrightarrow{\alpha_n} v_n$ through a trellis T of length n defines an n-tuple of branch labels $(\alpha_1, \alpha_2, \ldots, \alpha_n)$. The set of all such paths defines the block code associated with T. Conversely, it is always (trivially) possible to associate a trellis with any block code: for each codeword, simply create a path from the root to the goal passing through a unique set of intermediate states, and label each branch with the corresponding codeword component. Other than at the root and the goal ($i = 0$ or $i = n$), such a trellis will have as many states as codewords at every time index i. Our interest, of course, (as in [4, 5]) is to construct the *minimal* trellis associated with a given code, i.e., the trellis with the smallest possible number of states at each time index.

2.2 Binary Linear Block Codes

A binary linear block code $\mathscr{C}$ of length n and dimension k is a k-dimensional vector space of n-tuples over $GF(2)$. Such a code can be characterized as the row space of a $k \times n$ generator matrix G of rank k, i.e., as the set of all possible linear combinations of the rows of G (assumed to be linearly independent) with scalars drawn from $GF(2)$, and is often denoted as an $[n, k]$ code.

An equivalent characterization of an $[n, k]$ code is as the direct sum of the one-dimensional subspaces spanned by the rows of G. If the ith row of G is denoted by $\mathbf{g}_i$ and the set distinct scalar multiples of $\mathbf{g}_i$ by the elements of $GF(2)$ is denoted by $\langle \mathbf{g}_i \rangle$, then

$$\mathscr{C} = \langle \mathbf{g}_1 \rangle \oplus \langle \mathbf{g}_2 \rangle \oplus \cdots \oplus \langle \mathbf{g}_k \rangle \tag{1}$$

where $\oplus$ denotes the direct sum of sets. Such a characterization is useful as it allows one to define a trellis for $\mathscr{C}$, as will become apparent in the sequel.

2.3 Constructing Trellises

We define the *span* (or support interval) of a codeword $\mathbf{c} \in \mathscr{C}$ as the smallest interval outside of which the components of $\mathbf{c}$ are strictly zero. A nonzero codeword $\mathbf{c} = (c_1, \ldots, c_n)$ with span $[a, b]$ is said to *start* at position a and *end* at position b, where $1 \leq a \leq b \leq n$; this means that $c_a \neq 0$, $c_b \neq 0$, but $c_i = 0$ if $i < a$ or $i > b$ (by convention, $c_0 = c_{n+1} = 0$). The span of the all-zero codeword $\mathbf{0}$ is the empty interval denoted by $[\,]$. The span length $L(\mathbf{c})$ of a codeword $\mathbf{c}$ is defined as the cardinality of its span, i.e., if $\mathbf{c}$ is nonzero with span $[a, b]$, then $L(\mathbf{c}) = b - a + 1$, while $L(\mathbf{0}) = 0$.

The utility of the concept of codeword span arises when one considers the minimal trellis for a one-dimensional code. Consider an $[n, 1]$ code $\langle \mathbf{g} \rangle$ over $GF(2)$, generated by some nonzero n-tuple $\mathbf{g}$. Such a code has two codewords: $\{\mathbf{0}, \mathbf{g}\}$. If $\mathbf{g}$ has span $[a, b]$, $b > a$, then the minimal trellis for $\langle \mathbf{g} \rangle$ is as shown, for a binary code, in Fig. 1.

Next we provide labels for the states in the minimal trellis of an $[n, 1]$ code $\langle \mathbf{g} \rangle$. A scheme for labeling states is said to be well-defined if, at every time index i, distinct states receive distinct labels. We provide a well-defined set of state labels by labeling states with elements of $GF(2)$. Every state through which the all-zero codeword $\mathbf{0}$ passes is labeled with "0," thereby defining the zero state at each time index i. Nonzero states are labeled with "1."

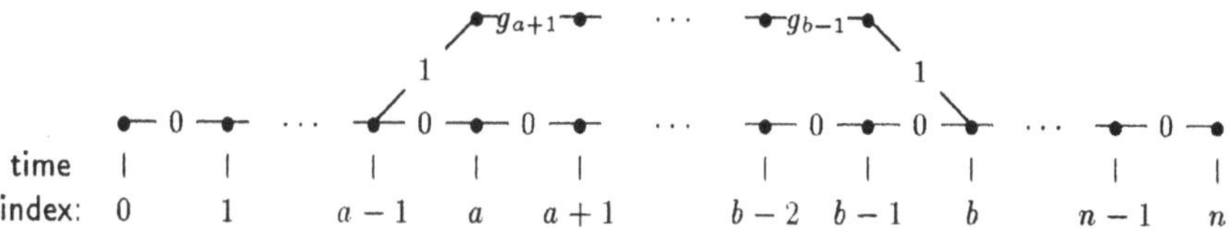

Fig. 1. Minimal trellis for a binary $[n, 1]$ code with generator having span $[a, b]$.

Having constructed and provided state labels for the minimal trellis of an arbitrary $[n, 1]$ code $\langle \mathbf{g} \rangle$, we are now in a position to construct and provide state labels for a (not necessarily minimal) trellis of the $[n, k]$ code defined as a direct sum. We simply construct the minimal trellis for each of the codes $\langle \mathbf{g}_i \rangle$, $1 \leq i \leq k$, and then form the "product" of these trellises. A similar trellis product construction can be used to build the trellis for a multilevel code, as in, e.g., [7].

The "product" $T = T_1 \times T_2$ of two length n "component" trellises with binary branch labels is defined as follows. Denote by $V_1(i)$ and $V_2(i)$ the set of state labels for trellises T_1 and T_2 at time index i, respectively. (Because the state label sets are well defined, we do not distinguish between the set of states and the set of state labels.) The number of states at time index i is denoted by $|V_1(i)|$ and $|V_2(i)|$, respectively. At each time index i, T will have a number of states equal to the product $|V_1(i)| \cdot |V_2(i)|$ of the number of states in the component trellises at that time index. A well-defined set of labels for these states is obtained by

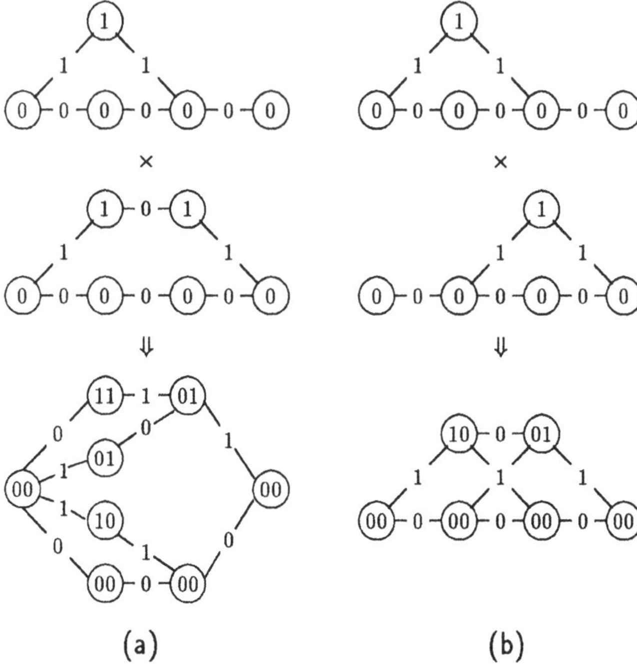

Fig. 2. Trellis diagrams for a binary [3,2] code with state label circled: (a) non-minimal trellis obtained from $\langle 110\rangle \oplus \langle 101\rangle$; (b) minimal-trellis obtained from $\langle 110\rangle \oplus \langle 011\rangle$.

assigning distinct elements of the Cartesian product $V_1(i) \times V_2(i)$ to the distinct states. The branches of T and their labels are obtained by forming the Cartesian product of the branches in the component trellises. If $B_1(i)$ and $B_2(i)$ denote the set of branches connecting states at time $i-1$ with states at time i in trellis T_1 and T_2, respectively, then T will have branches $B_1(i) \times B_2(i)$, in which an element

$$(s_1(i-1) \xrightarrow{\alpha_1} s_1(i), s_2(i-1) \xrightarrow{\alpha_2} s_2(i))$$

is interpreted as a branch

$$(s_1(i-1), s_2(i-1)) \xrightarrow{\alpha_1+\alpha_2} (s_1(i), s_2(i))$$

in the trellis for T.

Fig. 2 shows two examples of trellises obtained for the binary $[3,2]$ single parity check code. The two examples were obtained from two different generator matrices,

$$G_a = \begin{bmatrix} 110 \\ 101 \end{bmatrix}, \quad G_b = \begin{bmatrix} 110 \\ 011 \end{bmatrix},$$

both of which generate this code. The two trellises were obtained from product of the minimum trellises for the $[3,1]$ codes defined by the row vectors in the generator matrices. The trellis of Fig. 2(b) is the minimal trellis for the [3, 2] code.

It is important to note that the trellis obtained depends on the choice of generator matrix for the code. Any generator matrix is "trellis-oriented" in the sense that the trellis product construction described here gives rise to a trellis for the code it generates; however, the term "trellis-oriented generator matrix" [5] will be reserved for a generator matrix that gives rise to the minimal trellis. Thus G_b is a trellis-oriented generator matrix, while G_a is not.

2.4 Iterative Trellis Construction

Now we describe a method for computing the set of successor states and corresponding branch labels, starting from an arbitrary state in a linear block code trellis. (The algorithm is easily modified to compute the set of predecessor states, as might be required in a software implementation of the Viterbi algorithm.) By applying the basic algorithm iteratively, starting from the root node, the entire trellis diagram for a linear block code can be constructed.

All states are labeled with a k-tuple with components from $GF(2)$. The root node is labeled with the all-zero k-tuple. We first consider trellises with no "parallel transitions," i.e., with no generator having a span of the form $[a, a]$.

Let $\{\mathbf{g}_1, \ldots, \mathbf{g}_k\}$ be a set of generators for a binary $[n, k]$ code. A generator with span $[a, b]$ is said to contribute a *divergence* at time index $a - 1$ and a *convergence* at time index $b - 1$. Let the set $D(i) = \{d_1, d_2, \ldots, d_{|D(i)|}\}$ index the diverging generators at time i, i.e., $\{\mathbf{g}_{d_1}, \mathbf{g}_{d_2}, \ldots, \mathbf{g}_{d_{|D(i)|}}\}$ is the set of diverging generators at time index i. If no generators diverge at time index i, then $D(i)$ is empty. Note that $D(i)$ is easily pre-computed from the given generator matrix.

Consider a state v at time index i with state label $(s_1, s_2, \ldots, s_k)$. The state v will have $2^{|D(i)|}$ successors. An "intermediate" label $(s'_1, \ldots, s'_k)$ for each such state can be obtained by setting $s'_j = s_j$ for all $j \notin D(i)$, i.e., by copying the "non-diverging" component of the state label to each successor state. Each successor state is then uniquely labeled by choosing a different $|D(i)|$-tuple over $GF(2)$ to fill in the "diverging" component of the state label.

The branch label can be obtained by multiplying the intermediate state label (considered as a row vector) with the $i+1$st column of the generator matrix. The "final" state label for each successor state is then obtained by "masking out" (setting to zero) state label positions corresponding to the converging generators; i.e., those generators that contain only zeros in the future.

Parallel transitions are caused by generators having a span of the form $[i, i]$. In the case of parallel transitions, the above branch label computation can be eliminated; instead, the state v is connected to each of its successor states with 2 "parallel" branches, each labeled with 0 and 1, respectively.

By pre-computing $D(i)$ and the masks for computation of the final state label, an efficient algorithm results. Furthermore, if the generator matrix is in a special form (for example, systematic form), computation of branch labels can also be simplified for certain time indices.

2.5 Minimal Trellises

Note that, depending on the generator matrix used, the trellis construction described above does not necessarily result in a minimal trellis. Here we describe how to transform a generator matrix to "trellis-oriented" form, so that applying the above algorithm indeed *does* result in the minimal trellis.

The concept of codeword span defines an equivalence relation on any block code: two codewords are deemed equivalent if they have the same span. This equivalence relation partitions the code into disjoint equivalence classes called span classes, with codewords in each class having the same span, and codewords in different classes having different spans. Certain span classes—called atomic classes—are elementary in the sense that their elements cannot be expressed as linear combinations of codewords from $\mathscr{C}$ all having strictly smaller span length, i.e., they cannot be linearly "split" into smaller pieces—hence the terminology. The elements of an atomic class are said to be atomic codewords. The atomic codewords can be taken to be representatives for the "granules" of [8].

For a binary linear block code the following properties hold (see [3] for details):

1. atomic classes have distinct starting and ending positions;
2. codewords from different atomic classes are linearly independent; in fact, an $[n, k]$ code has exactly k atomic classes and representatives from each class generate the code;

A trellis-oriented generator matrix for a code $\mathscr{C}$ is obtained by taking one generator from each atomic class. Applying the trellis-construction algorithm of Section 2.3 to the trellis-oriented generator matrix results in the minimal trellis for $\mathscr{C}$. The key point is that a given generator matrix is in trellis-oriented form *if and only if* no two rows either start in the same position or end in the same position. Conversion to this form is easily accomplished via elementary row operations as described below.

It is easy to see that the set of spans in a trellis-oriented generator matrix for a given code is unique, since all atomic codewords that belong to the same atomic class have the same span and the trellis-oriented generator matrix is formed by choosing a representative from each atomic class. However, if any atomic class contains more than 1 codeword, the trellis-oriented generator matrix will not be unique.

Example: A trellis-oriented generator matrix for the [8,4] Reed-Muller code may be written as:

$$T = \begin{bmatrix} 1\,1\,1\,1\,0\,0\,0\,0 \\ 0\,1\,0\,1\,1\,0\,1\,0 \\ 0\,0\,1\,1\,1\,1\,0\,0 \\ 0\,0\,0\,0\,1\,1\,1\,1 \end{bmatrix}.$$

The atomic class of span [2,7] contains two codewords:

$$\left\{ \begin{matrix} 0\,1\,0\,1\,1\,0\,1\,0 \\ 0\,1\,1\,0\,0\,1\,1\,0 \end{matrix} \right\},$$

either of which may be used in the second row of T.

The process of converting an arbitrary generator matrix to trellis-oriented form can be achieved in two steps of Gaussian elimination, so that the rows of the resulting generator matrix satisfy the condition that no two generators start in the same position or end in the same position.

In the first step, using elementary row operations, the given matrix is transformed to "row echelon form," in which the first nonzero element in each row is a 1, and the leading 1 in any row appears in a column to the right of any leading 1 in a preceding row. The starting position of each row in the generator matrix is determined by this procedure.

In the second step, starting with the nth column, Gaussian elimination is used to ensure that no two generators end in the same position, thereby determining the ending position of each row in the generator matrix. However, in order not to perturb the row echelon form achieved in the first step, the only row operation allowed in the second step is "cancellation above," i.e., replacing a row above a given row by the sum of itself and the given row. Furthermore, only those rows whose ending positions have not yet been determined may be involved in these row operations. It is necessary, therefore, to maintain a list of rows that may (or, equivalently, may not) be involved in these row operations.

After these two steps, the given matrix is transformed into an equivalent canonical trellis-oriented form.

2.6 Trees

The essential difference between a tree and a trellis is that paths corresponding to different codewords never merge and pass through the same node, once they have diverged. It is easy to see, therefore, that construction of a code tree follows the same rules as the construction of a trellis with the only exception being that converging generators are ignored and masking of "intermediate" state labels is not performed.

As mentioned before, sequential decoders use code trees rather than trellises. However, the knowledge of the code trellis may also be used in sequential algorithms to eliminate paths in the stack that pass through the same state [9]. This allows one to reduce the stack size in some decoder implementations.

3 Sequential Decoding with Variable Bias-Term

The method for the construction of a code tree outlined in the previous section allows one to apply to block codes techniques used for decoding convolutional codes. For many block codes, algebraic decoding methods are not known. This compares with the "universality" of the decoding algorithms to be described, which offer a soft-decision decoding method for an arbitrary block code—a task generally unachievable by algebraic decoders.

3.1 The Basic Algorithm

Since the Viterbi algorithm has a high computational complexity for trellises with a large number of states, one would like to consider alternative decoding schemes. The natural candidate that (on average) provides reduced complexity is the sequential stack (Zigangirov-Jelinek) algorithm. (Different realizations are described in [6].) The conventional ZJ algorithm keeps a stack of partial paths through the tree ordered according to their accumulated metric. At each decoding step, the decoder extends the path positioned at the top of the stack. The decoder outputs a codeword if this path cannot be further extended, i.e., if the node at the top of the stack has reached depth n in the tree.

The conventional stack algorithm employs the Fano metric as the branch metric (assuming binary input, Q-ary output):

$$M\left(r_i \mid v_i\right) = \log_2 \frac{P\left(r_i \mid v_i\right)}{P\left(r_i\right)} - R, \tag{2}$$

where r_i is the received symbol, v_i is the transmitted bit and R is the code rate (the ratio k/n for block codes). Let us call the term R in Equation 2 the *bias* term. The bias term accounts for how deep the extended path is advanced into the tree.

We want to find an analog of (2) for soft-decision decoding. We assume binary ± 1 transmission over a Gaussian channel. The (soft) demodulator output corresponding to a transmitted codeword is denoted by $\mathbf{r} = (r_1, r_2, \ldots, r_n)$. The squared Euclidean distance between a candidate code word $\mathbf{v} = (v_1, v_2, \ldots, v_n)$ and the received word is

$$d^2(\mathbf{v}, \mathbf{r}) = \sum_{i=1}^{n} ||r_i - v_i||^2.$$

A branch metric is given by

$$M\left(r_i \mid v_i\right) = || \, r_i, \, v_i \, ||^2 - B_i, \tag{3}$$

where we call B_i a bias term (which may be constant for all positions i). Comparing (2) and (3), we note that in sequential algorithms the accumulated metric resulting from (3) behaves differently from (2) for all $B_i \leq || \, r_i, \, v_i \, ||^2$. In particular, the stack is ordered according increasing (rather than decreasing) metric for (3), i.e., the path of the smallest accumulated metric is positioned at the top of the stack. A zero bias term ($B_i = 0$) in (3) leads to an extensive search through the code tree for the maximum-likelihood sequence. A positive constant bias term for all i reduces the computational complexity (by limiting the search) but the algorithm becomes suboptimal, in general.

We would like to find a "good" expression for the bias term B_i. A lower bound for the accumulated metric of the maximum-likelihood code word can be determined as follows. Again, assuming ± 1 transmission, let

$$\Delta_i^2 = \min\{(r_i + 1)^2, (r_i - 1)^2\}$$

denote the *smallest possible* Euclidean distance between a code symbol and the received symbol in position i. Then, for every codeword $\mathbf{v}$,

$$d^2(\mathbf{v},\mathbf{r}) \geq \sum_{i=1}^{n} \Delta_i^2.$$

Then,

$$\begin{aligned} 0 &\leq d^2(\mathbf{v},\mathbf{r}) - \sum_{i=1}^{n} \Delta_i^2 \\ &= \sum_{i=1}^{n} (\|v_i - r_i\|^2 - \Delta_i^2) \end{aligned}$$

Fig. 3 illustrates how we make use of this lower bound on cost. The total estimated "cost" associated with any particular path through the tree is given the sum of (a) the (actual) cost through the already explored portion of the tree and (b) the lower bound on cost through the unexplored portion of the tree. For example, the estimated cost of opening the root node of the tree is thus given by $m(0) = \sum_{i\geq 1} \Delta_i^2$. Let $m(s)$ be the cost estimate associated with a state s in the tree and let b^2 be the actual cost (squared Euclidean distance) associated with a branch connecting state s to its successor state s^+. Then our cost estimate for s^+ is given by

$$m(s^+) = m(s) + (b^2 - \Delta_i^2)$$

where i is the depth of s^+ in the tree. We take the quantity $b^2 - \Delta_i^2$ as a modified branch metric in our decoding algorithm.

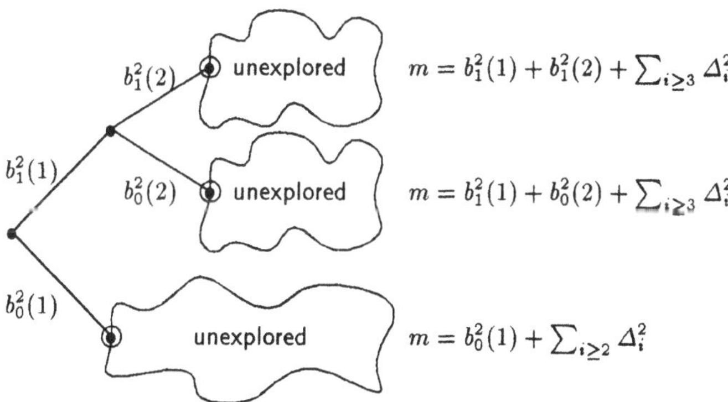

Fig. 3. A lower bound on the total cost to be accumulated in the unexplored future is given by $\sum \Delta_i^2$. The metric associated with the circled nodes is indicated by m.

The quantity Δ_i^2 is associated with the bias term B_i. The decoder places the node in the stack according this new "cost" metric. A node does not get extended while other nodes with smaller accumulated "cost" metric are above it in the stack. It is easy to see that the decoder with the "cost" metric will always output the maximum-likelihood path. Moreover, the decoder will open no more nodes than a decoder with no cost estimation of opening a node ($B_i = 0$). This is due to the fact that the cost estimate term $\sum_{i=1}^{n} \Delta_i^2$ reduces the accumulated Euclidean distance metric along the maximum-likelihood path faster than along any incorrect path.

The introduction of such a cost estimate function is known as algorithm A^* in the AI literature (see, for example, [10]). A cost estimate function can assume many different forms, even if we require the algorithm to be maximum-likelihood. Knowing certain properties of the code (for example, the Hamming weight distribution) one may suggest a more sophisticated (but more computationally complex) cost estimate function (see, for example, [11]). The cost estimate function presented here is a particularly simple cost estimate function, which does not relay on any particular code properties.

The variable bias term metric is obtained by subtracting Δ_i^2 from all branch "Euclidean distances." If we do this, any branch labelled "0" (corresponding to the transmission of -1) at time index i is assigned the branch metric

$$M_0(i) = \begin{cases} 0 & \text{if } r_i \leq 0 \\ 4r_i & \text{if } r_i > 0 \end{cases}$$

Similarly, branches labelled "1" (corresponding to the transmission of $+1$) at time index i are assigned the metric

$$M_1(i) = \begin{cases} 0 & \text{if } r_i \geq 0 \\ -4r_i & \text{if } r_i < 0 \end{cases}$$

Note that the factors of "4" can be omitted without losing the ML property.

All of this is equivalent to changing the bias term B_i in (3) from branch to branch as the decoder proceeds along the trellis — hence we call this calculated metric a *variable bias term* metric.

3.2 Algorithm Enhancements

Additional reduction in the computational complexity may be achieved by sorting the received information according its reliability (generally, this becomes feasible only for large codes since sorting entails re-computation of the trellis–oriented generator matrix for each permutation). For BPSK (± 1) modulation, we may take Δ_i^2, defined above, as a reliability measure, where lower Δ_i^2 means higher reliability. Then the received data is sorted in order of increasing D_i, and the trellis–oriented generator matrix for this code permutation is re–computed. The simulation results for decoders with data sorting are presented in the next section.

For high SNR the decoder is likely to open (explore) the minimally possible number of nodes in the tree diagram. However, with decreasing SNR more nodes are extended and for block codes with large n and k stack overflows are likely to occur. Practical strategies to handle such situations follow from simulation results such as those shown in Figure 7. From the Figure one can determine the probability that the ML path submerges into the stack below a certain depth. If one assumes the procedure of pushing out the bottom paths in case of stack overflows (the probability of stack overflow decreases linearly with the stack size and equals zero if the stack size is greater than 2^k) then with certain probability one can guarantee that the ML path remains in the stack. This seems to be a good strategy for intermediate SNR (Figure 7). Sorting of received data allows further decrease in the probability that the ML path will be pushed out of the stack.

Sorting is not the only plausible strategy that improves the decoder performance. An alternative method employs "fast push-throughs." Suppose we permute the code positions to obtain an equivalent code in the systematic form. Then the tree structure in the first k positions will contain two successors of every tree node and will represent a $[k, k]$ "free code". Our primary concern is time-efficiency. The most time-consuming operations are the insertion of new nodes and the updating of branch labels. The updating of branch labels involves a costly modulo–2 vector multiplication of the current state label with the corresponding column of the generator matrix. With a generator matrix in systematic form, we can save the computation of the branch labels during the k first time units corresponding to the k divergences. Indeed, for the newly formed nodes the branch labels are both 0 and 1. In order to reduce the number of stack "push" operations, we propose the following strategy: after each divergence we systematically push the two obtained nodes except after the k-th divergence where we decide first to extend the two nodes until we reach the terminal node in the tree. We then push only the path with smaller accumulated metric, the other one being certainly not the ML path.

The "fast push-through" algorithm can be easily extended to a limited–time search algorithm. Suppose our decoder is required to output a codeword within a limited period of time. If the noise level is high, the decoder might not be able to compute the ML path. In this case it is required to output its best estimate of the ML path. Again we assume that the generator matrix in systematic form is available. Then on receiving data, the decoder advances quickly to position k (since the tree is that of the "free code") and then pushes through as in the previous algorithm. This is the "first order" estimate of the transmitted codeword. If the accumulated metric of this path is less than the metric of any contending path, then the decoder outputs the path as the decoded codeword. Otherwise, it continues decoding, using the fast push-through algorithm and updating its estimate of the ML codeword whenever possible. Decoding proceeds until the ML codeword is found, or the decoding time limit is exceeded, in which case the decoder outputs its current best estimate.

Note that any nodes in the stack falling below a pushed codeword can effectively be eliminated, since they will never rise to the top of the stack. This leads to a further slight modification of the basic algorithm.

4 Simulation Results

Simulation results are shown in Figures 4 – 9. Two codes were simulated: the binary (24,12) extended Golay code, and the binary (128,64) extended BCH code.

Figs. 4, 5, and 8 measure the computational complexity associate with decoding the two codes. These graphs were obtained by counting elementary computations. By elementary computation we mean integer additions and the comparisons of two integers. In computing the branch and accumulated metrics, we used floating-point additions and comparisons; however, these operations are also counted as single elementary computations, since they are usually implemented with integer arithmetic in practice.

As another measure of computational complexity, we have also plotted the distribution of the number of nodes opened (explored) by the decoder as a function of SNR, for the Golay code in Fig. 6, and for the BCH code in Fig. 9. From the distributions of opened nodes of Figures 6 and 9 one can conclude that these distributions are *Pareto* as one would expect from a sequential decoder. In fact, since the number of computations for the stack algorithm depends on a particular realization, the comparison with the Viterbi algorithm becomes more meaningful if one compares the number of opened nodes in either case. The number of opened nodes in stack algorithm can be significantly less than in the VA even at a low SNR. The operation that contributes most to the computational complexity of the stack algorithm is the stack "push" operation (insertion into the stack according the accumulated metric), which depends on the number of opened nodes stored in the stack.

Plotted in Fig. 7 is the distribution of maximum stack depth for the maximum-likelihood codeword in the Golay decoder. This graph shows how deep the maximum likelihood codeword sinks in the stack at various SNR values. This data is useful in selecting a stack size for the decoder. One can see from the figure that, even in the presence of stack overflows, the maximum-likelihood path is likely not to sink too far in the stack. This will allow one to devise a strategy to handle stack overflows: bottom paths can be pushed out of the stack with the maximum-likelihood path retained in the stack with high probability, if the stack size is chosen appropriately.

Finally, one can see from the figures the effect of sorting the data according to reliability. The computational cost of sorting is not included in the figures; but, for large codes, will be a small fraction of the total cost of decoding. For most codes, sorting pays off in the medium SNR range. This is intuitively clear since in low SNR most received symbols are unreliable and in high SNR most symbols are reliable. Only in medium SNR can rearranging symbols according to their reliability lead to a search of a smaller portion of the tree and thus to computational savings.

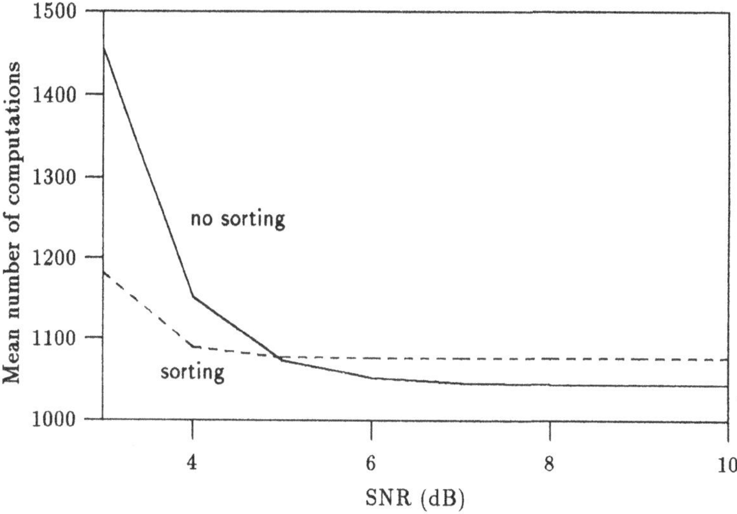

Fig. 4. Mean number of computations for decoding of the (24,12) extended Golay code using the variable bias-term stack algorithm.

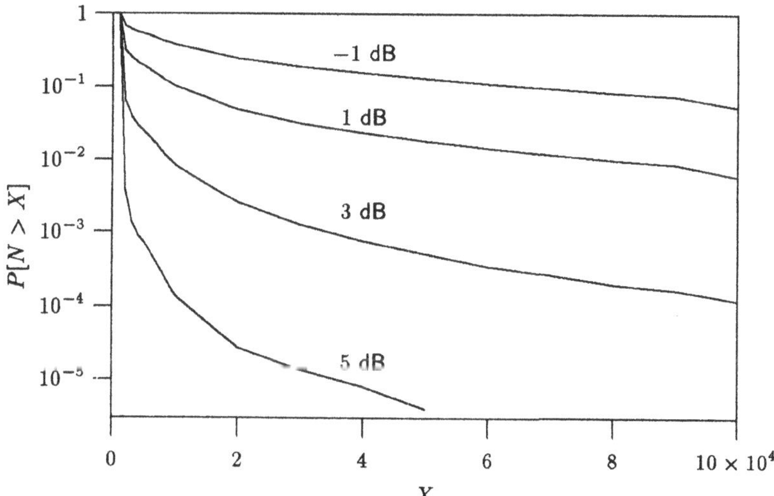

Fig. 5. Distribution of the number of computations for the (24,12) extended Golay code using the variable bias-term stack algorithm.

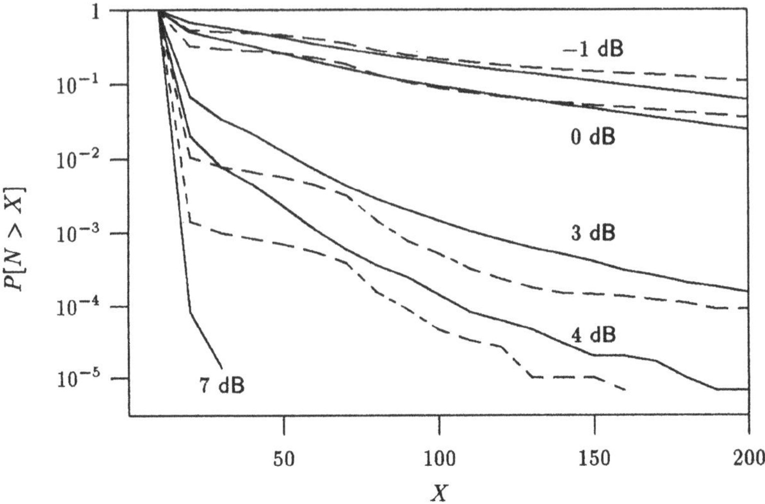

Fig. 6. Distribution of opened nodes for the (24,12) extended Golay code. (Solid—no sorting, dashed—with sorting.)

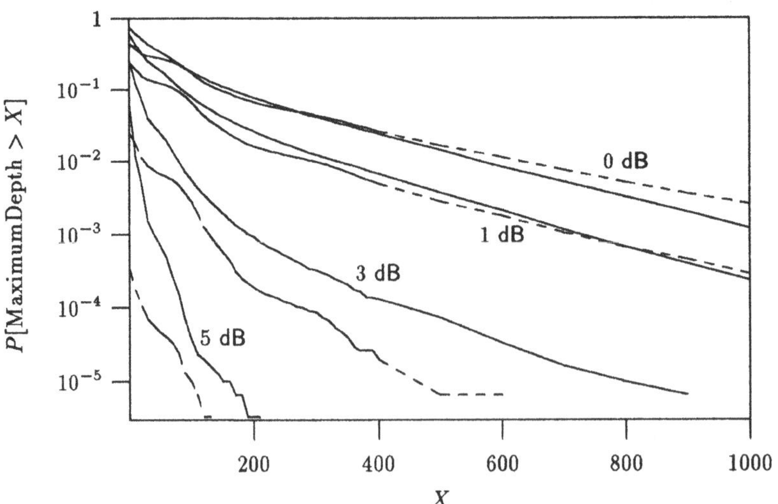

Fig. 7. Maximum depth of the ML path in the stack for (24,12) extended Golay code. (Solid—no sorting, dashed—with sorting)

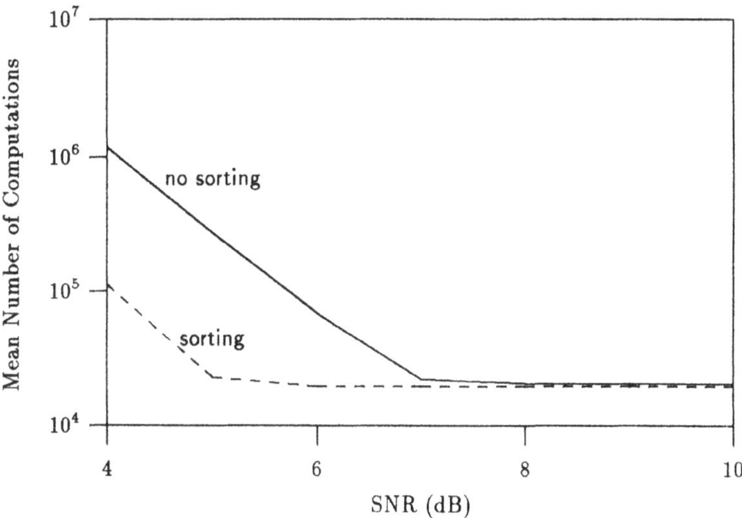

Fig. 8. Mean number of computations for decoding the (128,64) extended BCH code using the variable bias-term stack algorithm.

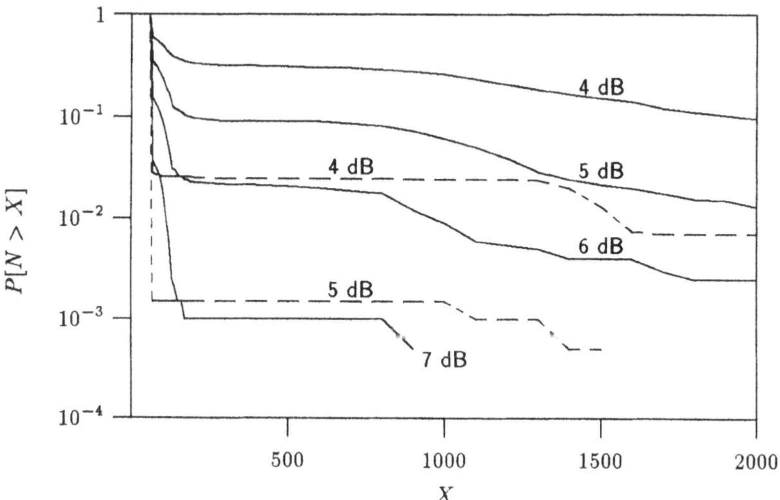

Fig. 9. Distribution of opened nodes for (128,64) extended BCH code. (Solid—no sorting, dashed—with sorting.)

5 Conclusions

In this paper we have presented an efficient iterative algorithm for the construction of minimal trellises (and the corresponding trees) for an arbitrary binary linear code. The starting point for these constructions is the generator matrix for the code. By transforming the generator matrix to trellis-oriented form—in which the rows of the generator matrix include representatives from each atomic class—the minimal trellis is obtained. A tree is then easily constructed from the given minimal trellis.

We have devised a maximum-likelihood sequential decoder with a variable bias-term metric to search the resulting tree. On average, for large codes and for intermediate to high SNRs, the soft-decision decoder with variable bias term performs a much smaller number of operations than a decoder based on the Viterbi algorithm. Various enhancements, such as sorting of data and "fast push-throughs" can be used to tailor the decoder performance to particular practical demands. The result is a general and flexible soft-decision decoder for binary linear block codes.

References

1. J. K. Wolf, "Efficient maximum-likelihood decoding of linear block codes using a trellis," *IEEE Trans. on Inform. Theory*, vol. IT-24, pp. 76–80, 1978.
2. J. L. Massey, "Foundation and methods of channel encoding," in *Proc. Int. Conf. Inform. Theory and Systems*, vol. 65, (Berlin), Sept. 1978.
3. F. R. Kschischang and V. Sorokine, "On the trellis structure of block codes." Submitted to *IEEE Trans. on Inform Theory*, Aug. 1993.
4. D. J. Muder, "Minimal trellises for block codes," *IEEE Trans. on Inform. Theory*, vol. 34, pp. 1049–1053, Sept. 1988.
5. G. D. Forney, Jr., "Coset codes II: Binary lattices and related codes," *IEEE Trans. on Inform. Theory*, vol. 34, pp. 1152–1187, Sep. 1988.
6. J. B. Anderson and S. Mohan, "Sequential coding algorithms: A survey and cost analysis," *IEEE Trans. on Commun.*, vol. COM-32, pp. 169–176, Feb. 1984.
7. A. R. Calderbank, "Multilevel codes and multistage decoding," *IEEE Trans. Commun.*, vol. 37, pp. 222–229, 1989.
8. G. D. Forney, Jr. and M. D. Trott, "The dynamics of group codes: State spaces, trellis diagrams and canonical encoders," *IEEE Trans. on Inform. Theory*, vol. 39, 1993. (To appear.).
9. D. Haccoun and M. J. Ferguson, "Generalized stack algorithms for decoding convolutional codes," *IEEE Trans. on Inform. Theory*, vol. IT-21, pp. 638–651, Nov. 1975.
10. N. J. Nilsson, *Principles of Artificial Intelligence.* Tioga Publishing Company, 1980.
11. Y. S. Han, C. R. P. Hartmann, and C. C. Chen, "Efficient maximum-likelihood soft-decision decoding of linear block codes using algorithm A*," in *Proc. IEEE Inter. Sympos. Inform. Theory*, (San Antonio, TX), 1993.

Permutation Decoding Using Primitive Elements as Multipliers

Tho Le-Ngoc Ming Jia Anader Benyamin-Seeyar

Department of Electrical and Computer Engineering
Concordia University, Montreal, Quebec, Canada H3G 1M8

Abstract: Permutation decoding employs a very simple combinational logic circuit for error detection and correction. In this paper, the topic of permutation decoding using primitive elements of a prime field as multipliers is addressed in order to increase the capability of the well known (T, U) permutation decoding method in decoding cyclic codes of prime length. Since only error positions are involved in the analysis, the results are applicable to cyclic codes over $GF(2^q)$.

1 Introduction

A variation of error trapping decoding, known as permutation decoding, was introduced by MacWillians [1]. In order to correct all error patterns of weight t or less with permutation decoding, it is necessary to determine a set of code-preserving permutations which can map every uncorrectable error pattern of basic error-trapping decoding onto one with at least k consecutive error-free digits, where t is the error correcting capability of a (n, k) cyclic code.

There are several code-preserving permutations for every cyclic code. Let C be a cyclic code over the Galois field $GF(q)$. If $c(x)$ is a code polynomial, then

$$c(x) = \sum_{j-0}^{n-1} b_j x^j$$

where b_j is a symbol from $GF(q)$. The well known (T, U) permutation is defined as follows:

$$T^{\beta} c(x) \equiv \sum_{j=0}^{n-1} b_j x^{j+\beta} \quad mod\ x^n - 1,$$

and

$$U^{i} c(x) \equiv \sum_{j=0}^{n-1} b_j x^{p^i j} \quad mod\ x^n - 1,$$

where p is the characteristic of $GF(q)$. The (T, U) permutation maps each code word onto another, provided $(n, q) = 1$ [1]. For the binary cyclic codes, $p = 2$, so

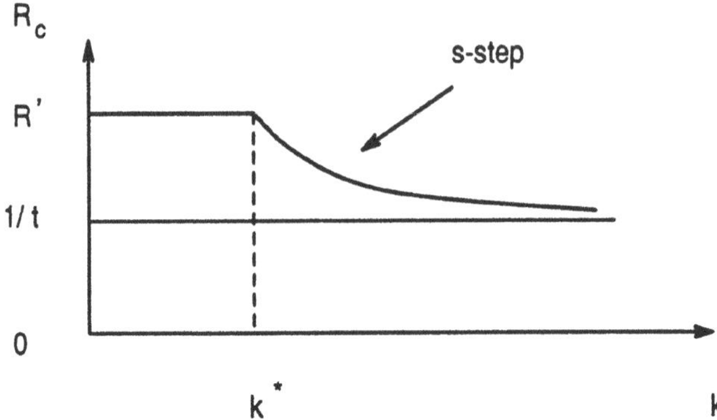

Figure 1: The relation between R_c, k and s

when n is odd, the code is invariant under group (U) permutation. If the errors are trapped with $i = 0, 1, \ldots, s < \gamma$, where γ is the least integer and $2^\gamma = 1 \; mod \; n$, then we say that the code is $(s+1)$ step permutation-decodable (PD). Some results on the capability of (T, U) permutation method have been found [2-6]. The main results can be illustrated by Figure 1 and the following points can be concluded:

1. When $k \rightarrow \infty$, the code rate of s-step permutation-decodable codes, R_c, decreases and approaches that of the error-trapping decodable codes.
2. When $k \rightarrow k_+^*$, R_c approaches $1/(t_o - 2 + 2/t_o)$ and $1/(t_o - 2)$ for $t_o = 5$ and $t \geq 7$, respectively, where t_o is an odd number.
3. For each additional step, the increase in k^* is more than doubled. In the region of $k < k^*$, the code length n does not keep decreasing at the rate of $\Delta n = (\Delta k) t_o$; therefore, the increase in code rate R_c is very slow and R_c is around the value R'.
4. Suppose that the code rate of a s-step permutation-decodable code is R_{cs}, then $\Delta R_{cs} = R_{c(s+1)} - R_{cs}$ increases as the permutation steps s increases until k^* is large enough and $k < k^*$. Therefore, there exists an optimal step which makes the largest improvement in the code rate of PD codes. Given a t-error correcting $(n, k, 2t+1)$ cyclic code, then the optimum step s can be determined.

For many individual cyclic codes it is also possible to find other code-preserving permutations [7]. However, the selection of a set of permutations for given values of n, k, and t remains unknown.

As we can see from the above results, the performance of (T, U) permutation method is bounded by R'. In this paper, we consider permutation decoding which uses primitive elements of a prime field as multipliers to increase the code rate of permutation-decodable codes in the region of $k \leq k^*$.

2 Using Primitive Elements as Multipliers

It is also possible to decode cyclic codes by using permutations which may not preserve the code. However the decoder must have the ability to correct error patterns in the equivalent codes. In order to define the generator polynomial $g(x)$ by the equation

$$g(x) = \prod_{k \in K} (x - \alpha^k)$$

we must specify both α, a particular primitive nth root of unity, and K, the set of powers of α which are roots of $g(x)$. Although different codes arise from different choices of α, these different codes are equivalent in the sense that they are permutations of each other. Clearly, the decoder can correct an error pattern in any code when all errors lie in or can be shifted into the "parity" portion of the codeword.

Let us consider a binary cyclic code with length n equal to a prime number p, then the set $S = \{0, 1, 2, \ldots, p-1\}$ is a field of order p under *modulo* $- p$ addition and multiplication. In fact, the set S can be partitioned into subsets[1] which are invariant under U permutation. For example, for $p = 31$, these subsets are (0), (1, 2, 4, 8, 16), (3, 6, 12, 24, 17), (5, 10, 20, 9, 18), (7, 14, 28, 25, 19), (11, 22, 13, 26, 21), (15, 30, 29, 27, 23). The union of any number of invariant subsets is also invariant under U; and this limits the improvement of the permutation decoding technique. These subsets exist because 2 is a non-primitive element of the field $\{0, 1, 2, \ldots, p-1\}$. Because $(2^5 \ mod \ p = 31) = 1$, so the maximum number of possible permutation steps is 5, and the subsets contain at most 5 elements each.

Now we consider the use of a primitive element of the field $GF(p)$ as a multiplier. We define M_i permutation as:

$$M_i^s c(x) \equiv \sum_{j=0}^{n-1} b_j x^{i^s j} \quad mod \ x^p - 1.$$

which is equivalent to

$$x^j \longmapsto x^{i^s j} \quad mod \ x^p - 1, \qquad i, s \in S.$$

If M_i-permutation is used $(s-1)$ times to decode a certain code, then we say the code is s-step permutation-decodable.

Since $GF(p)$ is a prime field, there exist some primitive elements i with order $p-1$ which satisfy $i^{p-1} = 1$. So we can perform M_i permutation $(p-2)$ times. Because the field $GF(p)$ can not be partitioned into subsets which are invariant under M_i permutation, the capability of the permutation method is expected to be increased. As an example, for the code BCH $(31, 11, d = 11)$, if we use 12 (which is a primitive element) as a multiplier, it is a 6-step permutation-decodable (PD) code. However, it is not a PD code if we use U permutation, i.e., by using

[1]The subsets in the partition of S are called *cyclotomic cosets* [8]

the primitive element as the multiplier, the code rate in the region of $k \leq k^*$ is increased.

Because p is a prime number, every multiplier i $(1 \leq i \leq p-1)$ is relatively prime to p (the block length of the word), and the mapping M_i is an automorphism. When i is not the characteristic of $GF(q)$, the permutation will not preserve the code, but it may map one cyclic code into another equivalent cyclic code.

The condition on which a cyclic code will be mapped to another equivalent cyclic code is given in the following theorem.

Theorem 1 *If the generator $g_1(x)$ of a linear cyclic code C_1 is mapped to a codeword of another equivalent cyclic code C_2, then C_1 is mapped to C_2.*

Proof[2]: Suppose $g_i(x)$ are the generator polynomials of $C_i (i = 1, 2)$. If $g_1(x)$ is mapped to a codeword of C_2, then $g_2(x)|M_j(g_1(x))$. For $x^h g_1(x)(1 \leq h \leq k)$, it is mapped to the codeword

$$\begin{aligned}(x^h \cdot g_1(x))^j \quad mod \ (x^n - 1) &= x^{hj}(g_1(x))j \quad mod \ (x^n - 1) \\ &= x^{hj}((g_1(x))^j \quad mod \ (x^n - 1)) \quad mod \ (x^n - 1) \\ &= x^{hj} \cdot M_j(g_1(x)) \quad mod \ (x^n - 1)\end{aligned}$$

which is the hjth cyclic shift of $M_j(g_1(x))$ and it is also a codeword of C_2. Clearly this mapping preserves addition and multiplication. If $x^{uj} = x^{vj}$ $(mod\ x^p)$, then $(u - v)j = 0\ mod\ p$. Since p is a prime, j is relatively prime to p, this implies that $i - j = 0\ mod\ p$ or $x^u = x^v$. Hence this mapping is an isomorphism, and the above condition holds.

Q. E. D.

3 An Illustrative Example

Now let us consider the BCH $(31, 16, d = 7)$ code. Suppose the code C_1 is generated by

$$g_1(x) = x^{15} + x^{11} + x^{10} + x^9 + x^8 + x^7 + x^5 + x^3 + x^2 + x + 1$$

Then, with M_{12} permutation, g_1 is mapped into a code word

$$c(x) = x^29 + x^27 + x^25 + x^24 + x^22 + x^15 + x^12 + x^8 + x^5 + x^3 + 1$$

which is divisible by

$$g_2(x) = x^{15} + x^{13} + x^{11} + x^8 + x^6 + x + 1.$$

[2]It was proven that if C is a binary cyclic code of length n , then the permutation M_i maps C on to another binary cyclic code C' or onto C itself [9].

Therefore, M_{12} maps the code C_1 into another BCH code C_2 which is generated by $g_2(x)$. Similarly, with M_{12} permutation, C_2 is mapped into other BCH codes which are generated by

$$g_3(x) = x^{15} + x^{14} + x^{12} + x^{11} + x^{10} + x^8 + x^6 + x^4 + x^3 + x^2 + 1$$

$$g_4(x) = x^{15} + x^{14} + x^{13} + x^{12} + x^{10} + x^8 + x^7 + x^6 + x^5 + x^4 + 1$$

successively. Therefore we can use the primitive element 12 as a multiplier to decode this code. This code is found to be 4 - step permutation decodable. The comparison of U and M_{12} permutation for decoding the BCH codes of length 31 is summarized in the following table.

BCH Code	Multiplier	Steps (PD or NPD)
(31, 16, $d = 7$)	2	5 - step PD
	12	4 - step PD
(31, 11, $d = 9$)	2	4 - step PD
	12	3 - step PD
(31, 11, $d = 11$)	2	NPD
	12	6 - step PD
(31, 6, $d = 15$)	2	4 - step PD
	12	3 - step PD

Table I: A comparison between U and M_{12} permutation of decoding BCH codes of length 31.

4 Implementation of U and M_i Permutation Decoders

There are two ways to realize the permutation decoding of a certain cyclic code: Parallel processing and Serial processing. For s-step permutation decodable codes, s syndrome registers are needed in parallel processing. As shown in Figure 2, the received vector is shifted to the s syndrome registers at the same time, so the time needed to decode the code is almost the same as the basic error-trapping decoder. The block diagram of a permutation decoder using serial processing is given in Figure . In this case, only one syndrome register is need, but the decoding time is approximately equal to the number of steps needed to decode the code multiplied by the decoding time of a parallel processing permutation decoder.

If parallel processing is applied for fast decoding, the M_i permutation decoder is simpler than the U permutation decoder because the former requires a smaller number of required syndrome registers than the latter.

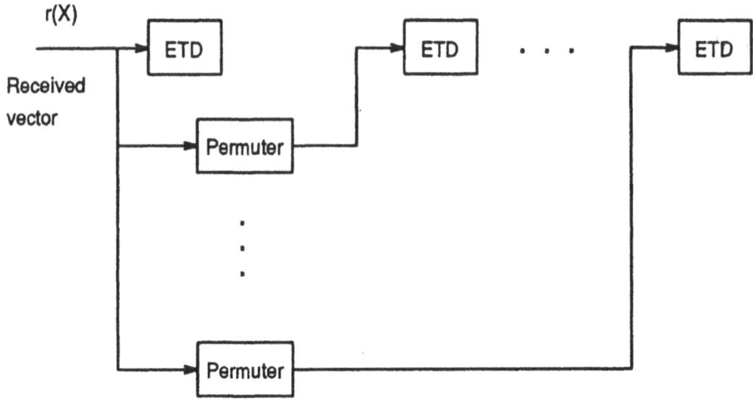

Figure 2: Permutation decoder using parallel processing

If serial processing is used for decoding, then the M_i permutation decoder is slightly more complex than the U permutation decoder because the M_i permutation may not preserve the code, and, therefore, a logical circuit is needed to control the feedback of the syndrome register. However the decoding time of a serial M_i-permutation decoder is shorter than that of a serial U-permutation decoder. The syndrome register of the M_{12} decoder for the BCH (31, 16, $d=7$) code is shown in Figure 4.

5 Remarks

1. In some cases, the capability of the permutation decoding method is expanded. For example, the BCH (31, 11, $d=11$) code is not permutation decodable by (T, U) permutation, but it is a 6-step permutation decodable code by (T, $M_1 2$) permutation.

2. M_i permutation decoder is simpler with parallel processing than U permutation decoder, it is also faster in both parallel and serial processing decoding. However, in serial processing, since M_i permutation may not preserve the code, an additional logical circuit is needed to control the feedback in the syndrome register of the M_i permutation decoder, this makes the M_i permutation decoder slightly more complex than the U permutation decoder.

3. When t increases, more permutation steps are required. The maximum number of required syndrome registers equals the number of existing equivalent codes. After the code is permuted so many times by the primitive multiplier, it will be mapped back to the original one.

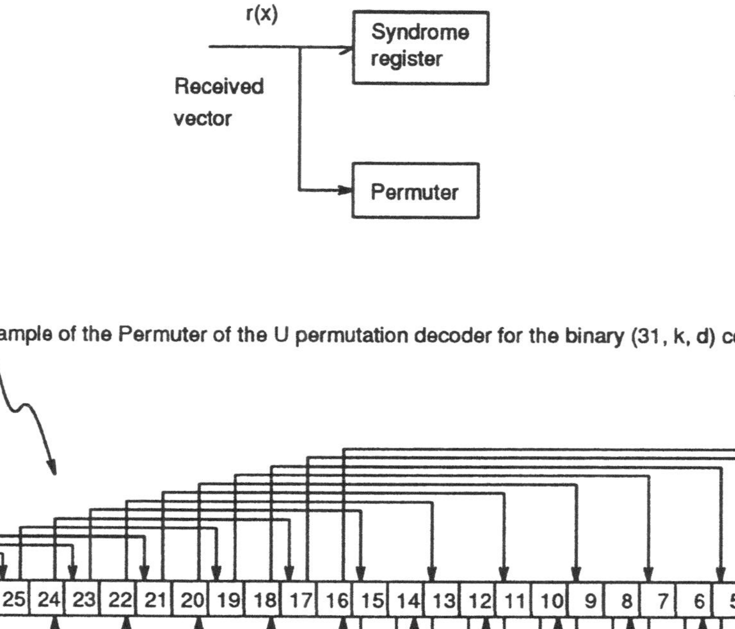

Figure 3: Permutation decoder using serial processing

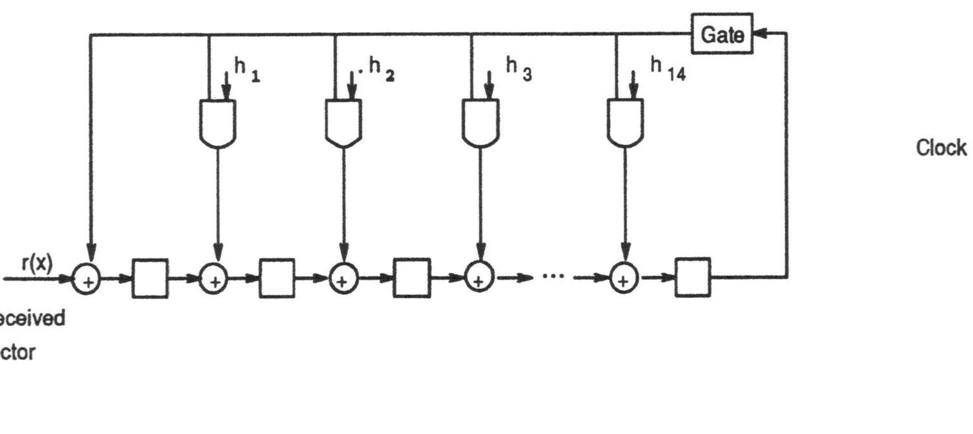

$h_1 = \bar{c}_2$ $h_2 = \bar{c}_1$ $h_3 = \bar{c}_1$ $h_4 = c_2$ $h_5 = \bar{c}_1 \bar{c}_2 + c_1 c_2$ $h_6 = c_1 + c_2$

$h_7 = \bar{c}_1 \bar{c}_2 + c_1 c_2$ $h_8 = 1$ $h_9 = \bar{c}_1 \bar{c}$ $h_{10} = \bar{c}_1 + c_2$ $h_{11} = \bar{c}_1 + \bar{c}_2$

$h_{12} = c_2$ $h_{13} = c_1$ $h_{14} = c_2$

Figure 4: Syndrome register of the M_{12} decoder for the BCH (31, 16, d=7) code

References

[1] F. J. MacWilliams, "Permutation Decoding of Systematic Codes," *Bell Syst. Tech. J.*, vol. 43, pp. 485-505, 1964

[2] A. Benyamin-Seeyar, S. S. Shiva, and V. K. Bhargava, "Capability of the Error-Trapping Cyclic Code," *IEEE Trans. Inform. Theory*, vol. 32, No. 2, pp. 166-180, Mar. 1986.

[3] M. Jia, A. Benyamin-Seeyar, and T. Le-Ngoc, "Exact Lower Bounds on the Code Length of Three-Step Permutation-Decodable Cyclic Codes," *1991 IEEE International Symposium on Information Theory*, Budapest, Hungary.

[4] M. Jia, A. Benyamin-Seeyar, and T. Le-Ngoc, "On the Capability of (T, U) Permutation Decoding Method," *1991 Canadian Conference on Electrical And Computer Engineering*, Quebec, Canada.

[5] S. G. S. Shiva and K. C. Fung, "Permutation Decoding of Certain Triple-Error-Correcting Binary Codes," *IEEE Trans. Inf. Theory,* IT-8, pp 444-446, May 1972.

[6] P. W. Yip, S. G. S. Shiva and E. L. Cohen, "Permutation Decodable Binary Cyclic Codes," *Electronic Letters*, Vol. 10, pp. 467-468, October 1974.

[7] W. W. Peterson, *Error-Correcting Codes*, The MIT Press, 1972.

[8] F. J. MacWillians and N. J. A. Sloane, *The Theory of Error-Correcting Codes*, North-Holland, Amsterdam, 1977.

[9] Vera Pless, *Introduction to the Theory of Error-Correcting Codes*, John Willey and Sons, New York, 1982.

Overflow Constraint in Hybrid Nodes with Movable Boundary Scheme

F. Ghazi-Moghaddam[1], I. Lambadaris[2] and J.F. Hayes[1]

[1] Department of Electrical and Computer Engineering, Concordia University, Montréal, Québec

[2] Department of Systems and Computer Engineering, Carleton University, Ottawa, Ontario

Abstract. The heterogeneous nature of traffic sources into the Broadband Integrated Services Digital Network (B-ISDN) Communication systems requires the switching centers to allocate resources dynamically. The strategy used to allocates the resources among these streams should consider their particular requirements. We are considering a system with two input traffic, one with high priority which can tolerate some loss, and the lower priority which cannot experience any loss. The switching scheme is a movable boundary system with overflow traffic. This system bounds the observed delay by the overflow traffic by restricting their access.

1 Introduction

The evolution in hybrid switching systems and the B-ISDN technology has started a new era in telecommunications. The demand to integrate different traffic sources with their distinct characteristics has prompted scientists to develop schemes with resource sharing and dynamic bandwidth allocation. [1]

One of the main goals of ISDN systems has always been the ability to provide instantaneous connection between users via voice, data and finally video imagery. [2] Thus bandwidth on demand, is one of the requirements of the B-ISDN systems.

A number of studies have aimed to find schemes which make resource sharing possible without a large degradation of service. [3] One of these methods is to use a movable boundary scheme to use the voice empty slots for the queued data packets. A particular configuration of this scheme was investigated in [4] which type-I traffic upon finding no empty slots could be redirected into the queue (see figure 1). This scheme will make resource sharing possible without any loss probability for type-I traffic.

However, the apparent problems with this scheme are:

- a relatively large overflow traffic could result into extremely long delays seen by the type-II traffic.
- Type-I traffic may require the delay to be bounded (e.g. real time voice and video traffic)

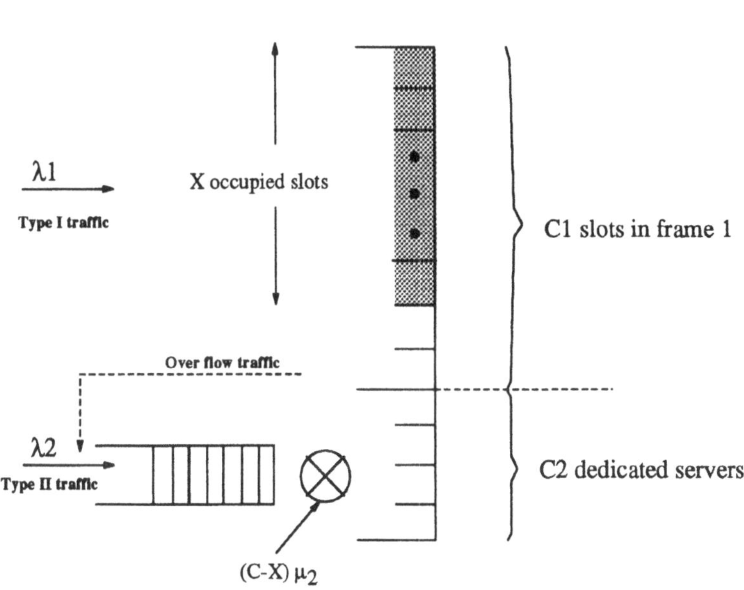

Fig. 1. Movable Boundary system with queuable overflow traffic

The present model describes a system which restricts the overflow arrivals depending on the number of customers present in the queue. We thus maintain an upper bound for the delay observed by the overflow traffic. The difference in the transition digram of this model and the previous one is shown in figure 2.

1.1 Model Description

The model considered here, is described by the following parameters;

- Poisson arrival into the frame 1 of C1 slots
- Poisson arrival of type-II traffic
- Movable voice boundary
- Exponential service times μ_1, μ_2
- Preemptive priority of type-I arrival
- Overflow arrivals from frame 1 into the queue if all C_1 slots are occupied
- The overflow arrival can only join if there are less than K customers in the queue otherwise those cells are lost

2 Matrix Geometric Method

As an exact analysis of this system we use the versatile approach by M.F. Neuts, namely matrix geometric techniques [5]. The transition states, form a two dimensional markov chain which is described by a quasi birth-death rate matrix, Q.

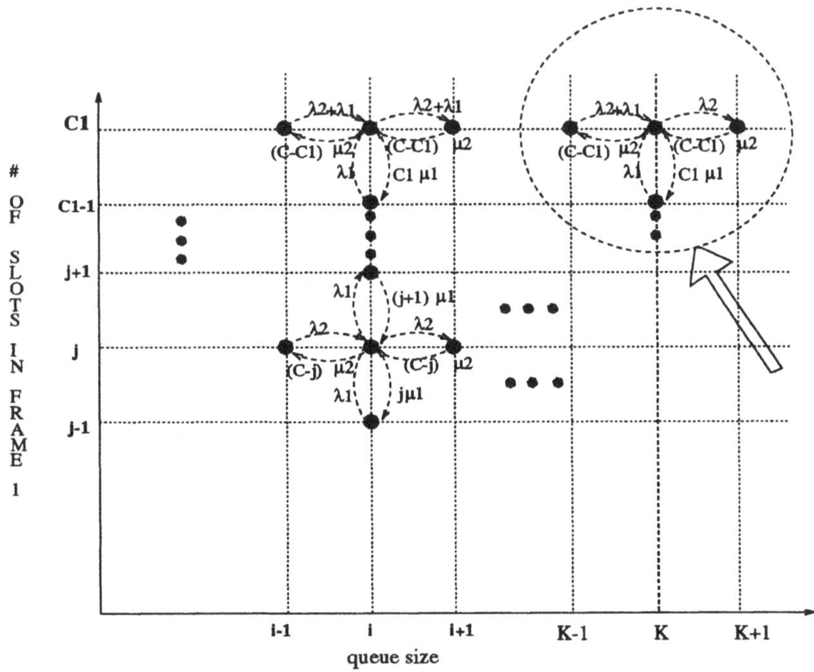

Fig. 2. Transition Diagram

Elements i, j of Q correspond to the number of customers in the queue and the number of occupied slots in C_1 frame.

The Matrix G is the key factor in this method. It is obtained from a first passage argument which states that a transition from level i to level 0 has to go through all intermediate levels namely $i-1$, $i-2$, etc.

$$G(z,s) = \sum_{i=1}^{\infty} \int_0^{\infty} e^{-sx} d\tilde{G}^{[1]}(i;x) z^i \tag{1}$$

for $|z| \leq 1$, Re $s \geq 0$

$$G(z,s) = \sum_{i=0}^{\infty} A_i(s) G^i(z,s) \tag{2}$$

The equation 2 is the key equation in MG and represents the number of customers served during a busy period and the busy period duration.

$$G = G(z,s)\mid_{z=1,s=0} = \sum_{i=0}^{\infty} A_i G^i$$

from which we can find its invariant probability vector $\mathbf{g} = \mathbf{g}G$

$$\mu = \left.\frac{dG(z)}{dz}\right|_{z=1} \mathbf{e}$$

$$= (I - G + \mathbf{eg})[I - A + (\mathbf{eg})\beta]^{-1}\mathbf{e} \tag{3}$$

where μ is the mean number of customers served during a busy period

Another quantity of interest is the matrix K which is related to the probability of returning to $\mathbf{0}$ from phase j to j'

$$K = K(z,s)\,|_{z=1,s=0} \; = \sum_{i=0}^{\infty} B_i\; G^i \tag{4}$$

As usual $\mathbf{k}K \;=\; \mathbf{k}$ and $\mathbf{ke} = \mathbf{1}$ and we define k^* as the row-sum mean of $K(z)$;

$$k^* = \left.\frac{d}{dz}K(z)\right|_{z=1} \mathbf{e}$$

$$= \left[\sum_{k=1}^{\infty} B_k \sum_{j=0}^{k-1} G^j\right] \mu \; + \; \mathbf{e} \tag{5}$$

then $\mathbf{x_0}$ can be obtained by using

$$\mathbf{x_0} \;=\; \frac{k}{k\,k^*} \tag{6}$$

Once $\mathbf{x}_0$ has been calculated the Ramaswami recursion can be applied to determine the stationary probabilities for other queue sizes.

$$\mathbf{x}_i = \left[\mathbf{x}_0 \bar{B}_i + \sum_{j=1}^{i-1} \mathbf{x}_j \bar{A}_{i+1-j}\right] (I - \bar{A}_1)^{-1} \tag{7}$$

where

$$\bar{A}_k = \sum_{i=k}^{\infty} A_i G^{i-k}$$

$$k \geq 0$$

$$B_k = \sum_{i=k}^{\infty} B_i G^{i-k}$$

In this particular case, Q is a rate matrix thus giving rise to the following equation;

$$R^2 A_2 \;+\; RA_1 \;+\; A_0 \;=\; \mathbf{0} \tag{8}$$

which is a result of $\mathbf{x}Q = \mathbf{0}$ and $\mathbf{x}\mathbf{e} = \mathbf{1}$. In order to apply the matrix geometric method, Uniformization of the rate matrices is required.

$$\begin{aligned} \hat{B}_0 &= I + \frac{1}{v} B_0 \\ \hat{A}_1 &= I + \frac{1}{v} A_1 \\ \hat{A}_0 &= \frac{1}{v} A_0 \\ \hat{A}_2 &= \frac{1}{v} A_2 \end{aligned} \tag{9}$$

which results into:

$$R = R^2 \hat{A}_2 + R\hat{A}_1 + \hat{A}_0 \tag{10}$$

A positive recurrent markov chain results in a matrix R which has all its eigenvalues inside the unit disk. Then $\mathbf{x}_0$ can be found using the set of equations;

$$\begin{aligned} \mathbf{x}_0 \, (B_0 + RA_0) &= \mathbf{0} \\ \mathbf{x}_0 \, (I - R)^{-1} \, \mathbf{e} &= 1 \end{aligned} \tag{11}$$

Vector $\mathbf{x}_0$ corresponds to the stationary probability of having a queue size of zero and the state of the frame-I being one of the $\{C_1 + 1\}$ possible cases.

$$\mathbf{x}_0 = [x_{00}, x_{01}, \ldots, x_{0C_1}] \tag{12}$$

and It can be used to determine other probabilities;

$$\mathbf{x}_n = \mathbf{x}_0 \, R^n \qquad \text{for } n \geq 0 \tag{13}$$

And as usual the average queue length is given by

$$\begin{aligned} \bar{N} &= \sum_{i=1}^{\infty} i \, \mathbf{x}_i \, \mathbf{e} \\ &= \mathbf{x}_0 \sum_{i=1}^{\infty} i \, R^i \, \mathbf{e} \\ &= \mathbf{x}_0 \, R \left((I - R)^{-1} \right)^2 \, \mathbf{e} \end{aligned} \tag{14}$$

In order to formulate the main blocks of the Q matrix, we need to define the intermediate matrices, $\tilde{A}_0$, $\tilde{A}_1$, and $\tilde{A}_2$. Also larger matrices of the size $(k+1) * (C_1 + 1)$ namely $\hat{B}_0$, $\hat{A}_0$, $\hat{A}_1$, and $\hat{A}_2$. In reality these superblocks of matrices are the mapping of three-dimensional state space. Once the matrices are known the same procedure is applied. Hence,

$$R^2 \hat{A}_2 + R\hat{A}_1 + \hat{A}_0 = \mathbf{0} \tag{15}$$

A result of $\mathbf{x}Q = \mathbf{0}$ and $\mathbf{x}\mathbf{e} = \mathbf{1}$. Applying uniformization again, we obtain the corresponding matrices, $\hat{B}_0$, $\hat{A}_1$, $\hat{A}_0$, $\hat{A}_2$ which results into:

$$R = R^2 \hat{A}_2 + R\hat{A}_1 + \hat{A}_0 \tag{16}$$

It is evident that the resulting $\hat{\mathbf{x}}_0$ is a supervector containing;

$$\hat{\mathbf{x}}_0 = [\mathbf{x}_0, \mathbf{x}_1, \ldots, \mathbf{x}_K] \tag{17}$$

where $\mathbf{x}_0 \ldots \mathbf{x}_K$ are given by (12).

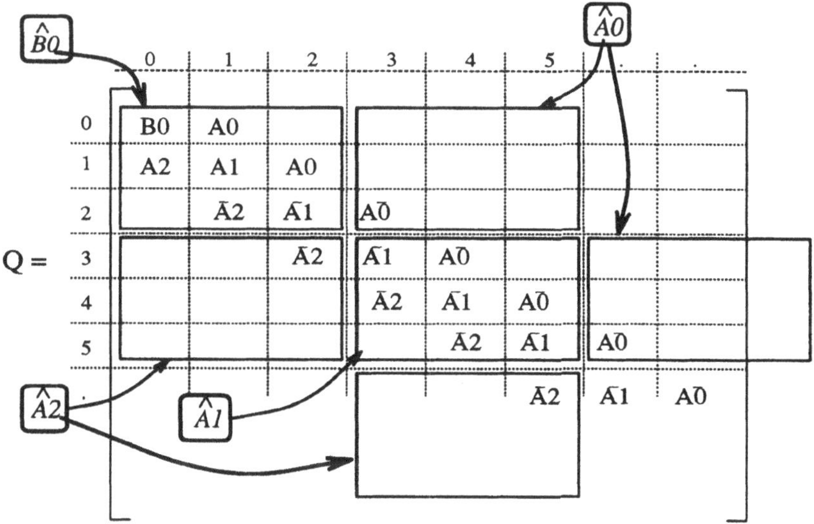

Quasi Birth-Death Process

Fig. 3. The Matrix Geometric Block partitioned structure

The same procedure is used to obtain $\hat{\mathbf{x}}_0$ and the subsequent stationary probabilities i.e. $\hat{\mathbf{x}}_i$. The quasi-birth-death structure of the transition matrix helps reduce the numerical complexities in determining the invariant probability vectors.

$$\hat{\mathbf{x}}_{\mathbf{i}} = \hat{\mathbf{x}}_{\mathbf{0}} R^i \tag{18}$$

The resulting data agrees with the simulation, however both taking considerable amount of time. Also by considering the limiting cases such as $k = 0$ or $k \to \infty$, we could examine the results.

3 Approximate Method

Matrix Geometric provides an exact numerical method, however as the size of the matrices increases, the time required to perform these calculation, rapidly increases. Another factor is the total arrival rate into the buffer; as $\rho \rightarrow 1$, the number of iterations required to obtain matrix G or equivalently matrix R exponentially increases. Thus if there is a need for determining the behaviour of the system in fast varying environment, the MG method would be severely limited. The approximate methods offered here, have proven to be close to the MG and the simulation results, yet requiring a fraction of the time needed by the other methods.

The main procedure is outlined here, however the simplified results are presented.

- Approximate the process using
  - moments matching of the arrival rate
  - moments matching of the service rate
  - using IPP [6] to approximate the overflow
- apply the Z-transform to find a closed-form expression for the mean number of packets in the queue.

A two state markov chain is used to approximate the original model and following the approach by Yechiali and Naor as in [7], we find the Z-transform of the system. Then the root z_0 which is located in [0,1) will determine various quantities such as $\bar{G_1}$ and $\bar{G_2}$ to determine the average population related to state 1 or state 2.

The moments matching techniques use the following quantities to match the statistical properties of the original process to the 2-state approximate one. These quantities are:

- Mean μ
- Variance v
- Third moment μ_3
- The Integral of the Covariance function $\int r(t)\,dt$

The integral of the covariance function is used in;

$$\lim_{T\to\infty} \frac{Var[n(T)]}{E[n(T)]} = 1 + \frac{2}{m}\int_0^\infty r(t)\,dt \tag{19}$$

We define the time constant τ_c which is used in an exponential approximation to the covariance function.

$$\tau_c = \frac{1}{v}\int_0^\infty r(t)\,dt \tag{20}$$

$$r_e(t) = ve^{-t/\tau_c} \tag{21}$$

These relationships may be inverted to find the required parameters from the known characteristics of the system.

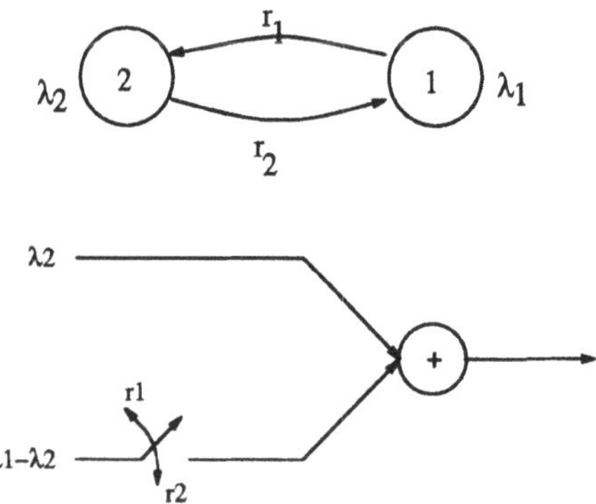

Fig. 4. MMPP source represented as a superposition of an IPP process and a Poisson process

$$r_1 = \frac{1}{\tau_c(1+\eta)} \tag{22}$$

$$r_2 = \frac{\eta}{\tau_c(1+\eta)} \tag{23}$$

$$\lambda_1 = m + \sqrt{\frac{v}{\eta}} \tag{24}$$

where

$$\eta = 1 + \delta/2\left[\delta - \sqrt{4+\delta^2}\right] \tag{25}$$

and

$$\delta = \frac{\mu_3 - 3mv - m^3}{v^{3/2}} \tag{26}$$

Once the two-state approximation is achieved we use the method introducod by Yechiali and Naor to obtain the Z-transform of $G_1(z)$ and $G_2(z)$. The balance equations which describe the system are:

$$\begin{aligned}
P_{10}(\lambda_1+\eta_1) &= P_{20}\eta_2 + P_{11}\mu_1 && m=0\\
P_{1m}(\lambda_1+\eta_1+\mu_1) &= P_{2m}\eta_2 + P_{1,m+1}\mu_1 + P_{1,m-1}\lambda_1 && 0<m<K\\
P_{1K}(\lambda_2+\eta_1+\mu_1) &= P_{2m}\eta_2 + P_{1,m+1}\mu_1 + P_{1,m-1}\lambda_1 && m=K\\
P_{1m}(\lambda_2+\eta_1+\mu_1) &= P_{2m}\eta_2 + P_{1,m+1}\mu_1 + P_{1,m-1}\lambda_2 && m>K
\end{aligned}$$

After some algebraic manipulation, and using the root to the characteristic polynomial $g(z)$, z_0 we reach the following results

$$p_1 = \eta_2 / (\eta_1 + \eta_2) \quad (27)$$
$$p_2 = \eta_1 / (\eta_1 + \eta_2) \quad (28)$$
$$\bar{\lambda} = p_1 \lambda_1 \ + \ p_2 \lambda_2 \quad (29)$$
$$\bar{\mu} = p_1 \mu 1 \ + \ p_2 \mu 2 \quad (30)$$
$$\bar{\mu} - \bar{\lambda} > 0 \quad (31)$$

It is also possible to consider a matrix geometric approach for the two-state approximate chain as a check for the Z-transform. As usual, resorting to numerical methods involves more computation and time. The following diagrams correspond to the different methods of approximation used for this case.

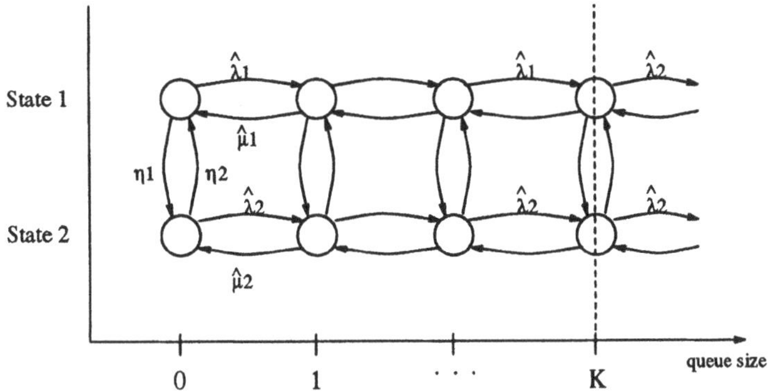

Fig. 5. The approximate model for IPP and arrival rates moment matching

Applying Z-transform to the two state approximate diagram, we get the following results:

- following the same procedure we get a characteristic function $g(z)$ whose root has to be found ($0 \leq z_0 \leq 1$ for a stable system i.e. $\lambda_2 < \bar{\mu}$)
- We obtain a set of linear equations which have to be solved to specify the system completely ($P_{10}, P_{11} \ldots, P_{1,K-1}$ are needed to find close form solution for $G_1(z)$ and $G_2(z)$)

The moment matching of the arrival rates results into a situation illustrated in figure 6. The procedure to obtain the $G_1(z)$ and $G_2(z)$ requires these probabilities to completely describe the system

$$P_{10}, P_{11}, \ldots, P_{1,K-2}, P_{1,K-1}$$
$$P_{10}, P_{11}, \ldots, P_{1,K-2}, P_{1,K-1}$$

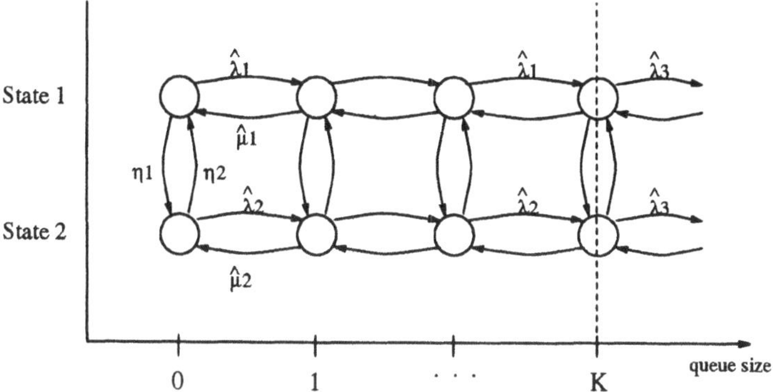

State 1 : Overflow traffic changes the arrival rate
State 2 : Overflow source is idle

Fig. 6. The approximate model for the service rates moment matching

Obviously the condition for stability becomes $\lambda_3 < \bar{\mu}$ since the arrival rates λ_1 and λ_2 are transient and do not contribute to the steady state arrival rate.

4 Results

As expected the three approximate method showed small differences in predicting the average delay for the packets in the buffer compared to the results of the matrix geometric. However the error in results are generally small enough to justify the use of the the approximate method.

It is however of importance to choose a threshold K which provides the acceptable loss probability for type-I traffic. This probability is shown for varying λ_2. The loss probability is calculated using the Matrix Geometric method as well as the approximate ones. This probability is simply

$$\begin{aligned} Pr\{\text{loss}\} &= Pr\,\{\text{Overflow, queue size} \geq k\} \\ &= Pr\{\text{Blocking type-I}\}\,(1 - Pr\{n < k\}) \end{aligned}$$

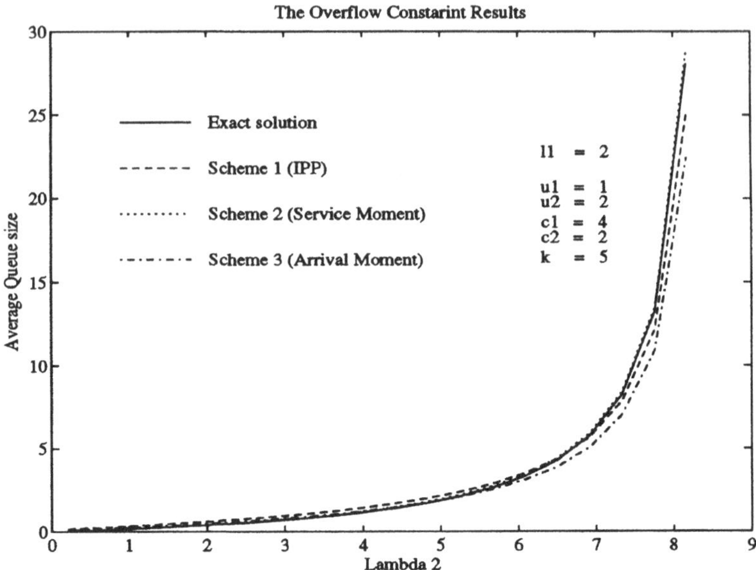

Fig. 7. Results of the approximate model versus the matrix geometric

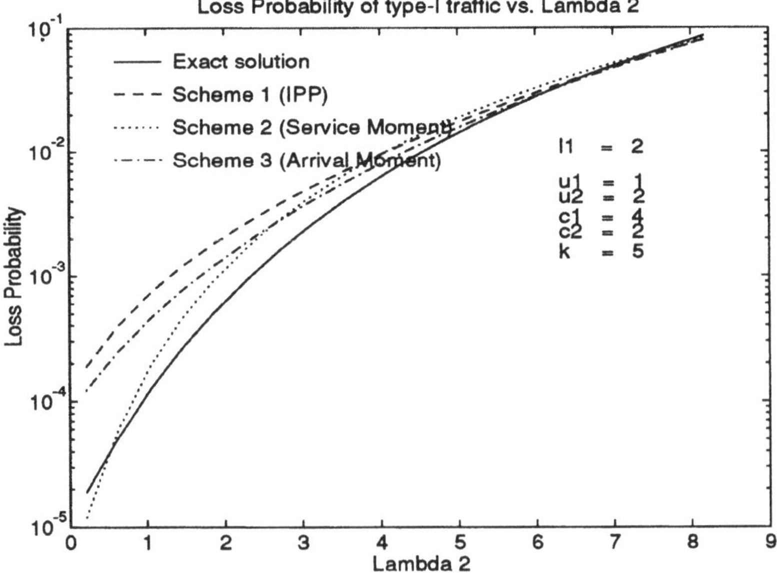

Fig. 8. The Loss Probability of type-I Traffic vs. Lambda 2

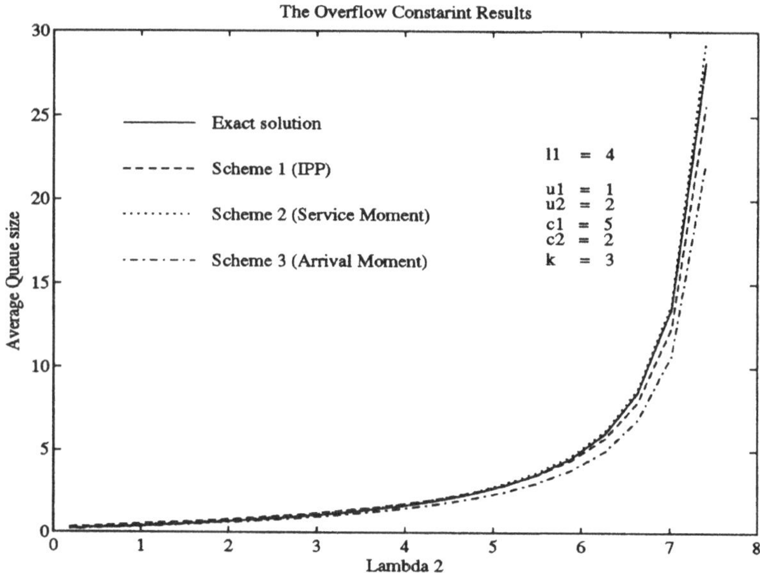

Fig. 9. Results of the approximate model versus the matrix geometric

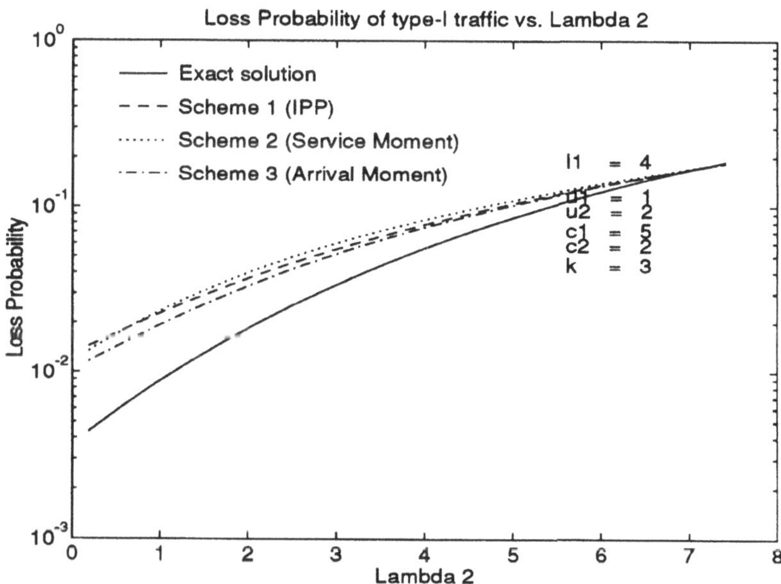

Fig. 10. The Loss Probability of type-I Traffic vs. Lambda 2

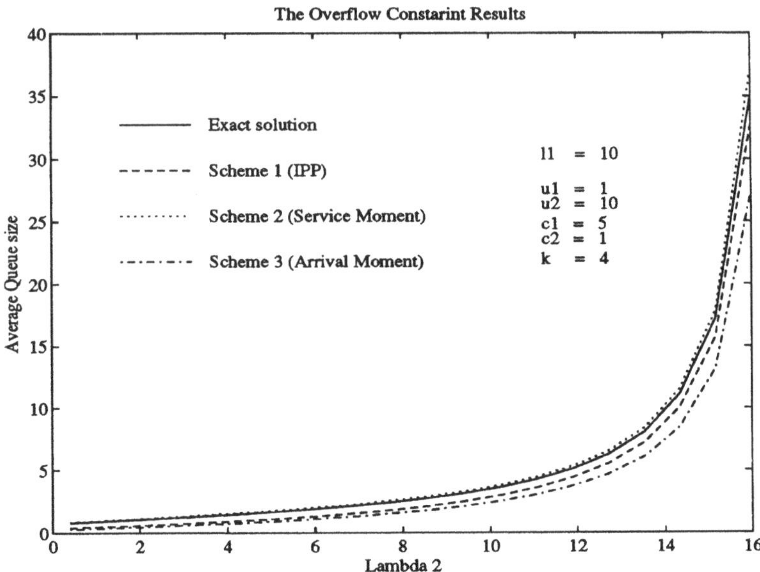

Fig. 11. Results of the approximate model versus the matrix geometric

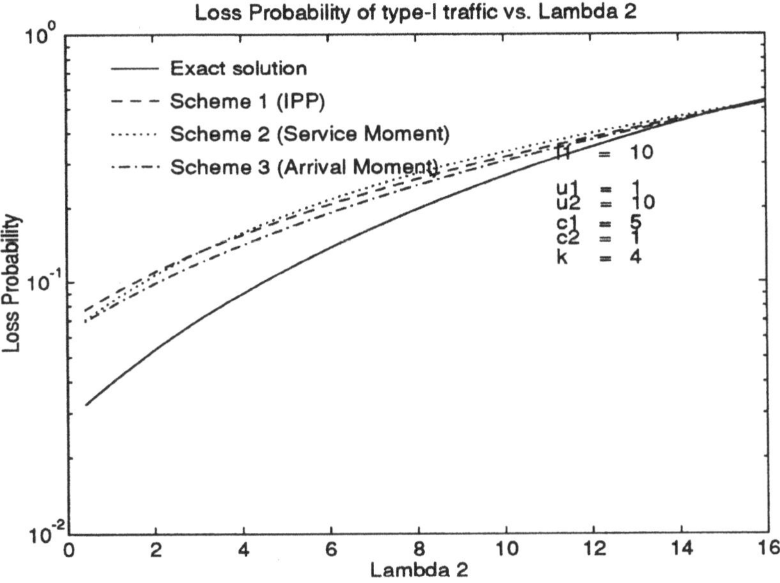

Fig. 12. The Loss Probability of type-I Traffic vs. Lambda 2

References

1. H. Delic and P. Papantoni-Kazakos. An optimal policy for competitive processing of high and low priority arrivals. *Int'l J. Digital and Analog Commun. Systems*, 4:209–224, 1991.
2. B. Kraimeche and M. Schwartz. Analysis of traffic control strategies in integrated service network. *IEEE Trans. Commun.*, COM-33(10):1085–1093, October 1985.
3. H. Delic and P. Papantoni-Kazakos. A class os scheduling policies for mixed media data with renewal arrival processes. *IEEE Trans. Automatic Control*, 38:455–459, March 1993.
4. I. Lambadaris, X. Chen, J.F. Hayes, and F.G. Moghaddam. Performance evaluation of a circuit-switched node with queuable overflow traffic. *Proceedings of Int'l Conf. Commun. in Chicago*, pages 516–520, June 1992.
5. Marcel F. Neuts. *Matrix Geometric Solutions in Stochastic Models.* Johns Hopkins University Press, Baltimore, Maryland, 1981.
6. A. Kuczura. The interrupted poisson process as an overflow process. *Bell Syst. Tech. J.*, 52(3):437–448, March 1973.
7. U. Yechiali and P. Naor. Queuing problems with heterogeneous arrivals and service. *Operations Research*, 19:722–734, 1971.

Statistical Analysis of Wavelet Transform Coded Multi-channel Signals for Transmission over ATM Packet Networks

Mehmet Zeytinoğlu[1] and Yen Chuan Hsu[2]

[1] Department of Electrical and Computer Engineering, Ryerson Polytechnic University, Toronto, Ontario M5B 2K3, Canada
[2] School of Electronic and Manufacturing Engineering, University of Westminster, London W1M 8JS, United Kingdom

Abstract. In this paper we investigate wavelet transform based coding of multi-channel (stereophonic) digital audio signals. Our goal is the derivation and implementation of efficient coding and interpolation methods in the context of data transmission over asynchronous transfer mode (ATM) networks. The coding algorithm is designed to exhibit robustness with respect to packet losses due to traffic congestion. An important concept is that of "layered coding" where the output of the encoder is divided into cells of varying significance such that when the network is congested cells of lower priority will be dropped. The decoder at the receiver must be capable to implement packet substitution to maintain reconstructed signal quality at an acceptable level. The high degree of correlation between co-channel signals provides significant advantages for layered coding and packet substitution methods applicable to multi-channel audio signals. We first decompose the wideband stereophonic signal into subbands which we analyze with respect to their relative energy and then separate into high- and low-priority data streams. We perform statistical analysis of subband signals generated through multi-resolution decomposition of wideband stereophonic signals We study the performance of the proposed algorithm under complete low priority cell loss conditions and report on preliminary results of these simulations.

1 Introduction

In the last decade digital coding became the dominant format for all types of audio signals. However, the data rate resulting from coding high-fidelity digital audio signals is simply too high for many transmission channels and storage media. Consequently, in recent years data reduction of digital audio signals has received much attention. In applications of digital audio, multi-channel signal processing is also becoming a increasingly desirable feature for realistic sound stage presentation. This however, further increases the data rate. Present data reduction algorithms are essentially monophonic where the algorithm is applied independently to each channel signal. In this paper we investigate wavelet transform based signal coding for multi-channel digital audio signals in the context of data transmission over asynchronous transfer mode (ATM) packet networks. The goal is the derivation and implementation of efficient multi-channel source coding and interpolation methods which are robust with respect to packet losses due to traffic congestion.

ATM is now accepted by CCITT as the preferred transfer mode for the broadband integrated services digital network (BISDN) [1], [2]. Diverse information services such as video, voice, data can all be transported via ATM. Such a network based on service independence can be characterized by:

- **flexibility:** advances in the coding efficiency and signal processing will certainly reduce the bandwidth requirements of existing services. The ATM network which is based on packet transmission and can support variable bit-rate coding can accommodate these changes in the years to come.
- **efficiency:** all network resources are available to every service having access to the network, therefore optimal statistical sharing of the resources is easily accomplished.
- **universality:** only one network needs to be designed and consequently, the overall cost of the system can be significantly reduced.

ATM is a packet oriented transfer mode based on fixed length cells. Each cell consists of an *information field* which is kept relatively small and a *header* with a very limited function. The main header function is the identification of the virtual connection which allows proper routing of each packet admitted to the network. Due to its limited functionality, processing of header information in ATM nodes is simple and can be done at high speeds. This results in very low processing and queuing delays. To further guarantee fast processing in the network, no error protection or flow control is implemented on a link-by-link basis. If a link introduces an error during the transmission thereby causing loss of packets, no special action will be taken on that link to correct the error. This protection can be omitted, since the links in the network are of high quality. With proper resource allocation the network can guarantee a controlled number of queue overflows which are the main source of packet losses in the network. As a direct consequence of these operational characteristics the ATM network can achieve packet loss probability in the order of 10^{-8} to 10^{-12}.

The "cell" oriented transport mechanism and the high data transmission capacities of the ATM networks allow variable bit-rate (VBR) coding and can support real-time services. However, ATM networks are susceptible to buffer overflow because of network congestion which may result in cell loss, and can severely degrade service quality. When the traffic design of ATM networks is based on the average value of a VBR source encoder, buffer overflow is likely to occur at those times when the bit rate exceeds its average. In times of network congestion, the network may have to discard some ATM cells to maintain the guaranteed capacity. Therefore, CCITT has adapted the solution with 2 priorities in a single virtual connection: high-priority (HP) for cells within the negotiated throughput and cells with lower priority (LP) which are subject to discarding, depending on the network conditions. A specific bit is defined in the header to indicate the cell priority. Cell priority control leads to the concept of *layered coding* where high and low priority is assigned to different parts of the coded signal [3]. Thus, the emergence of ATM based broadband data transmission services requires a fresh look at source coding strategies applicable for multimedia services such as HDTV and wideband audio signals.

In this paper we consider multi-resolution decomposition based source coding of wideband stereo audio signals. The multi-resolution decomposition is implemented by a wavelet transform (WT). In Section 2, we introduce the WT and a WT based subband decomposition. The relation between the WT and quadrature mirror filterbank structures is also discussed. Section 3 addresses layered coding in the context of stereo audio signals. We present statistical analysis results of stereo wideband and subband signals and then proceed to develop a layered coding scheme based on WT and energy distribution properties of the subband signals. Section 4 presents simulation results. We conclude by discussing current research directions.

2 Source Coding for ATM Networks

Section 1 introduced basic operational characteristics of ATM networks and briefly discussed how source coding methods should be modified to address the special challenges presented by the ATM networks. From an implementation point of view the following source coding characteristics are most significant:

- compatibility of ATM networks with VBR source coding algorithms;
- layered coding of source data to maintain service quality in the case of cell losses;
- missing cell recovery/substitution techniques to restore service quality if the low priority data stream is lost due to traffic congestion in the network.

For wideband audio signals the state of the art in source coding is based on subband coding. Among the most successful of the current low bit-rate coders is the subband coding method that has been employed by the ISO/MPEG audio standard [4], MUSICAM [5] and PASC [6]. These algorithms attempt to eliminate source redundancy by first separating the wideband signal into subbands through the use of a filterbank, whose output is then analyzed to establish the temporal and spectral masking properties as specified by a psycho-acoustic model of the human auditory system.

While the above subband coding techniques rely on the psycho-acoustic model for efficient source coding and bit allocation, other simpler methods have been implemented with success. One such method is based on the subband energy distribution in which the total number of bits are allocated to the individual subbands according to their respective energy content [7]. The energy distribution method is simple to implement and therefore will be basis of source coding applied in this study.

2.1 Wavelet Transform based Multi-resolution Decomposition

Wavelet transforms (WT) recently have been proposed as a new multi-resolution decomposition tool [8]. The WT has the capability of variable time-frequency localization and therefore is a natural match to some of the properties of audio-visual information. Consequently, WTs have enjoyed an increasing popularity in signal coding [9-10], including some research aimed specifically to WT based image coding in the context of ATM network transmission [11-13]. In the sequel we briefly summarize the most pertinent characteristics of the WT and then proceed to describe the specifics of the WT used in this particular study. We consider the WT applied to square-integrable functions defined over the set of real numbers. Consider a sequence of closed subspaces in $\mathcal{L}^2(\mathbb{R})$:

$$\begin{array}{c} \ldots V_2 \subset V_1 \subset V_0 \subset V_{-1} \ldots \\ \longleftarrow \text{coarser} \qquad\qquad\qquad \text{finer} \longrightarrow \end{array}$$

Bases for V_m are built by means of dilations and translations of a *scaling function* $\phi(t)$. Let W_m be the orthogonal complement of V_m in V_{m-1}, i.e. $W_m \oplus V_m = V_{m-1}$. Similarly, bases for W_m are built by means of dilations and translations of the *wavelet function* $\psi(t)$. Let

$$\begin{aligned} P_m : &\quad \text{projection from } \mathcal{L}^2(\mathbb{R}) \rightarrow V_m; \\ Q_m : &\quad \text{projection from } \mathcal{L}^2(\mathbb{R}) \rightarrow W_m; \end{aligned} \tag{1}$$

such that, for any $x \in \mathcal{L}^2(\mathbb{R})$:

$$\begin{aligned} P_m x &= x_m; \\ Q_m x &= d_m; \\ P_{m-1} x &= P_m x + Q_m x. \end{aligned} \tag{2}$$

Thus, by applying the first set of projections the WT can decompose $x(t)$ into the sum of a coarser signal, x_{-1}, and a detail signal, d_{-1}. This coarse approximation x_{-1} in turn can be further decomposed into a yet coarser signal, x_{-2}, and a detail signal, d_{-2}, at that resolution. One continues with this process until the required level of decomposition is achieved. For example, the WT of a square integrable function x at resolution level N is described by the set:

$$\{x_{-N}, d_{-N}, d_{-N+1}, \ldots, d_1\}. \tag{3}$$

The equation set also describes the multi-resolution decomposition of the function x, as versions of the function x at resolution levels $-N+1$, $-N+2$, ..., 1 can be generated using the set in (3) as the input of the reconstruction algorithm. For further discussion on reconstruction techniques see [10]. This discussion can be best illustrated by a block diagram which depicts the WT as a multi-resolution decomposition pyramid:

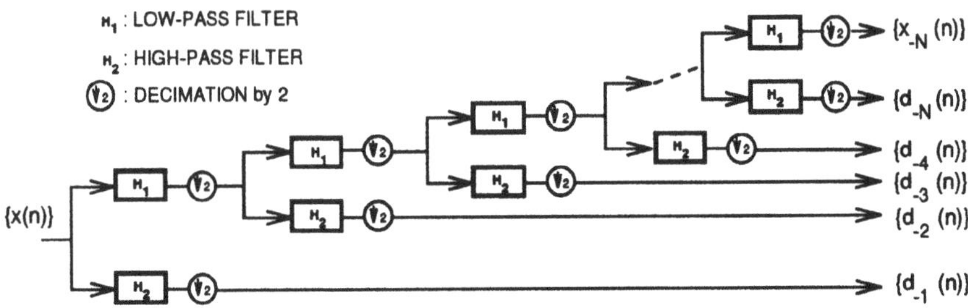

Fig. 1. Multi-resolution pyramid decomposition.

The low-pass and high-pass filters with impulse response functions $\{h_1(n)\}$, and $\{h_2(n)\}$, respectively, are related the scaling and to the wavelet function used by the WT through the following equations:

$$\begin{aligned}\phi(t) &= 2\sum_n h_1(n)\phi(2t-n);\\ \psi(t) &= 2\sum_n h_2(n)\phi(2t-n).\end{aligned} \tag{4}$$

One can show that compactly supported orthonormal wavelet bases imply a paraunitary 2-band FIR PR-QMF bank such that the discrete sequences $\{h_1, h_2\}$ constitute respectively the low- and high-pass filter impulse response sequences. Thus, the theory of WT based signal decomposition is intimately related to the well researched area of subband coding where the analysis and synthesis filters constitute a PR-QMF pair [10].

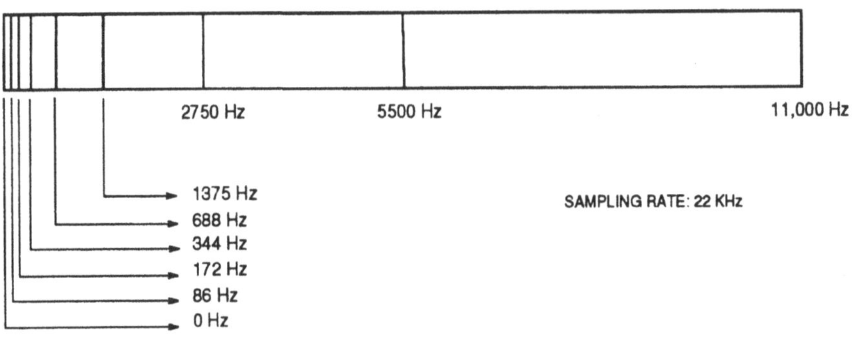

Fig. 2. 8-band multi-resolution decomposition.

In the context of this study we use an 8-band wavelet transform based on 32-tap FIR filters derived from the Daubechies wavelet function [14]. This wavelet function has desirable regularity properties that can be shown to correspond to the *maximally flat* magnitude response characteristics of the low-pass filter H_1. Furthermore, the orthonormal wavelet filter derived from the Daubechies wavelet function corresponds to the Binomial-QMF [11]. The signal decomposition used in this study has the subband structure shown in Figure 2.

2.2 Subband coding of Stereo Signals

Layered source coding. In this section we briefly outline the basic layered coding strategy where we first attempt to identify high and low priority data steams based on the energy distribution among the subbands. We use a short segment from Dvorak's *7th* Symphony sampled at 22 kHz. The segment consists of 65,536 samples representing approximately 3 seconds of sound data. The choice of sampling rate and number of samples was aimed at reducing the computational complexity, however results presented below are also representative of sound signals sampled at a more typical 44.1/48 kHz rate and are of longer duration. The block size is set at 512 samples per block. The lowest frequency subband [0-86 Hz] in Figure 2 is referred to as subband number 1, and the highest frequency subband [5.5-11 kHz] is referred to as subband number 8. Based on the Dvorak sound file and on the above set of system parameters we now proceed to illustrate energy distribution of the subband signals.

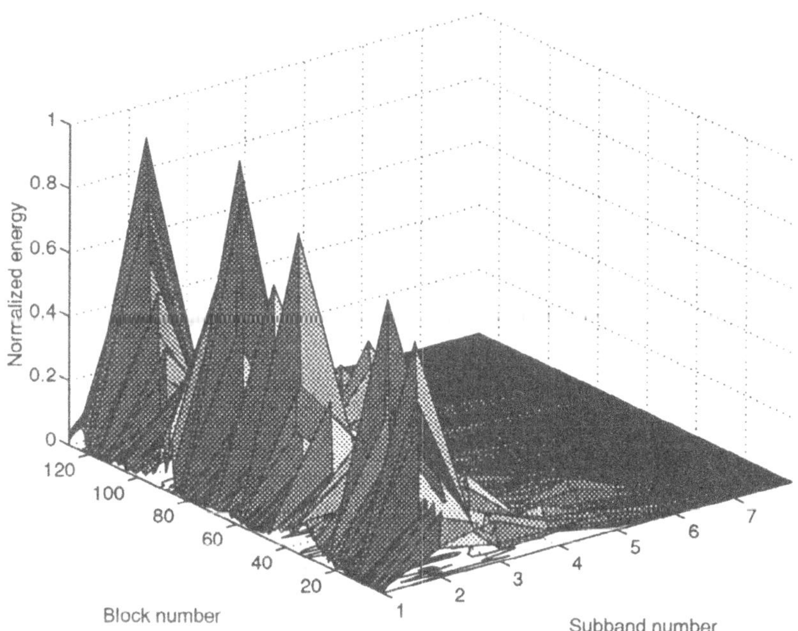

Fig. 3. Energy distribution

The energy distribution diagram depicted in Figure 3 confirms that for sound signals the signal energy is mostly concentrated at lower frequencies. A different visualization as depicted in Figure 4 yields a more striking illustration of the sound energy distribution.

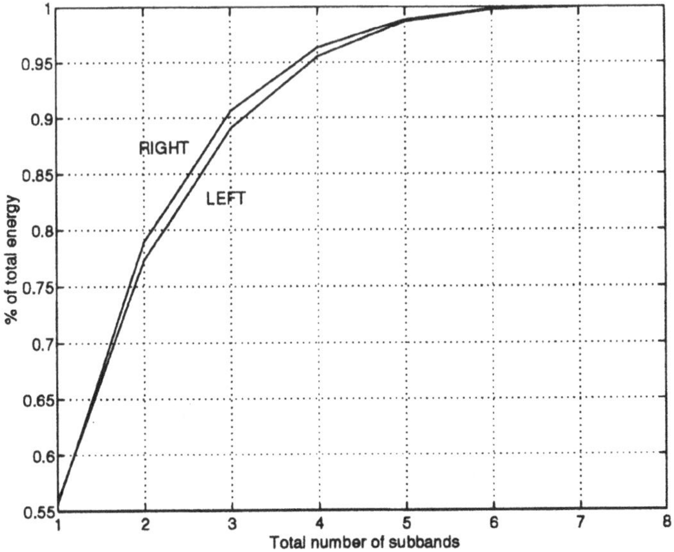

Fig. 4. Cumulative energy distribution

For each block of 512 samples we first apply the 8-band multi-resolution decomposition (based on the WT as described in the previous section) separately to left and right channel data. The individual subbands are then sorted according to their energy contents measured relative to the total energy of the entire block of 512 samples. The sorted subband energies are subsequently accumulated to determine the number of subband signals required to reach a fixed percentage of the total block energy. Looking at the two curves in Figure 4, we observe that on the average the four subbands with the largest relative energy content will yield approximately 95% of the total signal energy for that block.

The non-uniform energy distribution among the subbands will form the basis of layered source coding. The high energy subbands which together contain a certain percentage (α) of the total signal energy for that block will constitute the HP data stream. The threshold value α can be adjusted to yield a satisfactory balance between the sound quality and the increased HP data rate. Results presented in this section are based on the value $\alpha = 0.95$. This value is empirically determined and its optimum value may change depending on other signal parameters. While identifying the HP data stream solely on the basis of subband energy contents will not yield the same "optimality" as a masking based bit allocation and priority identification [15]; however, by carefully choosing the energy threshold α we can generate a simple algorithm that will allow testing of stereo signal coding strategies. Other works in the literature show that subband coding based on subband energy content yields subjective results comparable to masking based algorithms [7]. As researchers commented in [15] band-limiting the wideband signal to determine the HP data stream is not suitable for high-fidelity sound coding. This observation becomes particularly significant when one considers sound data which contains significant high frequency energy.

For typical audio signals most signal energy is concentrated at lower frequency bands. Consequently, from the multi-resolution decomposition block diagram depicted in Figure 2, we observe that on the average the high energy subbands have higher decimation factors compared to low energy subbands. Thus, identifying the HP data stream with most of the lower frequency bands generates highly desirable data rates for the HP and LP data streams.

In our experience working with the Dvorak and other audio files it is difficult to predict which subband signals constitute the HP data stream. However, the surface in Figure 3 which depicts the energy distribution as a function of time (block number) and frequency (subband number) is a fair representation for most audio signals. Other audio files which were examined during the course of this study indicate a "more uniform" energy distribution where the subbands that constitute the HP data stream are evenly distributed over the subbands 1–6. In these cases the two highest frequency bands (subbands 7 and 8) contribute only marginally to the total energy of the block and therefore with a very high probability these subbands are delegated to the LP data stream. The actual choice of subbands in the HP data stream changes from one audio file to the other as well as from one block to another within the same audio file. For the particular case where the energy threshold α is set to 0.9 we can state with a fair degree of certainty that that the HP subbands are selected among subbands 1–6 and the average value of the total number of subbands that constitute the HP data stream equals *four*.

These observations are important in establishing the exact HP/LP data rates generated by the layered coding algorithm. Assume the aggregate data rate for the original stereo audio signal is given by r bits per second. Let r_H and r_L represent respectively the HP and LP data rates which do include any data overhead due to side information such as energy scaling factors as such data constitutes only a very small fraction of the overall data rate. Each sequence of r_H and r_L values resulting from processing successive audio blocks of K samples constitute a realization of a random process whose statistics are dictated by the source file. These statistics – in particular the first and second order statistics – play an important role when the source negotiates a connection with the network [20]. Based on the assumption that for typical energy threshold levels subbands 7 and 8 constitute the LP data stream, typical values for r_H and r_L are $\frac{5}{8}r$ and $\frac{3}{8}r$, respectively.

Missing cell recovery techniques. For a well designed ATM network and network traffic control strategy the cell loss rate is low. However, the audio signal coding algorithm should address the recovery in the case of complete or bursty cell losses in the LP data stream. In particular, if real-time transmission and playback of the audio signal is envisioned, then the data rates precludes any retransmission possibility. Thus, the decoder faces the task of reconstructing the original data even in the case of cell losses. In this study, we concentrate on missing cell recovery techniques applicable to the LP data stream. This choice was motivated by the priority control strategy implemented by the network which results in extremely low HP cell loss probabilities. The issue of cell recovery methods for HP data stream will be addressed in a later study.

Among the most frequently used cell recovery techniques we can list *cell interpolation*, *zero-substitution* and *using the last set of samples* [3]. These methods have been used with varying degree of success. The suitability and applicability of any one of the above techniques is very much a function of the underlying signal type and the source coding method. Another alternative is to "pre-distort" the encoded signal in such a way so that any LP cell loss will have negligible effect on the decoder. The embedded DPCM [16] follows precisely this approach and has been shown to perform satisfactorily under quite adverse conditions [3].

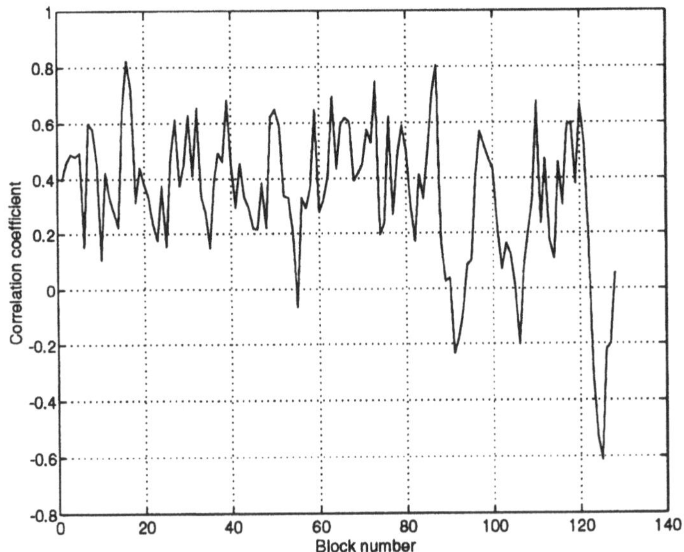

Fig. 5. Co-channel correlation for the wideband signal

The individual channels of a multi-channel wideband audio signal are highly correlated [17]. A number of researchers have attempted to exploit co-channel correlation for further improvement in source coding efficiency. These researchers devised "optimal" transformations applicable on the stereo signal such that the transformed co-channel signals are "nearly" orthogonal [18]. An undesirable side effect of matrixing is the phenomenon known as co-channel masking release where the quantization noise of the decoded stereo signal becomes audible even though it is in-perceptible when one listens to one channel at a time [19]. However, from the point of view of layered signal coding for ATM networks the co-channel correlation carries distinct advantages.

We now proceed to demonstrate the correlation properties between the left and right channel data which will be expressed in terms of the correlation coefficient of the wideband and subband signals. We also display the probability density function of the correlation coefficient between the left and right channel signals as a function of the subband number.

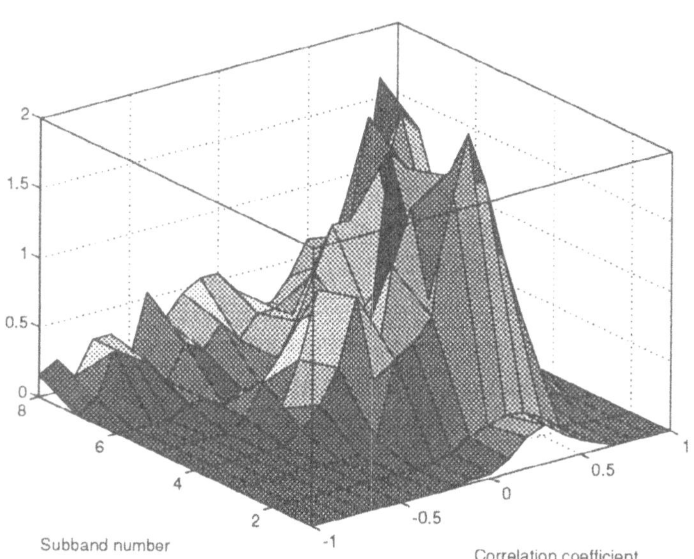

Fig. 6. Correlation coefficient pdf's across subbands

Observe that Figure 6 specifies an increasing co-channel correlation as one moves to higher frequency subbands in the multi-resolution decomposition of the wideband signal. This observation also establishes a framework for LP cell assignment and missing cell recovery as the LP data stream is more likely to contain data from higher frequency subbands where the co-channel correlation is also correspondingly high. Such an approach conforms with the optional *intensity stereo coding* specified in the MPEG standards draft. The MPEG documentation states:

> *"Psychoacoustic results indicate that at high frequencies (above about 2 kHz) the localization of the stereophonic image within a critical band is deter-*

mined by the temporal envelope and not by the temporal fine structure of the audio signal. The basic idea for intensity stereo coding is that for some subbands, instead of transmitting separate left and right subband samples only the sum-signal is transmitted, but with scale factors for both the left and right channels, thus preserving the stereophonic image."

We now define a source coding algorithm applicable to stereo wideband audio signals which are decomposed by an N-band WT. First, we introduce the following notation:

N : number of subbands;
K : number of samples in one block;
HP_k : high-priority data stream for the k*th* block;
LP_k : low-priority data stream for the k*th* block;
α : energy threshold;
M_n : normalized n*th* subband mono signal;
L_n : n*th* subband left channel signal;
R_n : n*th* subband right channel signal;
$\mathcal{E}_M$: total mono signal energy;
$\mathcal{E}_{Mn}$: n*th* subband mono signal energy;
$\mathcal{S}_{Ln}$: n*th* subband left channel scaling coefficient;
$\mathcal{S}_{Rn}$: n*th* subband right channel scaling coefficient;
$\mathcal{H}$: index set for the HP data stream;
$\overline{\mathcal{H}}$: index set for the LP data stream; (complement of $\mathcal{H}$ in N).

In this study *normalization* refers to scaling the coefficients in a given block such that the total signal energy for the samples becomes unity. $\mathcal{S}$ represents the multiplicative factor that will achieve this normalization.

Algorithm:

Step 1:

Separate the incoming data stream into non-overlapping blocks of K samples each. The right and left channel data of K samples each is decomposed into N subbands using the multi-resolution decomposition discussed in Section 2.1.

Step 2:

Compute the normalized mono signal $\{M_n,\ n = 1,\dots,N\}$, mono signal energy $\{\mathcal{E}_{Mn},\ n = 1,\dots,N\}$ and the left and right channel scaling coefficients $\{\mathcal{S}_{Ln}, \mathcal{S}_{Rn}\ n = 1,\dots,N\}$ for each subband.

Step 3:

Based on the set $\{\mathcal{E}_{Mn},\ n = 1, \ldots, N\}$ and a user specified energy threshold α determine the subband index set $\mathcal{H}$ such that the subbands pointed by the indices in $\mathcal{H}$ have a total signal energy that equals or exceeds $\alpha\mathcal{E}_M$.

Step 4:

Determine the high- and low-priority data streams as $LP_k = \{R_i,\ i \in \overline{\mathcal{H}}\}$ and $HP_k = \{R_i, L_i,\ i \in \mathcal{H};\ M_n, \mathcal{S}_{Ln}, \mathcal{S}_{Rn},\ n \in \overline{\mathcal{H}}\}$.

In a simpler algorithm the HP data stream may contain only those subbands pointed by the index set $\mathcal{H}$; then, no special missing cell recovery techniques need to be implemented. If the entire LP data stream is dropped due to traffic congestion within the ATM network, then the HP data stream is sufficient to reconstruct the audio data at an acceptable quality as all the subband signals that contains at least $\alpha\mathcal{E}_M$. As we remarked earlier such a coding strategy is superior to only retaining the lower frequency bands. Thus the proposed algorithm adapts the HP data stream to the non-stationary characteristics of the audio signal by changing the entries in the $\mathcal{H}$ from one block to the next. The basic algorithm is modified by including the normalized mono signal as part of the HP data stream. Such an approach introduces a degree of robustness with respect to LP cell loss possibility. In the case LP data cells are dropped by the network, the normalized mono signal, M_n, and the scaling coefficients, $\{\mathcal{S}_{Ln}, \mathcal{S}_{Rn}\}$, for each LP subband are used to generate re-scaled mono signals which will in turn be used in place of each missing LP subband signals. As we notice however, this robustness is achieved at the expense of increased HP data rate.

4 Results

We simulated the performance of the layered coding algorithm and measured the resulting SNR values under a variety of LP cell loss scenarios. The simulation results are based on four audio files sampled in stereo format at 32 kHz with sample amplitudes represented in 16-bit, uniform quantization format. Each audio file is 4.096 seconds long representing 256 blocks ($N = 256$) of 512 samples ($K = 512$). The audio files were sampled from Tchaikovsky's Violin Concerto in D op. 35, Tchaikovsky's 1812 overture and some generic harp music. We labelled these files Tch35, Tch1812 and Harp32, respectively. We used an 8-band wavelet transform based on 32-tap FIR filters derived from the Daubechies wavelet function for signal decomposition and reconstruction.

A significant characteristic of data transmission over ATM networks is LP cell losses seldom happen in isolation. In other words, a realistic simulation of traffic congestion in an ATM network should be based on "bursty" cell loss events of relatively long duration. The results presented in this section are based on complete LP data loss of 1.98 seconds duration representing approximately 50% of each test file. For this 1.98 second long segment the entire LP data stream is assumed to be dropped by the network during transmission. The decoder at the receiver implements the missing cell recovery technique based on scaled mono substitution method discussed earlier. The layered coding algorithm has been used at four different values of α, the energy threshold. Table 1 displays the left and right channel segmental SNR values for each test files.

Table 1. Segmental SNR (Right/Left) vs. α

File	Energy Threshold α			
	0.98	0.9	0.8	0.5
Tch35	16.9/16.1	9.3/10.5	8.1/8.2	6.1/6.1
Tch1812	18.0/17.8	12.2/13.6	9.6/11.1	7.1/8.4
Harp32	19.7/22.3	14.2/16.3	9.1/12.4	6.1/9.3

A value of 0.98 represents for most audio files a practical upper limit for the energy threshold parameter α. It is our experience that if α exceeds this limit the HP data rate r_H may equal the original data rate r for a significant number of processing blocks. Such an occurrence may violate the conditions which the source had negotiated at the connect time. Conversely, the source must have declared $\bar{r}_H \approx r$, which may may delay the *go-ahead* for network access during times of high traffic as the network may not guarantee transmission capacity at requested level of service. The choice of $\alpha = 0.98$ typically results in $\bar{r}_H = \frac{3}{4}r$ and $\bar{r}_L = \frac{1}{4}r$ which introduces the desired robustness (and the corresponding minimum acceptable sound quality) to the coding algorithm.

Table 2. MOS values vs. α

File	Energy Threshold α			
	0.98	0.9	0.8	0.5
Tch35	5	5	3	3
Tch1812	5	5	4	3
Harp32	5	4	2	2

5 = excellent; 4 = good; 3 = normal; 2 = poor; 1 = very poor.

The segmental SNR values provide an objective measure of the algorithm performance. However, most objective measurements do not correlate well with the subjective sound quality. A better assessment of the performance of the proposed layered coding algorithm is based on subjective listening tests. The quality of the sound files with the missing cell substitution method is measured using the *mean-objective-score* (MOS) values which ranks the resulting sound quality on a scale of 1 (=very poor) to 5 (=excellent). The MOS scores are summarized in Table 2.

The MOS values for the three sample audio files indicate that satisfactory sound quality can be achieved with value of the energy threshold variable α set to 0.9. While the MOS values represent only a subjective evaluation of the sound quality relative to changes in the value of the parameter α, we would like to emphasize that a minimum MOS value of 4 yields reconstructed sound quality that is virtually indistinguishable from the original.

5 Conclusions

The proposed algorithm for layered coding of stereo audio signals has an inherent simplicity as a direct consequence of the WT filterbank implementation. Among subband coding algorithms MPEG and MUSICAM rely on uniform bandwidth filterbank structures which do not provide the source encoder with sufficient resolution at lower frequencies in the order of critical bands as required by the underlying psycho-acoustic model. Therefore, the uniform bandwidth filterbank structures in these algorithms are supplemented by an FFT analyzer operating in parallel to the filterbank. This approach increases the computational complexity of the algorithm. The WT based subband decomposition/reconstruction on the other hand introduces a simpler framework for signal analysis and coding.

The coding algorithm of Section 3 together with the proposed missing cell recovery technique performs well in initial simulation and subjective evaluation tests. The initial performance assessment is based on a limited number of sound files. However, we expect the performance improvement will be higher for sound files with significant high frequency energy content.

A further direction of research is to combine the mono signal based missing cell recovery method with a masking based layered coding scheme which can be easier to integrate with the standardized MPEG coding algorithm. Currently we are working to have better understanding of the distortion mechanism due to replacing certain subband signals with the scaled mono signal. The distortion due to mono replacement has The results from this study will be the basis of an enhanced algorithm where the energy threshold is adaptively altered for each block to maintain the desired sound quality. Such an approach to layered coding is highly desirable as it will result to minimize the HP data rate with respect to the desirable sound quality.

Acknowledgements. This work was supported by grants from the National Science and Engineering Research Council (NSERC) and from the Office of Research and Innovation, Ryerson Polytechnic University.

References

1. Minzer, S.E, "Broadband ISDN and asynchronous transfer mode (ATM)," *IEEE Commun. Mag.*, 1989, 27, pp. 17-24.
2. M. de Prycker, *Asynchronous Transfer mode: Solution for broadband ISDN*. New York, Ellis Horwood, 1991.
3. N. Kitawaki et.al, "Speech coding technology for ATM networks," *IEEE Commun. Magazine*, January 1990, pp. 21-27.
4. Second draft of proposed standard on information technology of moving pictures and associated audio for digital storage media up to about 1.5 Mbits/s. *Document ISO/IEC JTC1/SC2/WG11 MPEG 90/001*, Sept. 1990.
5. Y.F. Dehery et al., "A MUSICAM source codec for digital audio broadcasting and storage," *Proc. ICASSP-91*, May 1991, pp. 3605-3608.
6. "How PASC Data Compression Works in Phillips Digital Compact Cassette," *Audio*, September 1991, pp. 32–39.
7. D.H. Teh, A.P. Tan and S.N. Koh, "Subband coding of high-fidelity quality audio signals at 128 kbps," *Proc. ICASSP-92*, 1992, pp. 197-200.
8. S. Mallat, "A theory for multiresolution signal decomposition: The wavelet representation," *IEEE Trans. Patt. Anal. Mach. Intel.*, pp. 674-693. July 1989.
9. P. Rioul and M. Vetterli, "Wavelets and signal processing," *IEEE Signal Proces. Mag.*, pp. 14-38, Oct. 1991.
10. A.N. Akansu and R.A. Haddad, *Multiresolution signal decomposition: transforms, subbands, and wavelets*. Boston, MA: Academic Press, 1992.
11. M.M. Lara-Barron and G.B. Lockhart, "Packet-based embedded encoding for transmission of low-bit-rate-encoded speech in packet networks," *IEE Proceedings-I*, vol. 139, no. 5, October 1992, pp. 482-487.
12. S. Zafar, Y.Q. Zhang and B. Jabbari, "Multiscale video representation using multiresolution motion compensation and wavelet decomposition," *IEEE Jour. on Sel. Areas in Commun.*, vol. 11, no. 1, pp. 24-35, January 1993.
13. S. Minami, "CCITT H.261 compatible mixed bit rate coding of video for ATM networks," *Proc. ICC-92*, 1992, pp. 537-543.
14. I. Daubechies, "Orthonormal bases on compactly supported wavelets," *Comm. Pure Appl. Math.*, pp. 909-996, 1988.
15. F. Hazu, I. Kuroda and T. Nishitani, "ATC-based hi-fi audio coding for ATM networks," *Proc. ICC-92*, 1992, pp. 57-61.
16. D.J. Goodman, "Embedded DPCM for variable bit rate transmission," *IEEE Trans. on Commun.*, COM-28, pp. 1040-1046, 1980.
17. D. Bauer and D. Seitzer "Statistical properties of high quality stereo signals in the time domain," *Proc. of ICASSP-89*, May 1989, pp. 2045-2048.
18. R.G. van der Waal and R.N.J. Veldhuis, "Subband coding of stereophonic digital audio signals," *Proc. of ICASSP-91*, May 1991, 3601-3604.
19. W.D.Th. ten kate et.al., "Matrixing of bit rate reduced audio signals," *Proc. ICASSP-92*, 1992, pp. 205-208.
20. Y.C. Hsu and M. Zeytinoğlu, "Subband statistics of layered coded, wideband stereo audio signal," *in preparation*.

Statistical Analysis of the Traffic Generated by the Superposition of N Independent Interrupted Poisson Processes

Faouzi Kamoun and M. Mehmet Ali

Department of Electrical and Computer Engineering, Concordia University
Montreal, Quebec, H3G 1M8

Abstract. This paper concentrates on the statistical characterization of the traffic generated by the superposition of N independent and homogeneous Interrupted Poisson processes. The superposed traffic is considered here as a candidate for modeling bursty and correlated arrival processes, such is the case for the aggregate packet arrival process at an ATM multiplexer. More precisely, we approximate the traffic generated by N' multimedia sources by the superposition of homogeneous interrupted Poisson processes, whose number , N, as well as their three parameters can be estimated using some statistical matching methods. In this paper, the main part of our analysis focuses on the appropriateness of the proposed model, from a statistical point of view through the theoretical investigation of its variability and correlation behaviors. In particular, we pay a special attention to a statistical problem that has not been yet investigated properly, namely the theoretical investigation of the dependence among the successive packet inter-arrival times in the superposition process. Based on some numerical examples, our analysis shows that as we increase the number of component processes, the burstiness of the superposition traffic decreases while the correlation among the packet inter-arrival times increases. We will also highlight a statistical matching method that can be applied to estimate the parameters of the proposed model.

1 Introduction

Several experimental studies have confirmed that the workload of an ATM multiplexer which is characterized by the aggregate packet arrival process, generated by many multiplexed sources, is a very complex process which exhibits a high degree of burstiness and correlation. The term *Burstiness* has often been used in the literature to characterize the relative variability of a given traffic when compared to that of a Poisson process. The term *Correlation* usually refers to the dependence that exists among successive packet inter-arrival times or to the dependence in the average packet arrival rates in successive time intervals. In this paper we consider a bursty and correlated process that can be used to model the traffic generated by the superposition of many independent ATM sources. More precisely, and as illustrated in Figure 1, we model the traffic generated by N' ATM sources by the superposition of N homogeneous ON/OFF sources of the type described in the next Section. What we focus on in this study, is mainly the appropriateness of the proposed process for modeling the workload of an ATM multiplexer, from a statistical point of view. Therefore we have organized this paper as follows:

In the next Section we will start by considering a single source model, namely an ON/OFF source, generating an Interrupted Poisson Process (IPP). Using renewal theory, we investigate the statistics of the traffic generated by such a source. In Section three we look at the pooled traffic generated by the superposition of N independent and homogeneous Interrupted Poisson Processes. Specifically, we derive an expression for the packet inter-arrival time distribution and then we carry a complete statistical analysis of the pooled process. The variability as well as the statistical correlation of the superposition traffic are dealt with in full details. In particular, and based on the generating function of the number of arrivals during a time interval, we have been able to suggest a method to compute the serial autocorrelation coefficients of arbitrary lag number in the pooled traffic; which is the main contribution of this paper. Computing the auto-correlation coefficients, ρ_i's, in a superposition process is a statistical problem that has not been properly investigated, mostly because of its difficulty. In particular, Enns [1] considered the superposition of a specific type of component processes and then derived a bound on the first serial covariance. Lawrance [2] used the joint distribution for the inter-arrival times to derive ρ_1 and ρ_2 for some particular superposition processes. For serial coefficients of order greater than two, the approach presented in [2] becomes quite involved because of the difficulty encountered in deriving the joint distribution of a large number of adjacent inter-arrival times. These problems, among others, have led many researchers (see for example [3,4]) to resort to extensive simulation experiments in order to estimate the ρ_i's. In Section four we consider an interesting statistical problem, namely that of estimating the equivalent number of sources, N, as well as the three single source parameters. In Section five we give a conclusion and some suggestions for future research.

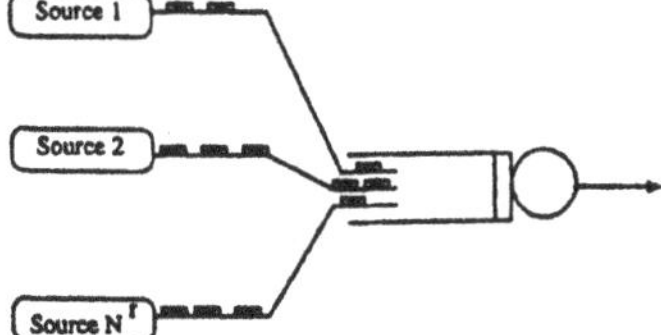

Fig1.a. ATM Multiplexer Loaded with N^f Sources

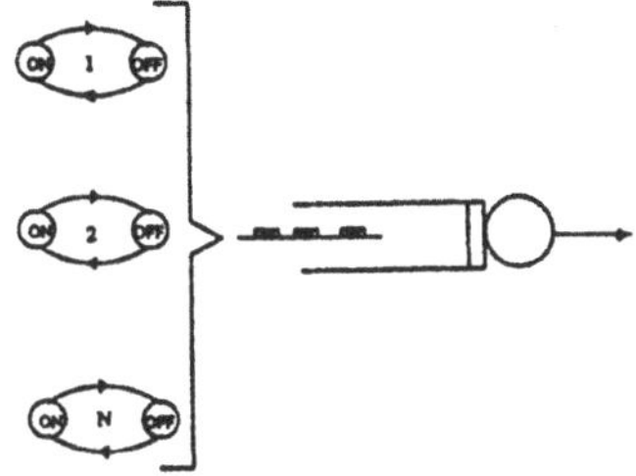

Fig1.b. Approximate Model (N Homogeneous IPP's)

Fig. 1. The Model for the Arrival Process at an ATM Multiplexer

2 The Single Source Model

2.1 Model Description

Here we consider an ON/OFF source, generating an Interrupted Poisson Process (IPP), as shown in Figure 2. When a source is ON (Burst mode), packets are generated according to a Poisson process of rate λ. When it is OFF (Silence mode) no packets are generated. We also assume that both ON and OFF periods are exponentially distributed with means $T_{on} = \alpha^{-1}$ and $T_{off} = \beta^{-1}$ respectively. The state probabilities of the burst state and silence state are given by:

$$\Pi_{on} = \frac{\beta}{\alpha+\beta} \quad ; \quad \Pi_{off} = \frac{\alpha}{\alpha+\beta}$$

This process can be viewed as a Poisson process, modulated by a random switch.

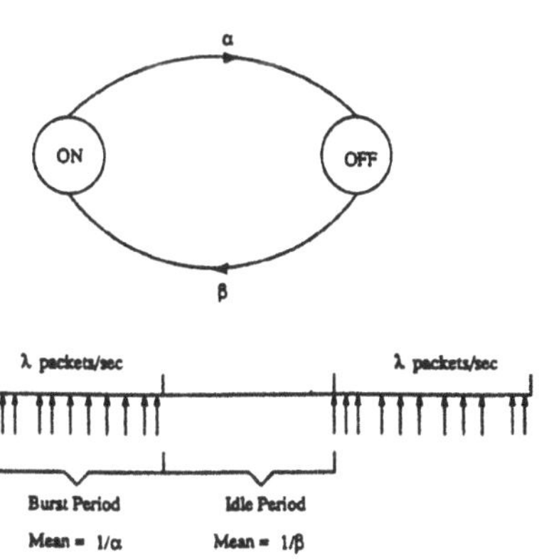

Fig. 2. The IPP as a Single Source Model

Since the burst durations are exponentially distributed, then the number of packets in a burst is geometrically distributed. Further, and because of the memoryless property of the geometric distribution, the packet stream generated by a single source is an ordinary renewal process [3,5]. From [6] the packet arrival process from a single source has a two-phase hyper-exponential inter-arrival time distribution given by:

$$F_X(t) = pr[X \le t] = k_1\left(1 - e^{-r_1 t}\right) + k_2\left(1 - e^{-r_2 t}\right) \tag{1}$$

where:

$$r_1 = \frac{1}{2}\{\lambda + \beta + \alpha + \sqrt{(\lambda+\beta+\alpha)^2 - 4\lambda\beta}\} \ , \ r_2 = \frac{1}{2}\{\lambda + \beta + \alpha - \sqrt{(\lambda+\beta+\alpha)^2 - 4\lambda\beta}\}$$

$$k_1 = \frac{\lambda - r_2}{r_1 - r_2} \ , \quad k_2 = 1 - k_1$$

Therefore:

$$f_X(t) = \frac{dF_X(t)}{dt} = k_1 r_1 e^{-r_1 t} + k_2 r_2 e^{-r_2 t} \tag{2.1}$$

$$\bar{F}(s) = L[f_X(t)] = \frac{k_1 r_1}{s + r_1} + \frac{k_2 r_2}{s + r_2} \tag{2.2}$$

where L (.) denotes the Laplace transform operator.

Throughout this paper, and unless otherwise specified, the single source parameters are given by $\alpha^{-1} = 352\,\text{ms}$, $\beta^{-1} = 650\,\text{ms}$ and $\lambda = 62.5$ packets/sec, which match the voice source model used in [3].

2.2 Statistical Charaterization

Generally speaking, there are two ways through which we can look at the packet stream generated by a single source. The first is based on the number of arrivals during a fixed time interval, while the other focuses on the intervals between packet arrivals. Without any loss of generality, we start by considering the sequence of inter-arrival times ($X_1, X_2, X_3, \ldots, X_n$). We assume that the arrival process is at least weakly stationary. Let

$\overline{X}$ and $Var(X)$ be the common mean and variance of the inter-arrival times which, from (2.2), are given by:

$$\overline{X} = \frac{k_1}{r_1} + \frac{k_2}{r_2} = \frac{\alpha+\beta}{\lambda\beta} \tag{3.1}$$

$$Var(X) = \frac{k_2(2-k_2)r_1^2 + k_1(2-k_1)r_2^2 - 2k_1k_2r_1r_2}{r_1^2 \cdot r_2^2} = \frac{2\lambda\alpha + (\alpha+\beta)^2}{(\lambda\beta)^2} \tag{3.2}$$

The squared coefficient of variation (SQV), $C^2 = \frac{Var(X)}{(\overline{X})^2}$, has often been used to characterize the burstiness of a given process and also to quantify, in a rough way, dispersion from the Poisson process for which $C^2 = 1$. For a single source, the squared coefficient of variation is given by:

$$C^2 = \frac{k_2(2-k_2)r_1^2 + k_1(2-k_1)r_2^2 - 2k_1k_2r_1r_2}{(k_1r_2 + k_2r_1)^2} = 1 + \frac{2\lambda\alpha}{(\alpha+\beta)^2} \tag{4}$$

It is important to note, at this stage, that although for the chosen parameters C^2 is relatively low (19.51), the single source model can be used to model a very bursty multimedia source. In fact, since:

$$C^2 = 1 + 2\lambda T_{on}\left(\frac{T_{off}}{T_{on}+T_{off}}\right)^2$$

then for fixed (λ, T_{off}), $C^2$is maximum when $T_{on} = T_{off}$, whereas for fixed (λ, T_{on}), C^2 increases with increasing values of T_{off}. Therefore, for a fixed λ, very large values for C^2 can be obtained by increasing the period of the source (Fig 3). In addition, for fixed burst and silence durations, the traffic's variability increases as the average packet arrival rate per burst (λ) gets larger. We also note that since $C^2 \geq 1$, then the traffic of a single IPP is over-dispersed relative to the Poisson process.

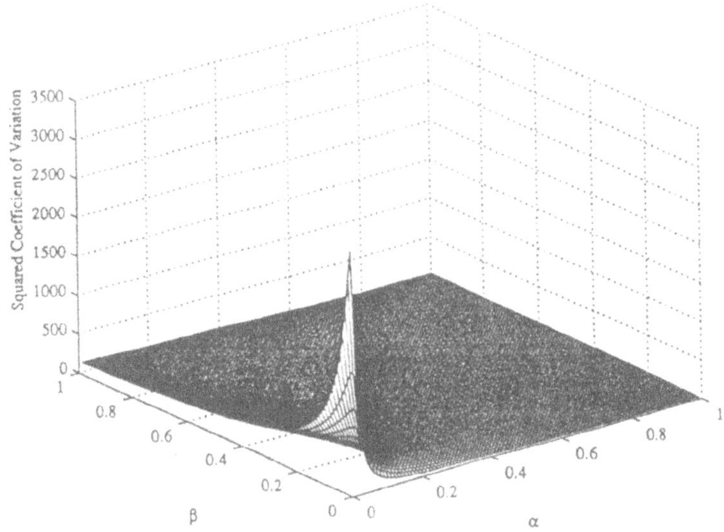

Fig. 3. Squared Coefficient of Variation ($\lambda = 62.5$)

Next, let T(n) be the variance of the time span from an arbitrary arrival to the nth successive arrival:

$$T(n) = Var(X_{i+1} + X_{i+2} .. + X_{i+n}) = nVar(X) + 2\sum_{j=1}^{n-1}\sum_{k=1}^{j} Cov(X_j, X_{j+k})$$

where $Cov(X_j, X_{j+k})$ is the autocovariance function of lag k. T(n) is a very important statistical characteristic of the packet arrival process since it embodies some of the covariance/correlation aspects of the cell stream. Associated with T(n) is the index of dispersion for intervals (IDI) J_n defined by:

$$J_n = \frac{T(n)}{n(\overline{X})^2} = C^2\left[1 + 2\sum_{j=1}^{n-1}\left(1 - \frac{j}{n}\right)\rho_j\right] \qquad (5)$$

where ρ_j is the autocorrelation coefficient of lag j. For the single source model, and because of the renewal property, J_n is constant and is equal to C^2 for all n's.

We next look to the counting process, N_t, of the number of packet arrivals during an interval $(0, t]$, which starts with, but does not include an arbitrary arrival. Let $M(t) = E[N_t]$ be the mean number of arrivals during time t. For a single source, and using renewal theory (see for example [7,8]) the Laplace transform of $M(t)$ is:

$$\tilde{K}(s) = \frac{\tilde{F}(s)}{s[1 - \tilde{F}(s)]} = \frac{1 + as}{s^2(1 + Ts)} \qquad (6)$$

where:

$$a = \frac{k_1 r_1 + k_2 r_2}{r_1 r_2}, \quad T = \frac{1}{r_1 k_2 + r_2 k_1}$$

Hence

$$M(t) = \Lambda t + \frac{1}{2}(C^2 - 1)\{1 - e^{-(r_1 k_2 + r_2 k_1)t}\} \qquad (7)$$

where $\Lambda = \frac{1}{\overline{X}}$ is the single source mean packet arrival rate. The variance $V(t) = Var(N_t)$ of the number of arrivals during $(0, t]$ is found [7,8] using:

$$V(t) = \frac{1}{\overline{X}}\int_0^t \Phi(n)dn \qquad (8.1)$$

where

$$\tilde{\Phi}(s) = L[\Phi(t)] = \frac{1}{s} + 2\tilde{K}(s) - \frac{2}{\overline{X}s^2} \qquad (8.2)$$

Substituting (6) in (8.2) and using (8.1), we get:

$$V(t) = C^2\Lambda t - \frac{\Lambda(C^2 - 1)}{k_1 r_2 + k_2 r_1} \cdot \{1 - e^{-(r_1 k_2 + r_2 k_1)t}\} \qquad (9)$$

Equations (7) and (9) show that the mean and variance time curves become linear as $t \to \infty$, and that while the displacement of the linear portion of $M(t)$ is always positive, that of $V(t)$ is always negative, leading to a positive covariance in the number of arrivals. The index of dispersion for counts (IDC), which is also used to characterize the variability of an arrival process, is given here by:

$$IDC(t) = \frac{V(t)}{M(t)} = \frac{C^2\Lambda t - \frac{\Lambda(C^2-1)}{k_1 r_2 + k_2 r_1} \cdot \{1 - e^{-(r_1 k_2 + r_2 k_1)t}\}}{\Lambda t + \frac{1}{2}(C^2 - 1)\{1 - e^{-(r_1 k_2 + r_2 k_1)t}\}} \qquad (10)$$

We next check for the correlation behavior of the number of arrivals in disjoint intervals. In particular let $C_1(t)$ be the covariance of the number of arrivals in two consecutive intervals of length t:

$$C_1(t) = \frac{V(2t)}{2} - V(t) = \frac{1}{2}\frac{\Lambda(C^2-1)}{(k_1r_2+k_2r_1)} \cdot \left(1-e^{-(r_1k_2+r_2k_1)t}\right)^2 \geq 0 \tag{11}$$

We can also define the associated autocorrelation function:

$$AC(t) = \frac{C_1(t)}{V(t)} = \frac{V(2t)}{2V(t)} - 1 \tag{12}$$

Equations (11) and (12) show that even though the packet arrival process from a single source is renewal, there is a positive correlation in the number of arrivals in consecutive intervals. In addition the asymptotic value of the covariance time curve , $C_1(\infty)$, is of practical interest since it equals to one half of the displacement of the linear Section of $V(t)$. This can be shown [8] by rewriting (9) in the form:

$$V(t) = V'(\infty)t - 2C_1(\infty) + \mu(t)$$

where $\mu(t) \to 0$ as $t \to \infty$.

Another way to look to the correlation properties of the number of arrivals is through the spectral and covariance density functions, which characterize the transient behavior of the second-order properties of the arrival process for very small time intervals. For this purpose consider the stationary counting process N_t' which is defined as the cumulative number of arrivals in an interval t, following an arbitrary selected point where the observation of arrivals begins. Associated with N_t' is the differential process $\Delta N_t'$ which represents the number of arrivals in $(t, t+\Delta t)$ where $\Delta t \to 0$. Define,. for this differential process, a covariance density function:

$$\psi(\tau) = \lim_{\Delta t \to 0} \frac{Cov\{\Delta N_t', \Delta N_{t+\tau}'\}}{(\Delta t)^2}$$

From Renewal theory, The spectral density function, $\Psi(w)$, of the single source counting process, N_t', is obtained by taking the Fourier transform of $\psi(\tau)$, giving:

$$\Psi(w) = \frac{1}{\pi\overline{X}}\left\{1 + \frac{F(jw)}{1-F(jw)} + \frac{F(-jw)}{1-F(-jw)}\right\} \qquad 0 \leq w \leq +\infty$$

$$= \frac{1}{\pi\overline{X}}\left\{1 + \frac{2k_1k_2(r_1-r_2)^2}{w^2+(r_1k_2+r_2k_1)^2}\right\} \qquad 0 \leq w \leq +\infty$$

The corresponding covariance density function can be then obtained by taking the inverse Fourier transform, giving:

$$\psi_+(\tau) = \frac{r_1r_2}{\pi(k_1r_2+k_2r_1)}\delta(\tau) + \frac{k_1k_2r_1r_2(r_1-r_2)^2}{\pi(k_1r_2+k_2r_1)^2}e^{-(r_1k_2+r_2k_1)\tau} \qquad ;\tau \geq 0$$

$$= \frac{\lambda\beta}{\pi(\alpha+\beta)}\delta(\tau) + \frac{\lambda^2\alpha\beta}{\pi(\alpha+\beta)^2}e^{-(\alpha+\beta)\tau}$$

where $\delta(\tau)$ is the Dirac delta function.

In particular it can be shown [8] that the spectral contribution at the origin ($w = 0$) is π^{-1} times the asymptotic slope ($V'(\infty) = \frac{C^2}{\overline{X}}$) of the variance-time curve. Furthermore, the spectrum $\Psi(w)$ decreases monotonically from $\Psi(0) = \frac{C^2}{\pi\overline{X}}$ to $\Psi(\infty) = \frac{1}{\pi\overline{X}}$, and when the process reduces to Poisson, the spectrum becomes flat, taking on the value $\frac{1}{\pi\overline{X}}$.

The $M(t)$, $V(t)$, $IDC(t)$, $C_1(t)$ $AC(t)$ and $\tilde{\Psi}(w)$ curves for the single voice source are also plotted as shown in Figures 4, 5, 6, 7, 8 and 9. A visual inspection of the autocorrelation-time curve (Figure 8) shows a rapid increase during a time span equivalent to approximately one and a half times of the mean cell inter-arrival time. The curve then decreases and the time required for the correlation to be neglected (let us say when $AC(t) < 0.1$) is of the order of 1 second, which also corresponds to the mean source period.

3 Superposition of N Independent On/Off (IPP) Sources

In this Section we consider the pooled traffic that results when the packet streams of N independent and homogeneous ON/OFF sources, of the type described in Section 2.1, are superimposed. Since the packet stream from a single source is renewal, then we can use the superposition theory of renewal processes to describe the aggregate (pooled) traffic. In the sequel, and in order to distinguish those variables related to the single source model from those related to the superposition process, we will often add the index [p] to the latter. Let N_t^p be the number of packets in the pooled traffic at time t. Let $N_t^{(i)}$ be the number of packets generated by the ith source at time t. Then $N_t^p = \sum_{i=1}^{N} N_t^{(i)}$. We will start with the statistical properties of the counting process N_t^p since they are easier to derive. Because of the independence assumtion among the N sources, many of the stattistical curves of the counting process, N_t^p are just additive, giving:

$$M^p(t) = N \cdot M(t) \quad , \quad V^p(t) = N \cdot V(t) \quad , \quad C_1^p(t) = N \cdot C_1(t) \quad , \quad AC^p(t) = AC(t)$$

$$\psi_+^p(\tau) = N \cdot \psi_+(\tau) \quad , \quad \Psi^p(w) = N \cdot \Psi(w) \quad , \quad IDC^p(t) = IDC(t)$$

Hence the correlation behavior of the superposition counting process N_t^p is closely tied to that of the single source counting process, N_t. We also note that the superposition process embodies the same type of positive correlation that characterizes the average cell arrival rate in a multimedia environment.

We next look to the sequence $(Y_1, Y_2, ..Y_n)$ of inter-arrival times in the superposition process.

First we note that although these inter-arrival times are not independent, they are in fact identically distributed, with a common mean given by:

$$\overline{Y} = \frac{1}{\Lambda^p} = \frac{\overline{X}}{N} = \frac{1}{N} \cdot \left[\frac{k_1}{r_1} + \frac{k_2}{r_2} \right] \tag{13}$$

where Λ^p is the mean packet arrival rate in the pooled process.

3.1 The Packet Inter-arrival Time Distribution

Let $F_c(t)$ and $F_c^p(t)$ be the survivor functions of the packet inter-arrival times in the single source and in the pooled traffic, respectively. Then [3,9]:

$$F_c^p(t) = [F_c(t)]\,[1 - F_e(t)]^{[N-1]} \tag{14.1}$$

where

$$F_e(t) = \frac{1}{\overline{X}} \int_0^t F_c(u)\, du \tag{14.2}$$

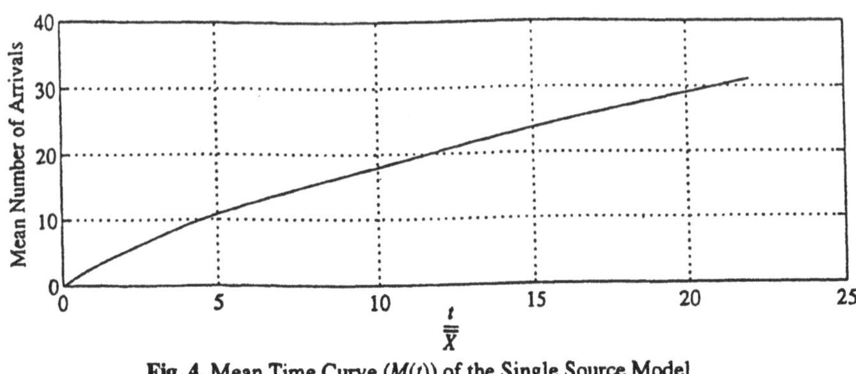

Fig. 4. Mean Time Curve ($M(t)$) of the Single Source Model

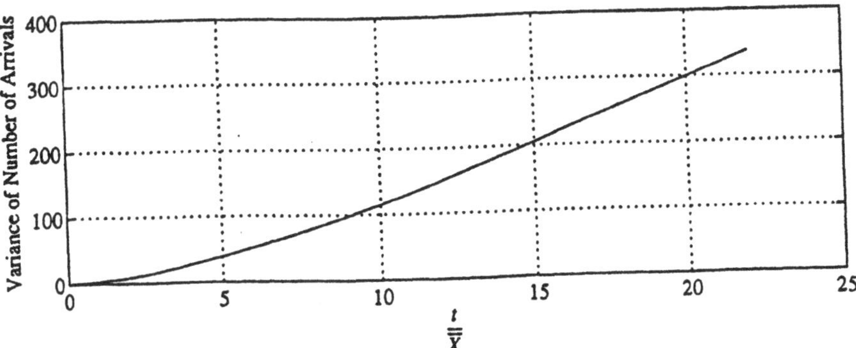

Fig. 5. Variance Time Curve $(V(t))$ of the Single Source Model

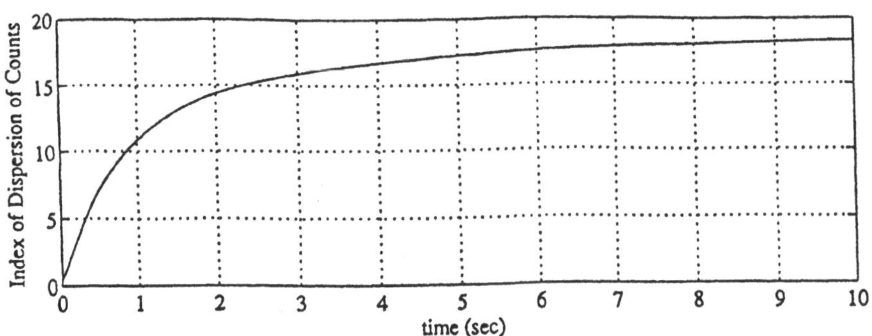

Fig. 6. Index of Dispersion of Counts for the Single Source Model

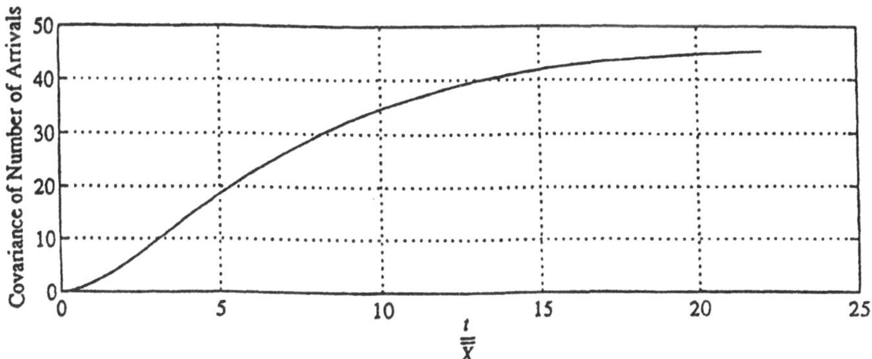

Fig. 7. Covariance Time Curve of Number of Arrivals

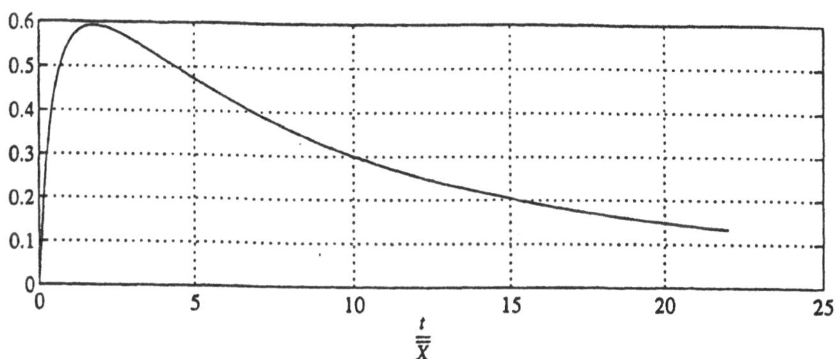

Fig. 8. Autocorrelation Function of Number of Arrivals

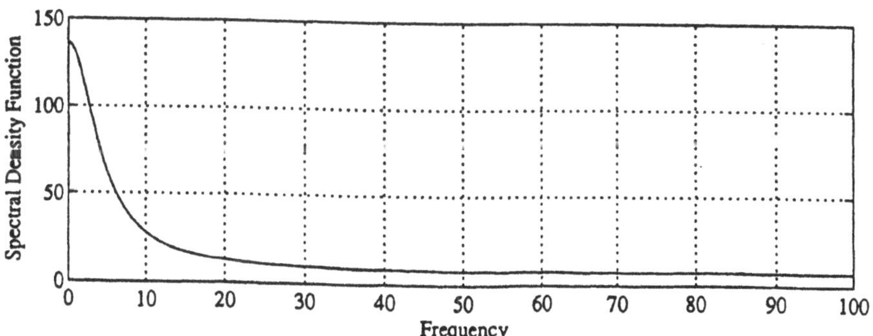

Fig. 9. Spectral Density Function of the Single Source Model

For our case, the survivor function of the packet inter-arrival time in the pooled traffic is given by:

$$F_c^p(t) = \left(k_1 e^{-r_1 t} + k_2 e^{-r_2 t}\right)\left[\frac{k_1 r_2 e^{-r_1 t} + k_2 r_1 e^{-r_2 t}}{k_1 r_2 + k_2 r_1}\right]^{N-1} \qquad (15)$$

In particular, we note that if we appropriately scale time by N, then the asymptotic survivor function, as $N \to \infty$, becomes exponential:

$$\lim_{N \to \infty} F_c^p\left(\frac{t}{N}\right) = \lim_{N \to \infty} (1 - \Lambda^p t)^{N-1} = e^{-\Lambda^p t}$$

This shows that over very short time intervals, the superposed packet arrival process is Poisson. This turns out to be a general result which holds when the component processes are renewal [8].

3.2 The Squared Coefficient of Variation

In this Section we investigate the relationship between the burstiness (variability) of the pooled traffic and the number of active sources, by deriving the squared coefficient of variation C_p^2 of the superposition traffic. From [3], the second moment of the aggregate packet inter-arrival time is given by:

$$\overline{Y^2} = 2 \int_0^\infty t F_c^p(t)\, dt \qquad (16)$$

After some mathematical manipulations (to avoid carrying the integration) it can be shown (see Appendix 1) that:

$$\overline{Y^2} = \frac{2}{N r_1 r_2 (k_1 r_2 + k_2 r_1)^{N-1}} \cdot \sum_{i=0}^{N} \binom{N}{i} \frac{(k_1 r_2)^i (k_2 r_1)^{N-i}}{i(r_1 - r_2) + N r_2} \qquad (17.1)$$

and therefore:

$$C_p^2 = \frac{Var(Y)}{(\overline{Y})^2} = \left\{\frac{2N r_1 r_2}{(k_1 r_2 + k_2 r_1)^{(N+1)}} \cdot \sum_{i=0}^{N} \binom{N}{i} \frac{(k_1 r_2)^i (k_2 r_1)^{N-i}}{i(r_1 - r_2) + N r_2}\right\} - 1 \qquad (17.2)$$

For the case of voice, and as shown in Figure 10, the squared coefficient of variation of the pooled process decreases monotonically with increasing number of sources.

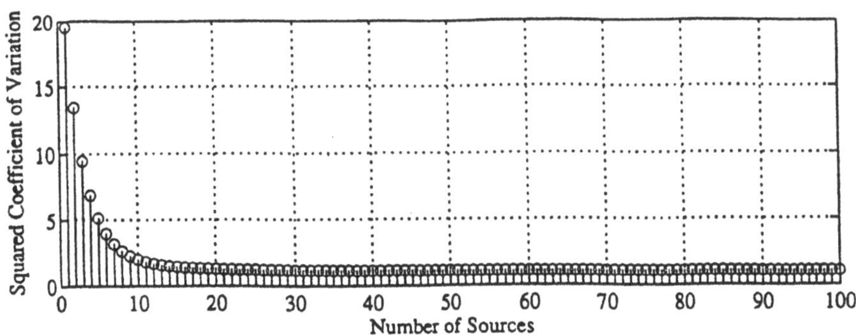

Fig. 10. Squared Coefficient of Variation of the Pooled Process

In the asymptotic limit, as $N \rightarrow \infty$, the squared coefficient of variation approaches that of the Poisson process (one). This shows that the burstiness of the pooled traffic decreases with increasing number of sources. This fact can be explained by the clustering effect that takes place in the pooled process and which decreases as the number of component processes gets larger. In fact as N increases, the average packet inter-arrival time ($\overline{Y}$) gets smaller and packets will have a tendency to arrive in clusters whose packet inter-arrival times are very close to $\overline{Y}$. However, as we will show shortly, one can not infer from the above observations that the pooled traffic becomes Poisson when the number of component processes becomes very large.

3.3 The Correlation Among Successive Intervals

Although a single source was modeled by a renewal process, it is well known that for $N > 1$, the pooled process is far from renewal because of the dependency among successive packet inter-arrival times. This dependency makes the superposition process suitable for modeling correlated arrival processes. For $N > 1$, the IDI of the pooled process:

$$J_n^p = \frac{Var(Y_{i+1}, \ldots, Y_{i+n})}{n(\overline{Y})^2} = C_p^2\left[1 + 2\sum_{j=1}^{n-1}\left(1 - \frac{j}{n}\right)\rho_j\right]$$

is no longer constant. To compute J_n^p, one has to evaluate the serial autocorrelation coefficients, ρ_i's, a problem which, as mentioned before, has not been properly investigated, mostly because of its difficulty.

In the sequel, and based on the Probability Generating Function (PGF) of the number of arrivals, $N_t'^p$, in the pooled traffic during time t (starting from an arbitrary time origin) we will propose a recursive scheme to compute all the ρ_i's for the proposed superposition model. Before doing so, we note that by writing the asymptotic matching between the IDI and the IDC curves in the superposition traffic, we get:

$$\lim_{n \rightarrow \infty} J_n^p = C_p^2\left[1 + 2\sum_{j=1}^{\infty}\rho_j\right] = \lim_{t \rightarrow \infty} IDC^p(t) = C^2$$

Hence the cumulative correlation of the superposition traffic is:

$$\chi_c(N) = \sum_{j=1}^{\infty}\rho_j = \frac{1}{2}\left[\frac{C^2}{C_p^2} - 1\right] \tag{18}$$

From (18) we can deduce that in the superposition of N homogeneous renewal processes, the cumulative correlation in the inter-arrival times depends merely on the relative variability of the single source traffic with respect to that of the superposition traffic. In addition since in Section 3.2, the variability of the superposition traffic is found to be decreasing with increasing number of sources then the cumulative correlation increases when the number of sources gets larger, reaching an asymptotic value, $\chi_c(\infty) = \frac{C^2-1}{2}$. Figure 11, illustrates this fact, for the case of voice traffic. In particular we note that it only takes few sources (around 80) before the $\chi_c(N)$ curve approaches its asymptotic limit of 9.75.

From the above we conclude that although the squared coefficient of variation approaches that of a Poisson traffic for a large number of sources, the increase in the cumulative correlation makes the superposition traffic deviate considerably from the Poisson process. It is clear from Figure 11 that, for a very large N, a Poisson approximation for the pooled process is hard to justify. This also shows that the component processes are not relatively sparse, a condition which must be satisfied if the pooled traffic (with $N \rightarrow \infty$) is to converge to a Poisson process [4].

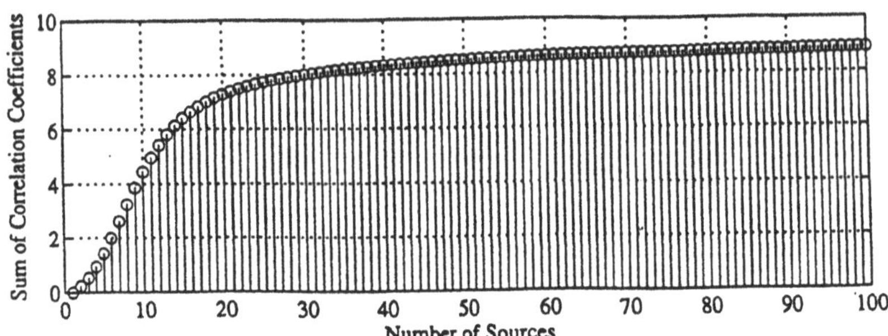

Fig. 11. Sum of Correlation Coefficients in the Pooled Process

Finally we look to the problem of finding the serial autocorrelation coefficients of arbitrary lag in the proposed superposition process, in an attempt to acquire a better understanding of type of dependence that governs the aggregate packet inter-arrival times.

Theoretically, the serial autocorrelation coefficients of arbitrary lag numbers can be computed using McFadden [10] identity:

$$\rho_i = \frac{E(Y_j Y_{j+i}) - E(Y_j)E(Y_{j+i})}{Var(Y)} = \frac{\overline{Y}}{Var(Y)} \cdot \left\{ \int_0^\infty p(i,t)dt - \overline{Y} \right\} \qquad (i = 1, 2, \ldots.) \quad (19)$$

where $p(i,t) = pr\left[N_t^{'p} = i\right]$.

Let $\Phi(z,t)$, $\Phi'(z,t)$ and $\Phi_p'(z,t)$ be the Generating functions of N_t, N_t' and $N_t^{'p}$ respectively. Let $\tilde{\Phi}(z,s)$, $\tilde{\Phi}'(z,s)$ and $\tilde{\Phi}_p'(z,s)$ be their corresponding Laplace transforms. From renewal theory, we know that:

$$\tilde{\Phi}(z,s) = \frac{1 - \tilde{F}(s)}{s[1 - z\tilde{F}(s)]} \qquad (20)$$

substituting (2.2) in (20), and taking the inverse Laplace transform, we obtain:

$$\Phi(z,t) = \frac{\gamma_1 - a}{\gamma_1 - \gamma_2} \cdot e^{-\gamma_1 t} + \frac{\gamma_2 - a}{\gamma_2 - \gamma_1} \cdot e^{-\gamma_2 t} \qquad (21.1)$$

where:

$$\gamma_1 = \frac{(r_1 + r_2) - (r_1 k_1 + r_2 k_2)z + \sqrt{(r_1 k_1 + r_2 k_2)^2 z^2 + 2(r_1 - r_2)(r_2 k_2 - r_1 k_1)z + (r_1 - r_2)^2}}{2} \qquad (21.2)$$

$$\gamma_2 = \frac{(r_1 + r_2) - (r_1 k_1 + r_2 k_2)z - \sqrt{(r_1 k_1 + r_2 k_2)^2 z^2 + 2(r_1 - r_2)(r_2 k_2 - r_1 k_1)z + (r_1 - r_2)^2}}{2} \qquad (21.3)$$

$$a = r_1 k_2 + r_2 k_1$$

Next the PGF of the counting process N_t' is related to that of the counting process N_t through the relationship:

$$\Phi'(z,t) = 1 + \frac{(z-1)}{\overline{X}} \int_0^t \Phi(z,u)du \qquad (22)$$

Substituting (21.1) in (22) and after some mathematical manipulations, we get:

$$\Phi'(z,t) = \frac{1}{a(\gamma_1 - \gamma_2)} \cdot \left[(\gamma_1\gamma_2 - a\gamma_2)e^{-\gamma_1 t} + (a\gamma_1 - \gamma_1\gamma_2)e^{-\gamma_2 t} \right]$$

where $\gamma_1\gamma_2 = r_1 r_2(1-z)$.
Because of the homogenousity and independence assumptions among the component processes, the PGF of $N_t^{'p}$, is readily obtained:

$$\Phi_p^{'}(z,t) = E\left[z^{N_t^{'p}}\right] = [\Phi^{'}(z,t)]^N$$

$$= \frac{1}{a^N(\gamma_1-\gamma_2)^N} \cdot \left[\sum_{j=0}^{N} \binom{N}{j} (\gamma_1\gamma_2 - a\gamma_2)^j (\gamma_1 a - \gamma_1\gamma_2)^{N-j} \cdot e^{-((\gamma_1-\gamma_2)j+\gamma_2 N)t}\right] \tag{23}$$

The integral expression in the McFadden identity (19), which also has the interpretation of the mean inter-arrival time, starting with an arbitrary time origin [8], can be evaluated from the relationship:

$$\int_0^\infty p(i,t)dt = \left[\frac{1}{i!} \cdot \frac{\partial^i \tilde{\Phi}_p^{'}(z,s)}{\partial z^i}\right]_{z=0,\, s=0^+} \tag{24}$$

where $\tilde{\Phi}_p^{'}(z,0^+)$ is readily obtained by evaluating the Laplace transform of (23) at $s = 0^+$, giving:

$$\tilde{\Phi}_p^{'}(z,0^+) = \frac{1}{a^N(\gamma_1-\gamma_2)^N} \cdot \left[\sum_{j=0}^{N} \frac{\binom{N}{j}(\gamma_1\gamma_2 - a\gamma_2)^j (\gamma_1 a - \gamma_1\gamma_2)^{N-j}}{((\gamma_1-\gamma_2)j + \gamma_2 N)}\right] \tag{25}$$

From the above discussion, we conclude that the main complexity in computing the ith autocorrelation coefficient resides in the evaluation of the ith derivative of (25) at $z = 0$. Using Leibniz's rule for successive differentiation, we have developed a simple recursive algorithm to compute all the serial autocorrelation coefficients. The details can be found in Appendix 2. For the chosen single source parameters, we have computed some serial correlation coefficients for different number of sources, as shown in Figures 12, 13 and 14. The corresponding IDI curves are also obtained as illustrated in Figure 15. These results show that for the case of voice the magnitude of the correlation coefficients gets larger as the number of component processes increases. This suggests that as the number of sources augments, burstiness is reduced due to the statistical smoothing effect while correlation increases.

4. Parameter Estimation of the Proposed Model

So far we have assumed that the number of sources, N, as well as the single source parameters (α, β and λ) are given. In this Section we consider a very interesting statistical problem, namely estimating the four parameters (α, β, λ and N) from the observation of the actual ATM traffic mix. Since it is very hard to derive a likelihood function of the observed ATM traffic under the proposed model, our approach will be based on some statistical matching of the characteristics of the proposed model, expressed in terms of α, β, λ and N, with that of the real observed ATM traffic. Our main goal behind the selection of some specific fitting parameters is to capture the correlation behavior of the real ATM traffic. In addition since these fitting parameters should be measured with reasonable efforts, we have based our estimation technique on the counting, rather than on the inter-arrival time process [8]. The following four statistical characteristics have been selected in order to estimate the parameters (α, β, λ and N) of the proposed model. These are:

- The mean packet arrival rate:

$$\Lambda^p = \lim_{t\to\infty} \frac{E[N_t^p]}{t} = N\lambda\frac{\beta}{\alpha+\beta} \tag{26.1}$$

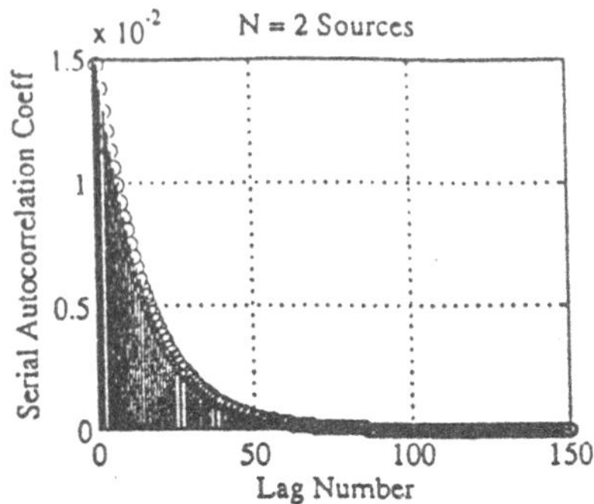

Fig. 12. Serial Autocorrelation Coefficient (N=2)

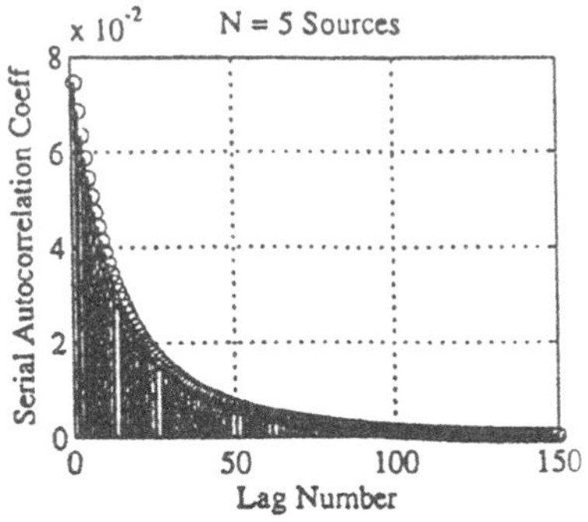

Fig. 13. Serial Autocorrelation Coefficient (N=5)

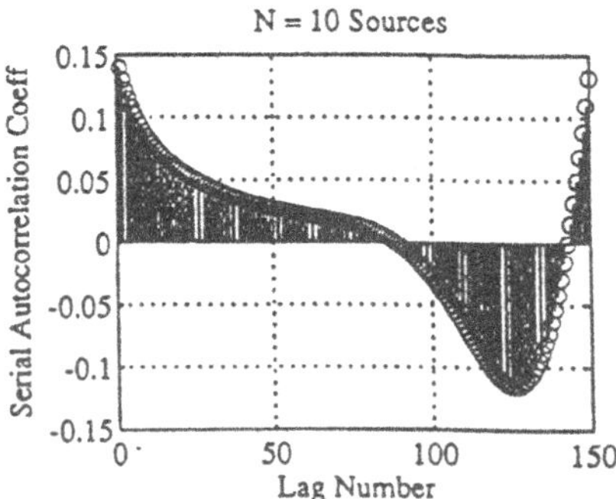

Fig. 14. Serial Autocorrelation Coefficient (N=10)

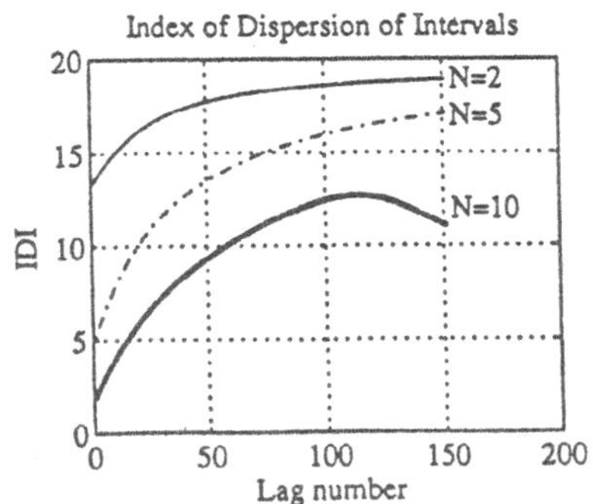

Fig. 15. Index of Dispersion of Intervals

- The long-term variance to mean ratio of the number of arrivals:

$$IDC^{p}(\infty) = C^{2} = 1 + \frac{2\lambda\alpha}{(\alpha+\beta)^{2}} \tag{26.2}$$

- The long-term covariance of the number of arrivals between two adjacent intervals:

$$C_{1}^{p}(\infty) = N\lambda^{2}\frac{\alpha\beta}{(\alpha+\beta)^{4}} \tag{26.3}$$

which can be also estimated from the intercept of the linear portion of the variance-time curve.

- The ratio, B, of the maximum burst cell rate to the overall average cell rate:

$$B \approx \frac{[N\lambda]}{\left[N\lambda\frac{\beta}{\alpha+\beta}\right]} = 1 + \frac{\alpha}{\beta} \tag{26.4}$$

By equating the measured and theoretical values of the above four characteristics, the model's parameters can be easily derived by solving the system of equation:

$$N\frac{\lambda\beta}{\alpha+\beta}=\tilde{a}\ ,\ 1+\lambda\frac{2\alpha}{(\alpha+\beta)^2}=\tilde{b}\ ,\ N\lambda^2\frac{\alpha\beta}{(\alpha+\beta)^4}=\tilde{c}\ ,\ 1+\frac{\alpha}{\beta}=\tilde{d} \tag{27}$$

giving:

$$\lambda=\frac{1}{4}\frac{\tilde{a}\tilde{d}(\tilde{b}-1)^2}{\tilde{c}(\tilde{d}-1)}\ ,\ \alpha=\frac{1}{2}\frac{\tilde{a}(\tilde{b}-1)(\tilde{d}-1)}{\tilde{c}\tilde{d}}\ ,\ \beta=\frac{1}{2}\frac{(\tilde{b}-1)\tilde{a}}{\tilde{c}\tilde{d}}\ ,\ N=\left[\frac{4\tilde{c}(\tilde{d}-1)}{(\tilde{b}-1)^2}\right]^{\pm} \tag{28}$$

where $[x]^{\pm}$ refers to the nearest integer to x, while $\tilde{a}$, $\tilde{b}$, $\tilde{c}$ and $\tilde{d}$ are the corresponding parameters in the actual ATM superposition traffic. These parameters can be estimated, using standard traffic measurement techniques (see for example [8], chapter 5). At the present stage, we are running some simulation tests to confirm the validity of the proposed model and evaluate the accuracy of the approximation.

5. Conclusion

In this paper we have considered a correlated analytical model for the packet arrival process to an ATM multiplexer. The model consists of approximating the aggregate packet arrival process by the superposition of N homogeneous and identical Interrupted Poisson Processes. A detailed statistical analysis of the superposition process has shown that our proposed model is very adequate since it captures most of the correlation and burstiness behaviors that characterize multimedia traffic. In particular we attributed a special attention to the dependence among the packet inter-arrival times in the superposition process. We have also addressed the statistical estimation problem of the proposed model's parameters, based on some statistical matching methods. The exact queuing analysis of the corresponding NG_i/D/1/K queue is left for future research.

Appendix 1 Computing the Second Moment of the Packet Inter-arrival Times

Recall from (16), that the second moment of the inter-arrival times is given by:

$$\overline{Y^2}=2\int_0^\infty tF_c^p(t)dt=2\int_0^\infty t\left(k_1e^{-r_1t}+k_2e^{-r_2t}\right)\left[\frac{k_1r_2e^{-r_1t}+k_2r_1e^{-r_2t}}{k_1r_2+k_2r_1}\right]^{N-1}dt \tag{29}$$

Let $\zeta(t)=1-F_e(t)$, where $F_e(t)$ is as defined in (14.2). Let $Z(s)=L\{\zeta(t)\}$ and $\bar{F}_c^p(s)=L\{F_c^p(t)\}$. In particular we have:

$$\zeta(t)=\frac{1}{(k_1r_2+k_2r_1)^N}\sum_{i=0}^{N}\binom{N}{i}(k_1r_2)^i(k_2r_1)^{N-i}e^{-[(r_1-r_2)i+Nr_2]t} \tag{30}$$

We can avoid performing the integration in (29) by noting that since:

$$F_c^p(t)=-\frac{\overline{X}}{N}\cdot\frac{d\zeta(t)}{dt} \tag{31.1}$$

$$\bar{F}_c^p(s)=-\frac{\overline{X}}{N}\cdot sZ(s) \tag{31.2}$$

then

$$\overline{Y^2}=-2\left[\frac{d\bar{F}_c^p(s)}{ds}\right]_{s=0}=\frac{2\overline{X}}{N}[Z(s)]_{s=0} \tag{32}$$

Taking the Laplace of transform (30) and substituting in (32), we obtain the second moment expression, as given in equation (17.1).

Appendix 2 A Recursive Algorithm for the Computation of All serial autocorrelation Coefficients

First let us rewrite (25) in the form:

$$\tilde{\Phi}_p'(z,0^+) = \frac{1}{a^N}\sum_{j=0}^{N}\binom{N}{j}\frac{A_j \cdot B_j}{C \cdot D_j}$$

where: $A_j = (r_1 r_2(1-z) - a\gamma_2)^j$, $B_j = (a\gamma_1 - r_1 r_2(1-z))^{N-j}$

$$C = (\gamma_1 - \gamma_2)^N \text{ and } D_j = (\gamma_1 - \gamma_2)j + \gamma_2 N$$

In what follows, whenever a superscript (i) is appended to a symbol, it will refer to that symbol's i^{th} derivative.

A.1. Computing the ith derivative of γ_1 and γ_2

To start with, we note that it would be convenient to write (21.2) in the form:

$$\gamma_1 = \frac{(r_1+r_2)-(r_1k_1+r_2k_2)z+\sqrt{mz^2+bz+c}}{2}$$

where $m = (r_1k_1+r_2k_2)^2$, $b = 2(r_1-r_2)(r_2k_2-r_1k_1)$, $c = (r_1-r_2)^2$

From the above, we have $\gamma_1(0) = r_1$ and $\gamma_1^{(1)}(0) = -r_1k_1$. Next let:

$$y = \frac{\sqrt{mz^2+bz+c}}{2} \tag{33}$$

Then $\forall i \geq 2$ $\qquad \gamma_1^{(i)} = y^{(i)}$

From (30) we have:

$$y^2 = \frac{mz^2+bz+c}{4}$$

Differentiating both sides twice with respect to z, we get:

$$yy^{(1)} = \frac{mz}{4}+\frac{b}{8} \tag{34}$$

$$[y^{(1)}]_{z=0} = \frac{r_2k_2-r_1k_1}{2} \quad , \quad \gamma_1^{(2)} = y^{(2)} = \frac{\frac{m}{4}-(y^{(1)})^2}{y}$$

By applying Leibniz's rule and taking the $(i-1)^{th}$ derivative $(i>2)$ on both sides of (34), all remaining higher order derivatives of γ_1 can be recursively computed:

$$\forall i \geq 3; \qquad \gamma_1^{(i)} = y^{(i)} = -\frac{\sum_{k=0}^{i-2}\binom{i-1}{k}y(i-k-1)\cdot y(k+1)}{y}$$

For γ_2 we have: $\qquad [\gamma_2]_{z=0} = r_2$, $[\gamma_2^{(1)}]_{z=0} = -r_2k_2$

$$\forall i \geq 2; \qquad \gamma_2^{(i)} = \gamma_1^{(i)}$$

A.2. Computing the ith derivative of A_j, B_j, C and D_j

First let $P = r_1r_2(1-z) - a\gamma_2$ and let $Q = \gamma_1 a - r_1r_2(1-z)$. Then we have:

$$P^{(1)} = -r_1r_2 - a\gamma_2^{(1)} \quad \text{and} \quad Q^{(1)} = a\gamma_1^{(1)} + r_1r_2$$

$\forall i \geq 2$
$$P^{(i)} = -a\gamma_2^{(i)} \quad \text{and} \quad Q^{(i)} = a\gamma_1^{(i)}$$

Since $A_j = P^j$ then its first derivative is:

$$A_j^{(1)} = j \cdot P^{(1)}P^{j-1}$$

while the remaining higher derivatives are recursively obtained by noting since:

$$A_j^{(1)}P = jA_jP^{(1)}$$

then taking the $(i-1)^{th}$ derivative on both side, and once again using the Leibniz's rule, gives: $\forall i \geq 2$:

$$A_j^{(i)} = \frac{1}{P}\{\, j[P^{(i)}A_j + P^{(1)}A_j^{(i-1)}] - P^{(i-1)}A_j^{(1)} + \sum_{k=1}^{i-2}\binom{i-1}{k}[jP^{(k+1)}A_j^{(i-k-1)} - P^{(k)}A_j^{(i-k)}]\,\}$$

Using a similar approach we get the following recursive relationships for the derivatives of B_j, C and D_j:

$$B_j = Q^{N-j}$$

$$B_j^{(1)} = (N-j)Q^{N-j-1}Q^{(1)}$$

$\forall i \geq 2$
$$B_j^{(i)} = \frac{1}{Q}\{(N-j)\,[Q^{(i)}B_j + Q^{(1)}B_j^{(i-1)}] - Q^{(i-1)}B_j^{(1)}$$
$$+ \sum_{k=1}^{i-2}\binom{i-1}{k}[(N-j)Q^{(k+1)}B_j^{(i-k-1)} - Q^{(k)}B_j^{(i-k)}]\}$$

Next:

$$C^{(1)} = N(\gamma_1^{(1)} - \gamma_2^{(1)})(\gamma_1 - \gamma_2)^{N-1}$$

Let $\kappa = \gamma_1 - \gamma_2$. Then $\forall i \geq 2$:

$$C^{(i)} = \frac{1}{\kappa}\left\{N(\kappa^{(i)}C + \kappa^{(1)}C^{(i-1)}) - \kappa^{(i-1)}C^{(1)} + \sum_{k=1}^{i-2}\binom{i-1}{k}[N\kappa^{(k+1)}C^{(i-k-1)} - \kappa^{(i-k-1)}C^{(k+1)}]\right\}$$

Finally:

$\forall i \geq 1$
$$D_j^{(i)} = (\gamma_1^{(i)} - \gamma_2^{(i)})j + N\gamma_2^{(i)}$$

A.3. Computing the ith derivative of $U_j = A_j \cdot B_j$, $V_j = C \cdot D_j$ and $R_j = \frac{U_j}{V_j}$

A straight forward application of Leibniz's rule yields the following relationships:

$$U_j^{(i)} = A_j^{(i)}B_j + A_jB_j^{(i)} + \sum_{k=1}^{i-1}\binom{i}{k}A_j^{(i-k)}B_j^{(k)} \quad , \quad V_j^{(i)} = C^{(i)}D_j + CD_j^{(i)} + \sum_{k=1}^{i-1}\binom{i}{k}C^{(i-k)}D_j^{(k)}$$

$$R_j^{(i)} = \frac{U_j^{(i)} - R_jV_j^{(i)} - \sum_{k=1}^{i-1}\binom{i}{k}V_j^{(i-k)}R_j^{(k)}}{V_j}$$

A.4. Algorithm for Computing the Autocorrelation Coefficients

Using the results of Sections A.1, A.2 and A.3, the first p^{th} serial autocorrelation coefficients are computed using the following recursive algorithm (note that all derivatives are evaluated at $z = 0$):

Begin
Enter the single source parameters (α, β and λ) *and the number of sources, N.*
Compute r_1, r_2 and k_1 (k_2).
Find $\overline{Y}$ and $Var(Y)$ *using (13) and (17.1).*
Find the first derivative of γ_1, γ_2, y and C.
for j from 0 to N do
Compute the first derivative of A_j, B_j, D_j, U_j, V_j and R_j
end for
for i from 2 to p do
Compute $y^{(i)}$, $\gamma_1^{(i)}$, $\gamma_2^{(i)}$ *and* $C^{(i)}$.
for j from 0 to N do
Compute $A_j^{(i)}$, $B_j^{(i)}$, $D_j^{(i)}$, $U_j^{(i)}$, $V_j^{(i)}$ *and* $R_j^{(i)}$,
end for
end for
for i from 1 to p do
Compute:

$$\left[\tilde{\Phi}_p^{'(i)}(z,0^+)\right]_{z=0} = \frac{1}{a^N}\sum_{j=0}^{N}\binom{N}{j}R_j^{(i)}$$

Compute:

$$\rho_i = \frac{\overline{Y}}{Var(Y)} \cdot \left(\frac{\left[\tilde{\Phi}_p^{'(i)}(z,0^+)\right]_{z=0}}{i!} - \overline{Y}\right)$$

end for
end

References

1 E. G. Enns: A Stochastic Superposition Process and an Integral Inequality for Distributions with Monotone Hazard Rates , Aust. J. Statist., No.12, pp. 44-49, 1970.
2 A. J. Lawrance: Dependency of Intervals Between Events in Superposition Processes , J. R. Statist. Soc, No.2, pp. 307-315, 1973
3 K. Sriram and W. Whitt: Characterizing Superposition Arrival Processes in Packet Multiplexers for Voice and Data , IEEE. JSAC, Vol. SAC-4, No.6, Sept 1986.
4 E. Cinlar: Superposition of Point Processes. In: P. A. W. Lewis (ed): Stochastic Point Processes: Statistical Analysis, Theory, and Applications, Wiley-Interscience, pp. 549-606, 1972.
5 H. Kobayashi: Performance Issues of Broadband ISDN: Part I: Traffic Characterization and Statistical Multiplexing , ICCC 90, New Delhi, India, pp. 349-354.
6 A. Kuczura: The Interrupted Poisson Process as an Overflow Process , The Bell System Technical Journal, Vol.52, No.3, pp. 437-448, 1973.
7 D. R. Cox: Renewal Theory, London: Methuen, 1962.
8 D. R. Cox and P. A. Lewis: The Statistical Analysis of Series of Events, John Wiley & Sons, 1966.
9 W. Whitt: Approximating a Point Process by a Renewal Process : Two Basic Methods , Oper. Res, Vol 30, No.1, pp. 125-147, Jan-Feb, 1982.
10 J. A. McFadden: On the Lengths of Intervals in a Stationary Point Process , J. R. Statist. Soc. B, No.24, pp. 364-382, 1962.

Throughput Increase of ALOHA-based Systems with Multiuser Sequence Detection *

J. A. Fergus Ross[1] and Desmond P. Taylor[2]

[1] Communications Research Laboratory
McMaster University
1280 Main Street West
Hamilton, Ontario
Canada, L8S 4K1
e-mail: ross@comres2.Eng.McMaster.CA
[2] Dept. of Electrical and Electronic Engineering
University of Canterbury
Private Bag 4800
Christchurch, New Zealand

Abstract. Multi-user detectors have been suggested to improve the performance of CDMA based random-access systems [1, 2, 3, 4]. It is shown that with the maximum-likelihood (ML) multi-user sequence detector (MUSD) [1], spreading codes are often unnecessary in the AWGN channel; the MUSD can separate interfering packets of data using the same symbol pulse. A bit error probability of 10^{-4} is attained at signal-to-noise ratios of 12 to 30 dB provided the packets are offset slightly. The offset may be considered as a probable outcome in many systems, with failure (collision) occurring when the offset is too small. Alternatively, the desired offset may be designed into the system in a more controlled way. A generalization of ALOHA is given where 'slot'-collisions are replaced by 'subslot'-collisions. Throughputs obtained depend on the complexity available. For example, relative to a single user continuously transmitting, 91% throughput is achieved with 4 subslots, which requires an 8-state Verdú-Viterbi trellis detection algorithm, while 159% throughput is achieved with 10 subslots, requiring 512 states. Potentially, systems with K-subslots obtain K carrier-sense channels when carrier-sensing is available for each subslot.

1 Introduction

The collision event is the main impediment to throughput in random-access channels (RAC). For example, in the ALOHA network, whenever two or more user's packets interfere, called a collision, retransmission is required by each user and is performed at a random later time. If each packet's duration is l seconds, there is an interval of $2l$ during which two users may transmit and interfere. A simple improvement occurs in slotted-ALOHA where users may only transmit

* This work is based on writings found in the Ph.D. thesis of the first author.

at the beginning of frame intervals, spaced in time by l, reducing the interference interval to l. The maximum throughput for slotted-ALOHA, with equal retransmission probabilities for all users, is 37%. In a sense, the ALOHA protocol is the most fundamental protocol for random-access channels, with many other systems adding to the basic transmit-when-necessary strategy to improve performance, such as Ethernet [5]. The viewpoint of this work is that if the collision rate can be reduced, the throughput of ALOHA-based systems may be increased.

Interfering signals may be separated by utilizing signal waveforms of mutually low cross-correlation, however, increased bandwidth is necessary to maintain a set of nearly orthogonal signals (dimensionality is proportional to the time-bandwidth product, WT). Alternatively, it is suggested here that with maximum likelihood detection the different users may use the same waveforms (or at least a small set relative to the spread-spectrum approach above) with the system exploiting the fact that the symbols are unlikely to arrive in near synchronism. Bandwidth expansion is not necessary, nor are spreading codes. The data is often recoverable if the colliding packets do not completely overlap.

The general model of interest is one of linearly combined signals corrupted by additive white Gaussian noise (AWGN). It is assumed that there are K users, each transmitting a packet of L binary symbols. The received signal $R_t(\mathbf{b})$, where $\mathbf{b}$ is the time-ordered transmitted bit-sequence of all users, is modelled by[3]

$$\begin{aligned} R_t(\mathbf{b}) &= \sum_{i=0}^{L-1} \sum_{j=1}^{K} b_j(i) s_j(t - iT - \tau_j) + n(t) \\ &= S_t(\mathbf{b}) + n(t) \end{aligned} \tag{1}$$

where $\mathbf{b} = (b(0), b(1), \ldots, b(L-1))$ with $b(i) = (b_1(i), b_2(i), \ldots, b_K(i))$ and $b_j(i) \in \{+1, -1\}$; $\|s_j\| = 1$; $s_j(t) = 0$, t outside $[0, T)$; $\tau_j \in [0, T)$, numbered such that $0 \le \tau_1 \le \cdots \le \tau_K < T$. The AWGN is modelled by $n(t)$.

To begin the investigation, consider the error probability of K evenly- offset binary users transmitting simultaneously with rectangular pulses of duration T. The first user begins transmission at $t = 0$, the second at $t = T/K$, the third at $t = 2T/K$, and so on, such that each is separated by T/K. This may be modelled by an ISI channel of length K using Forney's whitened matched-filter (WMF) approach [6].

The pulse autocorrelation function for the K overlapping users is,

$$\begin{aligned} R(D) &= \frac{1}{K} D^{-(K-1)} + \frac{2}{K} D^{-(K-2)} + \frac{3}{K} D^{-(K-3)} + \cdots + \frac{K-1}{K} D^{-1} \\ &\quad + D^0 + \frac{K-1}{K} D^1 + \cdots + \frac{1}{K} D^{K-1} \\ &= \frac{1}{K} D^{-(K-1)} (1 + 2D^1 + 3D^2 + \cdots + (K-1) D^{K-2} + \\ &\quad K D^{K-1} + (K-1) D^K + \cdots + D^{2K-2}) \end{aligned}$$

[3] This notation, commonly seen in the literature, follows that of Verdú.

$$= \frac{1}{K} D^{-(K-1)} (\sum_{i=0}^{K-1} D^i)^2 \tag{2}$$

Thus $R(D) = f(D)f(D^{-1})$ where

$$f(D) = \frac{1}{\sqrt{K}}(1 + D + D^2 + \cdots + D^{K-1}), \tag{3}$$

and $f(D)$ defines the white noise equivalent channel.

It is noted that according to Forney [6], the $1 + D$ channel has a minimum distance output error sequence of the form $\pm(1 - D^n)$, $n \geq 1$, caused by the input error sequence $\pm(1 + D + \cdots + D^{n-1})$, exactly the form of Eq. 3. It thus appears that the minimum distance output error sequence for the channels described by Eq. 3 is $\pm(1 - D^n)$ (reverse the roles of channel and input in Forney's argument). This also holds true by inspection. The constant factor, $\frac{1}{\sqrt{K}}$, can equivalently be considered to scale the noise variance at the output of the WMF by $K\sigma^2$. The minimum squared distance between output signals for the $\{+1, -1\}$ alphabet in the whitened equivalent channel is thus $(1 - (-1))^2 + ((-1) - 1)^2 = 8$. The SNR loss over binary signalling is, for $K > 1$,

$$\begin{aligned}\gamma(K) &= \frac{d^2_{min,overlap}/\sigma^2_{WMF}}{d^2_{min,binary}/\sigma^2} \\ &= \frac{8/K\sigma^2}{4/\sigma^2} \\ &= \frac{2}{K}\end{aligned} \tag{4}$$

or, in decibels,

$$\gamma_{dB}(K) \approx 3 - 10\log_{10} K. \tag{5}$$

Examples: The equivalent WMF is $\frac{1}{\sqrt{2}}(1 + D)$ for two users, or, alternatively, $1 + D$ followed by a white noise source of variance $2\sigma^2$. The closest squared distance is 8, thus there is no SNR loss relative to the single user system. The SNR loss over binary signalling for a ten user system is

$$\gamma_{dB}(10) = 3 - 10\log_{10} 10 \tag{6}$$

$$= -7 \tag{7}$$

It is noted that the magnitude of the autocorrelation of the Manchester pulse (used in coaxial cable Ethernet [5] systems) is always less than or equal to the magnitude of the autocorrelation of the rectangular pulse over the same duration. Maximum likelihood detection is thus expected to perform at least as well in Manchester coded systems.

A general algorithm for ML detection of interfering sequences of data originating from a multiple of users was developed by Verdú [1]. The algorithm requires 2^{K-1} states for binary data and does not require a whitened matched-filter. Motivated by the error-performance estimate of Eq. 5 and the efficiency of the Verdú algorithm, a proposal for the modification of ALOHA-like random-access systems is presented.

2 Proposal: Generalized ALOHA

The following proposal for the RAC is investigated. Active users must wait for frame intervals before transmission as in slotted-ALOHA. At the beginning of the frame each transmitting user randomly picks a frame offset of a fraction of a symbol period and then transmits the packet starting at that offset, Fig. 1. These slight offsets are called *subslots* and the proposal is named the *slotted-subslotted* modification. An advanced detection processor such as the ML detector of Verdú is assumed to separate the signals.

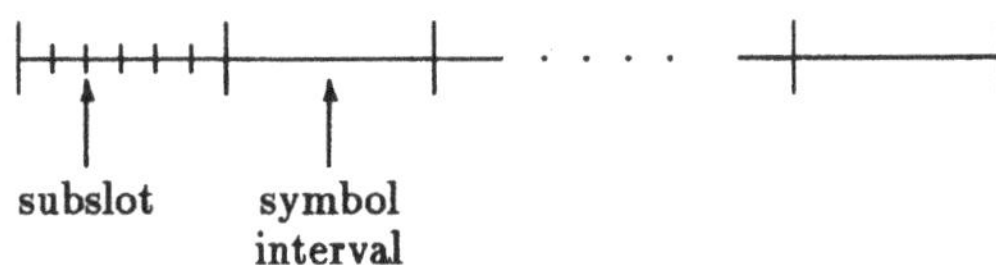

Fig. 1. Frame

2.1 Çarrier Systems

In carrier systems it is probably most reasonable to assume that carrier phases cannot be kept in synchronism, thus the signals $\{s_j\}$ become a function of phase, ϕ_j, and frequency, ω_j, ie., $\{s_j(t;\omega_j;\phi_j)\}$, where ϕ_j, uniform over $[0, 2\pi]$, models phase, and $\omega_j = \omega + \delta\omega_j$ models frequency offsets between oscillators.

The random parameters only serve to decrease the correlation amongst interfering signals. Not only does this improve performance but it also may allow the selection of less precise, and therefore less costly, oscillators. The additional effort placed on the receiver may be a reasonable tradeoff in many-to-one communications systems. One particular incoherent approach to demodulation is suggested below.

2.2 Demodulation: The Pilot Signal

A procedure must be devised to (1) determine the correlation coffficients of the correlation matrix G used by the MUSD [1] , and (2) obtain estimates of matched filter outputs for each offset symbol period. Theoretically, estimates of the signal parameters could be obtained through analysis of the received packet, but data would be lost during the time interval in which the number of users, signal pulse shapes, and carrier phases and frequencies were estimated. A more practical approach to demodulation is now discussed. The method is shown to be very effective under stable channel conditions, losing as little as 1dB in bit energy to ideal coherent recovery.

It is proposed that an interval prior to actual data transmission is reserved, called the prefix. The prefix is divided into N sub-intervals. A user signifies that it is transmitting in the jth subslot by first transmitting a pulse pilot signal in the jth sub-interval of the prefix . The pilot signal solves two other problems as shown in Fig. 2. The receiver may use the signals found in the prefix to determine the correlation matrix G used by the MUSD. The pilot signal is also used as a reference signal for a form of differential detection.

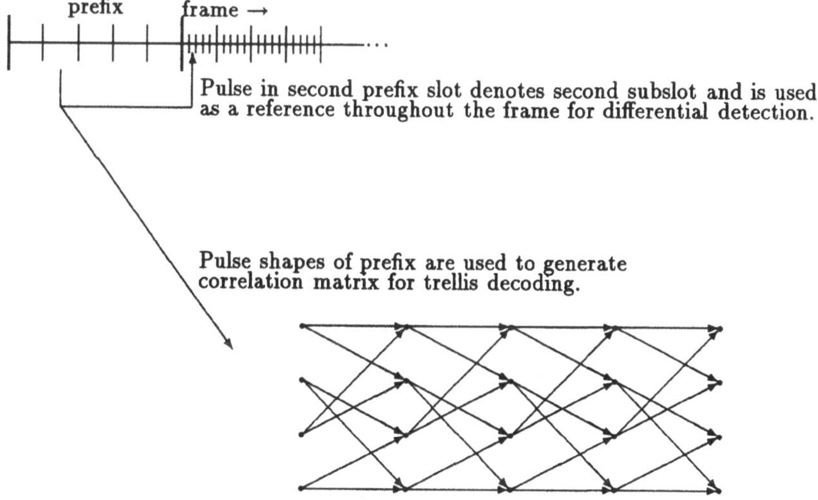

Fig. 2. Use of pilot signal in demodulation and detection.

It is assumed that the channel varies slowly with respect to the data rate, if at all. Then the pilot symbol may be used to provide a succession of estimates to the ideal matched filter outputs of each offset symbol period. For example, denote a particular pilot symbol by $s_p(t)$ and the noise waveform associated with it by $n_p(t)$. Suppose there are K users transmitting in a given frame, disturbed by noise $n(t)$. Then one estimate of a matched-filter output is

$$y = \int_T (s_p(t) + n_p(t))(\sum_{j=1}^{K} s_j(t) + n(t))\, dt. \tag{8}$$

It is insightful to rewrite this correlation into its component terms.

$$y = \int_T s_p(t) \sum_{j=1}^{K} s_j(t) + \int_T s_p(t) n(t)$$

$$+\int_T \sum_{j=1}^{K} s_j(t)n_p(t) + \int_T n_p(t)n(t) \tag{9}$$

Substitute

$$s_p(t) = \sqrt{a}\alpha_p(t), \;\; s_j(t) = \sqrt{b}\alpha_j(t) \tag{10}$$

where

$$\|\alpha\| = 1, \tag{11}$$

then Eq. 9 becomes

$$\begin{aligned} y = \sqrt{ab} + \sqrt{ab}\int_T \sum_{j \neq p} \alpha_p \alpha_j + \sqrt{a}\int_T \alpha_p(t)n(t) + \\ \sqrt{b}\int_T \sum_{j=1}^{K} \alpha_j(t)n_p(t) + \int_T n_p(t)n(t) \\ \approx \sqrt{ab} + c_p + \sqrt{a}n + \sqrt{b}n_p \end{aligned} \tag{12}$$

where c_p represents the summed projection of the interfering users, and n and n_p are the corresponding noise random variables. The fourth term of Eq. 9 is assumed neglible.

We concentrate on the symbol-by-symbol demodulation, leaving the task of processing c_p to the sequence detection algorithm. It is thus desirable to maximize the performance parameter

$$\kappa = \frac{ab}{a\sigma_n^2 + b\sigma_{n_p}^2} \tag{13}$$

for any channel use.

For example, with orthogonal signalling, $\sigma_{n_p}^2 = K\sigma_n^2$, and Eq. 13 becomes

$$\kappa = \frac{ab}{(a + Kb)\sigma_n^2} \tag{14}$$

Assuming finite power with the average energy per symbol normalized to one, and an L-symbol packet,

$$a + (L-1)b = L, \tag{15}$$

Eq. 14 can be maximized. This occurs at

$$b = \frac{L(L-1) - L\sqrt{K(L-1)}}{L^2 - 2L - LK + K + 1}. \tag{16}$$

Note it is common in single user channels to use differential detection between adjacent symbols. Setting $K = 1$, we find

$$b = \frac{L(L-1) - L\sqrt{L-1}}{(L-1)(L-2)} \tag{17}$$

maximizes the SNR parameter κ.

In the limit, as L increases, the 3dB loss from differential detection in single-user channels may be recovered. In a practical application the value of L is determined by the coherence time of the channel. If the channel is very stable and slowly varying relative to the data rate, a greater value of L may be used. On the other hand, very dynamic channels imply lesser L, or another modulation/reception scheme altogether. A plot of the maximum value of κ, Eq. 14, versus L is shown in Fig. 3, for $K = 1$ and normalized noise variance. A 1 dB improvement is achieved at $L = 11$ and 2 dB is recovered by $L = 60$ (2.5 dB by $L = 260$, not shown), for single user channels.

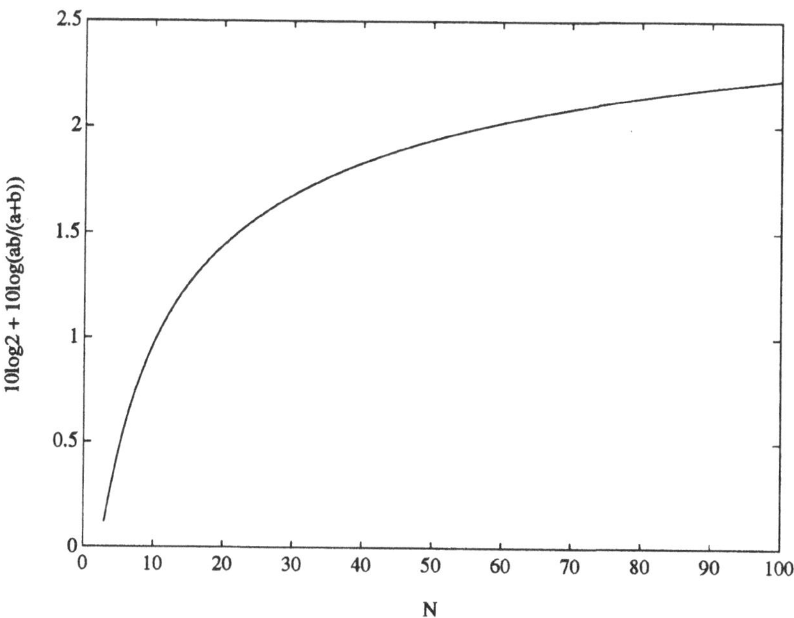

Fig. 3. Maximum values of κ at different L for $K = 1$

3 Simulation Results: P(e) Curves

A predetermined value for K must be chosen to set a and b since the number of users transmitting is unknown to the users in advance. Reference to pilot-symbol assisted demodulation in the simulation results are for Eq. 14 maximized with $K = 1$ and $L = 100$. There is an associated expected loss from optimum performance as the actual number of interfering users increases. Each user transmits 100 rectangular-pulse symbols per packet in the simulation results, unless stated otherwise. Note, SNR$=2Eb/N_0$ in the simulation.

A comparison of non-optimized ($a = b$) pilot symbol assisted detection/ demodulation to optimized pilot-symbol assisted detection/demodulation is shown in Fig. 4. The non-optimized pilot symbol detection is closest to standard differential detection. The benefit of the pilot-symbol is clear, even for the larger number of interfering users. There is a gain of 5 dB for two users and a gain of 9 dB for four users at P(e)=10^{-4}.

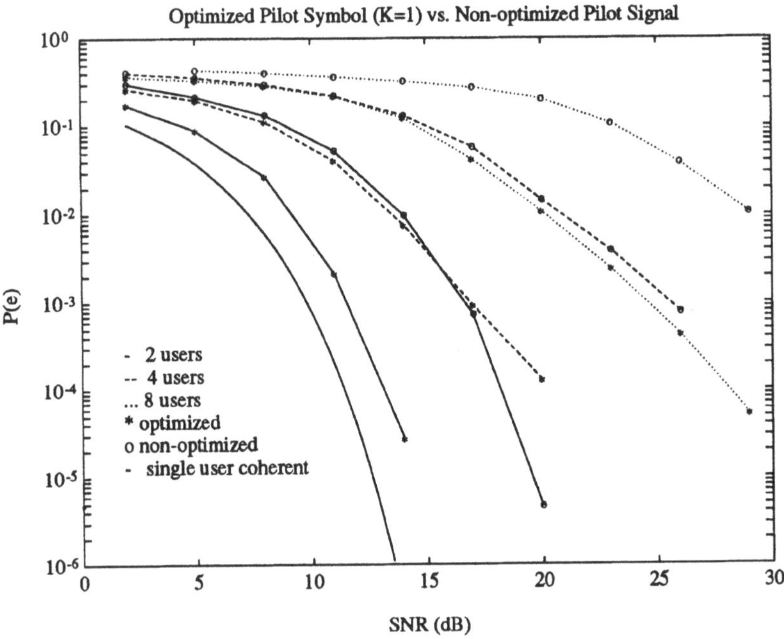

Fig. 4. Comparison of optimized pilot-assisted detection vs. non-optimized pilot-assisted detection

It appears to be difficult to design a system that could coherently demodulate the interfering users. Nevertheless, ideal results are shown as the limiting case. In Figs. 5 and 6 the performance of baseband, non-optimized and optimized pilot symbol detection, and coherent detection systems are shown for 2 and 4 users, respectively. Note the coherent detection system has a lower $P(e)$-curve than the baseband system as it benefits from the larger dimensionality associated with the random phase offsets. The optimized pilot-symbol results are about 2dB poorer than the coherent system while the non-optimized system suffers a loss of about 7dB. The optimized pilot-symbol approach suffers more loss in the 4-user case, about 6.4 dB at P(e)=10^{-4}, but is significantly more efficient than the non-optimized system which has lost more than 14 dB to the coherent results in this case.

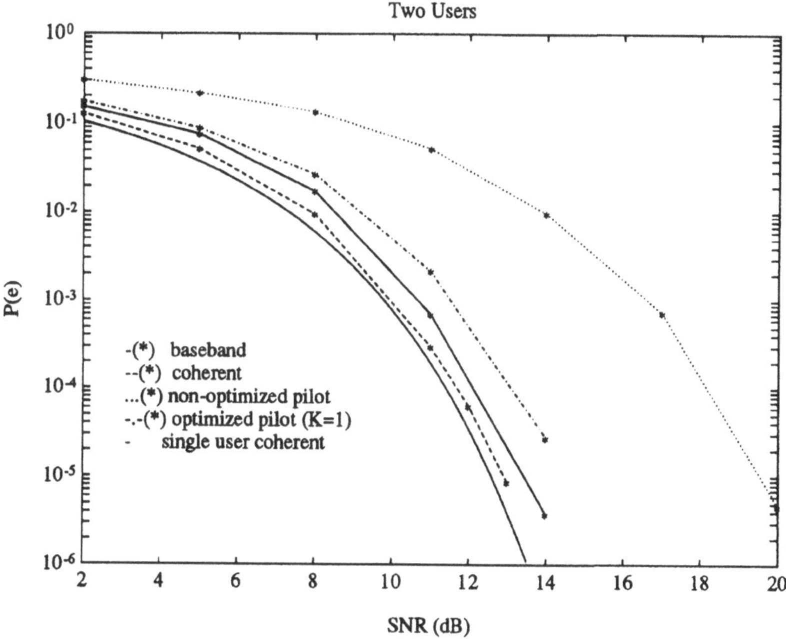

Fig. 5. Different detection methods, 2 users.

Simulation results of baseband signalling with the raised-cosine time-limited pulse of duration T were also obtained, Fig. 7. This pulse was chosen for its rapid squared-magnitude spectral rolloff which varies inversely with f^6 [7]. Compared to the error-performance of baseband rectangular pulse signalling (not shown), the more rounded pulse shape improves detection performance when there are under 7 evenly-spaced users since there is less correlation between the signals. In particular, the two-user system is almost identical in performance to a single-user system. The situation is reversed beyond 7 users as the autocorrelation function increases more rapidly for the raised-cosine pulse at smaller offsets.

A bandwidth and power comparison to more traditional signalling is useful. At baseband, the raised-cosine time-limited pulse of duration T_t has a 99% bandwidth of approximately $1.43/T_t$, while a 50% rolloff raised-cosine frequency pulse of symbol period T_f has a 99% bandwidth of about $0.55/T_f$. The time-limited pulse obtains 10 bits/T_t with $P(e) = 4 \times 10^{-5}$ at about 22dB SNR according to Fig. 7, or 7.14 bits/sec/Hz at 22dB SNR. A frequency-limited pulse could operate with a symbol period of $0.385T_t$ in the same bandwidth. The error-rate of 4×10^{-5} is attained by 8-PAM and 16-PAM at approximately 21dB and 25.8 dB SNR, respectively. Thus PAM achieves about 3.13 bits/symbol (taking logarithms into account) at 22dB SNR, obtaining 5.69 bits/sec/Hz. Without further coding, the fractionally-shifted pulse modulation appears to be more efficient.

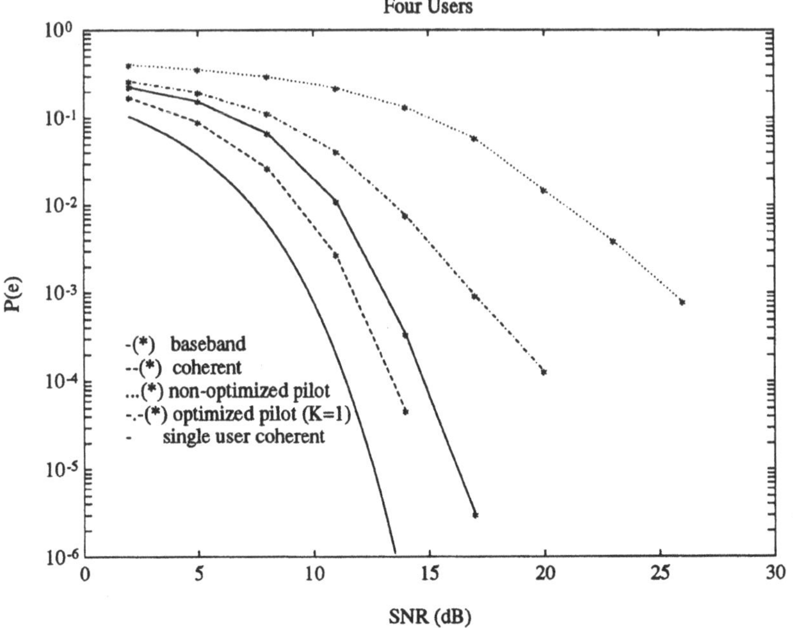

Fig. 6. Different detection methods, 4 users.

4 Throughput

Channel throughput highly depends on the communication protocol in operation. The actual choice of protocol is strongly influenced by the physical channel length. In a system of short physical channel length, it is possible for each receiver to monitor the channel and sense when other users have selected the same subslot. This method, commonly applied in multiple-access communications, is known as carrier-sense multiple-access with collision detection (CSMA/CD). If carrier sensing is available in a K-subslot system then effectively K CSMA/CD channels are obtained. The prefix interval would be particulary useful for resolving conflicts.

In a channel of longer physical length, such as a satellite link, carrier sensing may not be possible as the round-trip time delay may exceed the packet duration. In this case, two or more users selecting the same subslot will transmit unknowingly for the whole packet. Simulation trials showed that unfortunately all of the users' data is lost by the MUSD for this 'true' collision.

There are other channel environments that have not been considered in this work that may result in less catastrophic circumstances when users share the same subslot. This is left to future work. Nevertheless, the limiting throughput rate of a slotted-subslotted system subject to true collisions is much greater than slotted-ALOHA as will be shown. This can also be anticipated from the through-

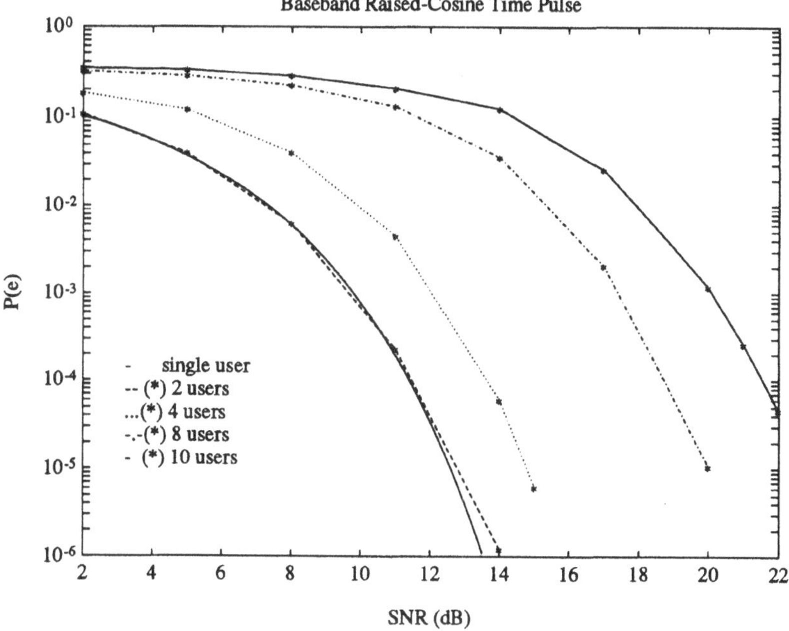

Fig. 7. P(e) curves for 2,4,8, and 10 overlapping users.

put versus load curves of ALOHA, slotted-ALOHA, and 10 subslot-subslotted ALOHA subject to true collisions, all under Poisson arrivals, shown in Fig. 8.

4.1 Stability Considerations

Up to this point the ALOHA protocol was mentioned as the basic protocol to which multi-user detection is suggested for improving throughput. This protocol in its purest form actually is unstable [8, 9]. For example, the curves of Fig. 8 show that throughputs tend to zero as loads increase. In the operation of these protocols, from time to time loads become heavy and cause a high collision rate. Consequently retransmissions rise, adding to the load and further reducing throughput. At some point the heavy load persists long enough to drive the protocol to a constant state of zero throughput. The loop between the transmitter and receiver which controls the rate of retransmission needs to be closed to assure stability. With optimal decentralized control, the throughput rate of e^{-1} for slotted ALOHA is achieved [10]. This method is considered below. Other techniques such as 'collision-resolution' algorithms which generate unequal retransmission probabilities amongst the set of users are also possible [11].

In decentralized control, information is provided to the users regarding the number of packets waiting for retransmission, termed 'backlog'. Three important theorems from the work of Ghez, Verdú, and Schwartz (GVS) [10, 12] are used

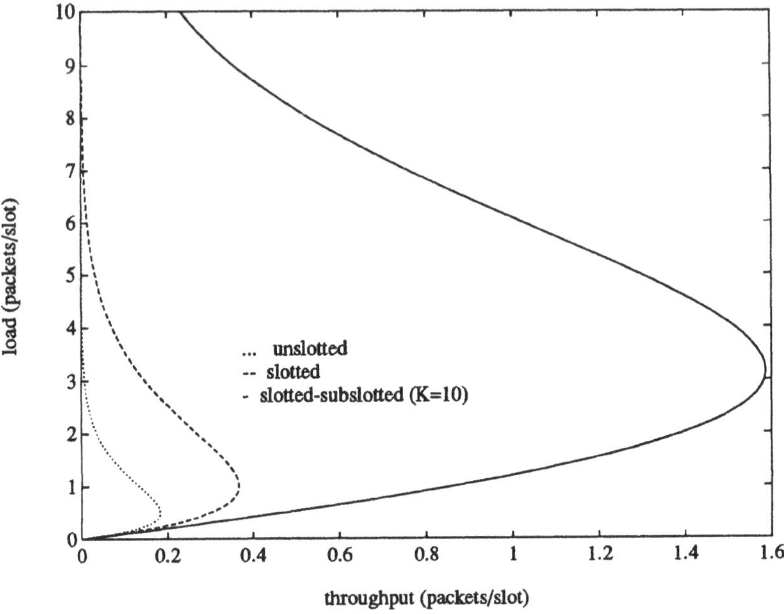

Fig. 8. Throughput vs. load comparison of pure, slotted, and slotted-subslotted ALOHA.

to show the instability of the open loop protocol and define the maximum stable throughput for a closed loop K-subslot system with decentralized control for both full backlog information and partial backlog information.

Define

$\epsilon_{nk} \stackrel{\text{def}}{=} \text{P}[k \text{ packets are correctly received} \mid n \text{ packets are transmitted}]$,

then the average number of successfully processed packets given n packets are received is

$$C_n \stackrel{\text{def}}{=} \sum_{k=1}^{n} k\epsilon_{nk}. \tag{18}$$

With the proposed K-subslot multi-user detection scheme, either all of the interfering packets are successfully separated or none are, giving

$$\epsilon_{nk} = \begin{cases} 0, & n \neq k,\ K \geq 1 \\ \frac{(K-1)!}{(K-n)!K^{n-1}}, & n = k,\ n \leq K \end{cases}$$

and

$$C_n = \frac{n}{K^{n-1}} \frac{(K-1)!}{(K-n)!} \quad \text{for } n \leq K. \tag{19}$$

The condition for open-loop stability with mean arrival rate λ and C_n defined in Eq. 18 is determined by

Theorem 1 GVS [12]. *If C_n has a limit $C = \lim_{n\to\infty} C_n$, then the system is stable for all arrival distributions such that $\lambda < C$ and is unstable for $\lambda > C$. This also holds if C is infinite: if $\lim_{n\to\infty} C_n = +\infty$, then the system is always stable.*

Since $C_n = 0$, $n > K$, there is no open-loop stable throughput for this system. Ultimately, retransmissions will dominate the channel and the throughput will go to zero. The instability of the ALOHA system is also noted by setting $K = 1$.

It is imperative that some state information be supplied to the users regarding the other users' channel requirements. It is assumed in ALOHA that the users are notified of the success/failure of their packet's reception. In addition, the protocol now considered requires that data is sent to all users regarding the status of the last frame whether or not they transmitted during it. This (usually incomplete) information will be used to estimate the backlog of packets waiting for retransmission which, in turn, will affect the transmission probability of new packets. Essentially, as the backlog increases, the new packet transmission probability is reduced, thus it is more likely for future transmissions to be successful. This procedure is known as decentralized control because the transmit probabilities are calculated by each user rather than at a central station. The following theorem of GVS is stated to define the maximum throughput.

Theorem 2 GVS [10]. *There exists a retransmission probability p_n^* that minimizes the expected backlog increase when the backlog is equal to n. With such a retransmission probability, the system is stable for $\lambda < \eta_c$ and unstable for $\lambda > \eta_c$, with*

$$\eta_c = \sup_{x\geq 0} e^{-x} \sum_{n=1}^{\infty} C_n \frac{x^n}{n!}. \tag{20}$$

The maximum throughput for decentralized control of a K-subslot system with C_n defined in Eq. 19 is now calculated from the result of Theorem 2 (GVS). Define

$$t(x) \stackrel{\text{def}}{=} e^{-x} \sum_{n=1}^{\infty} C_n \frac{x^n}{n!}. \tag{21}$$

Substituting C_n of Eq. 19 into Eq. 21 and differentiating with respect to x, we find

$$\frac{dt}{dx} = 1 + \left(\sum_{n=1}^{K-1} \frac{1}{n!} x^n K^{-(n-1)} \frac{(K-1)!}{(K-n+1)!} (K - n - n^2) \right) - \frac{x^K}{K^{K-1}}. \tag{22}$$

The positive zeros of this equation for $K = 1, 2, 3$, and 4 occur at $x = \sqrt{K}$. Plots (not shown) of Eq. 21 for greater K showed that $x = \sqrt{K}$ does indeed locate the maximum. For example, the maximum stable throughputs for K ranging from 1 to 20 are shown in Table 1. Note that even the two subslot case involving only a trellis with two states has a maximum stable throughput 60% greater than slotted-ALOHA. The maximum throughput is greater by more than a factor of 4 if the 10 subslot system is used.

No. of Subslots	Maximum Throughput
1	0.367879
2	0.586936
3	0.762424
4	0.913513
5	1.04834
6	1.17131
7	1.28512
8	1.39156
9	1.49193
10	1.58716
11	1.67797
12	1.76493
13	1.84850
14	1.92905
15	2.00688
16	2.08225
17	2.15538
18	2.22647
19	2.29567
20	2.36313

Table 1. Maximum Throughputs

A variety of feedback models less costly than providing complete backlog information have been proposed and analyzed in the recent literature. For example, Mahravari [13] discusses a ternary feedback system known as zero/success/collision (0,S,C) feedback, where all users are notified if the last frame resulted in a collision (C), a successful transmission (S), or no transmission (0). Alternatively, Mahravari combined the success and collision categories for something/nothing (S/N) feedback. Binary feedback is notable as it is the simplest possible. The binary feedback procedure of Ghez et al is assumed below [10]. Only an estimate S_t of the backlog X_t is known to the users. The users estimate the backlog from the reception of the binary variable, Z_t, which is 0 if slot t was empty and 1 if slot t was occupied.

This more practical result, under condition
$C0$: There exists $\theta > 0$ and B such that for all $n \geq 1, \sum_{k=1}^{n} e^{\theta k} \epsilon_{nk} \leq B$,
of GVS is found in the following

Theorem 3 GVS [10]. *Assume that there exists $A \in (0, +\infty)$ such that $t(A) = \sup_{x \geq 0} t(x)$, that the new packet arrivals $(A_t)_{t \geq 0}$ are exponential type with $\lambda < t(A)$, and that condition C0 holds. If $\alpha < 0$ and $\beta > 0$ verify the following conditions:*

$C1 : \beta > \lambda$

$C2 : \beta(1 - e^{-A}) + \eta_c - \lambda + \alpha e^{-A} = 0$

then the control algorithm

$$S_{t+1} = \max\{A, S_t + \alpha\overline{Z_t} + \beta Z_t\} \tag{23}$$
$$p_t = \begin{cases} 1, & S_t < A \\ A/S_t, & S_t \geq A \end{cases}$$

has maximum stable throughput equal to η_c.

The theorem shows, for example, that the simple closed-loop control algorithm for a 10-subslot system

$$p_t = \begin{cases} 1, & S_t < \sqrt{10} \\ \sqrt{10}/S_t, & S_t \geq \sqrt{10} \end{cases}$$
$$S_{t+1} = \max\{\sqrt{10}, S_t - .0221\overline{Z_t} + 2Z_t\}$$

achieves the maximum stable throughput listed in Table 1 of 1.587 packets/frame. Note, $\beta = 2$ and $\alpha = -.0221$ satisfy $C1$ and $C2$ for the 10 user subslot case.

5 Discussion

Low-level [4] protocols for the packet broadcasting RAC are designed to cope with the occurence of unpredictable collisions. The refinement of the frame to include subslots essentially moves the collision event of the RAC from the slot to the subslot. In this sense, any enhancement to the slotted-ALOHA protocol is also applicable to slotted-subslotted ALOHA.

The main practical drawbacks of the proposal are the need for finer timing to place signals in subslots and the complexity of the MUSD receiver. It is interesting to note that if timing, phase, and pulse shape estimates could be made, users could begin transmission at any time in the first symbol interval of the packet without being restricted to subslots. This would reduce the likelihood of a collision and consequently increase throughput.

Two systems that expose some main alternatives to the slotted-subslotted proposal must be mentioned. First, an ordinary ALOHA channel could simply be put in place with the reduced symbol period. Multi-level transmission with greater peak energy consumption is then required to maintain the maximum possible data rate. Second, the channel could be split into a number of frequency separated ALOHA channels. There would then be a tradeoff between the filtering requirements of the two pulses. For fair comparison, each user should be able to access any of these independent frequency channels at random, adding to transmitter cost. The receiver would need to scan all frequencies also. The cost of associated hardware must be determined before these competing systems can be fairly evaluated.

Acknowledgement

The first author is grateful to A. Yasotharan for his kind help during the development of this work and to the anonymous reviewers for their suggestions to improve the clarity of the presentation.

[4] Data Link or Network Layer depending on your point of view.

References

1. S. Verdú, "Minimum Probability of Error for Asynchronous Gaussian Multiple-Access Channels," *IEEE Trans. Inform. Theory*, vol. IT-32, pp. 85–96, Jan. 1986.
2. R. Lupas and S. Verdú, "Near-Far Resistance of Multiuser Detectors in Asynchronous Channels," *IEEE Trans. Commun.*, vol. COM-38, pp. 496–508, April 1990.
3. M. K. Varanasi and B. Aazhang, "Multistage Detection in Asynchronous Code-Division Multiple-Access Communications," *IEEE Trans. Commun.*, vol. COM-38, pp. 509–519, April 1990.
4. Z. Xie, C. K. Rushforth, and S. R. T., "Multiuser Signal Detection Using Sequential Decoding," *IEEE Trans. Commun.*, vol. COM-38, pp. 578–583, May 1990.
5. R. M. Metcalfe and D. R. Boggs, "Ethernet: Distributed Packet Switching for Local Computer Networks," *Commun. ACM*, vol. 19, pp. 395–404, July 1976.
6. G. D. Forney, "Maximum Likelihood Sequence Estimation of Digital Sequences in the Presence of Intersymbol Interference," *IEEE Trans. Inform. Theory*, vol. IT-18, pp. 363–378, May 1972.
7. J. G. Proakis, *Digital Communications*. New York: McGraw-Hill, Inc., 1983.
8. W. A. Rosenkrantz and D. Towsley, "On the Instability of the Slotted ALOHA Multiaccess Algorithm," *IEEE Trans. Automat. Contr.*, vol. 28, pp. 994–996, 1983.
9. S. M. Ross, *Introduction to Probability Models*. San Diego, CA.: Academic Press, Inc., 4th ed., 1989.
10. S. Ghez, S. Verdu, and S. C. Schwartz, "Optimal Decentralized Control in the Random Access Multipacket Channel," *IEEE Trans. Automat. Contr.*, vol. 34, pp. 1153–1163, Nov. 1989.
11. J. L. Massey, "Some New Approaches to Random-Access Communications," in *Performance '87, also in Multiple-Access Communications, N. Abramson, Ed. , pp. 354-368*, pp. 551–569, 1988.
12. S. Ghez, S. Verdu, and S. C. Schwartz, "Stability Properties of Slotted Aloha with Multipacket Reception Capability," *IEEE Trans. Automat. Contr.*, vol. 33, pp. 640–649, July 1988.
13. N. Mahravari, "Random-Access Communication with Multiple Reception," *IEEE Trans. Inform. Theory*, vol. IT-36, pp. 614–622, May 1990.

Auto-Regressive Transfer Function Identification Using Chaos

Xinping Huang
Applied Silicon Inc. Canada
220-2427 Holly Lane, Ottawa, Canada K1V 7P2
huang@grumpy.rdr.dreo.dnd.ca

Henry Leung
Radar Division, Defence Research Establishment Ottawa
Ottawa, Ontario, Canada K1A 0K2
leung@grumpy.rdr.dreo.dnd.ca

Abstract. Transfer function identification is an important issue in many areas such as process control and communications. A new method is proposed in this paper to identify a transfer function described by an auto-regressive (AR) model. It uses a chaotic sequence generated by a logistic map as an input and estimate channel parameters according to dynamics of the chaotic sequence. The new method outperforms the least square (LS) method with a white Gaussian input signal.

1. Introduction

A model which describes dynamic response of a system to an input is called a transfer function model. It can be illustrated by the block diagram in Figure 1, where the recorded signal r_t, the output y_t due to the input x_t and the observation noise n_t are related by the following equation:

$$r_t = y_t + n_t. \tag{1}$$

Among various models, the autoregressive (AR) model is one of the most popular model for such a system. It is described by

$$y_t = \sum_{i=1}^{p} a_i y_{t-i} + x_t \tag{2}$$

where p is the order, $a_i's$ are the parameters, of the system, respectively. Its transfer function is given by

$$H(z) = \frac{1}{\sum_{i=0}^{p} a_i z^{-i}}. \tag{3}$$

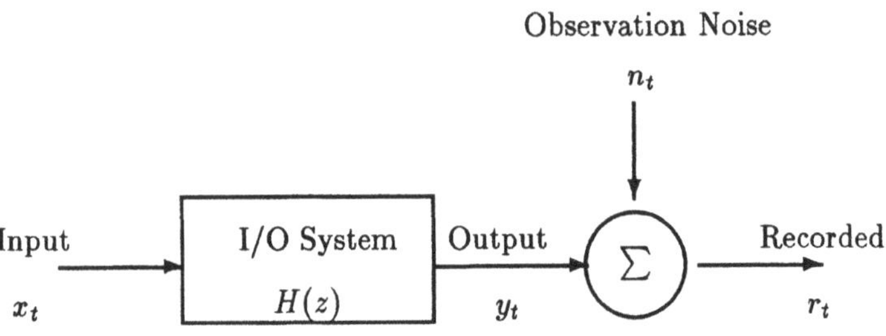

Figure 1: Illustration of input-output systems

Many methods have been proposed to identify the model. They can generally be classified into three catgories: (1). to measure the impulse response directly by applying an impulse to the input; (2). to measure the transfer function in the frequency domain by scanning an unmodulated carrier frequency; (3). to apply a pseudonoise (PN) sequence as an input and to infer the impulse response by using a correlation tehnique or various statistic-based criteria method [1]. The last method is especially useful to identify systems which can be parameterized in a close form of Eq. 2.

In this paper, a new method is proposed to identify the transfer function of Eq. 3. The method is "new" in twofold. Firstly, it uses a chaotic sequence instead of a random sequence. While a chaotic sequence is deterministic, it appears to be noise-like in a statistical sense. Secondly, it uses the dynamical property of the chaotic sequence for order selection and parameter estimation instead of classical estimation techniques such as maximum likelihood and minimum variances. Due to the deterministic nature of the chaotic sequence, the conventional statistic-based objective functions cannot exploit the information contained in the chaotic sequence completely.

The paper is organized as follows. In Section 2, the new system identification method is described. Computer simulations are performed in Section 3 to understand the effectiveness of the new method and the Least Square technique is used as a benchmark for performance comparisons. Conclusions and discussions are given in Section 4.

2. New Dynamic-Based Approach

Probably one of the most well known chaotic process is the logistic map [2] given by

$$x_{t+1} = 4x_t(1 - x_t) \tag{4}$$

This equation is deceptively simple, however, the time-evolution of x_t has an exceedingly rich behaviour. A realization of the logistic map is shown in Figure 2a. Although behaving irregularly in the time domain, the logistic map has remarkably simple characteristics in a 2-dimensional phase space defined by (x_t, x_{t+1}): all points fall on a simple curve defined by Eq. 4 as shown in Figure 2b. Such a simple functional relationship does not exist when the sequence is truly random: the points merely fill up the phase space instead. In addition, although it is deterministic process, the chaotic sequence is uncorrelated, as shown in Figure 3.

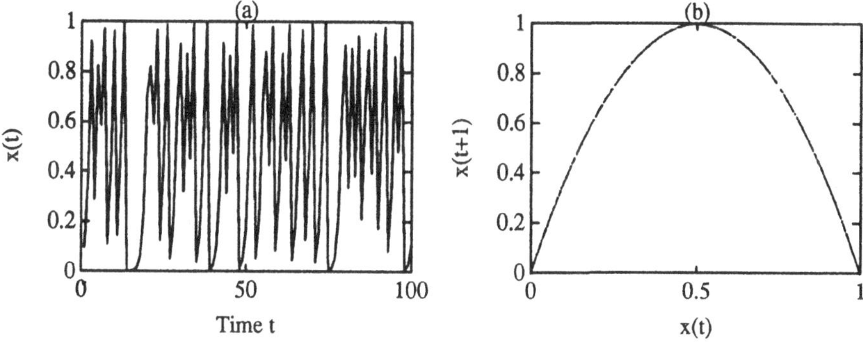

Figure 2: A realization of the logistic Map and its phase space presentation.

In the new method, we apply a chaotic sequence $\{x_t\}$ generated by Eq. 4 to the input of the system. In noiseless cases, we have

$$r_t = y_t. \tag{5}$$

Therefore, we may rewrite Eq. 2 as follows

$$r_t = a_1 r_{t-1} + a_2 r_{t-2} + \cdots + a_p r_{t-p} + x_t \tag{6}$$

x_t satisfies the following equation

$$x_t = r_t - a_1 r_{t-1} - a_2 r_{t-2} - \cdots - a_p r_{t-p} \tag{7}$$

If we construct an inverse filter $U(z)$ as

$$U(z) = \sum_{j=0}^{q} b_j z^{-j} \tag{8}$$

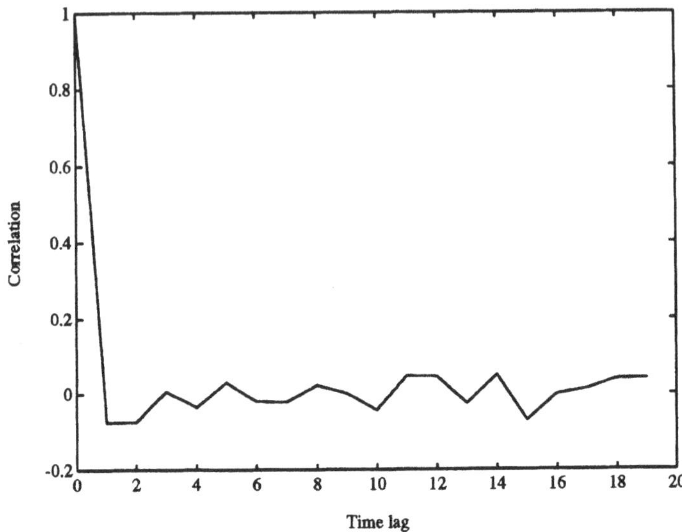

Figure 3: Autocorrelation function of the chaotic sequence generated by the logistic map.

where $q \geq p$, and apply it to r_t, we may obtain an estimate of x_t at the output of this inverse filter, i.e.,

$$\hat{x}_t = r_t + b_1 r_{t-1} + \cdots + b_q r_{t-q} \tag{9}$$

Comparing Eqs. 7 and 9, we can deduce that if we choose the inverse filter $U(z)$ such that

$$q = p \tag{10}$$

and

$$b_j = \begin{cases} 1 & j = 0 \\ -a_j & j = 1, \cdots p \\ 0 & j = p+1, \cdots, q \end{cases} \tag{11}$$

then the chaotic sequence can be recovered at the output of $U(z)$. In the other words, $\hat{x}_t$ should closely follow the recurrence relationship of Eq. 4 in the phase space as x_t does if the correct inverse filter is used. Therefore, the problem of transfer function identification can be reduced to finding the optimal inverse filter $U(z)$ such that differences between $\hat{x}_t$ and x_t in the phase space is minimized. After we obtain the inverse filter $U(z)$, we can easily derive the transfer function $H(z)$ from it according to Eq. 11.

Since the distribution of x_t in the phase space is known and fixed, the difference between $\hat{x}_t$ and x_t in the 2-dimensional phase space may be computed in the following way. Given $\hat{x}_t$, we form a two dimensional sequence:

$$(\hat{x}_1, \hat{x}_2), (\hat{x}_2, \hat{x}_3), \cdots, (\hat{x}_t, \hat{x}_{t+1}), \cdots$$

For the t-th point, $(\hat{x}_t, \hat{x}_{t+1})$, we compute the distance between this point and a point from x_t, denoted as $(x_t, 4x_t(1-x_t))$, which is defined as

$$e_t(x_t) \quad = \quad |\, (\hat{x}_t, \hat{x}_{t+1}) - (x_t, 4x_t(1-x_t)) \,| \tag{12}$$

The notation $e_t(x_t)$ implies that it is a function of x_t. Since we wish to examine the difference between $\hat{x}_t$ and x_t in the phase space, we choose the point (x_t, x_{t+1}) which is closest to $(\hat{x}_t, \hat{x}_{t+1})$ to compute the difference attributed to $(\hat{x}_t, \hat{x}_{t+1})$. It follows that

$$e_t = \min_{0<x_t<1} e_t(x_t) \tag{13}$$

The distance e_t represents the absolute error contributed by the t-th point $(\hat{x}_t, \hat{x}_{t+1})$. The next natural step is to sum all errors as follows

$$d_q = \frac{1}{L-1} \sum_{t=1}^{L-1} e_t^2 \tag{14}$$

where the subscript q denotes the test order and L denotes the length of sequence used in calculation.

The In the noiseless case, $\hat{x}_t$ will closely follow the recurrence relationship of Eq. 4 and d_q becomes zero. However, if the observation noise exists as in Eq. 1, a perfect fitting may not be available and we can only expect a best fit in the least square sense. Therefore, the optimal inverse filter can be obtained by minimizing d_q with respect to the model order q and parameter set $\{b_j\}$. Hence, the transfer function identification can be considered as choosing the order $\hat{p}$ and the parameter set $\{\hat{a}_j\}$ so that the error d_q is minimized, i.e.,

$$\{\hat{p}, \hat{a}_j\} = \arg \left\{ \min_{q, b_j} d_q \right\} \tag{15}$$

The method is termed as the chaotic method. In the following studies, we will examine how this method works and its effectiveness.

3. Simulation Studies

To illustrate how the chaotic method works and to understand its effectiveness, a second order model is constructed as the transfer function to be estimated:

$$H(z) = \frac{1}{1 - 0.195z^{-1} + 0.95z^{-2}} \tag{16}$$

We study issues of order selection and parameter estimation as well as estimation errors using the chaotic method with computer simulations. The least square (LS) technique which employs a white Gaussian sequence as input is used in performance comparisons. The observation noise is assumed to be absent in this studies. Its effect will be investigated in a separate work.

Case 1: Error surface

To understand how d_q distributes, the error surface defined as $\frac{1}{d_q}$ is plotted in Figure 4. 200 points are used in this simulation. It shows that the

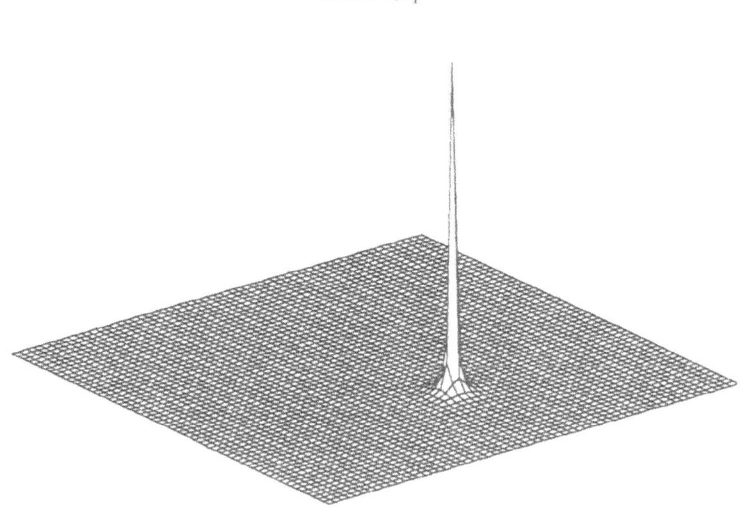

Figure 4: 2-D error surface $\frac{1}{d_p}$.

error surface is smooth and contains a unique sharp peak located at $a_1 = 0.195$ and $a_2 = -0.95$. The smoothness of the error surface makes it easier to search for the maxima and the sharp peak assure the accuracy of parameter estimation.

Case 2: Order selection and Parameter estimation

The issure of order selection and parameter estimation is examined in the following way. We start with a test model of order 1, estimate the parameter a_1 and compute d_1 according to Eq. 15. Then we proceed with order 2, 3 and 4. The results are listed in Table 1. It is observed that when the test order is below 2, d_q is relatively large. As the test order q increases, d_q decreases until $q = 2$ where d_q reaches an insignificant value. For $q > 2$, d_q remains at unchanged as shown in Figure 5. From this we can easily infer that the model is of order 2. The data in Table 1 also show that $\hat{a}_j$'s become essentially zero for $j > 2$, which suggests that we may determine the order by simply counting the number of nonzero parameters if we start with a test model of higher than the second order.

Order	$\hat{a}_1$	$\hat{a}_2$	$\hat{a}_3$	$\hat{a}_4$	d_q
1	0.05440				9.92E-03
2	0.19499	-0.94999			1.56E-10
3	0.19499	-0.94999	-5.49E-06		1.33E-10
4	0.19499	-0.94999	-8.02E-07	-7.89E-07	1.44E-10
True	0.195	0.195			0.0

Case 3: Estimation error

In the next simulation, we choose the mean squared error (MSE) as a measure of estimation error. It is defined as

$$\mathrm{MSE} = \frac{1}{p} \sum_{i=1}^{p} \mathrm{E}\left\{\hat{a}_i - a_i\right\}^2 \tag{17}$$

where $\mathrm{E}\{\cdot\}$ denotes the ensemble average, which is implemented by averaging over 100 independent simulations.

In Figure 6, two curves are shown. The solid line denotes MSE of the chaotic method while the dashed line represents MSE of the LS method with a white Gaussian sequence as input. Number of points used in the simulations ranges from 5 to 1000. It is noted that when number of points used is below 20, both methods perform relatively poor. Above 20, the chaotic method becomes better. When the number of points reaches about 50, MSE of the chaotic method drops rapidly to a level of 10^{-8}, while MSE of the LS method decreases slowly as the number of points increases. Above 50, MSE of the chaotic method tends to saturate at 10^{-9}, which is limited by numerical errors since only single precision is used in the simulation.

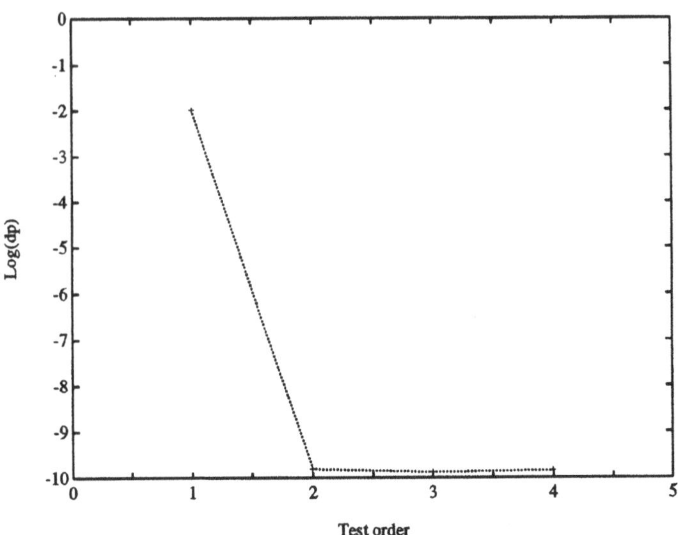

Figure 5: Min. d_p versus the test order q.

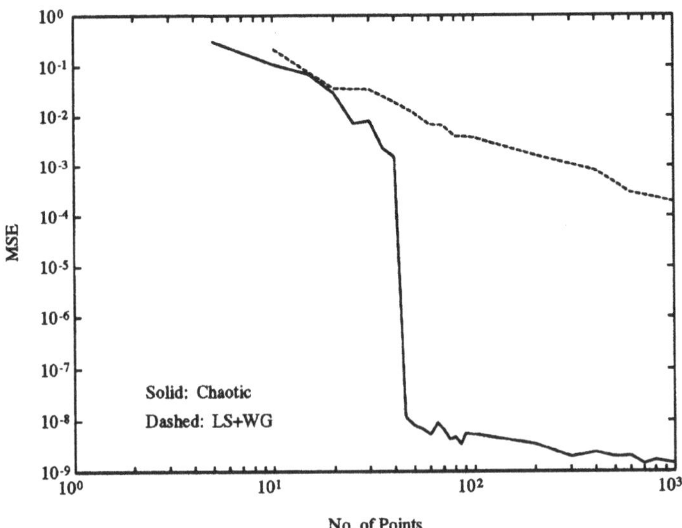

Figure 6: MSE in parameter estimates.

The large improvement in MSE is due to the fact that the new method is based on an inherently deterministic modelling of $\hat{x}_t$, while the LS method is based on a statistical modelling of $\hat{x}_t$. The latter obtains the true values of parameters only when the number of points approaches infinity. On the other hand, the new method has the potential of achieving the same level of performance with finite number of points as that achieved by the LS method with infinite number of points. This is major advantage of the new method developed in this paper.

4. Conclusions

In this paper, we have presented a new method, based on a dynamic property of chaos, to identifying the transfer function which is described by an AR model. A logistic map is introduced which generates white chaotic process as the input signal. The order and the parameters of the model can be determined jointly by matching the output to the logistic map. Simulation studies are carried out to understand usefulness of the new method. Compared with the LS method, it can obtain parameters accurately with only a short data record.

References

[1] G. Box, and G. Jenkins, *Time Series Analysis: Forecasting and Control*, revised edition, 1976, Holden-Day Inc.

[2] D. Ruelle, *Chaotic Evolution and Strange Attractors*, Cambridge Press, 1989.

A New Algorithm for Signal Classification Using Phase Information

R. J. Inkol

Defence Research Establishment Ottawa
National Defence, Ottawa, Ontario, Canada, K1A OZ4

R. H. Saper

Vantage Point International
Gloucester, Ontario, Canada, K1B 5C8

Abstract. An algorithm for the classification of pulsed signals having a phase or frequency modulated carrier is presented. First, the linear phase component from the unwrapped differential phase history of a pair of signals is modeled using linear regression. Statistical techniques are then used to determine if the phase residuals from the linear regression model are consistent with the hypothesis that both signals have the same modulation. The linear regression model also provides an accurate measurement of the carrier frequency offset. Simulation results for different modulations and signal-to-noise ratios are consistent with the theoretical performance.

1 Introduction

An important signal processing problem involves the comparison of an observed signal with a reference signal to decide if it is from the same class of sources. A special case occurs for bandpass signals generated by asynchronously gating a sinusoidal carrier. Such pulse waveforms are commonly employed in radar and navigation systems. This paper considers the signal classification problem for signals of the form

$$\begin{aligned} s_{ik}(t) = b_i(t)\cos(\omega_{ik}t + \phi_i(t) + \theta_{ik}) + n_{ik}(t), \\ b_i(t) > 0 \text{ when } 0 < t < \tau, \\ b_i(t) = 0 \text{ otherwise,} \end{aligned} \quad (1)$$

where $s_{ik}(t)$ is the observed real signal for the kth pulse from the class i. The pulse envelope has a shape defined by $b_i(t)$. $\phi_i(t)$ is the phase modulation, the initial phase θ_{ik} is a random variable with a uniform probability over a range of 2π radians, and

ω_{ik} is the carrier frequency. $n_{ik}(t)$ is additive zero mean Gaussian noise with a variance $\sigma^2_{\eta ik}$ and is uncorrelated for any nonidentical pair of signals. In practice, an observed signal will be shifted in time, but we will assume that the signal can be detected and its time-of-arrival shifted to be consistent with equation (1). Another assumption is that the carrier frequency is sufficiently large that the spectral spreading resulting from the amplitude modulation $b_i(t)$ does not significantly distort the phase information [1].

The problem we address here is to find some measure of the distance between the modulations $\phi_i(t)$ and $\phi_j(t)$ of a pair of signals, $s_{ik}(t)$ and $s_{jm}(t)$, that can be used to decide whether or not the signals belong to the same class.[1] This is not a trivial problem. The carrier frequency of each pulse may not be precisely known and could vary for members of an individual class. It is easily shown that the performance of techniques commonly employed for signal classification such as cross-correlation or the coefficient of similarity [2] are adversely affected by a carrier frequency offset between a pair of signals.[2] One solution to this problem is to first estimate ω_{ik} for each pulse and shift the carrier frequency to a common frequency. A more usual method of simplifying the signal representation is to perform frequency demodulation by differentiating the phase [3, 4]. This directly eliminates one parameter, the initial phase of the carrier, and the DC offset resulting from a carrier frequency offset can be estimated and removed. Signal classification can then be performed using the Euclidian distance between the time aligned and demodulated waveforms of each pair of pulses.

The use of frequency demodulation has several disadvantages. Algorithms commonly used for pattern recognition, such as cross-correlation, cannot be used to resolve classes of signals whose modulations $\phi_i(t)$ differ only by a scale factor. A more serious criticism is that frequency demodulation is equivalent to differentiating the phase and therefore results in a parabolic noise power spectrum for an input signal having white Gaussian noise. The noise power can be reduced by using a differentiator providing a smoothed estimate of the derivative, such as the Lanczos differentiator [5], or by additional lowpass filtering. However, the use of filters with lengthy impulse responses is undesirable for processing short duration pulse signals. The variance of the phase becomes large when the instantaneous signal-to-noise ratio is small and this corrupts the filtered phase data near the pulse edges.

Approaches involving the estimation of the modulation parameters of chirp signals

---

[1]Examples of practical situations where this problem is of interest include the comparison of a signal with a previously observed reference signal to determine if they are from the same class or to check if signals received simultaneously at spatially separated sites are from the same source.

[2]Doppler shift in the carrier frequency of the signal received by a radar system employing pulse compression results in a similar problem. Such systems often employ linear frequency modulation (FM) to obtain acceptable performance with a matched filter receiver. Other types of modulation such as non-linear FM or phase shift keying (PSK) modulation would require the correction of the doppler shift to ensure acceptable performance.

by fitting the phase to a polynomial have been extensively investigated [6, 7]. However, complex models may be required for common signals such as discrete phase or frequency modulation.

This paper describes a new technique for directly comparing the phase modulations of pairs of bandpass signals. A theoretical analysis of the performance is presented and confirmed by simulation results.

2. The Differential Phase Signal Comparison (DPSC) Algorithm

An important consideration in designing an algorithm for the classification of signals corresponding to the general model defined in equation (1) is to avoid assumptions or limitations concerning the types of phase or frequency modulation. Furthermore, the processing of phase data by filters having lengthy impulse responses is undesirable since this degrades the performance for short duration pulses.

The Differential Phase Signal Comparison (DPSC) algorithm avoids these problems by directly comparing the accumulated or unwrapped differential phase histories of a pair of signals, one of which is a template determined from *a priori* knowledge or previously received signals. The differential phase history of a pair of time aligned signals, $s_{ik}(t)$ and $s_{jm}(t)$, is given by

$$\phi_{ikjm}(t) = (\theta_{ik} - \theta_{jm}) + (\omega_{ik} - \omega_{jm})t + (\phi_i(t) - \phi_j(t)) + n_{ikjm}(t). \tag{2}$$

The first and second terms define the equation of a straight line with intercept $(\theta_{ik} - \theta_{jm})$ and slope $(\omega_{ik} - \omega_{jm})$. The third term is the difference in the phase modulations of the two signals and the last term represents a noise component which is dependent on the signal-to-noise ratios of the two signals.

The concept of the DPSC is to test $\phi_{ikjm}(t)$ for statistically significant departures from a linear model. An incidental advantage of this approach is that the slope of the linear model yields an accurate measurement of the difference in the carrier frequencies $(\omega_{ik} - \omega_{jm})$ of the pair of signals.

The DPSC is implemented in three steps. The first two steps are similar to an algorithm for measuring the differential Doppler shift of a signal received at two spatially separated locations [8]. The last step extends the previous work to provide statistical measures of the similarity of a pair of signals.

2.1 Computation of Differential Phase History

The computation of the differential phase history of a pair of signals, $s_{ik}(t)$ and $s_{jm}(t)$, involves forming their complex representations, $z_{ik}(t)$ and $z_{jm}(t)$, by quadrature demodulation. For high signal-to-noise ratios, the phase component of the conjugate product

$$z_{ik}(t)z^*_{jm}(t) = b_i(t)b_j(t)\,e^{j((\theta_{ik}-\theta_{jm}) + (\omega_{ik}-\omega_{jm})t + (\phi_i(t)-\phi_j(t)))}$$
$$+\, b_i(t)\,e^{j(\omega_{ik}t+\theta_{ik}+\phi_i(t))}n_{jm}^*(t)$$
$$+\, b_j(t)\,e^{-j(\omega_{jm}t+\theta_{jm}+\phi_j(t))}n_{ik}(t) \qquad (3)$$
$$+\, n_{ik}n^*_{jm}(t)$$
$$\approx \rho_{ij}(t)\,e^{j((\theta_{ik}-\theta_{jm}) + (\omega_{ik}-\omega_{jm})t + (\phi_i(t)-\phi_j(t)))}$$
$$+\, \eta_{ikjm}(t),$$

directly corresponds to $\phi_{ikjm}(t)$. Consequently, provided that the sampling period T is sufficiently small to avoid significant aliasing distortion, the discrete time differential phase history can be explicitly obtained by sampling $z_{ik}(t)$ and $z_{jm}(t)$ and computing

$$\phi_{ikjm}(nT) = U\left[\arctan\left[\frac{Re(z_{ik}(nT)\,z^*_{jm}(nT))}{Im(z_{ik}(nT)\,z^*_{jm}(nT))}\right]\right], \qquad (4)$$

where the symbol U represents phase unwrapping. Phase unwrapping can be carried out simply for high signal-to-noise ratios if the phase change does not exceed $\pi/2$ radians during one sample period [9]. More complex phase unwrapping algorithms [10] can provide better performance at low signal-to-noise ratios.

An alternative approach is to compute $\phi_{ikjm}(nT)$ as the difference between the separate time aligned and unwrapped phase histories obtained for each of $z_{ik}(nT)$ and $z_{jm}(nT)$. This doubles the number of phase unwrapping operations, but allows the variance of the phase data to be reduced by averaging the time aligned phase histories from a set of signals belonging to the same class.

The noise component of $z_{ik}(t)z^*_{jm}(t)$, $\eta_{ikjm}(t)$, has an expected value of zero and variance $\sigma^2_{\eta ikjm}(t) = b^2_{ik}(t)\sigma^2_{\eta jm} + b^2_{jm}(t)\sigma^2_{\eta ik} + \sigma_{\eta ik}\sigma_{\eta jm}$. Although $\eta_{ikjm}(t)$ is a random variable, it is dependent on the amplitudes of both signals. For large signal-to-noise ratios,[3] a simple geometric interpretation of its effect on the uncertainty $\sigma_{\phi ikjm}$ with which the phase can be measured yields the relationship [8]

$$\sigma_{\phi ikjm}(t) \approx \arctan\left[\frac{\sigma_{\eta ikjm}(t)}{\rho_{ij}(t)}\right]. \qquad (5)$$

The phase error is Gaussian with a mean of zero and a variance of $\sigma^2_{\phi ikjm}(t) = (2\gamma^2_{ikjm}(t))^{-1}$ where $\gamma^2_{ikjm}(t)$ is a composite signal-to-noise ratio defined as the harmonic mean of the signal-to-noise ratios of $z_{ik}(t)$ and $z_{jm}(t)$.

---

[3]The assumption of high signal-to-noise ratio (≥15 dB) is consistent with the need to perform signal detection with a low false alarm rate in practical applications.

2.2 Estimation of a Linear Model for $\phi_{ikjm}(t)$

The use of linear regression to compute a linear model for the first two terms in equation (2) is very reasonable. It provides the best linear unbiased estimate of the linear model if $\phi_i(t) = \phi_j(t)$. It is also a maximum likelihood estimator if $\phi_{\eta ikjm}(t)$ is Gaussian.[4] For a time interval NT, linear regression estimates of carrier frequency Δf_{ikjm} and initial phase offset $\hat{\phi}_{ikjm}(0)$ are given by [11]

$$\Delta f_{ikjm} = \frac{\sum_{n=0}^{N} n\,\psi_{ikjm}(nT)\,\phi_{ikjm}(nT) - \dfrac{\sum_{n=0}^{N} n\,\psi_{ikjm}(nT) \sum_{n=1}^{N} \psi_{ikjm}(nT)\,\phi_{ikjm}(nT)}{\sum_{n=0}^{N} \psi_{ikjm}(nT)}}{2\pi T\left[\sum_{n=0}^{N} n^2\psi_{ikjm}(nT) - \dfrac{\sum_{n=0}^{N} [n\,\psi_{ikjm}(nT)]^2}{\sum_{n=0}^{N} \psi_{ikjm}(nT)}\right]}, \tag{6}$$

$$\hat{\phi}_{ikjm}(0) = \frac{\sum_{n=0}^{N} \psi_{ikjm}(nT)\,\phi_{ikjm}(nT)}{\sum_{n=0}^{N} \psi_{ikjm}(nT)} - 2\pi\Delta f_{ikjm} \frac{\sum_{n=0}^{N} nT\,\psi_{ikjm}(nT)}{\sum_{n=0}^{N} \psi_{ikjm}(nT)}, \tag{7}$$

where $\Psi_{ikjm}(nT)$ is a weighting parameter whose purpose is to emphasize differential phase measurements having a low variance. Its optimal value is given by $1/\sigma^2_{\phi ikjm}(nT)$. Although *a priori* knowledge of $\sigma^2_{\phi ikjm}(nT)$ is unlikely to be available in practical situations, there are several reasonable approaches for setting $\Psi_{ikjm}(nT)$. If sets of signals for $z_{ik}(nT)$ and/or $z_{jm}(nT)$ are available, $\Psi_{ikjm}(nT)$ can be estimated from the variances computed for each $\phi_{\eta ikjm}(nT)$. Alternatively, using measurements of $\sigma_{\eta ik}$ and $\sigma_{\eta jm}$ outside the time intervals containing the pulses, reasonable weights for

---

[4]The assumption that the noise is Gaussian with zero mean is often reasonable in practical situations. If the noise consists of errors summed from many sources, it will tend to a Gaussian distribution in accordance with the Central Limit theorem regardless of the distributions associated with the individual sources of error.

high signal-to-noise ratios are given by the harmonic means of $|z_{ik}(nT)|^2/\sigma^2_{\eta ik}$ and $|z_{jm}(nT)|^2/\sigma^2_{\eta jm}$. The simple weighting obtained by setting $\Psi_{ikjm}(nT) = 1$ between the pulse endpoints and zero outside[5] is often satisfactory for pulses having an approximately constant amplitude over a large portion of their duration. This binary weighting significantly reduces computational cost and simplifies the performance analysis.

2.3 Statistical Tests of The Differential Phase History After Elimination of the Linear Phase Component

The similarity of the phase modulation of the pair of signals being processed is directly measured by the cost function defined as the weighted mean square error of the least squares fit

$$C_{ikjm} = \frac{\sum_{n=0}^{N} \Psi_{ikjm}(nT)\,[\phi_{ikjm}(nT) - \hat{\phi}_{ikjm}(0) - 2\pi\Delta f_{ikjm} nT]^2}{\sum_{n=0}^{N} \Psi_{ikjm}(nT)}. \quad (8)$$

For large N, the expected value of C_{ikjm} is approximately the weighted average of $((\phi_i(t) - \phi_j(t)) + \phi_{\eta ikjm})^2$. Consequently, C includes a signal-to-noise ratio dependent bias error which determines the ability to detect a given nonlinear phase component in a differential phase history. A different behaviour results if the denominator of equation (8) is replaced by a constant and a weighting proportional to the composite signal-to-noise ratio is used. The dependence of C on the signal-to-noise ratio will be reduced and its sensitivity to a given nonlinear component in a differential phase history will be approximately proportional to the signal-to-noise ratio.

C is affected by other sources of error including multipath propagation and quadrature demodulation errors. Their effect is often difficult to distinguish from a reduction in the signal-to-noise ratios.

The phase modulation of most signals has a lowpass spectrum whose bandwidth is small compared to the Nyquist frequency of a signal analysis system whose bandwidth is determined by the uncertainty in the carrier frequencies of the signals rather than their individual bandwidths. Consequently, statistical tests for serial (i.e., sample-to-sample) correlation between the phase residuals can effectively detect differences in the phase modulation of a pair of signals if error mechanisms

---

[5]The optimum processing duration with the binary weighting is likely to be dependent on the shape of the signal waveform and the transient behaviour of the quadrature demodulator. Consequently, it may be desirable to shorten the processing duration obtained using a given endpointing algorithm to ensure that good quality phase measurements can be obtained for most signals.

contributing significant correlated errors are absent.

Serial correlation of the phase residuals to the least squares fit implied by the linear regression can be determined by the von Neumann ratio [12] given by

$$VNR_{ikjm} = \frac{\sum_{n=1}^{N} [\phi_{ikjm}(nT) - \phi_{ikjm}(nT-T) - 2\pi\Delta f_{ikjm}T]^2/N}{\sum_{n=0}^{N} [\phi_{ikjm}(nT) - \hat{\phi}_{ikjm}(0) - 2\pi\Delta f_{ikjm}nT]^2/(N+1)}. \quad (9)$$

If the residuals have an independent Gaussian distribution, the VNR will have an expected value of two for large N. Its value will be reduced by serial correlation of the phase residuals. Consequently, a VNR near 2 has a simple interpretation; the differences in the phase modulations of the pair of signals cannot be distinguished from noise. In practical applications, sources of correlated phase errors such as multipath propagation, phase errors in the quadrature demodulator, and narrowband filtering of the signals[6] can degrade the usefulness of the VNR.

The use of both C and the VNR provides more information than if one parameter is used by itself. The following interpretations can be drawn if the fixed weighting $\Psi_{ikjm}(nT) = 1$ is used:

1. Low value of C and a high value of VNR (> ~ 1.2): unambiguous match.: the sum of $\sigma^2_{\phi ikjm}$ and the average value of $(\phi_i(t) - \phi_j(t))^2$ is small and the ratio of $\sigma^2_{\phi ikjm}$ to the average value of $(\phi_i(t) - \phi_j(t))^2$ is large.
2. Low values of C and the VNR: possible match.: $(\phi_i(t) - \phi_j(t))^2$ is not large, but is detectable.
3. High values of C and the VNR; inconclusive: the signal-to-noise ratio may be insufficient to detect a significant mismatch in the phase histories of the two signals.
4. High value of C and a low value of VNR: unambiguous mismatch: the average value of $(\phi_i(t) - \phi_j(t))^2$ is large.

Since C and the VNR are measures of how well pairs of signals match, they can be compared to appropriate thresholds to make an unambiguous match/mismatch decision if a signal pair matches. The optimal choice of threshold has a complex dependence on the type of weighting, the desired probabilities for correctly detecting matches and mismatches, and signal dependent parameters such as the size and variability of the signal population, and the signal-to-noise ratio. If sufficient *a priori* information is available, the optimal thresholds can be determined by Bayes or Neyman-Pearson criteria. However, in practical applications, the thresholds are often

[6] There is an implicit assumption that the signals are not processed by filters having bandwidths much less than the Nyquist bandwidth. Sample-to-sample correlation will be introduced if this condition is not satisfied

selected by empirical means. One approach is to examine histograms of C and the VNR for clusters that appear to correspond to pairs of signals with matching modulations and set the thresholds accordingly.

3 Performance Analysis

Important performance criteria for the DPSC algorithm include the statistical behaviour of Δf_{ikjm}, C, and the VNR. These can be readily analyzed given the assumptions that N is large, $\Psi_{ijkm}(nT) = 1$ between the pulse endpoints, and that the signal-to-noise ratios are constant over the time interval considered.

3.1 Two Signals Having Identical Phase Modulations

The frequency offset Δf_{ikjm} is known to be unbiased with variance [8]

$$var[\Delta f_{ikjm}] \approx \frac{12}{T^2 N^3 \gamma^2_{ikjm}}, \tag{10}$$

where N is the number of samples, and γ^2_{ikjm} is the composite signal-to-noise ratio.

For high signal-to-noise ratios, the phase residuals have a zero mean Gaussian distribution with variance $\sigma^2_{\phi ikjm}$ equal to $(2\gamma^2_{ikjm})^{-1}$. Consequently, C can be regarded as the mean value of the output signal from a square law detector operating on a signal $(\phi_i(t) - \phi_j(t))$ with additive Gaussian noise. It has an expected value and variance [13] given by

$$E[C_{ikjm}] \approx \frac{1}{2\gamma^2_{ikjm}}, \tag{11}$$

$$var[C_{ikjm}] \approx \frac{1}{4N\gamma^4_{ikjm}}. \tag{12}$$

The von Neumann ratio is normally distributed for large N [12] with

$$E[VNR_{ikjm}] = \frac{2N}{N-1}, \tag{13}$$

$$var[VNR_{ikjm}] \approx \frac{4N^2(N-2)}{(N+1)(N-1)^3}. \tag{14}$$

3.2 Two Signals Having Dissimilar Phase Modulations

For two signals having different phase modulations whose bandwidth is small relative to the Nyquist bandwidth, the residuals to the linear regression straight line will be correlated. In general, the estimates for the phase and frequency offsets will be biased and the variances will be increased. Since these effects can result in a reduction in the expected value of C, the exact performance analysis of C becomes complicated.[7] However, approximate theoretical results can be obtained for some simple cases by assuming that the bias errors in estimating the phase and frequency offsets are negligible.

If the difference in the phase modulations can be modeled as Gaussian noise having a standard deviation of σ_m, the results for E[C] and var[C] are similar to those of equations (12) and (13) except that $((2\gamma^2_{ikjm})^{-1}+\sigma_m^{\ 2})$ should be substituted for $(2\gamma^2_{ikjm})^{-1}$.

Another simple case occurs for phase modulations differing by a sinusoidal phase modulation having an amplitude A_p. C will have an expected value and variance [13] given by

$$E[C_{ikjm}] \approx \frac{A_p^2}{2} + \frac{1}{2\gamma^2_{ikjm}}, \tag{15}$$

$$var[C_{ikjm}] \approx \frac{A_p^4}{8} + \frac{A_p^2}{2\gamma^2_{ikjm}} + \frac{1}{4\gamma^4_{ikjm}}. \tag{16}$$

---

[7]This problem does not seriously limit the applicability of the DPSC. If the difference in the phase modulations of the two signals is large enough to introduce significant bias errors in the phase and frequency offsets obtained by equations (7) and (8), the DPSC usually has little difficulty in detecting them.

4 Simulation Results

The statistical behavior of Δf_{ikjm}, C and the VNR was investigated for various signal-to-noise ratios. Using a set of MATLAB macros, simulated pulses at an IF near 100 MHz were generated at a sampling rate of 400 MHz with 8-bit resolution. The signal models consisted of descriptions of the pulse shape, signal-to-noise ratio, and frequency modulation. Following quadrature demodulation using the algorithm described in [14], each pair of pulses was processed by the DPSC algorithm. The analysis was performed with weighting $\Psi(nT) = 1$ on those samples where the product of the magnitudes of both signals exceeded half its peak value. The actual number of samples processed thus fluctuated with the influence of noise as it would in a real system.

The test signals QP0-QP6 were rounded trapezoidal pulses having a total duration of 1.1 microsecond. The edges, constructed of halves of a raised cosine, had a transition time of 0.1 microsecond while the constant amplitude central portion had a duration of .9 microseconds. The reference pulse signal, QP0, had a 20 MHz/microsecond chirp frequency modulation with a 33 dB signal-to-noise ratio after quadrature demodulation. QP1-QP3 differed in having signal-to-noise ratios of 28 dB, 23 dB and 18 dB respectively. QP4 was similar to QP0, but had a 5 MHz carrier frequency offset. QP5 included a decaying sinusoidal frequency modulation having a maximum frequency deviation of 8 MHz superposed on the chirp modulation. QP6 had a 19.75 MHz/microsecond chirp rate (change of 0.25 MHz/microsecond in frequency deviation rate).

Figures 1 to 4 show typical differential phase histories for a QP0 reference pulse and QP3, QP4, and QP5 pulses. The smooth interpolating line drawn through the unwrapped differential phase points is plotted with crosses while the linear regression estimates of the phase are plotted with small circles. Figures 1 to 3 confirm that the differential phase histories of signals having the same modulation can be modeled by linear regression for different signal-to-noise ratios and carrier frequency offsets. Conversely, the results for Figure 4 demonstrate that a small difference in the modulations of a pair of signals can be discerned by comparing their differential phase history with the linear regression phase estimate, and detected by examining the values of C and the VNR.

In Figure 5, the phase residuals obtained with typical QP0 and QP6 pulses are plotted on an expanded scale. Although there is only a subtle difference in the chirp rate, the values of 0.235 and 0.0034 obtained for the VNR and C respectively provide unambiguous evidence that the signals do not have the same phase modulation. By itself, the observed value for C is more ambiguous since the same result would have been obtained if QP6 differed from the QP0 reference signal only in having a lower signal-to-noise ratio (20 dB instead of 33 dB).

Using a QP0 reference pulse, statistical results were generated for various signal-to-noise ratios using sets of 100 QP0, QP1, QP2 and QP3 pulses and summarized in Table 1. They are consistent with the results predicted by the performance analysis in Section 3.

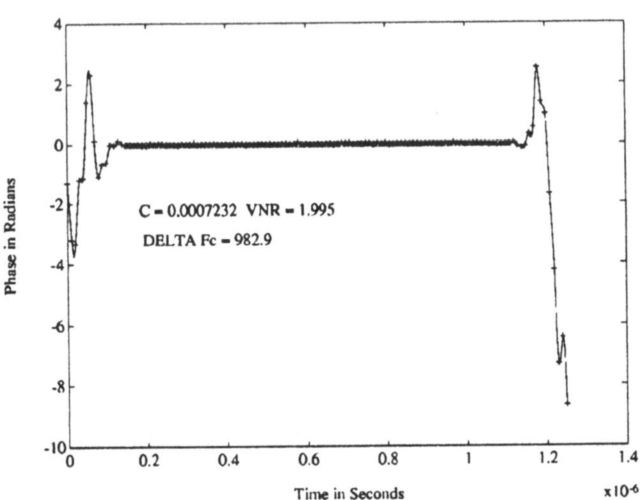

Fig. 1. Linear regression performed on differential phase history of two pulses having 33 dB and 28 dB signal-to-noise ratios. (QP0 and QP1). The differential phase shows little dispersion about the least squares straight line for the same modulation and high signal-to-noise ratios

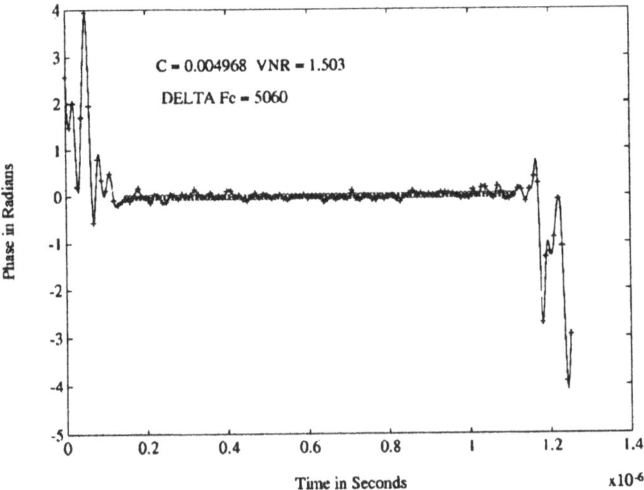

Fig. 2. Linear regression performed on differential phase history of two pulses having 33 dB and 18 dB signal-to-noise ratios (QP0 and QP3). The reduction in the signal-to-noise ratio in comparison with Figure 1 results in increased dispersion of the phase.

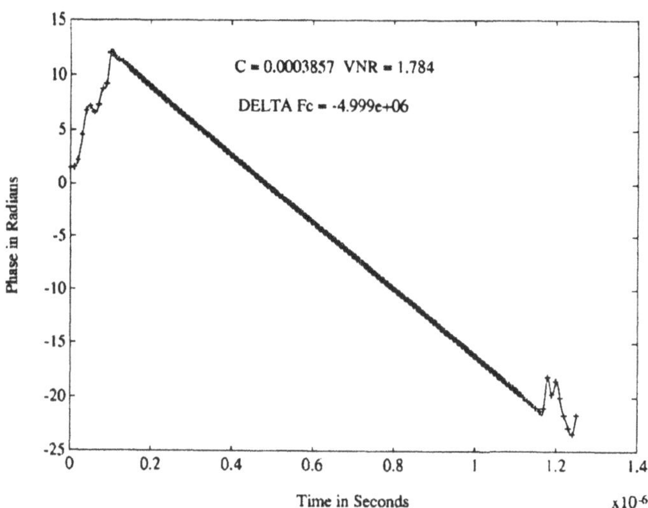

Fig. 3. Linear regression performed on differential phase of two pulses differing in carrier frequency by 5 MHz (QP0 and QP4). The carrier frequency offset does not significantly affect C and the VNR and is accurately measured by the slope of the straight line fitted to the differential phase history.

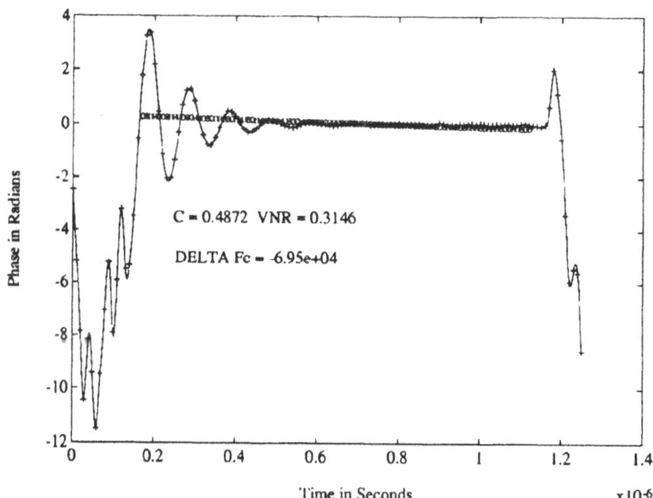

Fig. 4. Linear regression performed on differential phase of two pulses differing in modulation (QP0 and QP5). The difference in the modulations of the two signals is clearly visible and severely affects both C and the VNR. A small bias in the slope of the straight line fitted to the differential phase history is apparent.

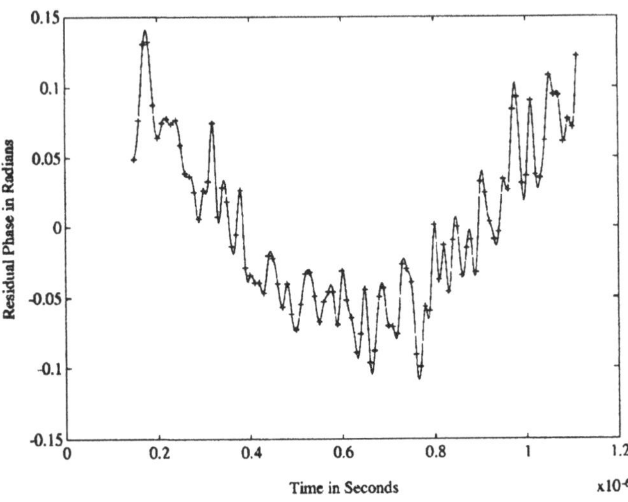

Fig. 5. Phase residuals from a linear regression performed on differential phase of two pulses differing in chirp rates by 0.25 MHz per microsecond (QP0 and QP6). The quadratic differential phase is recognizable on the expanded scale.

Table 1: Statistical results for the DPSC algorithm for pulses having different signal-to-noise ratios.

SNR (dB)	C (μ) ($rads^2$)	C (σ) ($rads^2$)	VNR (μ)	VNR (σ)	Δf_c(μ) (Hz)	Δf_c(σ) (Hz)
33	1.90e-04	2.97e-05	1.78	0.18	44.93	859.5
28	5.85e-04	7.98e-05	1.78	0.18	-83.82	1590
23	1.88e-03	2.89e-04	1.77	0.20	-74.61	2682
18	5.86e-03	8.86e-04	1.79	0.19	100.6	5621

Further simulations were carried out to compare the relative performances of C and the VNR and also to make a comparison of the DPSC algorithm with cross-correlation. Pairs of signals from two combinations of classes, S1-S1 and S1-C2 were used. The signals in the two classes differ in that the signals in S1 are unmodulated while those of C2 have a 0.35 MHz/microsecond chirp rate and the carrier frequencies of the two classes were selected to ensure the same expected value of the instantaneous frequency at the midpoint of the central one microsecond portion of the signals processed using the DPSC and cross-correlation.[8] In other respects, the signals from the two classes are similar. The total duration of each pulse was 1.6 microseconds with the edges, constructed of halves of a raised cosine, having transition times of 200 ns. The signal-to-noise ratio was 23 dB after quadrature demodulation.

The estimated probability distributions for C and the VNR, plotted in Figures 6 and 7, were obtained with 2,500 pairs of signals from each of the two combinations of classes, S1-S1 and S1-C2. They are consistent with the performance analysis presented in Section 3. The performance of C and the VNR are comparable; the two combinations of classes are clearly resolved by the VNR and C with over 95% of the matches and mismatches correctly determined using a fixed threshold.

The peak of the cross-correlation function was computed for each pair of signals as the sum of the conjugate products of the sampled and time aligned inphase and quadrature signals. The magnitude was normalized by 1/N where N was the number of samples processed.

The performance of the cross-correlation algorithm was very poor in comparison with that of the DPSC algorithm. The observed probability distributions obtained for the two combinations of classes, plotted in Figure 8, are very similar and the two combinations of classes cannot be resolved. Additional simulation results, involving shifts of the relative carrier frequencies of the two classes confirmed that cross-correlation was severely affected by carrier frequency offsets. With a 1 MHz offset in the relative carrier frequencies, the peak of the cross-correlation function has an expected value of only 0.32.

The relative performance of the DPSC algorithm and the Euclidian distance measured between pairs of frequency demodulated signals[9] was also investigated. For phase modulations $\phi_i(t)$ having a bandwidth small compared to the Nyquist bandwidth, the DPSC algorithm had a considerable performance advantage.

---

[8]This condition maximizes the cross-correlation for zero time lag between a pair of signals selected from two different classes.

[9]Frequency demodulation was performed using a filter having the impulse response {1,-1}. The variance of the demodulated signal would be reduced with a differentiator providing smoothing, but this would be undesirable for processing short duration signals.

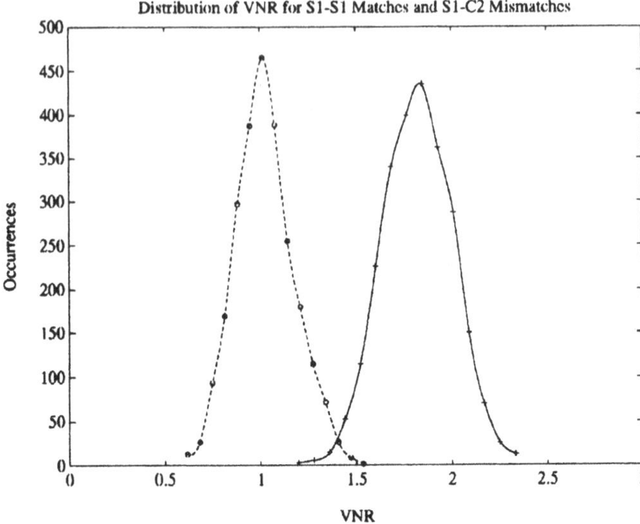

Fig. 6. Estimated probability distributions of VNR obtained using pairs of signals from two combinations of classes, S1-S1 and S1-C2.

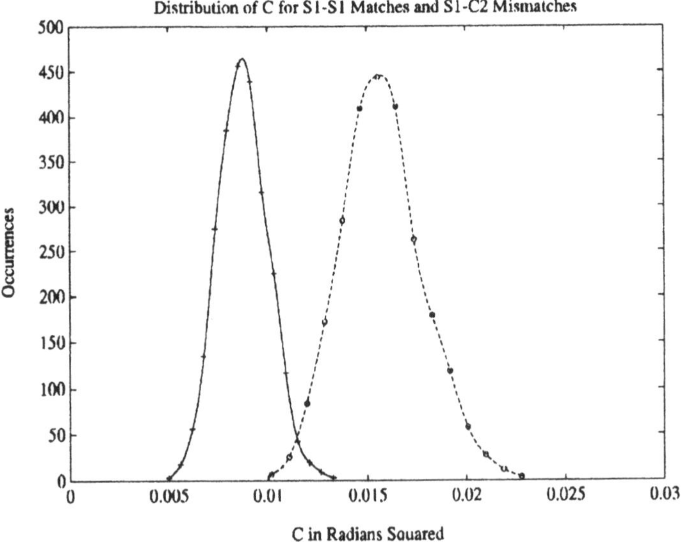

Fig. 7. Estimated probability distributions of C obtained for pairs of signals from two combinations of classes, S1-S1 and S1-C2.

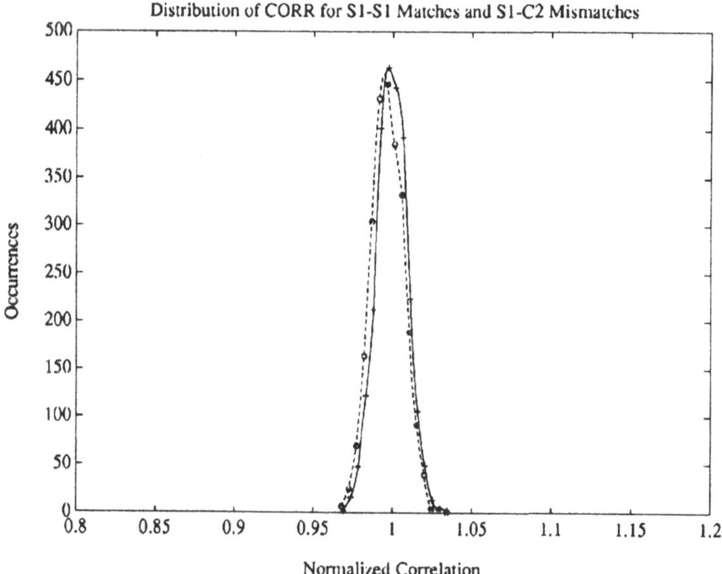

Fig. 8. Estimated probability distributions of peak values obtained from the normalized cross-correlation of pairs of signals from two combinations of classes, S1-S1 and S1-C2. Unlike the results shown in Figures 6 and 7, the distributions for matches and mismatches are nearly indistinguishable.

5 Conclusions

The similarity of the phase modulations of a pair of signals can be evaluated by testing their differential phase history for a statistically significant departure from a linear model. The proposed algorithm can process signals having an arbitrary duration and is particularly suitable for transient signals since filtering of phase data by differentiation/smoothing filters is not required. Its performance is unaffected by carrier frequency and phase offsets between the pair of signals. If the phase modulations of the pair of signals match and their noise is uncorrelated and Gaussian, it has the incidental advantage of providing a maximum likelihood estimate of the carrier frequency offset

Acknowledgements

The authors wish to acknowledge the assistance of Serge Martineau who was responsible for implementing the signal generator software used to obtain the simulation results. Also, the two anonymous referees provided comments that aided the revision of this paper.

References

1. A. Nutall: On the Quadrature Approximation to the Hilbert Transform of Modulated Signals. Proc. of the IEEE. 54, 1458 (1966)

3. G. Papadopoulis, K. Efstathiou: Implementation of an Instrument for Passive Recognition and Two-Dimensional Location Estimation of Acoustic Targets. Proc. IEEE Conference on Instrumentation and Measurement. 66 (1992)

3. H. L. Hirsch: Statistical Signal Characterization. Artech House 1992, pp. 125-134

4. I. L. Coat, P. C. Hill: The Application of Neural Networks to Radar Classification. Proc. IEE Colloqium Electronic Warfare Systems. 009, 4/1-6 (1991)

5. C. Lanczos: Applied Analysis. Englewood Cliffs N. J.: Prentice Hall, 1956, 315-326

6. P. M. Djuric, S. M. Kay: Parameter Estimation of Chirp Signals. IEEE Trans. Acoustics, Speech and Signal Processing. 38, 2118 (1990)

7. J. M. Dunn-Rogers: Digital Sampling Techniques for ESM Receivers. Proc. Military Microwave Conference. 19 (1990)

8. R. ·K. Otnes: Frequency Difference of Arrival Accuracy. IEEE Trans. on Acoustics, Speech, and Signal Processing. 37, 306 (1989)

9. N. Boutin: An Arctangent Type Wideband Demodulator with Improved Performance. IEEE Trans. Consumer Electronics. 38, 5 (1992)

10. K. Steiglitz, B. Dickinson: Phase Unwrapping by Factorization. IEEE Trans. Acoustics, Speech, and Signal Processing. 30, 984 (1982)

11. J.R. Green, D. Margerison: Statistical Treatment of Experimental Data. Amsterdam, Oxford, New York: Elsevier, 1978, pp.198-235

12. J. Johnston: Econometric Methods. New York: McGraw-Hill, 1963, pp. 249-250

13. W. B. Davenport Jr., W. L. Root: An Introduction to the Theory of Random Signals and Noise. New York: McGraw-Hill, 1958, pp 250-267

14. R. J. Inkol, R. H. Saper, G. Zhang, D. Al-Khalili: A Wide Bandwidth Digital Quadrature Demodulator for Electronic Warfare Receivers. Proc. NAECON. 313 (1993)

Efficient Algorithms for Fixed-Rate Entropy-Coded Vector Quantization*

A. K. Khandani*, P. Kabal[†,‡] and E. Dubois[†]

*Dept. of Elec. and Comp. Eng., University of Waterloo, Waterloo, Ont., N2L 3G1

[†]INRS–Telecommunications, 16 Place du Commerce, Verdun, PQ., H3E 1H6

[‡]McGill University, 3480 University, Montreal, PQ., H3A 2A7

Abstract: In quantization of any source with a nonuniform probability density function, the entropy coding of the quantizer output can result in a substantial decrease in bit rate. A straight-forward entropy coding scheme presents us with the problem of the variable data rate. A solution in a space of dimensionality N is to select a subset of elements in the N-fold cartesian product of a scalar quantizer and represent them with code-words of the same length. A reasonable rule is to select the N-fold symbols of the highest probability. For a memoryless source, this is equivalent to selecting the N-fold symbols with the lowest *additive* self-information. The *search/addressing* of this scheme can no longer be achieved independently along the one-dimensional subspaces. In the case of a memoryless source, the selected subset has a high degree of structure which can be used to substantially decrease the complexity. In this work, a dynamic programming approach is used to exploit this structure. We build our recursive structure required for the dynamic programming in a hierarchy of stages. This results in several benefits over the conventional trellis-based approaches. Using this structure, we develop efficient rules (based on aggregating the states) to substantially reduce the search/addressing complexities while keeping the degradation negligible.

1 Introduction

Consider the problem of quantizing a source with a nonuniform probability density function. If the dimensionality of the quantizer is not high enough, the entropy coding of the output can result in a substantial decrease in bit rate. A straight-forward entropy coding method presents us with the problem of variable data rate. Also, if the bit rate per quantizer symbol is restricted to be an integer, we are potentially subject to wasting up to one bit of data rate per quantizer output. A solution in a space of dimensionality N is to code the N-fold cartesian product of a scalar quantizer. To avoid having a variable data rate, one can select a subset of the N-fold symbols and represent them with code-words of the same length.

*This research was supported by a grant from the Canadian Institute for Telecommunications Research under the NCE program of the Government of Canada and also by a funding from INRS–Telecommunications.

In such a block-based source coding scheme, as some of the elements in the N-fold cartesian product space are not allowed, the *search* for the quantizer output and also the corresponding *addressing/reconstruction* processes can no longer be achieved independently along the one-dimensional (one-D) subspaces. The basic idea is to select the subset of points in such a way that these processes can be simplified.

One class of schemes are based on using a subset of points from a lattice (quantization lattice) bounded within the Voronoi region around the origin of another lattice (shaping lattice) [1]. In this case, the selected subset forms a group under vector addition modulo the shaping lattice.

Another class of schemes are based on selecting the N-fold symbols with the lowest additive self-information. This approach is traditionally denoted as the geometrical source coding [2], [3]. In this case, the selected subset has a high degree of symmetry which can be used to substantially reduce the complexity. A method for reducing the complexity of such a quantizer based on using a state diagram (with the states corresponding to the length of the code-words) is introduced by Laroia and Farvardin in [4]. Subsequently, Balamesh and Neuhoff in [5], introduce some complementary techniques to further reduce the complexity. In the present work, we introduce some more advanced techniques showing improvement with respect to the schemes of [4], [5].

We discuss a dynamic programming approach. The key point is to use the additive property of the self-information, in conjunction with the additive property of the distortion measure, to decompose the search/addressing into the lower dimensional subspaces. This decomposition avoids the exponential growth of the complexity. The core of the scheme, as in any problem of dynamic programming, is a recursive relationship. We build our recursive structure in a hierarchy of stages where each stage involves the cartesian product of two lower dimensional subspaces. This results in several benefits over the conventional trellis-based approach used in [4], [5]. By effectively quantizing the state space, we obtain suboptimum methods with low complexity and negligible performance degradation.

2 Basic Structure

Consider a memoryless source and a scalar quantizer composed of M partitions. In the N-fold cartesian product of this quantizer, we obtain M^N, N-D partitions. The final vector quantizer is selected as a subset of the N-D partitions composed of T elements. Each partition is represented by a code-word composed of $\lceil \log_2 T \rceil$ bits. The N-D reconstruction vectors are denoted as $\mathbf{r}_i, i = 0, \ldots, T-1$. For a given source vector $\mathbf{x}$, the quantization rule (decoding) is to find the reconstruction vector $\mathbf{r}_i$ which has the minimum square distance to $\mathbf{x}$, addressing is to produce the index i when $\mathbf{r}_i$ is selected, and reconstruction is to reproduce $\mathbf{r}_i$ from the index i.

Assume that the induced self-information and the expected value of the symbols mapped to the j'th one-D partition are equal to c_j and r_j, respectively. The self-information associated with a one-D point is considered as a cost associated with that point. The selection rule for the N-D symbols is to keep the points with the lowest overall additive cost. The N-D reconstruction vectors are obtained by concatenating the corresponding one-D reconstruction levels, namely r_j's. The search operation is formulated as:

$$\begin{array}{ll} \text{Minimize} & \sum_{i=0}^{N-1}(x_i - r_{j_i})^2 \\ \text{Subject to:} & \sum_{i=0}^{N-1} c_{j_i} \leq C_{\max} \end{array} \tag{1}$$

The immediate approach to solving (1) is to perform an exhaustive search.

For the addressing/reconstruction, we need a one-to-one mapping between the set of the code-words and the set of the integer numbers $0, \ldots, T-1$ such that the mapping (addressing) and its inverse (reconstruction) can be easily implemented. The immediate approach to obtain such a mapping is to use a look-up table.

In a high dimensional space, as the number of the symbols is usually quite high, one can not make use of the immediate approaches based on exhaustive search and lookup table. The main idea is to use the high degree of structure, which is mainly due to the symmetry of the problem in (1), to reduce the complexity of the involved operations.

3 Recursive merging of shells

If $F_N(C)$ denotes the set of N-D points of cost C (shell of cost C), we have the following recursive relationship:

$$F_N(C) = \bigcup [F_{N_1}(C_1) \otimes F_{N_2}(C_2)] \tag{2}$$

where $\otimes$ denotes the cartesian product, $N = N_1 + N_2$, and the union is computed over all the pairs (C_1, C_2) satisfying $C_1 + C_2 = C$. We refer to each cartesian product element in (2) as a *cluster*. We are specially interested in the case that $N_1 = N_2 = N/2$.

For a given input vector $\mathbf{x}$, by decoding of a shell we mean the process of finding the element of the shell which has the minimum distance to $\mathbf{x}$. Using (2), we can decode a shell recursively. To do this, $\mathbf{x}$ is split into two parts $\mathbf{x}_1$ and $\mathbf{x}_2$ of lengths N_1 and N_2. Assume that the nearest vectors of $F_{N_1}(C_1)/F_{N_2}(C_2)$ to $\mathbf{x}_1/\mathbf{x}_2$ are equal to $\hat{\mathbf{x}}_1/\hat{\mathbf{x}}_2$ with the minimum distances d_1/d_2. The nearest vector of $F_{N_1}(C_1) \otimes F_{N_2}(C_2)$ to $\mathbf{x}$ is equal to $(\hat{\mathbf{x}}_1, \hat{\mathbf{x}}_2)$ with the minimum distance $d_1 + d_2$. The minimum distance of a shell is equal to the smallest of the minimum distances of its clusters.

For $N_1 = N_2 = N/2$, if we know the minimum distance and the nearest vector for all the shells of the $N/2$-D subspaces, we can decode all the N-D shells. The lower is the number of shells in the $N/2$-D subspaces, the simpler will be the decoding process.

One can also use the recursive structure of the shells to develop an algorithmic addressing/reconstruction procedures. The basic idea is that the addressing within each cluster can be achieved independently along its lower dimensional shells. This results in the same decomposition principle as proposed for the first time in [6] and elaborated in [7], [8]. To complete the recursion, it remains to select a single cluster within a shell. This is achieved by arranging the clusters within a shell in a preselected order and assuming that the points in a higher order cluster have a larger label. Based on this ordering, a cluster is selected according to the range of the index and the corresponding residue with respect to the start of the range is used for the addressing within the cluster.

The procedure of recursive addressing becomes specially simple if all the cardinalities are restricted to be an integral power of two (integral bit rate). The key point behind the simplicity is as follows: Consider two sets of cardinalities 2^{c_1} and 2^{c_2}. The cartesian product of these sets is composed of $2^{c_1+c_2}$ elements. To address an element of the cartesian product, the input bit stream composed of $c_1 + c_2$ bits is split into two parts of lengths c_1 and c_2. Each part is subsequently used to select a point within one of the two sets. In other words, the address of a composite symbol can be easily obtained by concatenating the addresses of its constituents lower dimensional elements. The effect of merging is reflected through some additional bits which are stored in a block of memory.

4 Hierarchical dynamic programming

Dynamic programming is a multi-stage optimization procedure based on an inductive principle. It makes use of a recursive relationship to decompose a complicated problem into a sequence of easier subproblems. In the following, we introduce our approach to dynamic programming. As the schemes of [4] and [5] are also based on a dynamic programming, we have focused our explanation on a comparison between the methods.

The core of the idea in the schemes of [4], [5] is to use a state diagram with the transitions corresponding to one-D symbols. This results in a trellis composed of N stages. The states s and $s+c$ in two successive stages are connected by a link corresponding to the one-D symbol(s) of cost c. Consequently, the states in the nth stage, $n = 1, \ldots, N$, represent the accumulative cost over the set of the first n dimensions. The links connecting two successive stages are labeled by the corresponding distortions. Then, the viterbi algorithm is used to find the path of the minimum overall distortion through the trellis.

The straight-forward approach is to assign an independent state to each possible value of cost at a given stage. Let K denote the number of the distinct values of cost along a dimension. Number of distinct values of cost in N dimensions can be as large as:

$$C = \sum_{\sum_{i=0}^{K-1} n_i = N} \frac{N!}{\prod_i n_i!} \tag{3}$$

The general term in (3) represents the total number of N-tuples where the one-D symbol with the ith value of cost has occurred for n_i times. If two different combinations in (3) result in the same value of the additive cost, the corresponding states merge together. This is denoted as a *natural* merge.

Even for a moderate value of K, the number of distinct states in N-D (after the natural merge) can be impractically large. The solution is to synthetically *aggregate* distinct states into a smaller number. This is denoted as the *state-space quantization* and is the key point to the effectiveness of any dynamic programming approach. In [4], the self-information associated with the one-D symbols are rounded to rational numbers with a common denominator. In [5], to reduce the complexity with respect to [4], these are rounded to integer numbers.

Unlike [4] and [5] which are based on a component-by-component analysis, we use a hierarchy of stages where each stage involves the cartesian product of lower-dimensional subspaces. This approach is specially effective when the space dimensionality is equal to $N = 2^u$. In this case, the hierarchy is composed of u stages where the ith stage, $i = 0, \ldots, u-1$, is based on the (pair-wise) cartesian product of the 2^i-D subspaces (there are 2^{u-i} identical pairs of cartesian product in the ith stage). All our following discussions are based on this structure.

The immediate benefit of this approach is the possibility of using a parallel processing system. Another benefit is that this structure can be easily combined with the state diagram of a lattice (used to decode the lattice [9]). This provides a means to easily use the scheme in conjunction with a quantization lattice. More importantly, as we will see later, this approach provides the basis for an effective state-space quantization rule.

4.1 State-space quantization, aggregation of states

As already mentioned, a straight-forward approach results in a large number of distinct states (shells). The major question is how we can aggregate the shells into *macro-shells* while keeping the degradation negligible. Obviously, after aggregation, the points of the macro-shells are no more of the same cost (each macro-shell has a range of costs). Based on our hierarchy in an $N = 2^u$-D space, we consider the following recursive structure.

Recursive aggregation rule: The macro-shells in 2^i-D subspaces are composed of the union of the elements in the cartesian product of the 2^{i-1}-D macro-shells.

In devising a specific rule, we should keep the following three implicit objectives in mind:

1. As truncation is achieved by discarding some of the macro-shells, while the objective is to discard a given number of points of the highest cost, we should try to minimize the overlap between the range of the costs of different macro-shells.

2. The number of the macro-shells should be as small as possible. This suggests that we should try to put an equal number of points in different macro-shells. As we will see later, in the case that the macro-shells have

an equal number of points, the addressing is also much simpler than the general case.

3. Aggregation rule should be compatible with our recursive structure mentioned earlier.

Concerning the first objective of this list, the best approach is to partition the dynamic range of the cost into nonoverlapping segments. Then, each macro-shell is considered as the set of elements with the cost in one of these subranges. By appropriately selecting the subranges, one can even put an equal number of points in each macro-shell and satisfy the second objective. This sounds excellent, however, unfortunately, no recursive structure is known for this type of aggregation. As we will see later, by partitioning the space into macro-shells of increasing *average cost*, it is possible to remain compatible with our recursive structure. In the following, we propose two rules for the state-space quantization which partially fulfill the afore-mentioned objectives. In the first method, the aggregation is limited to the one-D subspaces. This is based on a similar approach as used for the first time in the cotext of constellation shaping in [10]. In the second method this is achieved sequentially in different stages of our hierarchy. As we will see later, the second method is specially effective and results in a simple addressing procedure.

4.2 Aggregation on a one-D basis, Macro-shells of identical sum of the indices

The effect of natural merging of shells is specially pronounced when the cost of the one-D shells is an affine function of their indices (cost of the ith shell is equal to $c_0 + i\Delta$). This results in a set of KN distinct shells in an N-D space where K is the number of one-D shells.

Based on this observation, in our first method, the one-D symbols are aggregated into K information macro-shells with a fixed spacing (increment in the self-information) Δ. The probabilities of the points in the ith macro-shell satisfy $0 < -\log_2 p \leq c_0$ for $i = 0$ and $c_0 + (i-1)\Delta < -\log_2 p \leq c_0 + i\Delta$, for $i = 1, \ldots, K-1$. Obviously, some of the one-D macro-shells may remain empty. The higher-dimensional macro-shells are considered as the set of the symbols with a fixed sum of the indices. This results in a recursive structure. The final subset is selected as the union of the N-D macro-shells with the sum of the indices less than a given value $L_{\max}$. This results in $\min[2^i K, L_{\max}]$ states in the ith stage of the hierarchy. This approximation method can be considered as a more general formulation for the schemes of [4] and [5] which are based on approximating the costs on a one-D basis.

From the three objectives in the afore-mentioned list, this method just fulfills the last one, namely the recursive structure. In the following, we introduce another method which is more compatible with these objectives.

4.3 Aggregation on a sequential basis, Macro-shells of increasing average costs and identical cardinalities

In our second method, the quantization of the state-space is based on a sequential aggregation of the macro-shells in the 2^i-D subspaces, $i=0,\ldots,u-1$. In other words, the state-space quantization is achieved gradually at different stages of the hierarchy. The subspaces involved at each stage of the hierarchy are partitioned into a number of macro-shells of increasing average costs and identical cardinalities. The key point is to approximate the costs of all the points within a given macro-shell by their average value.

Consider an $N=2^u$-dimensional space and assume that there are $K_i=2^{k_i}$ macro-shells in the $N_i=2^i$-D subspaces, $i=0,\ldots,u-1$. In the cartesian product of two of the N_i-D subspaces, we obtain K_i^2 clusters of equal volume. The clusters are arranged in the order of increasing average costs. A number equal to K_i^2/K_{i+1} of subsequent clusters are aggregated into a higher level ($2Ni=N_{i+1}$-D) macro-shell. Then, the whole process is repeated recursively. The final subset is obtained by keeping some of the N-D clusters of the lowest average cost.

Using macro-shells of *integral, equal bit rate* results in a specially simple addressing scheme. This is discussed in the following: Consider the case that the macro-shells in a given stage of our hierarchy, say at dimensionality N', are composed of 2^{c_1} elements. Also, assume that a higher level macro-shell (dimensionality $2N'$) is obtained by aggregating 2^{c_2} clusters in the two-fold cartesian product of the set of the N'-D macro-shells. The addressing of the $2N'$-D macro-shells requires $2c_1+c_2$ bits. The address of an $2N'$-D element is computed by concatenating the addresses of its constituent components in the N'-D macro-shells and concatenating the result with an additional c_2 bits which are selected as the label of the corresponding cluster within the $2N'$-D macro-shell.

For addressing in an $N=2^u$ dimensional space, all we need is a set of u memory blocks to store the components of each macro-shell in the cartesian product of the macro-shells of the lower dimensional subspaces. The ith addressing stage, $i=0,\ldots,u-2$, requires a lookup table with $2k_i\times 2^{2k_i}$ bits. The last stage requires $2k_{u-1}\times 2^{2k_{u-1}-r_s}$ bits where r_s denotes the redundancy associated with the selection of the final N-fold symbols as a subset of the cartesian product space. This is defined as the logarithm of the ratio of the employed number of points per dimension to the minimum necessary number of points per dimension. By using a relatively small number of macro-shells in lower dimensional subspaces and imposing an appropriate constraint on r_s, one can provide a tradeoff between performance and complexity.

4.4 Comparison with other methods

Figures (1), (2) show the Signal-to-Noise-Ratio (SNR) obtained by using our sequential aggregation rule in conjunction with an independent identically distributed (iid) Gaussian source. Table (1) presents a comparison between our method and the scheme of [4] in terms of performance and complexity. It is difficult to have a fair comparison with the scheme of [5] because in their case the space dimensionality is usually quite high which results in a longer delay.

References

[1] M. V. Eyuboglu and G. D. Forney, "Lattice and trellis quantization with lattice- and trellis- bounded codebooks—high-rate theory for memoryless sources," *IEEE Trans. Inform. Theory,* vol. IT-39, pp. 46–59, Jan 1993.

[2] D. J. Sakrison, "A geometrical treatment of the source encoding of a Gaussian random variable," *IEEE Trans. Inform. Theory,* vol. IT-14, pp. 481–486, May 1968.

[3] T. R. Fischer, "Geometric source coding and vector quantization," *IEEE Trans. Inform. Theory,* vol. IT-35, pp. 137–145, January 1989.

[4] R. Laroia and N. Farvardin, "A structured fixed-rate vector quantizer derived from variable-length scalar quantizer—Part I: Memoryless sources," *IEEE Trans. Inform. Theory,* vol. IT-39, pp. 851–867, May 1993.

[5] A. S. Balamesh and D. L. Neuhoff, "Block-constrained methods of fixed-rate, entropy coded, scalar quantization," submitted to *IEEE Trans. Inform. Theory,* Sept 1992.

[6] G. R. Lang and F. M. Longstaff, "A leech lattice modem," *IEEE J. Select. Areas Commun.,* vol. SAC-7, pp. 968–973, Aug. 1989.

[7] A. K. Khandani and P. Kabal, "Shaping multi-dimensional signal spaces—Part II: shell-addressed constellations," to appear in the *IEEE Trans. Inform. Theory,* Nov. 1993.

[8] A. K. Khandani and P. Kabal, "An efficient block-based addressing schemes for the nearly optimum shaping of multi-dimensional signal spaces," submitted to *IEEE Trans. Inform. Theory,* Aug. 1992.

[9] G. D. Forney, "Coset codes—Part II: Binary lattices and related codes," *IEEE Trans. Inform. Theory,* vol. IT-34, pp. 1152–1187, Sept. 1988.

[10] A. R. Calderbank and L. H. Ozarow, "Nonequiprobable signaling on the Gaussian channel," *IEEE Trans. Inform. Theory,* vol. IT-36, pp. 726–740, July 1990.

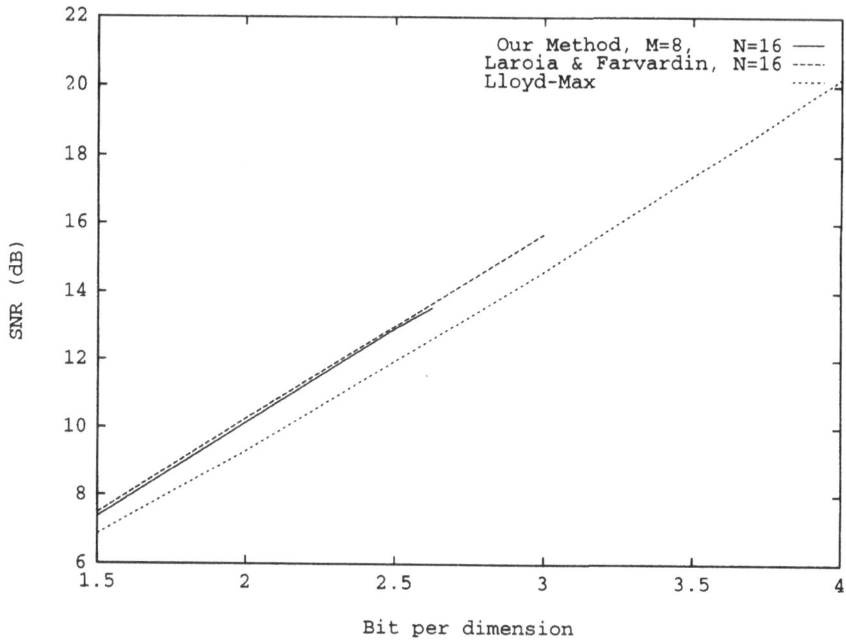

Figure 1: Quantization SNR for an iid Gaussian source, $N = 16$ (dimensionality), $M = 8$ (number of points per dimension) and $(k_1, k_2, k_3, k_4) = (3, 5, 6, 10)$.

Method	N	R	Memory	Computation	SNR (dB)
SMS	16	1.5	1.25 k	54 (33)	7.43
L-F	16	1.5	7.9 k	670	7.47
SMS	16	2.5	2.5 k	220 (97)	12.91
L-F	16	2.5	21.0 k	2240	13.00
SMS	32	3.5	14.3 k	1060 (290)	18.7
L-F	32	3.5	307 k	12500	18.8†

Table 1: Comparison between the method based on the sequential aggregation of shells (denoted by SMS) with the scheme of [4] (denoted by L-F). The memory size is in byte (8 bits) per N dimensions and the computational complexity is the number of additions/comparisons per dimension. The values inside parenthesis are the computational complexities of our method in the case of using a parallel processing system. (The value denoted by † is obtained using interpolation.).

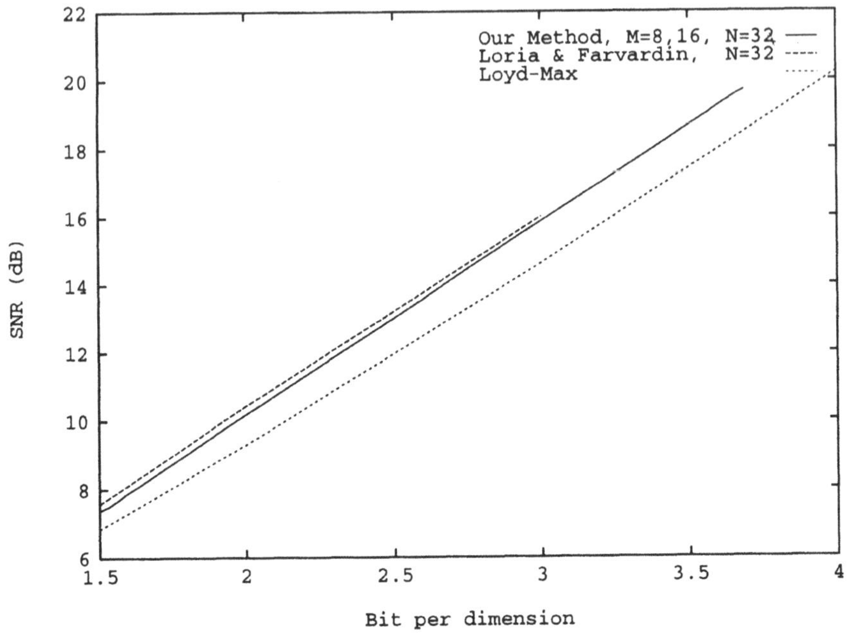

Figure 2: Quantization SNR for a Gaussian source, $N = 32$ (dimensionality), $M = 8, 16$ (number of points per dimension) and $(k_1, k_2, k_3, k_4, k_5) = (3, 5, 6, 7, 10), (4, 6, 7, 7, 10)$.

Lecture Notes in Computer Science

For information about Vols. 1–714
please contact your bookseller or Springer-Verlag

Vol. 715: E. Best (Ed.), CONCUR'93. Proceedings, 1993. IX, 541 pages. 1993.

Vol. 716: A. U. Frank, I. Campari (Eds.), Spatial Information Theory. Proceedings, 1993. XI, 478 pages. 1993.

Vol. 717: I. Sommerville, M. Paul (Eds.), Software Engineering – ESEC '93. Proceedings, 1993. XII, 516 pages. 1993.

Vol. 718: J. Seberry, Y. Zheng (Eds.), Advances in Cryptology – AUSCRYPT '92. Proceedings, 1992. XIII, 543 pages. 1993.

Vol. 719: D. Chetverikov, W.G. Kropatsch (Eds.), Computer Analysis of Images and Patterns. Proceedings, 1993. XVI, 857 pages. 1993.

Vol. 720: V.Mařík, J. Lažanský, R.R. Wagner (Eds.), Database and Expert Systems Applications. Proceedings, 1993. XV, 768 pages. 1993.

Vol. 721: J. Fitch (Ed.), Design and Implementation of Symbolic Computation Systems. Proceedings, 1992. VIII, 215 pages. 1993.

Vol. 722: A. Miola (Ed.), Design and Implementation of Symbolic Computation Systems. Proceedings, 1993. XII, 384 pages. 1993.

Vol. 723: N. Aussenac, G. Boy, B. Gaines, M. Linster, J.-G. Ganascia, Y. Kodratoff (Eds.), Knowledge Acquisition for Knowledge-Based Systems. Proceedings, 1993. XIII, 446 pages. 1993. (Subseries LNAI).

Vol. 724: P. Cousot, M. Falaschi, G. Filè, A. Rauzy (Eds.), Static Analysis. Proceedings, 1993. IX, 283 pages. 1993.

Vol. 725: A. Schiper (Ed.), Distributed Algorithms. Proceedings, 1993. VIII, 325 pages. 1993.

Vol. 726: T. Lengauer (Ed.), Algorithms – ESA '93. Proceedings, 1993. IX, 419 pages. 1993

Vol. 727: M. Filgueiras, L. Damas (Eds.), Progress in Artificial Intelligence. Proceedings, 1993. X, 362 pages. 1993. (Subseries LNAI).

Vol. 728: P. Torasso (Ed.), Advances in Artificial Intelligence. Proceedings, 1993. XI, 336 pages. 1993. (Subseries LNAI).

Vol. 729: L. Donatiello, R. Nelson (Eds.), Performance Evaluation of Computer and Communication Systems. Proceedings, 1993. VIII, 675 pages. 1993.

Vol. 730: D. B. Lomet (Ed.), Foundations of Data Organization and Algorithms. Proceedings, 1993. XII, 412 pages. 1993.

Vol. 731: A. Schill (Ed.), DCE – The OSF Distributed Computing Environment. Proceedings, 1993. VIII, 285 pages. 1993.

Vol. 732: A. Bode, M. Dal Cin (Eds.), Parallel Computer Architectures. IX, 311 pages. 1993.

Vol. 733: Th. Grechenig, M. Tscheligi (Eds.), Human Computer Interaction. Proceedings, 1993. XIV, 450 pages. 1993.

Vol. 734: J. Volkert (Ed.), Parallel Computation. Proceedings, 1993. VIII, 248 pages. 1993.

Vol. 735: D. Bjørner, M. Broy, I. V. Pottosin (Eds.), Formal Methods in Programming and Their Applications. Proceedings, 1993. IX, 434 pages. 1993.

Vol. 736: R. L. Grossman, A. Nerode, A. P. Ravn, H. Rischel (Eds.), Hybrid Systems. VIII, 474 pages. 1993.

Vol. 737: J. Calmet, J. A. Campbell (Eds.), Artificial Intelligence and Symbolic Mathematical Computing. Proceedings, 1992. VIII, 305 pages. 1993.

Vol. 738: M. Weber, M. Simons, Ch. Lafontaine, The Generic Development Language Deva. XI, 246 pages. 1993.

Vol. 739: H. Imai, R. L. Rivest, T. Matsumoto (Eds.), Advances in Cryptology – ASIACRYPT '91. X, 499 pages. 1993.

Vol. 740: E. F. Brickell (Ed.), Advances in Cryptology – CRYPTO '92. Proceedings, 1992. X, 593 pages. 1993.

Vol. 741: B. Preneel, R. Govaerts, J. Vandewalle (Eds.), Computer Security and Industrial Cryptography. Proceedings, 1991. VIII, 275 pages. 1993.

Vol. 742: S. Nishio, A. Yonezawa (Eds.), Object Technologies for Advanced Software. Proceedings, 1993. X, 543 pages. 1993.

Vol. 743: S. Doshita, K. Furukawa, K. P. Jantke, T. Nishida (Eds.), Algorithmic Learning Theory. Proceedings, 1992. X, 260 pages. 1993. (Subseries LNAI)

Vol. 744: K. P. Jantke, T. Yokomori, S. Kobayashi, E. Tomita (Eds.), Algorithmic Learning Theory. Proceedings, 1993. XI, 423 pages. 1993. (Subseries LNAI)

Vol. 745: V. Roberto (Ed.), Intelligent Perceptual Systems. VIII, 378 pages. 1993. (Subseries LNAI)

Vol. 746: A. S. Tanguiane, Artificial Perception and Music Recognition. XV, 210 pages. 1993. (Subseries LNAI).

Vol. 747: M. Clarke, R. Kruse, S. Moral (Eds.), Symbolic and Quantitative Approaches to Reasoning and Uncertainty. Proceedings, 1993. X, 390 pages. 1993.

Vol. 748: R. H. Halstead Jr., T. Ito (Eds.), Parallel Symbolic Computing: Languages, Systems, and Applications. Proceedings, 1992. X, 419 pages. 1993.

Vol. 749: P. A. Fritzson (Ed.), Automated and Algorithmic Debugging. Proceedings, 1993. VIII, 369 pages. 1993.

Vol. 750: J. L. Díaz-Herrera (Ed.), Software Engineering Education. Proceedings, 1994. XII, 601 pages. 1994.

Vol. 751: B. Jähne, Spatio-Temporal Image Processing. XII, 208 pages. 1993.

Vol. 752: T. W. Finin, C. K. Nicholas, Y. Yesha (Eds.), Information and Knowledge Management. Proceedings, 1992. VII, 142 pages. 1993.

Vol. 753: L. J. Bass, J. Gornostaev, C. Unger (Eds.), Human-Computer Interaction. Proceedings, 1993. X, 388 pages. 1993.

Vol. 754: H. D. Pfeiffer, T. E. Nagle (Eds.), Conceptual Structures: Theory and Implementation. Proceedings, 1992. IX, 327 pages. 1993. (Subseries LNAI).

Vol. 755: B. Möller, H. Partsch, S. Schuman (Eds.), Formal Program Development. Proceedings. VII, 371 pages. 1993.

Vol. 756: J. Pieprzyk, B. Sadeghiyan, Design of Hashing Algorithms. XV, 194 pages. 1993.

Vol. 757: U. Banerjee, D. Gelernter, A. Nicolau, D. Padua (Eds.), Languages and Compilers for Parallel Computing. Proceedings, 1992. X, 576 pages. 1993.

Vol. 758: M. Teillaud, Towards Dynamic Randomized Algorithms in Computational Geometry. IX, 157 pages. 1993.

Vol. 759: N. R. Adam, B. K. Bhargava (Eds.), Advanced Database Systems. XV, 451 pages. 1993.

Vol. 760: S. Ceri, K. Tanaka, S. Tsur (Eds.), Deductive and Object-Oriented Databases. Proceedings, 1993. XII, 488 pages. 1993.

Vol. 761: R. K. Shyamasundar (Ed.), Foundations of Software Technology and Theoretical Computer Science. Proceedings, 1993. XIV, 456 pages. 1993.

Vol. 762: K. W. Ng, P. Raghavan, N. V. Balasubramanian, F. Y. L. Chin (Eds.), Algorithms and Computation. Proceedings, 1993. XIII, 542 pages. 1993.

Vol. 763: F. Pichler, R. Moreno Díaz (Eds.), Computer Aided Systems Theory - EUROCAST '93. Proceedings, 1993. IX, 451 pages. 1994.

Vol. 764: G. Wagner, Vivid Logic. XII, 148 pages. 1994. (Subseries LNAI).

Vol. 765: T. Helleseth (Ed.), Advances in Cryptology – EUROCRYPT '93. Proceedings, 1993. X, 467 pages. 1994.

Vol. 766: P. R. Van Loocke, The Dynamics of Concepts. XI, 340 pages. 1994. (Subseries LNAI).

Vol. 767: M. Gogolla, An Extended Entity-Relationship Model. X, 136 pages. 1994.

Vol. 768: U. Banerjee, D. Gelernter, A. Nicolau, D. Padua (Eds.), Languages and Compilers for Parallel Computing. Proceedings, 1993. XI, 655 pages. 1994.

Vol. 769: J. L. Nazareth, The Newton-Cauchy Framework. XII, 101 pages. 1994.

Vol. 770: P. Haddawy (Representing Plans Under Uncertainty. X, 129 pages. 1994. (Subseries LNAI).

Vol. 771: G. Tomas, C. W. Ueberhuber, Visualization of Scientific Parallel Programs. XI, 310 pages. 1994.

Vol. 772: B. C. Warboys (Ed.),Software Process Technology. Proceedings, 1994. IX, 275 pages. 1994.

Vol. 773: D. R. Stinson (Ed.), Advances in Cryptology – CRYPTO '93. Proceedings, 1993. X, 492 pages. 1994.

Vol. 774: M. Banâtre, P. A. Lee (Eds.), Hardware and Software Architectures for Fault Tolerance. XIII, 311 pages. 1994.

Vol. 775: P. Enjalbert, E. W. Mayr, K. W. Wagner (Eds.), STACS 94. Proceedings, 1994. XIV, 782 pages. 1994.

Vol. 776: H. J. Schneider, H. Ehrig (Eds.), Graph Transformations in Computer Science. Proceedings, 1993. VIII, 395 pages. 1994.

Vol. 777: K. von Luck, H. Marburger (Eds.), Management and Processing of Complex Data Structures. Proceedings, 1994. VII, 220 pages. 1994.

Vol. 778: M. Bonuccelli, P. Crescenzi, R. Petreschi (Eds.), Algorithms and Complexity. Proceedings, 1994. VIII, 222 pages. 1994.

Vol. 779: M. Jarke, J. Bubenko, K. Jeffery (Eds.), Advances in Database Technology — EDBT '94. Proceedings, 1994. XII, 406 pages. 1994.

Vol. 780: J. J. Joyce, C.-J. H. Seger (Eds.), Higher Order Logic Theorem Proving and Its Applications. Proceedings, 1993. X, 518 pages. 1994.

Vol. 781: G. Cohen, S. Litsyn, A. Lobstein, G. Zémor (Eds.), Algebraic Coding. Proceedings, 1993. XII, 326 pages. 1994.

Vol. 782: J. Gutknecht (Ed.), Programming Languages and System Architectures. Proceedings, 1994. X, 344 pages. 1994.

Vol. 783: C. G. Günther (Ed.), Mobile Communications. Proceedings, 1994. XVI, 564 pages. 1994.

Vol. 784: F. Bergadano, L. De Raedt (Eds.), Machine Learning: ECML-94. Proceedings, 1994. XI, 439 pages. 1994. (Subseries LNAI).

Vol. 785: H. Ehrig, F. Orejas (Eds.), Recent Trends in Data Type Specification. Proceedings, 1992. VIII, 350 pages. 1994.

Vol. 786: P. A. Fritzson (Ed.), Compiler Construction. Proceedings, 1994. XI, 451 pages. 1994.

Vol. 787: S. Tison (Ed.), Trees in Algebra and Programming – CAAP '94. Proceedings, 1994. X, 351 pages. 1994.

Vol. 788: D. Sannella (Ed.), Programming Languages and Systems – ESOP '94. Proceedings, 1994. VIII, 516 pages. 1994.

Vol. 789: M. Hagiya, J. C. Mitchell (Eds.), Theoretical Aspects of Computer Software. Proceedings, 1994. XI, 887 pages. 1994.

Vol. 790: J. van Leeuwen (Ed.), Graph-Theoretic Concepts in Computer Science. Proceedings, 1993. IX, 431 pages. 1994.

Vol. 791: R. Guerraoui, O. Nierstrasz, M. Riveill (Eds.), Object-Based Distributed Programming. Proceedings, 1993. VII, 262 pages. 1994.

Vol. 792: N. D. Jones, M. Hagiya, M. Sato (Eds.), Logic, Language and Computation. XII, 269 pages. 1994.

Vol. 793: T. A. Gulliver, N. P. Secord (Eds.), Information Theory and Applications. Proceedings, 1993. XI, 394 pages. 1994.